BIOLOGICAL INDICATORS OF SOIL HEALTH

Biological Indicators of Soil Health

Edited by

C. Pankhurst, B.M. Doube
CSIRO Land and Water
Glen Osmond
Australia

and

V.V.S.R. Gupta
Cooperative Research Centre for Soil and Land Management
Glen Osmond
Australia

CAB INTERNATIONAL

CABI Publishing
Wallingford
Oxon OX10 8DE
UK

Tel: +44 (0)1491 832111
Fax: +44 (0)1491 833508
Email: cabi@cabi.org

CABI Publishing
10 E 40th Street, Suite 3203
New York, NY 10016
USA

Tel: +1 212 481 7018
Fax: +1 212 686 7993
Email: cabi-nao@cabi.org

ISBN 0 85199 158 0

First printed 1997
Reprinted 1998

Printed and bound in the UK by Biddles Ltd, Guildford and King's Lynn

Contents

Contributors

G.L. Bateman *IACR – Rothamsted, Harpenden, Hertfordshire AL5 2JQ, UK.*

C.D. Campbell *Macaulay Land Use Research Institute, Craigiebuckler, Aberdeen AB9 2QJ, UK.*

M.S. Cresser *Department of Plant and Soil Science, Cruickshank Building, University of Aberdeen, Aberdeen AB9 2UE, UK.*

R.P. Dick *Department of Crop and Soil Science, Oregon State University, Corvallis, Oregon 97331, USA.*

J.W. Doran *Soil/Water Conservation Research Unit, United States Department of Agriculture, Agricultural Research Service, University of Nebraska, Lincoln, Nebraska 68583, USA.*

B.M. Doube, *CSIRO Land and Water, Private Bag No. 2, Glen Osmond, South Australia 5064, Australia.*

E.T. Elliott *Natural Resource Ecology Laboratory, Colorado State University, Fort Collins, Colorado 80524, USA.*

L.F. Elliott *National Forage Seed Production Research Centre, United States Department of Agriculture, Agricultural Research Service, Oregon State University, Corvallis, Oregon 97331, USA.*

R. Gabbrielli *Dipartimento di Biologia Vegetale, Università di Frienze, Via Micheli 1, I-50121 Firenze, Italy.*

L.A. Glover *Department of Molecular and Cell Biology, Marischal College, University of Aberdeen, Aberdeen AB9 1AS, UK.*

P. Gremigni *Dipartimento di Biologia Vegetale, Università di Frienze, Via Micheli 1, I-50121 Firenze, Italy.*

V.V.S.R. Gupta *Cooperative Research Centre for Soil and Land Management, Private Bag No. 2, Glen Osmond, South Australia 5064, Australia.*

D. Hornby *IACR – Rothamsted, Harpenden, Hertfordshire AL5 2JQ, UK.*

K. Killham *Department of Plant and Soil Science, Cruickshank Building, University of Aberdeen, Aberdeen AB9 2UE, UK.*

J.M. Lynch *School of Biological Sciences, University of Surrey, Guildford, Surrey GU2 5XH, UK.*

S.J. Macnaughton *Microbial Insights, Inc., 201 Center Park Drive, Knoxville, Tennessee 37922-2105, USA.*

J.H. McCullum *Department of Land Resource Science, University of Guelph, Guelph, Ontario N1G 2W1, Canada.*

J.C.L. Meeussen *Macaulay Land Use Research Institute, Craigiebuckler, Aberdeen AB9 2QJ, UK.*

G.N. Mhatre *304 A, Kohinoor Tower, Bal Govindas Marg, Dadar, Bombay 400 028, India.*

M.H. Miller *Department of Land Resource Science, University of Guelph, Guelph, Ontario N1G 2W1, Canada*

K.M. Ophel-Keller *Cooperative Research Centre for Soil and Land Management, Private Bag No. 2, Glen Osmond, South Australia 5064, Australia.*

T. Pandolfini *Dipartimento di Biologia Vegetale, Università di Firenze, Via Micheli 1, I-50121 Firenze, Italy.*

C.E. Pankhurst, *CSIRO Land and Water, Private Bag No. 2, Glen Osmond, South Australia 5064, Australia.*

G.I. Paton *Department of Plant and Soil Science, Cruickshank Building, University of Aberdeen, Aberdeen AB9 2UE, UK.*

D.J. Rapport *Eco-Research Chair in Ecosystem Health, Faculty of Environmental Sciences, University of Guelph, Guelph, Ontario NIG 2W1, Canada.*

E.A.S. Rattray *Department of Plant and Soil Science, Cruickshank Building, University of Aberdeen, Aberdeen AB9 2UE, UK.*

M.M. Roper *Division of Plant Industry, CSIRO, Private Bag, Wembley, PO, Western Australia 6014, Australia.*

M. Safley *Biological Conservation Sciences Division, United States Department of Agriculture, Natural Resources Conservation Service, Post Box 2890, Washington DC 20013, USA.*

O. Schmidt *Department of Environmental Resource Management, Faculty of Agriculture, University College Dublin, Belfield, Dublin 4, Ireland.*

G.P. Sparling *Manaaki Whenua – Landcare Research, Private Bag 3127, Hamilton, New Zealand.*

N.M. van Straalen, *Vrije Universiteit, Department of Ecology and Ecotoxicology, De Boelelaan 1087, 1081 HV Amsterdam, The Netherlands.*

D.C. White *Center for Environmental Biotechnology, University of Tennessee, 10515 Research Drive, Knoxville, Tennessee 37932-2575, USA.*

G.W. Yeates *Manaaki Whenua – Landcare Research, Private Bag 11052, Palmerston North, New Zealand.*

Preface

Our ability to assess the health of soils and to identify key soil properties which can serve as indicators of soil health has become a major issue for land managers and for food and fibre producers throughout the world. The driving force behind this is the need to produce 'more' from our soils and to maintain increasing levels of production in the face of diminishing land resources resulting from expanding urbanization and land degradation. More than ever before, we now appreciate the wisdom of 'sustainable production' and realize that soils must be 'looked after' if they are to continue to produce an abundance of healthy foods. Hence the arrival of the concept of soil health and the desire to be able to assess and monitor it in some way.

Soil health, defined as 'the continued capacity of soil to function as a vital living system, within ecosystem and land-use boundaries, to sustain biological productivity, promote the quality of air and water environments, and maintain plant, animal, and human health' is a term that is used synonymously with soil quality, although many, including authors in this book, would argue for a distinction. The definition does, however, remind us that soils are living systems which contain vast assemblages of soil organisms which perform many of the functions that are critical to terrestrial life. These functions include the decomposition and recycling of nutrients from dead plant and animal tissues, the fixation of nitrogen, the maintenance of soil structure and the detoxification of pollutants. Often these functions are ignored and soil is regarded as an inanimate entity composed of minerals and chemicals. The key roles played by its living components are not recognized. Commonly, we are only reminded of the presence of or lack of specific soil biota only when a soilborne root disease wreaks havoc on a crop or when plants fail to grow through lack of an appropriate soil symbiont. The linkage between soil biota, soil health and the health of plants, animals and human beings is rarely considered.

The principles of soil conservation have been known for centuries and in many countries recognition of the dangers of soil degradation has prompted national soil conservation programmes. However, recent regional and global assessments of human-induced soil degradation (erosion, salinization, acidification, heavy metal pollution, organic matter decline) indicate that the productive capacity of millions of hectares of land continues to decline each year and warn us of ecological collapse of the world's productive soils. At a local level, we need to be able to assess how farming practices are affecting the capacity of the land to remain productive and how such practices are reducing or improving the health of the soil. The search for indicators which can be used as quantitative tools to assess the health of the soil has thus become a major challenge for both scientists and land managers. Indicators need to be robust and meaningful, and easy to measure and interpret. To date, emphasis has been given to physical and chemical soil properties as indicators of soil health, rather than to biological properties which are generally regarded as more difficult to measure, predict, or quantify. However, biological processes are intimately linked with the maintenance of soil structure and fertility and are potentially more sensitive to changes in the soil than indicators based on physical and chemical soil properties. Biological indicators therefore may provide an 'early warning' of system collapse and allow us to react before irreversible damage occurs. Although many problems surround the use of soil biological properties as indicators of soil health (e.g. the inherent temporal and spatial heterogeneity of soil organism populations and the unpredictable interaction of soil organisms with the climatic factors), modern technology is providing many new methodologies and approaches which may ultimately overcome some of these problems.

In 1994, a workshop entitled 'Soil Biota: Management in Sustainable Farming Systems' was held in Adelaide, Australia. The workshop was sponsored by the OECD Cooperative Research Project on Biological Resource Management, the CSIRO Division of Soils, The Cooperative Research Centre for Soil and Land Management (based in Adelaide) and by three Australian Research Corporations. This workshop was unique in that it focused attention on the soil biota and how it can be better managed to make agriculture more sustainable and less dependent on the use of non-renewable resources. The workshop also focused on opportunities for using the soil biota and soil biotic processes as biological indicators in farming systems. An outcome of the workshop was identification of the need for a detailed synthesis of current research of the soil biota and how it might be used as a component of an indicator package for the assessment and monitoring of soil health. This book is a response to that need for a synthesis. It contains 17 chapters, each prepared by authors who are internationally recognized for their knowledge and expertise in a particular area of soil/plant biology. There are four introductory chapters which address the concept of soil health, its relationship to ecosystem health and what is needed from biological indicators of soil health. These are followed by chapters which evaluate the potential of using different components of the soil biota and its activity as biological indicators. These cover soil microorganisms (including plant root pathogens), soil micro-, meso- and macrofauna, soil biodiversity, soil biotic processes and soil enzymes. In addition, two chapters address the development of new technologies which probe the composition and functioning of soil microbial communities, and two chapters review the use of plants as indicators of soil pollution. The final chapter is a synthesis that brings together the views and major points made by the authors of the volume and offers an analysis of the current status of the different biological indicators.

Throughout this volume the term 'bioindicator' is used whereas the term 'biological indicator' is used in the title. We do not differentiate between the two terms. The term 'bioindicator' is in common usage whereas the term 'biological indicator' is more technically correct.

C.E. Pankhurst
B.M. Doube
V.V.S.R. Gupta

1

Defining and Assessing Soil Health and Sustainable Productivity

J.W. Doran[1] and M. Safley[2]

[1] *Soil/Water Conservation Research Unit, United States Department of Agriculture, Agricultural Research Service, University of Nebraska, Lincoln, Nebraska 68583, USA;*
[2] *Biological Conservation Sciences Division, United States Department of Agriculture, Natural Resources Conservation Service, Post Box 2890, Washington DC 20013, USA*

Introduction

Increasing human populations, decreasing resources, social instability, and environmental degradation pose serious threats to the natural processes that sustain the global ecosphere and life on earth (Costanza *et al.*, 1992; Pearce and Warford, 1993). Agriculture, and society in general, is challenged to develop strategies for sustainability that conserve non-renewable natural resources such as soil, enhance use of renewable resources and are aligned with the natural processes that sustain life on earth. The problems of sustainability which we currently face are considered by some to result from an abandonment of ecological principles to produce human food and the acceptance of a cultural premise that places humankind as the ruler of the world, and therefore not subject to the laws of nature (Quinn, 1993). We often suffer from the delusion that we as humans can control nature when, in reality, the only thing we can control and manage is ourselves (Cline and Ruark, 1995). The challenge ahead in sustaining life on earth will require new vision, holistic approaches for ecosystem management and a renewed partnership between science and society. We must muster our cultural resources and 'put science to work' for both humanity and the natural ecosystems of which it is part and on which it depends.

We present the thesis that 'soil' is a dynamic, living resource whose condition is vital to both the production of food and fibre and to global balance and ecosystem function (Doran *et al.*, 1996). The quality and health of soils deter-

CAB INTERNATIONAL 1997. *Biological Indicators of Soil Health* (eds C.E. Pankhurst, B.M. Doube and V.V.S.R. Gupta)

mine agricultural sustainability (Acton and Gregorich, 1995), environmental quality (Pierzynski *et al.*, 1994) and, as a consequence of both – plant, animal and human health (Haberern, 1992). In its broadest sense, soil health can be defined as the ability of soil to perform or function according to its potential, and changes over time due to human use and management or to natural events. In this sense, soil health is enhanced by management and land-use decisions that weigh the multiple functions of soil and is impaired by decisions which focus only on single functions, such as crop productivity. In this chapter we present approaches to assessing the quality and health of soils and present the value of soil health to strategies for sustainable management of our natural resources.

Soil – a Vital and Finite Resource

Global function and sustainability

We enter the twenty-first century with greater awareness of our technological capability to influence the global environment and of the impending challenge for sustaining life on earth (Gore, 1993; Postel, 1994). Global climate change, depletion of the protective ozone layer, serious declines in species biodiversity, and degradation and loss of productive agricultural land are among the most pressing concerns associated with our technological search for a higher standard of living for an ever growing human population. Increasing worldwide concern for sustainable global development and preservation of our soil resources is reflected by numerous recent international conferences such as the United Nations Conference on Environment and Development (UNCED) in Rio de Janeiro, Brazil in 1992; the Soil Resilience and Sustainable Land Use Symposium in Budapest, Hungary in 1992; the Sustainable Land Management Conference in Lethbridge, Canada in 1993; and the International Congress of Soil Science in Acapulco, Mexico in 1994. Central to discussions at these conferences were the threats to sustainability posed by soil and environmental degradation associated with increasing intensity of land use and the search among increasing populations of the world for a higher standard of living. Sustainability of the energy and chemically intensive industrial agricultural model, which has enabled a two- to three-fold growth in agricultural output of many countries since World War II, is increasingly questioned by ecologists, earth scientists, and clergy (Jackson and Piper, 1989; Bhagat, 1990; Hillel, 1991 and Sagan, 1992).

Interest in evaluating the quality and health of our soil resources has been stimulated by increasing awareness that 'soil' is a critically important component of the earth's biosphere, functioning not only in the production of food and fibre but also in the maintenance of local, regional, and global environmental quality (Glanz, 1995). The thin layer of soil covering the surface of the earth represents the difference between survival and extinction for most land-based life. Like water, soil is a vital natural resource essential to civilization but, unlike water, soil is non-renewable on a human time scale (Jenny, 1980, 1984). Modern conservationists are quick to point out that, 'mismanagement and neglect can ruin the fragile resource and become a threat to human survival' (Lal and Pierce, 1991). This fact is supported by archaeological evidence showing that soil degradation was responsible for extinction or collapse of the Harappan civilization in western India, Mesopotamia in Asia Minor, and the Mayan culture in Central America (Olson, 1981).

Present-day agriculture evolved as we sought to control nature to meet the food and fibre needs of an increasingly urbanized society. With the development of modern chemistry during and after World War II, agriculturists often assumed a position of dominance in their struggle against a seemingly hostile natural environment, often failing to recognize the consequences of management approaches upon long-term productivity and environmental quality. Increased monocultural production of cash grain crops, extensive soil cultivation and greater reliance on chemical fertilizers and pesticides to maintain crop growth have resulted in two to three fold increases in grain yields and on-farm labour efficiency (Brown *et al.*, 1994; Northwest Area Foundation, 1994; Avery, 1995; Power and Papendick, 1985). However, these management practices have also increased soil organic matter loss, soil erosion, and surface and ground water contamination in the USA and elsewhere (Gliessman 1984; Hallberg, 1987; Reganold *et al.*, 1987). Motivations for shifting from input-intensive management to reduced external input farming include concern for protecting soil, human, and animal health from the potential hazards of pesticides; concern for protection of the environment and soil resources; and a need to lower production costs in the face of stagnant farm-gate receipts (Soule and Piper, 1992; US Department of Agriculture, 1980).

Past management of agriculture and other ecosystems to meet the needs of increasing populations has taxed the resiliency of soil and natural processes to maintain global balances of energy and matter. The quality of many soils in North America has declined significantly since grasslands and forests were converted to arable agriculture and cultivation was initiated (Campbell *et al.*, 1976). Mechanical cultivation and the production of continuous row crops has resulted in physical soil loss and displacement through erosion, large decreases in soil organic matter content, and a concomitant release of that organic C as carbon dioxide to the atmosphere (Houghton *et al.*, 1983). Within the last decade, inventories of the soil's productive capacity indicate severe degradation on well over 10% of the earth's arable land as a result of soil erosion, atmospheric pollution, cultivation, over-grazing, land clearing, salinization, and desertification (Sanders, 1992; World Resources Institute, 1992). Findings from a project of the United Nations Environment Program on 'Global Assessment of Soil Degradation' indicate that almost 40% of agricultural land has been adversely affected by human-induced soil degradation, and that more than 6% is degraded to such a degree that restoration of its original productive capacity is only possible through major capital investments (Oldeman, 1994). The quality of surface and sub-surface water has been jeopardized in many parts of the world by intensive land management practices and the consequent imbalance of C, N and water cycles in soil. At present, agriculture is considered the most widespread contributor to nonpoint source water pollution in the USA (CAST, 1992b; National Research Council, 1989). The major water contaminant in North America and Europe is nitrate-N; the principal sources of which are conversion of native to arable land use, animal manures and fertilizers. Soil management practices such as tillage, cropping patterns, and pesticide and fertilizer use are known to influence water quality. However, these management practices can also influence atmospheric quality through changes in the soil's capacity to produce or consume important atmospheric gases such as carbon dioxide, nitrous oxide and methane (CAST, 1992a; Rolston *et al.,* 1993). The present threat of global climate change and ozone depletion, through elevated levels of atmospheric gases and altered hydrological cycles, necessitates a better understanding of the influence of land management on soil processes.

Development of sustainable agricultural management systems has been complicated by the need to consider their utility to humans, their efficiency of resource use, and their ability to maintain a balance with the environment that is favourable both to humans and most other species (Harwood, 1990). We are challenged to develop management systems which balance the needs and priorities for production of food and fibre with those for a safe and clean environment. In the USA, the importance of soil quality in maintaining balance between environmental and production concerns was reflected by a major conclusion of a recent National Academy of Science report that, 'Protecting soil quality, like protecting air and water quality, should be a fundamental goal of national environmental policy' (National Research Council, 1993).

A recent call for development of a 'soil health index' was stimulated by the perception that human health and welfare is associated with the quality and health of soils (Haberern, 1992). However, defining and assessing soil quality or health is complicated by the fact that soils perform multiple functions in maintaining productivity and environmental well-being. Identifying and integrating the physical, chemical and biological soil attributes which define soil functions is the challenge (Rodale Institute, 1991; Papendick and Parr, 1992). Forums were held in Washington, DC, in the winter of 1995 to ensure that emphasis on maintaining the quality of our soil resources was included in the 1995 Farm Bill. Many agriculturists, scientists, politicians, and citizens recognize that maintaining the health and quality of soil should be a major goal of a sustainable society. An important question, however, is 'What defines a healthy or high quality soil and how might soil quality and health be maintained or improved through agricultural and land-use management?'

Defining soil quality and soil health

Soil is a dynamic, living, natural body that is vital to the function of terrestrial ecosystems and represents a unique balance between physical, chemical and biological factors. Soils form slowly, averaging 100 to 400 years per centimetre of topsoil, through the interaction of climate, topography, vegetation, and mineral parent material over time (Jenny, 1980; Lal, 1994). The major components of soil include inorganic minerals and sand, silt, and clay particles; reactive and stable forms of organic matter derived from decomposed soil biota; living organisms such as earthworms, insects, bacteria, fungi, algae, nematodes, etc. – in such a multitude that the numbers in a teaspoon (10 g) of soil can exceed the human population of the earth; water; and gases including O_2, CO_2, N_2, NO_x, and CH_4. Continual interchanges of molecules/ions between the solid, liquid and gaseous phases are mediated by physical, chemical, and biological processes in soil. The inorganic components of soil play a major role in retaining cations through ion exchange and non-polar organic compounds and anions through sorption reactions. Essential parts of the global C, N, P and S and water cycles occur in soil and soil organic matter is a major terrestrial pool for C, N, P and S; the cycling rate and availability of these elements is continually being altered by soil organisms in their constant search for food and energy sources.

The sun is the basis for most life on earth and provides radiant energy for heating the biosphere and for the photosynthetic conversion of carbon dioxide (CO_2) and water into food sources and oxygen for consumption by animals and other aerobic organisms. Most living organisms utilize oxygen to metabolize these food sources, capture their energy and recycle heat, CO_2, and water to the environment to begin this cycle again.

A simplified version of this 'Equation of Life' can be depicted as follows:

$$6CO_2 + 6H_2O + \text{ENERGY} \underset{\text{(heat)}\longleftarrow\text{(fuel)}}{\overset{\text{(radiant)}\longrightarrow\text{(food)}}{\rightleftharpoons}} C_6H_{12}O_6 + 6O_2$$

Photosynthesis (upper arrow); Decomposition and combustion (lower arrow)

The amount of CO_2 in the atmosphere is rather small and represents less than 0.04% of all gases present. If all combustion and respiration processes on earth were halted the plant life of the earth would consume all available CO_2 within a year or two (Lehninger, 1973). Thus, there is a fine balance between CO_2 production and utilization in the biosphere. Decomposition processes in soil play a predominant role in maintaining this balance. These processes are brought about by a complex web of organisms in soil, each group playing unique roles in the physical and chemical breakdown of organic plant and animal residues and reacting differently to a soil environment which is continually changing. Soils breathe and play a major role in transforming sunlight and stored energy and recycling matter through plants and animals. As such, living soils are vital to providing human food and fibre needs and in maintaining the ecosystems on which all life ultimately depends.

The concept of soil quality – soil function

Blum and Santelises (1994) describe a concept of sustainability and soil resilience based on six main soil functions – three ecological functions and three which are linked to human activity. Ecological functions include: (i) biomass production (food, fibre, and energy); (ii) the soil as a reactor which filters, buffers, and transforms matter to protect the environment, groundwater, and the food chain from pollution; and (iii) soil as a biological habitat and genetic reserve for many plants, animals, and organisms which should be protected from extinction. Functions linked to human activity include: (i) the soil as a physical medium, serving as a spatial base for technical and industrial structures and socio-economic activities such as housing, industrial development, transportation systems, recreation and refuse disposal; (ii) soil as a source of raw materials supplying water, clay, sand, gravel, minerals, etc.; and (iii) soil as part of our cultural heritage, containing palaeontological and archaeological treasures important to preserving the history of earth and humankind.

Our concepts of soil quality change as we become aware of the many essential functions soil performs in the biosphere, in addition to serving as a medium for plant growth, and as societal priorities change. In the late seventies, Warkentin and Fletcher (1977) discussed the evolution of soil quality concepts in intensive agriculture. The oldest and most frequently used concept was one of 'suitability for chosen uses', with emphasis on capability to support crop growth or engineering structures. This evolved to a 'range of possible uses' concept which is ecologically based and recognizes the importance of soil to biosphere function and its multiple roles in enhancing biological productivity, abating pollution, and even serving to enhance human health and aesthetic and recreational use of landscapes. Another stage in this evolution was development of the 'intrinsic value' concept of soil as a unique and irreplaceable resource, of value apart from its importance to crop growth or ecosystem function. As noted by Warkentin (1995), this view of soils is not widely explored by soil scientists but is held in various forms by naturalists and people who see a special relationship with the earth (Leopold, 1949). Historically soil has been used as a waste disposal system; it was conceived to be a biological incinerator destroying all the organic wastes deposited on or in it over time. However, in the 1960s and 1970s it became increasingly apparent that soils were receiving wastes of a type, and at a rate, that overwhelmed their assimilative capacity. This trend threatened soil function and called for a major responsibility by agriculturists in defining soil quality criteria (Alexander, 1971).

Quality of soil, as distinct from health, is largely defined by the ability of soil to perform various intrinsic and extrinsic functions. Quality is represented by a suite of physical, chemical, and biological properties that together: (i) provide a medium for plant growth and biological activity; (ii) regulate and partition water flow and storage in the environment; and (iii) serve as an environmental buffer in the formation and destruction of environmentally hazardous compounds (Larson and Pierce, 1991, 1994).

Soil serves as a medium for plant growth by providing physical support, water, essential nutrients and oxygen for roots. Suitability of soil for sustaining plant growth and biological activity is a function of physical properties (porosity, water holding capacity, structure and tilth) and chemical properties (nutrient supplying ability, pH, salt content, etc.), many of which are a function of soil organic matter content.

Soil plays a key role in completing the cycling of major elements required by biological systems (C, N, P, S, etc.), decomposing organic wastes and detoxifying certain hazardous compounds. The key role played by soils in recycling organic materials into CO_2 and water and degrading synthetic compounds foreign to the soil is brought about by microbial decomposition and chemical reactions. Ability of a soil to store and transmit water is a major factor controlling water availability to plants and transport of environmental pollutants to surface and ground water.

Much like air or water, the quality of soil has a profound effect on the health and productivity of any given biome and the environments and ecosystems related to it. However, unlike air or water for which we have quality standards, the definition and quantification of soil quality is complicated by the fact that it is not directly ingested or respired by humans and animals as are air and water. Soil quality is often thought of as an abstract characteristic of soils which can't be defined because it depends on external factors such as land use and soil management practices, ecosystem and environmental interactions, socioeconomic and political priorities, and so on. Historically, perceptions of what constitutes a good soil vary depending on individual priorities for soil function and intended land use. However, to manage and maintain our soils in an acceptable state for future generations, soil quality must be defined, and the definition must be broad enough to encompass the many functions of soil. In other words, as a natural body soil has importance and value in itself not necessarily as defined by its managed applications. These considerations led to the following definition: 'Soil quality is the capacity of soil to function, within ecosystem and land-use boundaries, to sustain biological productivity, maintain environmental quality and promote plant, animal and human health' (after Doran and Parkin, 1994).

Defining soil health

The terms soil quality and soil health are often used interchangeably in the scientific literature and popular press with scientists, in general, preferring soil quality and producers preferring soil health (Harris and Bezdicek, 1994). Some prefer the term soil health because it portrays soil as a living, dynamic organism that functions holistically rather than as an inanimate mixture of sand, silt and clay. Others prefer the term soil quality and descriptors of its innate quantifiable physical, chemical and biological characteristics. Much discussion at a recent soil health conference in the midwest USA centred on the importance of defining soil health (Soil Health: The Basis of Current and Future Production, Decatur, Illinois, December 7, 1994). In those discussions it was observed that efforts to define the concept of soil health have produced a polarization of attitudes concerning the term. On the one hand are those, typically speaking from outside agriculture, who view maintenance of soil health as an absolute moral imperative – critical to our very survival as a species. On the other hand is the attitude, perhaps ironically expressed most adamantly by academics, that the term is a misnomer – a viewpoint seated, in part, in fear that the concept requires value judgments which go beyond scientific or technical fact. The producers, and therefore society's management of the soil, are caught in the middle of these opposing views and the communication failures that result.

'Health' is defined as, 'the condition of an organism or one of its parts in which it performs its vital functions normally or properly' (Webster's Third New International Dictionary, 1986). The word is derived from the Old English word 'haelth' that was itself derived from the concept of 'whole'. Dr David White, a natural resource economist and speaker at the aforementioned soil health conference, proposed that any definition of soil health should: (i) reflect the soil as a living system; (ii) address all essential functions of soil in the landscape; (iii) compare the condition of a given soil against its own unique potential within climatic, landscape, and vegetation patterns; and (iv) somehow enable meaningful assessment of trends. It is interesting to note that with some modification, the definition of soil quality presented earlier could serve as a definition of soil health.

With consideration of the aforementioned factors, soil health can be defined as: 'the continued capacity of soil to function as a vital living system, within ecosystem and land-use boundaries, to sustain biological productivity, promote the quality of air and water environments, and maintain plant, animal and human health'. The challenge we face, however, is in quantitatively defining the state of soil health and its assessment using measurable properties or parameters. Unlike human health, the magnitude of critical indicators of soil health ranges considerably over dimensions of time and space.

For the remainder of this chapter the terms soil quality and soil health will be used synonymously. However, the term soil health is preferred in that it more clearly portrays the idea of soil as a living dynamic organism that functions in a holistic way depending on its condition or state rather than as an inanimate object whose value depends on its innate characteristics and intended use.

Assessment of Soil Quality and Health

Establishing an ongoing assessment of the condition and health of our soil resources is vital to maintaining the sustainability of agriculture and civilization. As discussed earlier, the failure of several earlier civilizations was sealed by their disregard for the health of finite soil resources. In today's energy- and technology-intensive world, the need for maintaining the health of our soil resources is imperative to sustaining productivity for increasing populations and in maintaining global function and balance. Assessment of soil quality and health is invaluable in determining the sustainability of soil and land management systems and in evaluating their long-term effectiveness. However, we need a framework for evaluation and standards of soil quality and health to identify problems in production areas, to make realistic estimates of sustainable food production, to monitor changes in environmental quality as related to agricultural management, and to assist government agencies in formulating and evaluating sustainable agricultural and other land-use policies (Granatstein and Bezdicek, 1992; Acton, 1993). Identification of appropriate indicators for soil health assessment is complicated by the fact that they must account both for multiple dimensions of soil function, such as productivity and environmental well-being, and the multiplicity of physical, chemical and biological factors which control biogeochemical processes; and their variation in intensity over time and space.

Use of indicators

Assessing soil health can be likened to a medical examination of humans in which certain measurements are taken of the quality of certain parameters as basic indicators of system function (Larson and Pierce, 1991). In a medical examination, the physician takes measurements of body system functions such as temperature, blood pressure, pulse rate, and perhaps certain blood or urine chemistries. The physician will also take note of visible, outward signs of health status. If these basic indicators are outside specific ranges, more diagnostic tests can be conducted to help identify the cause of the problem and find a solution. For example, excessively high blood pressure may indicate a potential for system failure (death) through stroke or cardiac arrest. Because one of the causes of high blood pressure may be improper diet, lack of exercise, or high stress level, the physician may request a secondary blood chemistry test for cholesterol, electrolytes, etc. Assessment of stress level as a causative factor for high blood pressure is less straightforward and generally involves implementing some change in lifestyle followed by periodic monitoring of blood pressure to assess change. This is a good example of using a basic indicator both to identify a problem and to monitor the effects of management on the health of a system.

Applying this human health analogy to soil health is fairly straightforward. Larson and Pierce (1991) proposed that a minimum data set (MDS) of soil parameters be adopted for assessing the health of world soils, and that standardized methodologies and procedures be established to assess changes in the quality of those factors. A set of basic indicators of soil quality and, therefore, health has not previously been defined, largely due to difficulty in defining soil quality and health, the wide range over which soil indicators vary in magnitude and importance, and disagreement among scientists and soil and land managers over which basic indicators should be measured.

Acton and Padbury (1993) defined soil quality attributes as measurable soil properties that influence the capacity of soil to perform crop production or environmental functions. Soil attributes are useful in defining soil quality criteria and serve as indicators of change in quality. Attributes that are most sensitive to management are most desirable as indicators and some such as soil depth, soil organic matter and electrical conductivity are often affected by soil degradation processes (Arshad and Coen, 1992).

To be practical for use by practitioners, extension workers, conservationists, scientists, and policy makers the set of basic soil quality/health indicators should be useful over a range of ecological and socioeconomic situations.
Indicators should:

1. Correlate well with ecosystem processes (this also increases their utility in process oriented modelling).
2. Integrate soil physical, chemical, and biological properties and processes and serve as basic inputs needed for estimation of soil properties or functions which are more difficult to measure directly.
3. Be relatively easy to use under field conditions and be assessable by both specialists and producers.
4. Be sensitive to variations in management and climate. The indicators should be sensitive enough to reflect the influence of management and climate on long-term changes in soil quality but not be so sensitive as to be influenced by short-term weather patterns.
5. Be components of existing soil data bases where possible.

The need for basic soil quality and health indicators is reflected in the question commonly posed by producers, researchers, and conservationists: 'What measurements should I make or what can I observe that will help me evaluate the effects of management on soil function now and in the future?'. Too often scientists confine their interests and efforts to the discipline with which they are most familiar. Microbiologists often limit their studies to soil microbial populations, having little or no regard for soil physical or chemical characteristics which define the limits of activity for microorganisms, plants, and other life forms. The proper approach in defining soil quality and health indicators must be holistic, not reductionistic. The indicators chosen must also be measurable by as many people as possible, especially managers of the land, and not limited to a select cadre of research scientists. Indicators should describe the major ecological processes in soil and ensure that measurements made reflect conditions as they exist in the field under a given management system. They should relate to major ecosystem functions such as C and N cycling (Visser and Parkinson, 1992) and be driving variables for process oriented models which emulate ecosystem function. Some indicators, such as soil bulk density, must be measured in the field so that laboratory analyses for soil organic matter and nutrient content can be better related to actual field conditions at time of sampling. Soil bulk density is also required for calculation of soil properties such as water-filled pore space (WFPS) which serves as an excellent integrator of physical, chemical and biological soil properties and aeration dependent microbial processes important to C and N cycling in soil (Doran *et al.*, 1990). Many

Table 1.1. A limited listing of soil attributes or properties which can be estimated from basic input variables using pedotransfer functions or simple models.

Soil attribute or property	Basic input variables	Reference
Cation exchange capacity	Org. C + clay type and content	Larson and Pierce, 1994
Water retension charac. (AWHC)	% sand, silt, clay, + org. C + BD	Gupta and Larson, 1979
Hydraulic conductivity	Soil texture	Larson and Pierce, 1994
Aerobic and anaerobic microbial activity	WFPS as calculated from BD and water content	Linn and Doran, 1984 Doran *et al.*, 1990
C and N cycling	Soil respiration (Soil temperature + WFPS)	Parkin *et al.*, 1996
Plant/microbial activity or pollution potential	Soil pH + EC	Smith and Doran, 1996
Soil productivity	BD, AWHC, pH, EC, and aeration	Larson and Pierce, 1994
Rooting depth	BD, AWHC, pH	Larson and Pierce, 1994
Leaching potential	Soil texture, pH, org. C (hydr. cond., CEC, depth)	Shea *et al.*, 1992

Abbreviations: AWHC, available water holding capacity; BD, soil bulk density; EC, soil electrical conductivity; WFPS, water-filled pore space.

basic soil properties are useful in estimating other soil properties or attributes which are difficult or too expensive to measure directly. A listing of these basic indicators and input variables and the soil attributes they can be used to estimate are given in Table 1.1.

Starting with the minimum data set (MDS) proposed by Larson and Pierce (1991), we have developed a list of basic soil properties (Table 1.2) which meets many of the aforementioned requirements of indicators for screening soil quality and health. Appropriate use of such indicators, however, will depend to a large extent on how well these indicators are understood with respect to the ecosystem of which they are part. Thus, interpretation of the relevance of soil biological indicators apart from soil physical and chemical attributes and their ecological relevance is of little value and, with respect to assessment of soil quality or health, can actually be misleading.

Data presented in a Science article (Reganold *et al.*, 1993) describing soil quality and financial performance of biodynamic and conventional farming management systems in New Zealand, are useful in illustrating this concern (Table 1.3). Our analyses, however, are not intended as criticisms of this published

Table 1.2. Proposed minimum data set of physical, chemical, and biological indicators for screening the condition, quality, and health of soil (after Doran and Parkin, 1994 and Larson and Pierce, 1994).

Indicators of soil condition	Relationship to soil condition and function; Rationale as a priority measurement	Ecologically relevant values/units; Comparisons for evaluation
Physical		
Texture	Retention and transport of water and chemicals; Modelling use, soil erosion and variability estimate	% Sand, silt, and clay; Less eroded sites or landscape positions
Depth of soil, topsoil, and rooting	Estimate of productivity potential and erosion; Normalizes landscape and geographic variability	cm or m; Non-cultivated sites or varying landscape positions
Infiltration and soil bulk density (SBD)	Potential for leaching, productivity, and erosivity; SBD needed to adjust analyses to volumetric basis	minutes per 2.5 cm of water and Mg cm^{-3}; Row and/or landscape positions
Water holding capacity (water retention characteristics)	Related to water retention, transport, and erosivity; Available H_2O: calculate from SBD, texture, and OM	% (Mg cm^{-3}), cm of available H_2O per 30 cm ; Precipitation intensity
Chemical		
Soil organic matter (OM) (total organic C and N)	Defines soil fertility, stability, and erosion extent; Use in process models and for site normalization	kg C or N ha^{-1}–30 cm; Non-cultivated or native control
pH	Defines biological and chemical activity thresholds; Essential to process modelling	Compared with upper and lower limits for plant and microbial activity
Electrical conductivity	Defines plant and microbial activity thresholds; Presently lacking in most process models	dS m^{-1}; Compared with upper and lower limits for plant and microbial activity
Extractable N, P, and K	Plant available nutrients and potential for N loss; Productivity and environmental quality indicators	kg ha^{-1}–30 cm; Seasonal sufficiency levels for crop growth
Biological		
Microbial biomass C and N	Microbial catalytic potential and repository for C and N; Modelling: early warning of management effects on OM	kg N or C ha^{-1}–30 cm; Relative to total C and N or CO_2 produced
Potentially mineralizable N (anaerobic incubation)	Soil productivity and N supplying potential; Process modelling; (surrogate indicator of biomass)	kg N ha^{-1}–30 cm d^{-1}; Relative to total C or total N contents
Soil respiration, water content, and temperature	Microbial activity measure (in some cases plants) Process modelling; estimate of biomass activity	kg C ha^{-1} d^{-1}; Relative microbial biomass actvity, C loss vs. inputs and total C pool

Table 1.3. Reported and ecologically relevant mean values of aggregated soil quality data for 0–20 cm layer of 16 biodynamic and conventional farms in New Zealand (after Reganold *et al.*, 1993).

Soil property	Biodynamic farms	Conventional farms	Ratio Bio./Conv.
Reported units and values			
0–5 cm Bulk density (Mg m^{-3})	1.07	1.15	0.93*
Topsoil thickness (cm)	22.8	20.6	1.11*
Carbon (%)	4.84	4.27	1.13*
Total N (mg kg^{-1})	4840	4260	1.14*
Mineralizable N (mg kg^{-1})	140.0	105.9	1.32*
Respiration (μl O_2 $h^{-1}g^{-1}$)	73.7	55.4	1.33*
Ratio: mineralizable N to C (mg g^{-1})	2.99	2.59	1.15*
Extractable P (mg kg^{-1})	45.7	66.2	0.69*
pH	6.10	6.29	0.97*
Ecologically relevant units and values			
0–20 cm Bulk density† (Mg cm^{-3})	1.2	1.3	0.92
Carbon (Mg C ha^{-1} –20 cm)	116.2	111.0	1.05
Total N (kg N ha^{-1} –20 cm)	11,616	11,076	1.05
Mineralizable N (kg N ha^{-1} –20 cm 14 d^{-1})	336	275	1.22
Respiration in lab (kg C ha^{-1} –20 cm d^{-1})	2275	1850	1.23
Ratio: mineralizable N to C	2.89	2.48	1.17*
Extractable P (excess)‡ (kg P ha^{-1} –20 cm)	110 (50)‡	172 (112)	0.63*
pH units above 6.0 lower limit	0.1	0.3	0.33

* Values differ significantly ($P < 0.01$); † Estimated, since data was only given for 0–5 cm depth; ‡ Threshold value for environmentally sound soil P level set at 60 kg P ha^{-1}.

work as the authors should be commended for their vision in choice of physical, chemical and biological indicators of soil quality. One point of discussion is the importance of expressing the results of soil quality tests on a volumetric rather than a gravimetric basis and in units for which ecological relevance can be readily ascertained. As illustrated in Table 1.3, the magnitude of differences in soil C, total N, respiration, and mineralizable N between management systems for samples expressed by weight of soil are 8–10% greater than where expressed on a volume basis using soil bulk density estimates. In cultivated systems, soil bulk density can vary considerably across the soil surface due to mechanical compaction and throughout the growing season due to reconsolidation of soil after tillage. Soil bulk density is also directly proportional to the mass of any soil component for a given depth of soil sampled. Where samples are taken in the field under management conditions of varying soil densities, comparisons made using gravimetric analyses will err by the difference in soil density at time of sampling. The observed differences due to management in the New Zealand study were statistically significant. However, since results were expressed on a gravimetric basis, they may not be valid nor ecologically relevant. In cases such as this, where values for soil bulk density at time of sampling are not available, the use of soil indicator ratios (in this case mineralizable N to C) can reduce errors of interpretation associated with use of results expressed on a weight basis. Reganold and Palmer (1995) recommend calculating soil measurements on a volume basis per unit of topsoil or solum depth for most accurate assessment of management effects on soil quality.

Choice of units of expression for soil quality indicators can also have an important bearing on determining the ecological relevance of measured values. In the New Zealand study, respiration of laboratory incubated soils from biodynamic farms averaged 73.7 ml O_2 $h^{-1}g^{-1}$, significantly greater (33%) than that from conventional farms. One interpretation of these results could be that the soils of the biodynamic farms are healthier since respiration was greater. However, if one assumes that for aerobic respiration a mole of oxygen is consumed for each mole of carbon dioxide produced, and the results are adjusted for soil density and expressed as kg C released per hectare per day, a different picture emerges. The quantities of C released in one day from both the biodynamic and conventional farms are incredibly high and represent 2.0 and 1.7%, respectively, of the total C pools of these surface soils. While the values for soil respiration from disturbed soils incubated in the laboratory only represent a potential for release of readily metabolizable C (labile C), the results clearly demonstrate that more may not be better and that high rates of respiration may be ecologically detrimental as they represent potentials for depletion of soil organic C with accelerated enrichment of the atmosphere with carbon dioxide. When expressed in ecologically relevant units, it becomes obvious that the respiration rates observed in this study are of limited use in evaluating the status of soil quality and health between these different farming management systems when used as the only indicator. Similar observations can be made for mineralizable N and extractable P. Levels of mineralizable N above that needed for crop production for biodynamic farms and extractable P levels above crop needs for conventional farms could represent a lower level of soil quality and health as a result of greater potential for environmental contamination through leaching, runoff, or volatilization losses. Specific upper limits for environmentally sound levels of soil P and N exist and are determined by local climatic, topographic, soil and management situations (Sharpley *et al.*, 1996). Again, an example that with respect to soil quality and health, more is not necessarily better and ecologically relevant units are needed for proper evaluation. Soil pH is another example of a soil quality attribute that must be referenced to a definable standard for upper and lower limits which are defined by the cropping system or biological processes of greatest ecological relevance. The above discussion serves to highlight the difficulty we have in interpreting the results of laboratory incubations and the need for in-field measurements of respiration and N cycling.

Indicators of soil quality and health are commonly used to make comparative assessments between agricultural management practices to determine their sustainability. However, the utility of comparative assessments of soil quality are limited because they provide little information about the processes creating the measured condition or performance factors associated with respective management systems (Larson and Pierce, 1994). Also, the mere analysis of soils, no matter how comprehensive or sophisticated does not provide a measure of soil quality or health unless the parameters are calibrated against designated soil functions (Janzen *et al.*, 1992).

Quantitative assessments

Quantitative assessments of soil quality and health will require consideration of the many functions that soils perform, their variations in time and space and opportunities for modification or change. Criteria are needed to evaluate the impact of various practices on the quality of air, soil, water and food resources. Soil quality and health can not be defined in terms of a single number, such as the 10 mg l^{-1} NO_3-N standard applied for drinking water, although such quantitative standards will be valuable to overall assessment. Assessments must consider specific soil functions being evaluated in their land use and societal contexts. Threshold values for key indicators must be established with the knowledge that these will vary depending upon land use, the specific soil function of greatest concern and the ecosystem or landscape within which the assessment is being made. For example, soil organic matter concentration is frequently cited as a major indicator of soil quality. Threshold values established for highly weathered Ultisols in the southeastern US indicate that surface soil organic matter levels of 2% (1.2% organic C) would be very good, while the same value for Mollisols developed under grass in the Great Plains, which commonly have higher organic matter levels, would represent a degraded condition limiting soil productivity (Fig. 1.1). As pointed out by Janzen *et al.* (1992) the relationship between soil quality indicators and various soil functions does not always comply to a simple relationship increasing linearly with magnitude of the indicator, as is commonly thought. Simply put, bigger is not necessarily better.

Soil quality and health assessments will have to be initiated within the context of societal goals for a specific landscape or ecosystem. Examples include establishing goals such as enhancing water quality, soil productivity, biodiversity, or recreational opportunities. When specific goals have been established or are known, then critical soil functions needed to achieve those goals can be agreed upon, and the criteria for assessing progress toward achieving those goals can be set. Periodic assessments of soil quality and health with known indicators, thresholds, and other criteria for evaluation will then make it possible to quantify soil quality and health.

To accomplish such goals, several approaches for assessing soil quality have been proposed (Acton and Padbury, 1993; Doran and Parkin, 1994; Karlen *et*

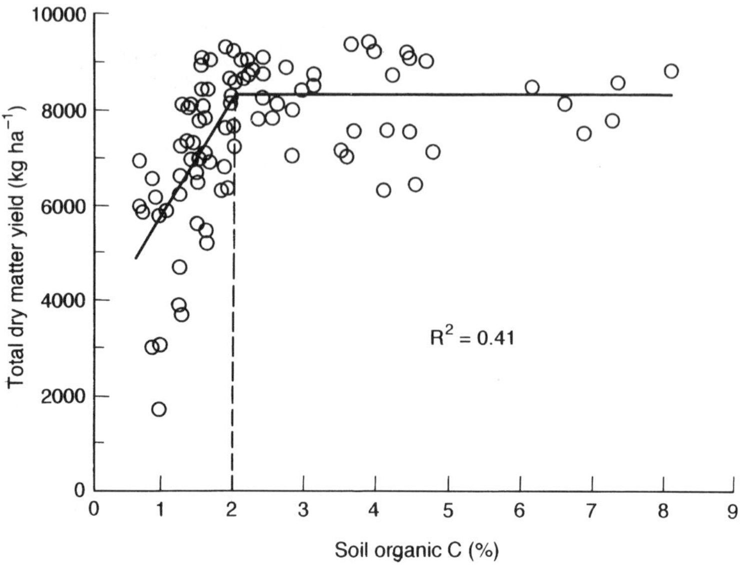

Fig. 1.1. Relationship between organic C concentration in the surface 0–15 cm of soil and soil productivity as determined by total dry matter yield at dryland site in Alberta, Canada in 1991 (after Janzen *et al.*, 1992; with permission).

al., 1994; Larson and Pierce, 1994). A common attribute among all these approaches is that soil quality is assessed with respect to specific soil functions. Larson and Pierce (1994) proposed a dynamic assessment approach in which the dynamics, or change in soil quality, of a management system is used as a measure of its sustainability. They proposed the use of a minimum data set of temporally variable soil properties to monitor changes in soil quality over time. They also proposed the use of pedotransfer functions (Bouma, 1989) to estimate soil attributes which are too costly to measure and to interrelate soil characteristics in evaluation of soil quality (Table 1.1). Simple computer models are used to describe how changes in soil quality indicators impact important functions of soil, such as productivity. An important part of this approach is the use of statistical quality control procedures to assess the performance of a given management system rather than its evaluation by comparison to other systems. This dynamic approach for assessing soil quality permits identification of critical parameters and facilitates corrective actions for sustainable management.

Karlen and Stott (1994) presented a framework for evaluating site-specific changes in soil quality. In this approach they define a high quality soil as one that: (i) accommodates water entry; (ii) retains and supplies water to plants; (iii) resists degradation; and (iv) supports plant growth. They described a procedure by which soil quality indicators which quantify these functions are identified, assigned a priority or weight which reflects relative importance, and are scored using a systems engineering approach for a particular soil attribute such as resistance to water erosion. Karlen *et al.* (1994) also demonstrated the utility of this approach in discriminating changes in soil quality between long-term crop residue and tillage management practices.

Doran and Parkin (1994) described a performance-based index of soil quality that could be used to provide an evaluation of soil function with regard to the major issues of: (i) sustainable production; (ii) environmental quality; and (iii) human and animal health. They proposed a soil quality index consisting of six elements:

$$SQ = f\ (SQE1, SQE2, SQE3, SQE4, SQE5, SQE6)$$

where soil quality elements are: SQE1 = food and fibre production; SQE2 = erosivity; SQE3 = groundwater quality; SQE4 = surface water quality; SQE5 = air quality; and SQE6 = food quality.

One advantage of this approach is that soil functions can be assessed based on specific performance criteria established for each element, for a given ecosystem. For example, yield goals for crop production (SQE1); limits for erosion losses (SQE2); concentration limits for chemicals leaching from the rooting zone (SQE3); nutrient, chemical and sediment loading limits to adjacent surface water systems (SQE4); production and uptake rates for gases that contribute to ozone destruction or the greenhouse effect (SQE5); and nutritional composition and chemical residue of food (SQE6). This list of elements is restricted primarily to agricultural situations but other elements such as wildlife habitat quality could be easily added to expand the applications of this approach.

This approach would result in soil quality indices computed in a manner analogous to the soil tilth index proposed by Singh *et al.* (1990). Weighting factors are assigned to each soil quality element, with relative weights of each coefficient being determined by geographical considerations, societal concerns and economic constraints. For example in a given region, food production may be the primary concern and elements such as air quality may be of secondary importance. If such were the case, SQE1 would be weighted more heavily than SQE5. Thus this framework has an inherent flexibility in that the precise functional relationship for a given region, or a given field, is determined by the intended use of that area or site, as dictated by geographical and climatic constraints as well as socioeconomic concerns.

Assessment of soil quality and health is not limited to areas used for crop production. Forests and forest soils are important to the global C balance as related to C sequestration and atmospheric levels of carbon dioxide. Soil organic matter and soil porosity, as estimated from soil bulk density, have recently been proposed among international groups as major soil quality indicators in forest soils (Richard Cline, personal communication). Criteria for evaluating rangeland health have recently been suggested in a National Research Council (1994) report which describes new methods to help classify, inventory and monitor rangelands. Rangeland health is defined as the degree to which the integrity of the soil and the ecological processes of rangeland ecosystems are sustained. Assessment of rangeland health is based on the evaluation of three criteria: (i) degree of soil stability and watershed function; (ii) integrity of nutrient cycles and energy flows; and (iii) presence of functioning recovery mechanisms.

Value of qualitative/descriptive assessments

The concept of soil health is in many ways farmer-generated and rooted in observational field experiences which translate into descriptive properties such as its look, feel, resistance to tillage, smell, presence of biota, etc. Harris and Bezdicek (1994) conclude that farmer-derived descriptive properties for assessing soil health are valuable for: (i) defining or describing soil quality/health in meaningful terms; (ii) providing a descriptive property of soil quality/health; and (iii) providing a foundation for developing and validating an analytical component of soil health based on quantifiable chemical, physical, and biological properties that can be used as a basis for management and policy decisions. Unfortunately, the potential contributions of indigenous farmer knowledge to management of soil quality/health throughout the world has not been fully utilized (Pawluk *et al.*, 1992).

Descriptive soil information is not commonly used in scientific literature dealing with characterization of soil quality/health. However, Arshad and Coen (1992) indicate that many soil attributes can be estimated by calibrating qualitative observations against measured values and recommend that qualitative (descriptive) information should be an essential part of soil quality monitoring programmes. Visual and morphological observations in the field can be used by both producers and scientists to recognize degraded soil quality caused by: (i) loss of organic matter, reduced aggregation, low conductivity, soil crusting and sealing; (ii) water erosion, as indicated by rills, gulleys, stones on the surface, exposed roots, uneven topsoil; (iii) wind erosion as indicated by ripple marks, dunes, sand against plant stems, plant damage, dust in air, etc. ; (iv) salinization, as indicated by salt crust and salt-tolerant plants; (v) acidification and chemical degradation, as indicated by growth response of acid-tolerant and -intolerant plants and lack of fertilizer response; and (vi) poor drainage and structural deterioration, as indicated by standing water and poor or chlorotic plant stands.

Doran *et al.* (1994a,b) stressed the importance of holistic management approaches which optimize the multiple functions of soil and conserve soil resources and support strategies for promoting soil quality and health. They proposed use of the basic set of soil quality and health indicators (Table 1.2) to assess soil health in various agricultural management systems. However, while many of these key indicators are extremely useful to specialists (i.e. researchers, consultants, extension staff and conservationists) many of them are beyond the expertise of the producer to measure (Hamblin, 1991). In response to this

Table 1.4. Sustainable management strategies for building soil quality and health and associated indicators which are assessable by producers.

Strategy	Indicators
Conserve soil organic matter through maintaining balance in C and N cycles where inputs × outputs	Direction/change in organic matter levels with time; OM potential within soil, climate, and cropping patterns; Both visual and analytical measures; Soil infiltration/water-holding capacity
Minimize soil erosion through conservation tillage and increased soil cover (residue, cover crops, green fallow, etc)	Visual signs (gullies, rills, dust, etc.); Surface soil characteristics: (depth of topsoil, organic matter content/texture, infiltration rate)
Substitution of renewable for non-renewable resources through less reliance on synthetic chemicals, use of conservation tillage, and greater use of natural balance and diversity (crop rotation, legume cover crops, etc.)	Crop growth characteristics (yield, N content, colour, rooting); Soil and water nitrate levels; Soil physical condition/compaction; Input costs
Move toward management systems which coexist more with and less dominate natural systems through optimizing productivity needs with environmental quality	Crop growth characteristics (yield, N content, colour, vigour); Soil and water nitrate levels; Synchronization of N availability with crop needs during year

dilemma, Doran (1995) presented strategies for building soil quality and health which also included generic indicators which are measurable by and accessible to producers within the time constraints imposed by their normally hectic and unpredictable schedules (Table 1.4). Soil organic matter, crop appearance and erosion were ranked by farmers in the Northern US Corn and Dairy Belt as the top three properties for describing soil health (Romig *et al.*, 1995).

Soil Assessment – Need for Producer/Specialist Interaction

Integration of soil health concepts into farm management

At a time when agriculture must address environmental degradation due to certain yield promoting practices against ever increasing demands for both greater and better-distributed food supplies, the concept of soil health can be a useful communication device in meeting present and future world needs. Stewardship of the soil resource that enhances soil quality and health while allowing for acceptable long-term production levels is in everyone's best interest and satisfies what has been called the 'Ecocentric' notion of the Common Good (Stauber, 1994). Soil management practices must now be evaluated for their impacts across the temporal scale – short, middle and long-term, as well as across the landscape, to be truly sustainable (Swift *et al.*, 1991).

Producers around the globe receive advice from many sources about recommended production practices. Unfortunately, much of this advice is often aimed at relatively short-term (1 or 2 years) economic gains to their operation, rather than on long-term resource conservation (Stauber, 1994). Additionally, advice may be value-laden, or linked to agribusiness sales, such as soil tests performed by private companies which may indicate need for unnecessary chemical fertilizers and pesticides (Cramer, 1986; Soule and Piper, 1992). Management recommendations are often developed for regions which may encompass a wide variation in soil type, topography and resource availability. In such cases, practices which are appropriate for experimental conditions may be inappropriate on a large portion of the individual farms to which they are recommended. To begin the move toward site-specific best management practices, tests for soil quality indicators should be developed as meters for gauging both the short and long-term effects of various production practices on soil health. Soil quality tests that yield results uncoupled from value judgments will allow both land stewards and researchers to evaluate production practices objectively under a wide range of conditions, to identify those that are truly improving soil health. Clearly, value judgment is always likely to be necessary to reconcile the need for food production with the need to maintain soil in a near-natural state. Tests which accurately measure impacts on soil quality of various options will help make the consequences of those options more apparent. If tests are made to be used by producers and other land stewards, production practices will not only be efficiently tailored for individual situations, but researchers will have a many fold increase in the information available to better understand soil processes.

The concept of soil health can be a key tool for educating producers about some of the less obvious potentials for soil degradation due to poor management. There is some evidence that a concern for soil health may lead land stewards to production practices that indeed improve some soil characteristics. Van Kooten *et al.* (1990) found in southwestern Saskatchewan that farmer concern for soil quality was in fact correlated to production practices which improved soil physical parameters. The authors found, however, that in areas of deep topsoil, farmers were less likely to be seriously concerned with soil quality, which pinpoints the need to emphasize the long-term vision of soil health.

Technology transfer

Producers and land managers need practical tools which they can use to determine the effectiveness of their management practices on soil health and sustainable production. Traditional research has identified management practices that conserve the soil resource, protect air and water quality, or maximize crop yields. However, development of sustainable management strategies that maintain soil quality and health and balance production needs with environmental concerns require new research approaches and on-site evaluation to confirm the specific applications of general strategies across the range of climatic, soil, economic and social conditions experienced by agriculture. Facilitating producer participation in the research process is essential to the development of practical production systems and assessment approaches which address both the needs of producers and society in general. Indicators of soil health and practical assessment tools are essential to forming this necessary partnership between producers and the technical community. National indicators of some aspects of soil quality and health are likely to be established within the next decade as a means of monitoring the state of our soil resources. It is imperative that these indicators be useful to producers, especially if incentives or regulations based on soil quality or health are enacted.

To include producers as active participants in on-site assessment of soil quality/health, tools and methodologies used by researchers must be adapted to be easily accessible to the producers themselves (Sarrantonio *et al.*, 1996). Tests should be simple to perform, require little in the way of expensive equipment, and give rapid results. Additionally, tests should be able to measure soil indicators that are meaningful to the producers' understanding of soil and its processes, and that give results that are reliable, accurate within an acceptable range, and are easily understood and used. A soil quality kit is currently being developed by USDA-ARS to help producers, researchers, conservationists, environmentalists and consultants assess the health and quality of soil and facilitate technology transfer (Cramer, 1994). The kit provides on-site capability for assessment of many of the potential indicators of soil quality and health (see Table 1.2) such as soil pH, electrical conductivity, soil and water nitrate levels, soil density, water infiltration, water-holding capacity, soil water content, water-filled pore space, soil temperature and soil respiration. The kit provides producers and agricultural specialists with the necessary means for a cursory assessment of the complex suite of physical, chemical and biological factors that comprise soil quality/health. Tests facilitate on-site identification of soil resource condition and its degree of degradation. Currently the cost of the test kit is under $250, yet results obtained with this kit compare well with standard laboratory procedures that are more time consuming and costly (Liebig *et al.*, 1996). The utility of this test kit is currently being evaluated by conservationists, researchers, extension educators, environmental monitors (EPA-EMAP), and producers at locations in Australia, Canada, Cuba, Greece, Honduras, India, New Zealand, Poland, Moldova, Russia and in over 50 locations the United States. Preliminary results suggest the kit is useful to specialists in fostering appreciation for the complexity of soil, in bridging across disciplinary boundaries, and facilitating assessment of soil quality and health. However, the overall procedure for on-site assessment of soil quality and health was found too complicated and time consuming for practical use by producers. The kit is best used by producers as a tool kit from which specific tests can be used as needed to assess soil quality and health. Also, compilation of a practical manual for the test kit, similar to that included with the 'Sustainability Kit' produced in Australia (Powell and Pratley, 1991), would greatly aid utility of this test kit and interpretation of results.

Practical tools for soil quality and health assessment by producers must aid their comprehension of the concept of soil health and be useful to them within the context of their normal work routines (after Nowak in Leopold Letter, 1995). Knowledge of soil for most producers is largely limited to that which they gain through their sensory experiences in working the soil with agricultural implements and watching plant growing conditions during the growing season. Clues producers most often use to differentiate soils include soil colour (largely organic matter), the workability of soil (structure and compaction), wetness or dryness of soil (drainage, storage and infiltration capacity) and topsoil texture and depth (indicators of soil erosion and production potential). Crop yield and input costs are indicators which producers most often rely upon to assess the short-term sustainability of their management practices. Inclusion of other tools for rapid assessment of efficiency of resource use such as quick tests for soil and water nitrate levels, adequacy of plant growth and N content and synchronization of soil nitrogen supplies with crop plant needs will facilitate development of reduced input management systems and management strategies for long-term sustainability (Table 1.4).

Conclusions

Soil is a finite and dynamic living resource that acts as an interface between agriculture and the environment and is vital to global function. Soil health can be defined as the continued capacity of soil to function as a vital living system, within ecosystem and land-use boundaries, to sustain biological productivity, promote the quality of air and water environments, and maintain plant, animal and human health. Advantages to giving value to soil health and its assessment include: (i) importance as a resource for evaluation of land-use policy; (ii) use in identification of critical landscapes or management systems; (iii) use in evaluation of practices that degrade or improve the soil resource; and (iv) utility in identifying gaps in our knowledge base and understanding of sustainable management.

To assure the sustainability of agricultural management systems, producers and land managers must be included as active participants in the quantitative and qualitative assessment of soil quality and health. Present research and education needs critical to assessment and enhancement of soil quality/health include:

1. Coordinated development of standards for soil quality/health, by national and local agencies and interest groups involved in agriculture, the environment, resource conservation, and economics, to assess sustainability changes with time. This requires establishment of reference guidelines and thresholds for indicators of soil quality/health that enable identification of relationships between soil measures and soil function which permit valid comparisons across variations in climate, soils, landuse, topography and management systems. This will also require identification of appropriate scales of time and space for assessment of soil quality/health and development of standardized protocols for sampling, processing and analysis.
2. Development of practical approaches and tools for on-site assessment of soil quality/health by farmers, researchers, extension workers, conservationists and environmental monitors that can also be used by resource managers and policy makers to determine the sustainability of land management practices.

We are beginning to realize that soil health, by its broadest definition, is inseparable from issues of sustainability. The challenge before us is to develop holistic approaches for assessing soil quality and health that are useful to producers, specialists, and policy makers in identifying agricultural and land-use management systems which are profitable and will sustain our soil resources for future generations.

References

Acton, D.F. (1993) *A Program to Assess and Monitor Soil Quality in Canada: Soil Quality Evaluation Program Summary (Interim).* Centre for Land and Biological Resources Research Contribution No. 93–49. Research Branch, Agriculture Canada, Ottawa.

Acton, D.F. and Gregorich, L.J. (1995) *The Health of our Soils: Toward Sustainable Agriculture in Canada.* Agriculture and Agri-Food Canada, Ottawa.

Acton, D.F. and Padbury, G.A. (1993) A conceptual framework for soil quality assessment and monitoring. In: Acton, D.F. (ed.) *A Program to Assess and Monitor Soil Quality in Canada: Soil Quality Evaluation Program Summary (Interim).* Centre for Land and Biological Resources Research Contribution No. 93–49. Research Branch, Agriculture Canada, Ottawa, pp. 2–7.

Alexander, M. (1971) Agricultures responsibility in establishing soil quality criteria. In: *Environmental Improvement – Agricultures Challenge in the Seventies.* National Academy of Sciences, Washington, DC, pp. 66–71.

Arshad, M.A. and Coen, G.M. (1992) Characterization of soil quality: Physical and chemical criteria. *American Journal of Alternative Agriculture* 7, 12–16.

Avery, D.T. (1995) *Saving the Planet with Pesticides and Plastic: The Environmental Triumph of High-Yielding Farming.* Hudson Institute Inc., Indianapolis.

Bhagat, S.P. (1990) *Creation in Crisis.* Brethren Press, Elgin, Illinois, 173 pp.

Blum, W.E.H. and Santelises, A.A. (1994) A concept of sustainability and resilience based on soil functions. In: Greenland, D.J. and Szabolcs, I. (eds) *Soil*

Resilience and Sustainable Land Use. CAB International, Wallingford, UK, pp. 535–542.

Bouma, J. (1989) Using soil survey data for quantitative land evaluation. *Advances in Soil Science* 9, 177–213.

Brown, L.R., Kane, H. and Roodman, D.M. (1994) Vital signs 1994: The trends that are shaping our future. *Worldwatch Institue.* W.W. Norton & Co., New York, p. 27.

Campbell, C.A., Paul, E.A. and McGill, W.B. (1976) Effect of cultivation and cropping on the amounts and forms of soil. In: Rice, W.A. (ed.) *Proceedings of the Western Canada Nitrogen Symposium.* Alberta Agriculture, Edmonton, Canada, pp. 9–101.

CAST (1992a) *Preparing US Agriculture for Global Climate Change.* Council for Agricultural Science and Technology, Ames, Iowa. Task force report no. 119.

CAST (1992b) *Water Quality: Agriculture's Role.* Council for Agricultural Science and Technology, Ames, Iowa. Task force report no. 120.

Cline, R.G. and Ruark, G.A. (1995) Management of forest soils. In: (Lal, R., Kimble, J., Levine, E. and Stewart, B.A. (eds) *Advances in Soil Science; Soil Management and Greenhouse Effect.* Lewis Publishers, Boca Raton, Florida, pp. 365–371.

Constanza, R., Norton, B.G. and Haskell, B.D. (1992) *Ecosystem Health: New Goals for Environmental Management.* Island Press, Washington, DC.

Cramer, C. (1986) *The Farmer's Fertilizer Handbook.* Regenerative Agricuture Association, Emmaus, Pennsylvania.

Cramer, C. (1994) Test your soils health: A three part series. *New Farm: Magazine of Regenerative Agriculture.* Jan., 17–21; Feb., 40–45; May/June, 46–51.

Doran, J.W. (1995) Building Quality Soil. In *Proceedings of the 1995 Alberta Conservation Tillage Workshop on 'Opportunities and Challenges in Sustainable Agriculture'.* Red Deer, Alberta, Canada.

Doran, J.W. and Parkin, T.B. (1994) Defining and assessing soil quality. In: Doran, J.W., Coleman, D.C., Bezdicek, D.F. and Steward, B.A. (eds) *Defining Soil Quality for a Sustainable Environment.* Soil Science Society of America Special Publication no. 35, Madison, Wisconsin. pp. 3–21.

Doran, J.W. and Werner, M.R. (1990) Management and soil biology. In: Francis, C.A., Flora, C.B. and King, L.D. (eds) *Sustainable Agriculture in Temperate Zones.* John Wiley & Sons, New York. pp. 205–230.

Doran, J.W., Mielke, L.N. and Power, J.F. (1990) Microbial activity as regulated by soil water-filled pore space. In: *Transactions of the 14th International Congress of Soil Science,* Kyoto, Japan, pp. 94–99.

Doran, J.W., Sarrantonio, M. and Jankke, R. (1994a) Strategies to promote soil quality and soil health. In: Pankhurst, C.E., Doube, B.M., Gupta, V.V.S.R. and Grace, P.R. (eds) *Soil Biota: Management in Sustainable Farming Systems.* CSIRO Press, Melbourne, Australia, pp. 230–237.

Doran, J.W., Varvel, G.E. and Culley, J.B.L. (1994b) Tillage and residue management effects on soil quality and sustainable land management. In: Wood, R.C. and Dumanski, J. (eds) *Proceedings of an International Workshop on Sustainable Land Management* Volume 2. Agricultural Institute of Canada, Ottawa. pp. 59–74.

Doran, J.W., Sarrantonio, M. and Liebig, M. (1996) Soil health and sustainability.

In: Sparks, D.L. (ed.) *Advances in Agronomy,* Vol. 56. Academic Press, New York.

Glanz, J.T. (1995) *Saving Our Soil: Solutions for Sustaining Earth's Vital Resource.* Johnson Books, Boulder, Colorado.

Gliessman, S.R. (1984) An agroecological approach to sustainable agriculture. In: Jackson, W., Berry, W. and Colman, B. (eds) *Meeting the Expectations of the Land,* North Point Press, San Francisco, California. pp. 160–171.

Gore, A. (1993) *Earth in the Balance: Ecology and the Human Spirit.* Houghton Mifflin Co., New York.

Granatstein, D. and Bezdicek, D.F. (1992) The need for a soil quality index: local and regional perspectives. *American Journal of Alternative Agriculture* 7, 12–16.

Gupta, S.C. and Larson, W.E. (1979) Estimating soil water retention characteristics from particle size distribution, organic matter percent, and bulk density. *Water Resources Research* 15, 1633–1635.

Habern, J. (1992) Viewpoint: a soil health index. *Journal of Soil Water Conservation* 47, 6.

Hallberg, G.R. (1987) Agricultural chemicals in ground water: extent and implications. *American Journal of Alternative Agriculture* 2, 3–15.

Hamblin, A. (1991) *Environmental Indicators for Sustainable Agriculture.* Report of a National Workshop, published by LWRRDC and GRDC, Australia. 96 pp.

Harris, R.F. and Bezdicek, D.F. (1994) Descriptive aspects of soil quality/health. In: Doran, J.W., Coleman, D.C., Bezdicek, D.F. and Stewart, B.A. (eds) *Defining Soil Quality for a Sustainable Environment.* Soil Science Society of America Special Publication no. 35, Madison, Wisconsin. pp. 23–35

Harwood, R.R. (1990) A history of sustainable agriculture. In: Edwards, C.A., Lal, R., Madden, P, Miller, R.H. and House, G. (eds) *Sustainable Agricultural Systems,* Soil Water Conservation Society of America, Ankeny, Iowa, pp. 3–19.

Hillel, D.J. (1991) *Out of the Earth: Soil and the Sustainability of Civilization.* The Free Press, New York.

Houghton, R.A., Hobbie, J.E., Melillo, J.M., Moore, B., Peterson, B.J., Shaver, G.R. and Woodwell, G.M. (1983) Changes in the carbon content of terrestrial biota and soils between 1860 and 1980: a net release of CO_2 to the atmosphere. *Ecological Monographs* 53, 235–262.

Jackson, W. and Piper, J. (1989) The necessary marriage between ecology and agriculture. *Ecology* 70, 1591–1593.

Janzen, H.H., Larney, F.J. and Olson, B.M. (1992) Soil quality factors of problem soils in Alberta. In: *Proceedings of the Alberta Soil Science Society Workshop.* Lethbridge, Alberta, Canada.

Jenny, H. (1984) The making and unmaking of fertile soil. In: Jackson, W., Berry, W. and Colman, B. (eds) *Meeting the Expectations of the Land.* North Point Press, San Francisco, California. pp. 44–55.

Jenny, H. (1980) The soil resource: origin and behavior. *Ecological Studies* 37, Springer-Verlag, New York.

Karlen, D.L. and Stott, D.E. (1994) A framework for evaluating physical and chemical indicators of soil quality. In: Doran, J.W., Coleman, D.C., Bezdicek, D.F. and Stewart, B.A. (eds) *Defining Soil Quality for a Sustainable Environment.*

Soil Science Society of America Special Publication no. 35, Madison, Wisconsin. pp. 53–72.

Karlen, D.L., Wollenhaupt, N.C., Erbach, D.C., Berry, E.C., Swan, J.B., Eash, N.S. and Jordahl, J.L. (1994) Crop residue effects on soil quality following 10-years of no-till corn. *Soil and Tillage Research* 31, 149–167.

Lal, R. (1994) Sustainable land use systems and soil resilience. In: Greenland, D.J. and Szabolcs, I. (eds) *Soil Resilience and Sustainable Land Use.* CAB International, Wallingford, UK, pp. 41–67.

Lal, R. and Pierce, F.J. (1991) The vanishing resource. In: Lal, R. and Pierce, F.J. (eds) *Soil Management for Sustainability,* Soil and Water Conservation Society, Ankeny, Iowa. pp. 1–5.

Larson, W.E. and Pierce, F.J. (1991) Conservation and enhancement of soil quality. In: *Evaluation for Sustainable Land Management in the Developing World, Vol. 2: Technical Papers.* IBSRAM Proceedings No. 12 (2), International Board for Soil Research and Management. Bangkok, Thailand.

Larson, W.E. and Pierce, F.J. (1994) The dynamics of soil quality as a measure of sustainable management. In: Doran, J.W., Coleman, D.C., Bezdicek, D.F. and Stewart, B.A. (eds) *Defining Soil Quality for a Sustainable Environment.* Soil Science Society of America Special Publication no. 35, Madison, Wisconsin. pp. 37–51.

Lehninger, A.L. (1973) *Bioenergetics: The Molecular Basis of Biological Energy Transformations,* 2nd edn W.A. Benjamin, Inc., Menlo Park, California.

Leopold, A. (1949) *A Sand County Almanac and Sketches Here and There.* Oxford University Press, New York.

Leopold Letter (1995) Perspectives on soil health: a discussion among scientists and farmers. March issue. Leopold Center for Sustainable Agriculture, Ames, Iowa.

Liebig, M.A., Doran, J.W. and Gardner, J.C. (1996) Evaluation of a field test kit for measuring selected soil quality indicators. *Agronomy Journal* 88, 683–686.

Linn, D.M. and Doran, J.W. (1984) Effect of water-filled pore space on carbon dioxide and nitrous oxide production in tilled and nontilled soils. *Soil Science Society American Journal* 48, 1267–1272.

National Research Council (1989) *Alternative Agriculture: Committee on the Role of Alternative Farming Methods in Modern Production Agriculture.* National Academy Press, Washington, DC.

National Research Council (1993) *Soil and Water Quality: An Agenda for Agriculture.* National Academy Press, Washington, DC.

National Research Council. (1994) *Rangeland Health: New Methods to Classify, Inventory, and Monitor Rangelands.* National Academy Press, Washington, DC.

Northwest Area Foundation (1994) *A Better Row to Hoe: The Economic, Environmental, and Social Impact of Sustainable Agriculture.* Northwest Area Foundation, St. Paul, Minnesota.

Oldeman, L.R. (1994) The global extent of soil degradation. In: Greenland, D.J. and Szabolcs, I. (eds) *Soil Resilience and Sustainable Land Use.* CAB International, Wallingford, UK, pp. 99–118.

Olson, G.E. (1981) Archaeology: Lessons on future soil use. *Journal of Soil Water Conservation* 36, 261–264.

Papendick, R.I. and Parr, J.F. (1992) Soil quality – The key to a sustainable agriculture. *American Journal of Alternative Agriculture* 7, 2–3.

Parkin, T.B., Doran, J.W. and Franco-Vizcaino, E. (1996) Field and laboratory tests of soil respiration. In: Doran, J.W. and Jones, A.J. (eds) *Methods for Assessing Soil Quality,* Soil Science Society of America, Special Publication no. 49, Madison, Wisconsin.

Pawluk, R.R., Sandor, J.A. and Tabor, J.A. (1992) The role of indigenous knowledge in agricultural development. *Journal of Soil Water Conservation* 47, 298–302.

Pearce, D. and Warford, J. (1993) *World Without End.* World Bank, Washington, DC.

Pierzynski, G.M., Sims, J.T. and Vance, G.F. (1994) *Soils and Environmental Quality.* Lewis Publishers, CRC Press, Boca Raton, Florida.

Postel, S. (1994) Carrying capacity: Earth's bottom line. In: Brown, L.R. (ed.) *State of the World, 1994.* W.W. Norton & Co., New York, pp. 3–21.

Powell, D. and Pratley, J. (1991) *Sustainability Kit Manual.* Centre for Conservation Farming, Charles Sturt University–Riverina, Wagga Wagga, Australia.

Power, J.F. and Papendick, R.I. (1985) Organic sources of nutrients. In: Engelstad, O.P. (ed.) *Fertilizer Technology and Use,* 3rd edn. Soil Science Society of America, Madison, Wisconsin, pp. 503–520.

Quinn, D. (1993) *Ishmael.* Bantam Books, New York.

Reganold, J.P. and Palmer, A.S. (1995) Significance of gravimetric versus volumetric measurements of soil quality under biodynamic, conventional, and continuous grass management. *Journal of Soil Water Conservation* 50, 298–305.

Reganold, J.P., Elliott, L.F. and Unger, Y.L. (1987) Long-term effects of organic and conventional farming on soil erosion. *Nature* 330, 370–372.

Reganold, J.P., Palmer, A.S., Lockhardt, J.C. and Macgregor, A.N. (1993) Soil quality and financial performance of biodynamic and conventional farms in New Zealand. *Science* 260, 344–349.

Rodale Institute. (1991) *Conference report and abstracts, International Conference on the Assessment and Monitoring of Soil Quality.* Rodale Press, Emmaus, Pennsylvania.

Rolston, D.E., Harper, L.A., Mosier, A.R. and Duxbury, A.R. (1993) *Agricultural Ecosystem Effects on Trace Gases and Global Climate Change.* American Agronomy Society Special Publication 55, Madison, Wisconsin, 206 pp.

Romig, D.E., Garlynd, M.J., Harris, R.F. and McSweeney, K. (1995) How farmers assess soil health and quality. *Journal of Soil Water Conservation* 50, 229–236.

Sagan, C. (1992) To avert a common danger. *Parade Magazine,* March 1, pp. 10–14.

Sanders, D.W. (1992) International activities in assessing and monitoring soil degradation. *American Journal of Alternative Agriculture* 7, 17–24.

Sarrantonio, M., Doran, J.W., Liebig, M.A. and Halvorson, J.J. (1996) On-farm assessment of soil health. In: Doran, J.W. and Jones, A.J. (eds) *Methods for Assessment of Soil Quality.* Soil Science Society of America, Special Publication no. 49, Madison, Wisconsin.

Sharpley, A., Daniel, T.C., Sims, J.T. and Pote, T.H. (1996) Determining environmentally sound soil phosphorus levels. *Journal of Soil Water Conservation* 51, 160–166.

Shea, P.J., Mielke, L.N. and Nettleton, W.D. (1992) Estimation of relative pesticide leaching in Nebraska soils. *Research Bulletin* 313-D, Agricultural Research Division, IANR, University of Nebraska, Lincoln, Nebraska.

Singh, K.K., Colvin, T.S., Erbach, D.C. and Mughal, A.Q. (1990) Tilth index: an approach towards soil condition quantification. Meeting paper No.90–1040. American Society of Agricultural Engineers, St. Joseph, Missouri.

Smith, J.L. and Doran, J.W. (1996) Measurement and use of pH and electrical conductivity for soil quality analyses. In: Doran, J.W. and Jones, A.J. (eds) *Methods for Assessment of Soil Quality*. Soil Science Society of America, Special Publication no. 49, Madison, Wisconsin.

Soule, J.D. and Piper, J.K. (1992) *Farming in Nature's Image*. Island Press, Washington, DC.

Stauber, K.N. (1994) The future of agriculture. *American Journal of Alternative Agriculture* 9, 9–15.

Swift, M.J., Kang, B.T., Mulongoy, K. and Woomer, P. (1991) Organic matter management for sustainable soil fertility in tropical cropping systems. In: *Evaluation for Sustainable Land Management in the Developing World, Vol. 2: Technical Papers*. IBSRAM Proceedings No. 12 (2), International Board for Soil Research and Management, Bangkok, Thailand.

US Department of Agriculture. (1980) *Report and Recommendations on Organic Farming*. US Government Printing Office, Washington, DC, 94 pp.

Van Kooten, G.C., Weisensel, W.P. and Chinthammit, D. (1990) Valuing trade-offs between net returns and stewardship practices: The case of soil conservation in Sasketchewan. *American Journal of Agricultural Economics* 72, 104–113.

Visser, S. and Parkinson, D. (1992) Soil biological criteria as indicators of soil quality: soil microorganisms. *American Journal of Alternative Agriculture* 7, 33–37.

Warkentin, B.P. (1995) The changing concept of soil quality. *Journal of Soil Water Conservation* 50, 226–228.

Warkentin, B.P. and Fletcher, H.F. (1977) Soil quality for intensive agriculture. In: *Proceedings of an International Seminar on Soil Environment and Fertility Management in Intensive Agriculture,* Tokyo, Japan, pp. 593–598.

Webster's Third New International Dictionary (1986) 3rd edn. Gove, P.B (ed.), G and C Merriam Company, Publishers, Springfield, Massachusetts.

World Resources Institute (1992) *New York Times,* March 3, 1992.

Soil Health: its Relationship to Ecosystem Health

2

D.J. Rapport[1], J. McCullum[2] and M.H. Miller[2]

[1]*Eco-Research Chair in Ecosystem Health, Faculty of Environmental Sciences, University of Guelph, Guelph, Ontario N1G 2W1, Canada;* [2]*Department of Land Resource Science, University of Guelph, Guelph, Ontario N1G 2W1, Canada*

Introduction

This paper introduces the concept of ecosystem health and its relationship to soil health. Ecosystem health is an emerging integrative science dealing with the health of regional systems (Rapport, 1995a). Akin to other health sciences, e.g. human health, the concept does not conjure up a single model, or a single method. Rather it speaks to a cluster of concepts and methods contributing to the goal of promoting viable life systems. Clearly the focus is early detection of pathologies, assessments, i.e. diagnosis of causal factors, and options for intervention and rehabilitation. Added to this list, importantly, and often neglected, ought to be a concerted effort on preventive strategies.

Both the concept of 'ecosystem' and of 'health' are definitely ambiguous. The concept of ecosystem suffers from the lack of definitive and concrete boundaries, and the concept of health suffers from the implicit value judgements that must enter into a human assessment of the condition of the system of interest. Added to this of course is the problem of complex dynamics which characterizes most ecosystems at a variety of temporal and spatial scales. With the exception of the boundary problem, these problems are also common to other health sciences, and while definitions in these domains continually evolve, this fact of life by itself should not stand in the way of progress in methods and actions.

With respect to the boundary problem, clearly there are a number of suitable demarcations for specific types of analyses. For example, drainage basins and sub-basins are rather unambiguously defined and appropriate for analysis of many aquatic and terrestrial ecosystems. Other classifications based on topography, climate, land use and dominant vegetation have been constructed and

 Biological Indicators of Soil Health (eds C.E. Pankhurst, B.M. Doube and V.V.S.R. Gupta)

utilized in environmental assessments and management at regional, national and international levels (Friend and Rapport, 1991; UNSO, 1991). These varying geographies suggest that the choice of boundaries, although to some extent arbitrary, is feasible on a pragmatic basis. The guiding principle is one of suiting the geography to the problem being addressed.

For purposes of this chapter, we will define ecosystem health with three major considerations in mind: firstly the ecosystem should be free of 'ecosystem distress syndrome (EDS)' (Rapport *et al.*, 1985; Rapport, 1989; Costanza, 1992; Rapport, 1995b). EDS comprises a group of signs by which ecosystem breakdown is generally recognized. For terrestrial ecosystems, these signs include leaching of soil nutrients, reduced species diversity, shifts in species composition to opportunistic species, reduced productivity, increased pest and disease loads, among others (Rapport *et al.*, 1985; Rapport, 1995a). Secondly, the ecosystem ought to be self-sustaining. In human dominated systems, this should mean with a minimal subsidy. A corollary to this, particularly relevant in agroecosystems and to some degree in managed forest systems, is that the degree of subsidy should not be increasing to maintain a given yield, particularly fertilizer and fossil fuel use, i.e. non-renewable resources. Thirdly, the ecosystem should not be punitive to surrounding systems. For example, a healthy managed forest ecosystem should not cause injury to the fisheries in the drainage basin; a healthy agroecosystem should not add a pollution or nutrient burden to its drainage. The first two requirements have clear analogies to human health, while the third is a more unique property of ecosystem health.

Associated with each of these considerations, methods for assessment of ecosystem health are evolving. These methods are diverse and involve palaeoecological approaches to portray the longer term history of environmental change and its relationship to human settlement and activity patterns (Smol, 1995); quantitative methods for assessing changes in the structure of biotic communities associated with stress (Hansen, 1995; Karr, 1995); and methods to identify the flow of services provided by the ecosystem to humans and changes in the quality and quantity of ecological services in response to economic activity (Cairns and Pratt, 1995; Rapport, 1995c). Yet there is no denying that many of these methods have a long way to go. The question of validation of indicators, for example, is particularly pertinent to our present discussion. Clearly what is required, just as in the health sciences, are indicators which have been shown in practice to be highly reliable and informative about the nature of dysfunctions, probable causes, or probable remedies. While much attention has been given to selection of indicators of ecosystem condition, and various schemes have been devised to classify indicators, for example into those that are useful as 'early warning' indicators of particular dysfunctions, those that are useful as 'general screening', those that are useful in 'risk assessment', it is as of yet a neglected area to validate the choice of indicators (Rapport, 1995c). What is becoming apparent however, from such anecdotal information that may be gleaned from various case studies (Rapport, 1989; Hilden and Rapport, 1993; Whitford, 1995), is that single indicators are insufficient: it is syndromes not signs or symptoms that can be relied upon for the evaluation of the health of ecosystems.

Ecosystems being functionally comprised of many components, both abiotic and biotic, naturally depend upon the well-functioning of sub-systems. It is precisely here that connective links can be established between indicators of sub-system function and indicators of ecosystem health. But the job is not easy. One has to bear in mind that some sub-system functions are more vital than others for the larger system, and that the sub-system-to-the-whole relations cross scales, and that always poses conceptual and analytical difficulties, particularly when one takes into account that the whole is more than the sum of the parts! Emergent properties are pervasive in nature (Funtowicz and Ravetz, 1994)!

While the subject of this volume is bioindicators of soil health, much of this paper is dedicated to setting out the context for assessing soil health from an ecosystem health viewpoint. Bioindicators may play a role in this assessment, but we believe it is critical to grasp the conceptual lessons which ecosystem health provides before any adoption of particular indicators or types of indicators can truly be useful.

An Eco-health Approach to Soils with Reference to Agroecosystems

In examining the soil sub-system from an eco-health perspective two realities must be considered: (i) soil is a vital component in the functioning of ecosystems accounting for virtually all decomposition processes and a significant proportion of energy flow, and; (ii) soil is a vital economic resource required for agriculture, silviculture, supply of many raw materials and indirectly as a platform for most infrastructural development (Blum and Sandelises, 1994). These two realities can appear to conflict, with unsustainable exploitation of the soil considered necessary for robust economic health and ecological health being identified with practices which provide uneconomically low levels of production. The unique facility of the eco health perspective is its ability to incorporate the associated sets of values from each perspective into a useful assessment, or diagnostic, framework which reconciles these seeming conflicts.

Agroecosystems are an excellent example of apparent conflicts between the ecological and economic roles of soils, and the discussion which follows will focus on soil health in agroecosystems. The farmer also has multiple roles. As producer, the farmer is dependent on sufficient production to maintain an adequate income and lifestyle, yet is also dependent on maintaining the satisfactory functioning of the agroecosystem, especially the soil sub-system, to continue to produce (farmer as steward). Agricultural management can be seen to affect all aspects of ecosystem health as set out in the definition above. In many agricultural soils degradation occurs, such as erosion or loss of structure, that may ultimately lead to destabilization of the ecosystem (Bormann and Likens, 1979) and EDS. Agricultural systems are necessarily subsidized with energy, nutrients and other agrochemicals (Gliessman, 1990). These subsidies, together with other aspects of agricultural management, affect the functioning of the agroecosystem and have the potential to affect surrounding systems. Yet the ills that may be wreaked on agricultural soils often have their roots in a striving for increased yield and economic output; an agroecosystem which forces farmers off the land because they can't make a living cannot be considered healthy. Economic viability must include the cost of production: for example maximum production from any crop variety cannot generally be achieved using the soil medium, but with hydroponics (Miller *et al.*, 1989). If maximum production per plant were the only criterion for healthy agriculture all crops would be grown hydroponically. However the cost of hydroponic production is much greater than soil-based farming. The farmer effectively uses certain ecological properties of the soil to economically subsidize the cost of production.

Fertilizer quandaries

The complexity of evaluating ecosystem health, and the need for reconciling ecological functioning with economic production in agroecosystems is well-illustrated by the current debate over control of nitrate leaching and the associated potential for groundwater contamination, and phosphate run-off causing eutrophication. Nitrogen and phosphorus are essential crop nutrients. There is a strong argument that nitrogen fertilizer is a necessary component of many farming systems, at least if economic viability is to be maintained. Phosphorus is also critical for adequate crop growth and phosphorus fertilization is seen as necessary to ensure sufficient production to meet economic needs. However, fertilizers can have negative consequences on ecological processes in the agroecosystem and on humans and other species living within the agroecosystem, as well as on adjoining ecosystems. There has been a serious public concern about pollution effects of phosphorus, such as eutrophication by phosphates, since the early 1970s, as evidenced by activities of the Pollution from Land Use Activities Reference Group (PLUARG) set up by the International Joint Commission on the Great Lakes (e.g. Coote *et al.*, 1982). More recently, the issue of nitrate contamination of groundwater with its attendant negative human health effects on those reliant on the contaminated water has attracted attention and public debate.[1]

A number of possible responses to fertilizer impacts on the environment can be readily prescribed. The most obvious is to decrease the amount of fertil-

[1] Note that nitrate contamination can result both from synthetic fertilizers and organic forms such as manure or legume cover crops.

izer input to the system as a way of controlling loss, including leaching to groundwater. In general, excessive leaching of nitrates occurs only when nitrogen availability exceeds crop nitrogen requirement (Addiscott *et al.*, 1991). In some cases fertilizer inputs are much greater than those recommended for a healthy economic return and can easily be decreased without significantly affecting yield (Legg *et al.*, 1989; Schepers *et al.*, 1991). In others, even application of fertilizer as recommended may result in unhealthy nitrate build-up in groundwater (Power and Broadbent, 1989). Here both choices result in an unhealthy situation: loss of production due to below-minimum fertilizer inputs results in unhealthy agroecosystems from an economic perspective, while raising nitrate concentration in groundwater above safe levels creates a danger to human health and may disrupt ecological processes, such as by causing eutrophication through increased nutrient levels.

Haynes (1994) has suggested the method and timing of cultivation and disposal of crop residues seems the most effective way to manage nitrogen loss in agroecosystems. Reductions in nitrate leaching might be achieved through increased accuracy of fertilizer application in space and time based on crop needs and site-specific available nutrient levels in soil. Minimizing the period where the soil is bare and/or devoid of active, nitrate-capturing roots can also minimize nitrate leaching (Addiscott *et al.*, 1991). Over-winter nitrate losses generally accounted for all nitrate remaining in soil after harvest, in a study of Ontario farming systems (Kowalenko, 1989). Planting of winter crops has been shown to decrease over-winter nitrate leaching in Denmark (Jensen *et al.*, 1994). Minimizing nitrate left in soil at the end of the cropping season and/or inclusion of a winter crop in the rotation may reduce over-winter nitrate leaching.

Another approach to reduce output of nutrients to surrounding systems is to change tillage and residue management practices. Adoption of no-till or minimum tillage systems decreases soil erosion (e.g. Karlen *et al.*, 1994; Pimental *et al.*, 1995), and attendant loss of adsorbed phosphorus (Logan, 1993). However there is a strong body of evidence that such systems also have increased development of biopores due to earthworm activity and rooting channels which may lead to increased nitrate loss through leaching (e.g. Cattle *et al.*, 1994). In addition, minimum tillage systems often require greater use of pesticides than conventional tillage (Sprague, 1986). They also generally require a greater degree of management skill and time in order to produce results comparable to conventional systems (Ball *et al.*, 1994), introducing an additional economic cost and greater potential for decreased yield and loss of nutrients due to human error. These effects underline the need to treat the system of interest, be it field, farm, watershed or biome, as an interlinked set of processes and components where a change to any part may affect any other part and the linkages between them.

Certain crop rotations present a greater risk of contributing to environmental degradation through fertilizer loss to the environment than others (e.g. Barry *et al.*, 1993), bringing the element of spatial and temporal scale into the discussion. Should nitrate leaching be minimized at the field scale or at a higher level of aggregation, such as the watershed, where a number of different rotations may individually be above or below a selected threshold but collectively meet the agreed standard? Temporally it may be possible for a rotation over a period of years to meet an average annual nitrate leaching standard while exceeding it in certain years. It may seem obvious that there is no single correct approach to the question of scale. From an ecosystem perspective, management at the watershed level makes sense, but this does not make the contamination of an individual farm wells less likely, nor any less of a disaster for the farmer involved. In general, management occurs at the farm scale, while factors affecting management decisions occur at many scales: fertility at the field scale; policy at the regional, provincial and national scales; economic forces at the local to global scales.

In assessing the health of an agroecosystem the impact of fertilizers is an important consideration and the assessment is never completely straightforward. Is a system which controls phosphorus output at the expense of allowing nitrate leaching healthier than one which has higher phosphorus output and lower nitrate loss? The answer will have to trade off production, nitrogen input and phosphorus input, and will clearly involve human values. The nitrate issue exemplifies the need for a holistic approach which incorporates values, such as the maximum acceptable nitrate concentration in groundwater, in determining ecosystem health.[2] A decision on the acceptable level of risk for any hazard must be based on all the associated costs and benefits including: health, economic, policy and social values (Gentile and Slimak, 1992).

Finally, it should be obvious that the role of nitrates and other fertilizers in agroecosystem health is but one component of a much larger set of factors which determine the system's total health picture. There is a great deal of scope for interaction of factors in different and complex ways and it is this scope that makes ecosystem health assessment challenging, controversial and absolutely critical to the future of humanity.

Values and objectives

A discussion of soil health in an ecosystem health context must include a consideration of values, since the two are difficult to separate (O'Neill *et al.*, 1992). Evaluation of ecosystem condition is only meaningful relative to societal values, which must be assessed through broad public consultation, although it must be kept in mind that this does not mean that any environmental situation should be considered healthy simply because it has sufficient public support (Rapport,

[2] The maximum acceptable nitrate concentration can never be determined solely on a scientific basis, since different concentrations will result in different outcomes depending on exposure, individual health and genetic history, etc.

1995a). In the case of managed ecosystems such as agroecosystems, 'values' include human-set objectives for the system in question. In agroecosystems, as with other systems, objectives will vary with different perspectives. For the farmer, explicit objectives for agroecosystems and their soils relate largely to quantity and quality of yield. Policy-makers may be interested in production levels of various commodities as well as soil loss, tillage systems or other soil-related public policy issues. Regulators will largely be interested in compliance with environmental standards and guidelines.

Of course objectives for soils in agroecosystems must be realistic, taking into account the inherent soil properties which affect the potential of the system, as well as the properties which can be affected by management. Inherent characteristics – such as texture, topography, depth, drainage and mineral composition resulting from parent material – change little over years and decades (although they can be altered with sufficient investment and/or inputs), and so set limits on objectives which management can reasonably expect to achieve. On the other hand, fertility characteristics can be modified with chemical and organic inputs over periods of days. Some economically 'unhealthy' soils can become 'healthy' high producers with judicious application of lime and/or particular nutrients. When setting objectives for the system, managers must have a clear understanding of limitations posed by inherent characteristics of the system in question as well as the range of management practices, and associated costs, available to modify the system.

It is also difficult to separate soil-related objectives from more general agroecosystem objectives. Most agroecosystem objectives will rely on soil and soil properties to fully or partially achieve them. For example a less explicit and less precise objective for a farm family might relate to their quality of life. Quality of life includes a number of components, but is definitely affected by income. Income is partially, but not completely affected by productivity and crop yield, which is in turn affected by soil. From an ecosystem health perspective it is important to identify objectives for the entire system early on, prior to defining what specific soil-related objectives might be.

Soil quality indicators

Much of the discussion related to soil health deals with indicators of soil quality; in fact the terms are used interchangeably in the scientific literature (Harris and Bezdicek, 1994). Soil quality relates to the capability of soil for production or provision of other services beneficial to humans such as pollution attenuation (Doran and Parkin, 1994). Soil quality indicators, such as those used by the US EPA's Environmental Monitoring and Assessment Program's (EMAP) Agroecosystems component, generally include a range of site-specific measures such as pH, texture, organic matter content or microbial biomass with a strong focus on the biophysical (Meyer *et al.*, 1992). They provide a framework within which to assess the ability of the soil to meet human objectives.

While the implication of the term 'quality' is of an absolute measure allowing one soil to be compared to another, it is clear that the quality of a soil to meet two different objectives is very unlikely to be the same for both. Soil quality to produce corn will not be the same as for coffee, and may in turn be very different from its quality to buffer acid precipitation. No soil can be best, or worst, at all things. As discussed above, objectives for the system must be set in order to guide the choice of indicators for assessing soil quality.

However, we suggest that interchanging the terms 'soil quality' and 'soil health' is a fundamental misuse of the two concepts. Soil quality, with its emphasis on meeting human needs, includes inherent soil characteristics which may assist or constrain the soil's ability to support the achievement of human objectives. Such inherent characteristics could be considered somewhat equivalent to genetic potential and should not be included in soil health. Soil health should strictly include only those characteristics which can be affected by management at scales relevant to managers, e.g. over years, not centuries. The fact that a soil is too steep, or too stony, to be economically farmed cannot be sufficient reason to assess that soil as unhealthy. Nor is it correct to say that soil health is simply a subset of soil quality, as is explained below; soil health includes factors which may be unrelated to the achievement of human objectives.

Another area of emphasis in the soil quality literature relates to the development of indices of soil quality (e.g. Parr *et al.*, 1992). The use of indices has been considered in the ecosystem health literature and the ability of aggregate indexes, i.e. formulaic or algorithmic calculations which condense values of a number of indicators into one overall value, to accurately measure ecosystem health or detect changes, has been questioned and/or treated sceptically by many investigators, although the value of suites of indicators is emphasized (e.g. Milne, 1992). Karr, who developed the much respected Index of Biotic Integrity for aquatic ecosystems, has advocated a multi-metric approach, but warned that each test of which an index is composed must have a sound theoretical and empirical basis, and that aggregation of test results into a single measure results in a loss of information about specific system attributes (Karr, 1992). Rapport *et al.* (1985) emphasize that detection of EDS requires a number of different metrics.

In adopting an ecosystem health perspective with respect to soils, indicators of productivity or other objectives form an important part of a group of indicators of ecosystem health. However, soil quality/health cannot necessarily be equated with agroecosystem health, or even with health of the soil sub-system. Agroecosystem health requires freedom from EDS; minimal, non-increasing subsidy per unit yield and prevention of negative impacts to surrounding systems. Soil quality indicators address the second consideration and, in some cases, aspects of the third, but have little direct relation to the first. There is a set of soil ecological indicators which addresses these gaps, and which intersects the set of soil quality indicators. Further, the soil sub-system is directly linked to the rest of the agroecosystem, so an assessment of its health cannot be made focusing solely on soil-related indicators or biophysical indicators. For example there can be human health impacts from nitrate in groundwater – there is a relationship with soil health but it is an indirect one. The assessment is incomplete without the socioeconomic component. We now turn to examine the two sets of indicators mentioned above, namely indicators relating to economic objectives and indicators relating to ecological objectives. Finally we consider how these two elements of soil health relate to agroecosystem health.

Soil health indicators

The number of possible soil health indicators is large, and the choice daunting. The principles of ecosystem health, and particularly EDS, show that suites of indicators are the most reliable approach to assessing ecosystem health, while single indicators or indices can be misleading. This maxim does not make the selection of indicators easier though. In general, we seek indicators which are mechanistically linked to human-set objectives for the system and/or its ecological health (Bernstein, 1992). A slightly different perspective is used in the EMAP approach (Heck *et al.*, 1993). They explicitly recognize the need to determine societal values, from which they derive assessment endpoints which reflect these values. These endpoints must be unambiguous, have social or biological relevance and be quantifiable. The quantifiable measurements associated with the assessment endpoints are EMAP's indicators.

The discussion of indicators below is not exhaustive by any means. Rather it is intended to discuss a few basic indicators from an agroecosystem health perspective.

Indicators of productivity and other human-set objectives

There is a broad range of proposed agricultural soil quality indicators. In terms of productivity, perhaps the best set of general screening indicators relates to yield trends over time under a consistent management system, i.e. rotation, tillage type, fertilizer regime, etc. (Powlson and Johnston, 1994) with a decrease indicating a loss of health. Large year-to-year variations in yield have been shown to be a prelude to complete crop failure (Woodward, 1993). Yield integrates an array of factors, some related to soil, so it is important to consider the impact of non-soil factors, particularly climatic factors, on yield changes before drawing conclusions regarding soil health. There are a number of other objectives which managers may also seek to assess, biodiversity for instance, for which indicators other than yield would be more appropriate.

Once a trend of declining health has been identified, the difficult work of diagnosing the cause begins. Diagnosing decreasing productivity often begins with an analysis of availability of water and nutrients in conjunction with pH, and there may be specific plant symptoms which indicate such problems. Toxic effects or soil-borne pathogens may also affect productivity. Changes in management practices, such as tillage type, can affect yield directly (e.g. Ball *et al.*, 1994; Carter, 1994). Some of these causes/symptoms occur over short timescales, certainly within a growing season, and can often be corrected with one-time management intervention. In some cases the factors may interact, for example aluminium toxicity is associated only with pH below 5.5. A difficulty in using these short-term changes as indicators is their transience and the possibility of other transient factors, notably climate, creating 'noise' which makes the cause–effect relationship to indicator response difficult to interpret.

Other changes in soil may be more chronic, responding slowly over years and decades to agricultural practices such as tillage, cropping and harvest, residue management or agrochemical use and include: soil erosion, organic matter content, structure, in-soil species diversity and population sizes, and factors related to infiltration and water-holding capacity. For example there is evidence that organic matter will decline when a previously uncultivated soil is cropped, eventually reaching a new equilibrium depending on the amounts of carbon and nitrogen returned to it by the specific farming system (Swift, 1994). The new equilibrium is not necessarily reached quickly though. Soil at Rothamsted, UK took 130 years to reach an equilibrium with respect to percent carbon under a continual, specific farming system (Powlson and Johnston, 1994). Other work suggests that such an equilibrium may not be reached in all agricultural systems (Bird and Rapport, 1986). Battiston *et al.* (1987) have shown that soil erosion decreased corn yield minimally until a particular soil depth threshold was crossed (which varied for differing soil types) at which point average yield losses were 59% (range from 16 to 80%) on these severely eroded soils. Soil structure and aggregation – particularly pore size distribution, continuity and stability – may be the most critical determinant of productivity but is difficult to measure in a single, consistently meaningful, way (e.g. Kay, 1990).

Finally, long term changes in soil can occur over centuries or millennia. These include topographic changes due to geomorphological processes and of particular importance for agriculture, loss of clay minerals and associated cation exchange capacity due to natural weathering processes over many thousands of years (Chesworth *et al.*, 1983). These latter processes result in decreasing fertility and ultimately barrenness unless new sources of rock are added to the system, but are also accelerated by one or two orders of magnitude by agriculture (Chesworth *et al.*, 1983). These changes occur far too slowly to provide useful indications of stress on the agroecosystem.

The characteristics of soil, discussed above, which respond over the medium term may be the most useful for indicating soil health, partly because they integrate many different processes affecting soil, including abiotic, biotic and management and partly because they result from the combined effects of management over time. Aggregation and structure are affected by a combination of abiotic processes such as wetting/drying cycles, biotic processes such as bacterial gels, and management practices such as tillage, and their interactions (Coleman *et al.*, 1994; Foster, 1994). More transient responses, such as nitrogen concentration, change quickly and can easily be modified by management so that the symptom is masked. Nitrogen fertilization can have an immediate effect on nitrogen concentration in soil. Similarly very slow processes, which result in what we have called inherent characteristics, are of little value in assessing the health of the agroecosystem.

The subject of this volume is biological indicators of soil health. Our discussion so far has concentrated on other, more conventional indicators, but the biotic element of soil, and indeed of any ecosystem, is by definition an essential component. Biota are both essential to soil processes and dependent on soil as habitat. Bioindicators generally include microbial indicators such as biomass and activity (Pankhurst, 1994; Turco *et al.*, 1994) or ratio of bacteria to fungi (Coleman *et al.*, 1993; Swift and Anderson, 1993), faunal indicators, especially nematodes and earthworms (Meyer *et al.*, 1992; Linden *et al.*, 1994), foodweb structure and soil species biodiversity (Pankhurst and Lynch, 1994).

A problem with biological indicators of soil quality is that it is often difficult to determine what quality is being indicated other than the actual object or process measured (Duxbury and Nkambule, 1994). In addition, many of the processes respond very quickly to environmental changes such as precipitation, temperature, fertilization or tillage. For example Hassink *et al.* (1991) found that climatic events occurring during the growing season, rather than agricultural practices, determine microbial dynamics. Stork and Eggleton (1992) summed up their review of soil biological criteria for soil quality by supporting the conclusion that the major hindrance to the use of bioindicators is '. .a lack of understanding of the links between gross soil processes and the structure of soil organism communities.' Karlen and Stott (1994) stated that biological indicators of soil quality are being evaluated but currently none have been agreed on and recommended for routine measurement. In short, bioindicators have not yet been directly linked to the human-set objectives (other than biodiversity) which are fundamental criteria against which the health of the soil, and the ecosystem, must be evaluated.

Indicators of soil ecological health

Ecological health is often considered to be embodied in ecological function. Measurements of ecological function involve basic ecosystem functions such as nutrient cycling or energy transfer which result from the interaction of many components (O'Neill, 1991). Gliessman (1990) stated that all ecosystem functions, including energy flow and nutrient cycling, are disturbed in agroecosystems. Swift and Anderson (1993) hypothesized that while 'in natural ecosystems the internal regulation of function is substantially a product of plant biodiversity through flows of energy, nutrients and information, this form of control is progressively lost under agricultural intensification so that ultimately the only integrated ecosystems function is invested in the below-ground sub-systems, regulated predominantly by chemical inputs of industrial origin.' In other words, the functioning of the soil sub-system may be the key to understanding the health of agroecosystems from an ecological perspective.

Soil plays an essential role in nutrient cycling via decomposition. In unmanaged ecosystems, nutrients tend to be conserved and losses are small. In temperate forest ecosystems nutrient conservation is largely achieved through control of water as it moves through the system: kinetic energy of precipitation is absorbed by leaves and the litter layer, chemistry of rainfall is altered at the leaf, stem, litter and soil interfaces and flow-through by evapotranspiration (Bormann and Likens, 1979; Likens, 1992). In particular, control of erosion may be the most critical factor in preventing destabilization of unmanaged ecosystems, with rates of erosion in second growth forests among the lowest recorded for humid ecosystems (Bormann and Likens, 1979). It would seem that indicators related to control of water and to erosion would have a relationship to nutrient cycling and to ecological health in agroecosystems as well. Such indicators would include infiltration vs. run-off rates and hydraulic conductivity, pore size distribution and pore continuity and stability, measures related to aggregation and structure as well as direct measures of soil erosion and loss of soil depth. They should also include measures related to the nutrient cycle function of the agroecosystem: nutrient conservation and leakage from the system in relation to inputs, sources, sinks and losses.

Decomposition would not occur without soil biota, so intuition suggests bioindicators should play a significant role in the assessment of soil ecological health. However, the same arguments which applied to bioindicators of soil quality pertain here. Anderson (1994) stated that management practices which reduce complex biological interactions in soils are not inherently unstable, but depend on the ecological and agronomic context. It appears that much more work is required before there is an immediate understanding of the relationship between the various bioindicators currently being investigated and soil ecological health. For example, it is not enough to say that microbial activity is critical to soil health when levels of such activity vary by orders of magnitude throughout a growing season in response to climatic factors and seasonal plant growth patterns, or even fertilizer application (Hassink *et al.*, 1991; Grace *et al.*, 1994). Some form of standard must be set, including ranges of healthy levels and thresholds, to allow the assessment of a measurement or series of measurements. However, it must be recognized that the development of absolute standards for assessment of ecosystem health may not be scientifically possible, underlining the need for suites of indicators which, taken together, can provide the required information. It may also be possible to monitor indicators over time to assist in determining whether a healthy or unhealthy trend exists. In general, the criticisms of bioindicators of soil ecological health are the same as for bioindicators of soil quality. The idea appears to make sense but more work is required, especially in the area of validation, before its time will truly have come.

Indicators of both productivity and ecological health

We have argued that there are at least two intersecting sets of indicators of soil health: those related to productivity and other human-set objectives and those related to soil ecology. We have also argued that productivity can sometimes appear to be in conflict with ecology. For purposes of comparison among indicators, and to identify processes where the dual roles of soils apparently conflict, it will be most useful to concentrate on indicators which fall into the overlap area between the two sets. Our analysis suggests that measures related to erosion and soil structure are among those which measure both aspects and so may be the most useful in assessing soil health. Undoubtedly there are other measures, possibly including bioindicators, which also fulfil this dual evaluative role.

Conclusions

In this paper we have suggested some linkages between soil health and ecosystem health and potential bridges to the area of bioindicators. We have taken note of the various time scales on which soils change, and the kinds of changes that directly relate to goals of ecosystem health, and those more associated with soil health. We view changes at intermediate time scales in soil properties, particularly those that are sensitive to management practices, as the most appropriate for gaining measures of soil health that relate to ecosystem health.

Ecosystem health indicators emphasize two elements:

1. There is the need for a group of indicators – syndromes, not symptoms (more technically signs) are to be sought.
2. Indicators fall into a number of categories. Some are useful for diagnostic purposes (identifying the potential causes of particular dysfunctions), some are useful for general screening purposes (overall assessments of health), and some for risk assessment (to evaluate potential losses which may not yet have surfaced, as a result of particular types of stresses impacting the environment).

In this discussion, we have placed emphasis on the importance of both biophysical and socioeconomic conditions to provide a spectrum of indicators for assessing ecosystem health.

The role of bioindicators of soil health within this context seems clear enough. There is an apparent relationship between ecosystem health and biodiversity within soils, community composition of soil organisms, nutrient cycling and productivity. However the nature of this relationship remains rather speculative. What is needed are methods to assess the sensitivity of such measures for ecosystem health evaluation. If bioindicators of soil health are to become a tool of practitioners in establishing guidelines for soil conditions and in recommending rehabilitation, it must be shown that such indicators, collectively, provide valid information. To accomplish this is no small task!

Presently, statistical methods, employing classical statistics are inadequate for situations wherein replicates are not appropriate or not possible. New methodologies will need to be developed whereby one can confirm the utility of such groups of indicators as suggested here. These methods would need to establish that where indicators indicate poor ecosystem health then, in the absence of interventions, there will be further declines in health and these can be directly related to the loss of ecosystem services and management options (Rapport, 1995a,b). Conversely, and necessary in parallel with standard medical practice, if indicators are to be used for rehabilitating the health of ecosystems, one needs validation and confidence that the appearance of certain groups of indicators is indeed a significant sign of recovery. These challenges are not only in the areas of soil health, but are general challenges as the entire area of ecosystem health advances to a firmer footing in practice and applications.

Acknowledgements

This chapter represents a collaboration between the Eco-Research Chair Program at the University of Guelph and the Agro-Ecosystem Health Project at the University of Guelph. The authors are grateful to the Tri-Council for their support under grants to the Eco-Research Chair Program at the University of Guelph, and the Agro-Ecosystem Health Project at the University of Guelph. The Tri-council is comprised of the Social Science and Humanities Research Council, the National Sciences and Engineering Research Council and the Medical Research Council.

References

Addiscott, T.M., Whitmore, A.P. and Powlson, D.S. (1991) *Farming, Fertilizers and the Nitrate Problem.* CAB International, Wallingford, UK, 170 pp.

Anderson, J.M. (1994) Functional attributes of biodiversity in land use systems. In: Greenland, D.J. and Szabolcs, I. (eds) *Soil Resilience and Sustainable Land Use.* CAB International, Wallingford, UK, pp. 267–290.

Ball, B.C., Lang, R.W., Robertson, E.A.G. and Franklin, M.F. (1994) Crop performance and soil conditions on imperfectly drained loams after 20–25 years of conventional tillage and direct drilling. *Soil and Tillage Research* 31, 97–118.

Barry, D.A.J., Goorahoo, D. and Goss, M.J. (1993) Estimation of nitrate concentrations in groundwater using a whole farm nitrogen budget. *Journal of Environmental Quality* 22, 767–775.

Battiston, L.A., Miller, M.H. and Shelton, I.J. (1987) Soil erosion and corn yield in Ontario. 1. Field evaluation. *Canadian Journal of Soil Science* 67, 731–745.

Bernstein, B.B. (1992) A framework for trend detection: coupling ecological and management perspectives. In: McKenzie, D.H., Hyatt, E.D. and McDonald, V.J. (eds) *Ecological Indicators*: Volume 2. Elsevier, London, pp. 1101–1114.

Bird, P.M. and Rapport, D.J. (1986) *State of the Environment Report for Canada.* Canadian Government Publishing Centre, 264 pp.

Blum, W.E.H. and Santelises, A.A. (1994) A concept of sustainability and resilience based on soil functions: the role of ISSS in promoting sustainable land use. In: Greenland, D.J. and Szabolcs, I. (eds) *Soil Resilience and Sustainable Land Use.* CAB International, Wallingford, UK, pp. 535–542.

Bormann, F.H. and Likens, G.E. (1979) *Pattern and Process in a Forested Ecosystem.* Springer-Verlag, New York, 253 pp.

Cairns, J. and Pratt, J.R. (1995) The relationship between ecosystem health and delivery of ecosystem services. In: Rapport, D.J., Gaudet, C. and Calow, P. (eds) *Evaluating and Monitoring the Health of Large-scale Ecosystems.* Springer-Verlag, Heidelberg, pp. 63–76.

Carter, M.R. (1994) A review of conservation tillage strategies for humid temperate regions. *Soil and Tillage Research* 31, 289–301.

Cattle, S.R., Koppi, A.J. and McBratney, A.B. (1994) The effect of cultivation on the properties of a Rhodoxeralf from the wheat/sheep belt of New South Wales. *Geoderma* 63, 215–225.

Chesworth, W., Macias-Vazquez, F., Acquaye, D. and Thompson, E. (1983) Agricultural alchemy: stones into bread. *Episodes* 1, 3–7.

Coleman, D.C., Hendrix, P.F., Beare, M.H., Cheng, W.X. and Crossley, D.A., Jr (1993) Microbial and faunal interactions as they affect soil organic matter dynamics in subtropical ecosystems. In: Paoletti, M.G., Foissner, W. and Coleman, D.C. (eds) *Soil Biota, Nutrient Cycling and Farming Systems.* Lewis Publishers, Boca Raton, pp. 1–14.

Coleman, D.C., Hendrix, P.F., Beare, M.H., Crossley, D.A., Jr, Hu, S. and van Vliet, P.C.J. (1994) The impacts of management and biota on nutrient dynamics and soil structure in sub-tropical agroecosystems: impacts on detritus food webs. In: Pankhurst, C.E., Doube, B.M., Gupta, V.V.S.R. and Grace, P.R. (eds) *Soil Biota: Management in Sustainable Farming Systems.* CSIRO Press, Melbourne, pp. 133–143.

Coote, D.R., MacDonald, E.M., Dickinson, W.T., Ostry, R.C. and Frank, R. (1982) Agriculture and water quality in the Canadian Great Lakes Basin: 1. Representative agricultural watersheds. *Journal of Environmental Quality* 11, 473–481.

Costanza, R. (1992) Toward an operational definition of ecosystem health. In: Costanza, R., Norton, B.G. and Haskell, B.D. (eds) *Ecosystem Health: New Goals for Fnvironmental Management.* Island Press, Washington, DC, pp. 239–256.

Doran, J.W. and Parkin, T.B. (1994) Defining and assessing soil quality. In: Doran, J.W., Coleman, D.C., Bezdicek, D.F. and Stewart, B.A. (eds) *Defining Soil Quality for a Sustainable Environment.* SSSA Special Publication No. 35. Soil Science Society of America, Madison, Wisconsin, pp. 3–21.

Duxbury, J.M. and Nkambule, S.V. (1994) Assessment and significance of biologically active soil organic nitrogen. In: Doran, J.W., Coleman, D.C., Bezdicek, D.F. and Stewart, B.A. (eds) *Defining Soil Quality for a Sustainable Environment.* SSSA Special Publication No. 35. Soil Science Society of America, Madison, Wisconsin, pp. 125–146.

Foster, R.C. (1994) Microorganisms and soil aggregates. In; Pankhurst, C.E., Doube,

B.M., Gupta, V.V.S.R. and Grace, P.R. (eds) *Soil Biota: Management in Sustainable Farming Systems.* CSIRO Press, Melbourne, pp. 144–155.

Friend, A.M. and Rapport, D.J. (1991) Evolution of macro-information systems for sustainable development. *Ecological Economics* 3, 59–76.

Funtowicz, S. and Ravetz, J.R. (1994) Emergent complex systems. *Futures* 26, 568–582.

Gentile, J.H. and Slimak, M.W. (1992) Endpoints and indicators in ecological risk. In: McKenzie, D.H., Hyatt, E.D. and McDonald, V.J. (eds) *Ecological Indicators,* Volume 2. Elsevier, London, pp. 1385–1397.

Gliessman, S.R. (1990) Agroecology: researching the ecological basis for sustainable agriculture. In: Gliessman, S.R. (ed.) *Agroecology: Researching the Ecological Basis for Sustainable Agriculture.* Springer-Verlag, New York, pp. 3–10.

Grace, P.R., Ladd, J.N. and Skjemstad, J.O. (1994) The effect of management practices on soil organic matter dynamics. In: Pankhurst, C.E., Doube, B.M., Gupta, V.V.S.R. and Grace, P.R. (eds) *Soil Biota: Management in Sustainable Farming Systems.* CSIRO Press, Melbourne, pp. 162–171.

Hansen, P.D. (1995) Assessment of ecosystem health: development of tools and approaches. In: Rapport, D.J., Gaudet, C. and Calow, P. (eds) *Evaluating and Monitoring the Health of Large-scale Ecosystems,* Springer-Verlag, Heidelberg, pp. 195–217.

Harris, R.F. and Bezdicek, D.F. (1994) Descriptive aspects of soil quality/health. In: Doran, J.W., Coleman, D.C., Bezdicek, D.F. and Stewart, B.A. (eds) *Defining Soil Quality for a Sustainable Environment,* SSSA Special Publication No. 35. Soil Science Society of America, Madison, Wisconsin, pp. 23–35.

Hassink, J., Oude Voshaar, J.H., Nijhuis, E.H. and Van Neen, J.A. (1991) Dynamics of the microbial populations of a reclaimed-polder soil under a conventional and reduced-input farming system. *Soil Biology and Biochemistry* 23, 515–524.

Haynes, R.J. (1994) Impact of management practices on nutrient cycling. In: Pankhurst, C.E., Doube, B.M., Gupta, V.V.S.R. and Grace, P.R. (eds) *Soil Biota: Management in Sustainable Farming Systems.* CSIRO Press, Melbourne, pp. 172–181.

Heck, W.W., Campbell, C.L., Neher, D.A. and Munster, M.J. (1993) An agroecosystem monitoring and assessment program for sustainable agriculture. *Proceedings of the 12th Annual Organic Agriculture Conference,* University of Guelph, Guelph, Ontario, Canada.

Hildén, M. and Rapport, D.J. (1993) Four centuries of cumulative impacts on a Finnish river and its estuary: an ecosystem health approach. *Journal of Aquatic Ecosystem Health* 2, 261–275.

Jensen, C., Stougaard, B. and Ostergaard, H.S. (1994) Simulation of nitrogen dynamics in farmland areas of Denmark (1989–1993). *Soil Use and Management* 10, 111–118.

Karlen, D.L. and Stott, D.E. (1994) Evaluating physical and chemical indicators of soil quality. In: Doran, J.W., Coleman, D.C., Bezdicek, D.F. and Stewart, B.A. (eds) *Defining Soil Quality for a Sustainable Environment.* SSSA Special Publication No. 35. Soil Science Society of America, Madison, Wisconsin, pp. 53–72.

Karlen, D.L., Wollenhaupt, N.C., Erbach, D.C., Berry, E.C., Swan, J.B., Eash, N.S.

and Jordahl, J.L. (1994) Crop residue effects on soil quality following 10 years of no-till corn. *Soil and Tillage Research* 31, 149–167.

Karr, J.R. (1992) Ecological integrity: protecting earth's life support systems. In: Costanza, R., Norton, B.G. and Haskell, B.D. (eds) *Ecosystem Health: New Goals for Environmental Management,* Island Press, Washington, DC, pp. 223–238.

Karr, J.R. (1995) Using biological criteria to protect ecological health. In: Rapport, D.J., Gaudet, C. and Calow, P. (eds) *Evaluating and Monitoring the Health of Large-scale Ecosystems,* Springer-Verlag, Heidelberg, pp. 137–152.

Kay, B.D. (1990) Rates of change of soil structure under different cropping systems. *Advances in Soil Science* 12, 3–50.

Kowalenko, C.G. (1989) The fate of applied nitrogen in a Fraser Valley soil using ^{15}N in field microplots. *Canadian Journal of Soil Science* 69, 825–833.

Legg, T.D., Fletcher, J.J. and Easter, K.W. (1989) Nitrogen budgets and economic efficiency: a case study of southeastern Minnesota. *Journal of Production Agriculture* 2, 110–116.

Likens, G.E. (1992) *The Ecosystem Approach: Its Use and Abuse.* In: Kinne, O. (ed.) *Excellence in Ecology Series, No. 3.* Ecology Institute, Oldendorf/Luhe, Germany, 166 pp.

Linden, D.R., Hendrix, P.F., Coleman, D.C. and van Vliet, P.C.J. (1994) Faunal indicators of soil quality. In: Doran, J.W., Coleman, D.C., Bezdicek, D.F. and Stewart, B.A. (eds) *Defining Soil Quality for a Sustainable Environment.* SSSA Special Publication No. 35. Soil Science Society of America, Madison, Wisconsin, pp. 91–106.

Logan, T.J. (1993) Agricultural best management practices for water pollution control – current issues. *Agriculture, Ecosystems and Environment* 46, 223–231.

Meyer, J.R., Campbell, C.L., Moser, T.J., Hess, G.R., Rawlings, J.O., Peck, S. and Heck, W.W. (1992) Indicators of the ecological status of agroecosystems. In: McKenzie, D.H., Hyatt, E.D. and McDonald, V.J. (eds) *Ecological Indicators.* Volume 1, Elsevier, London, pp. 629–658.

Miller, M.H., Walker, G.K., Tollenaar, M. and Alexander, K.G. (1989) Growth and yield of maize (*Zea mays*) grown outdoors hydroponically and in soil. *Canadian Journal of Soil Science* 69, 295–302.

Milne, B.T. (1992) Indicators of landscape condition at many scales. In: McKenzie, D.H., Hyatt, E.D. and McDonald, V.J. (eds) *Ecological Indicators.* Volume 2, Elsevier, London, pp. 883–895.

O'Neill, R.V. (1991) The systems approach to environmental assessment. In: Cairns, J. Jr and Crawford, T.V. (eds), *Integrated Environmental Management.* Lewis Publishers, Chelsea, Michigan, pp. 39–51.

O'Neill, R.V., Hunsaker, C.T. and Levine, D.A. (1992) Monitoring challenges and innovative ideas. In: McKenzie, D.H., Hyatt, E.D. and McDonald, V.J. (eds) *Ecological Indicators.* Volume 2, Elsevier, London, pp. 1443–1460.

Pankhurst, C.E. (1994) Biological indicators of soil health and sustainable productivity. In: Greenland, D.J. and Szabolcs, I. (eds) *Soil Resilience and Sustainable Land Use.* CAB International, Wallingford, UK, pp. 331–351.

Pankhurst, C.E. and Lynch, J.M. (1994) The role of the soil biota in sustainable agriculture. In: Pankhurst, C.E., Doube, B.M., Gupta, V.V.S.R. and Grace, P.R.

(eds) *Soil Biota: Management in Sustainable Farming Systems*. CSIRO Press, Melbourne, pp. 3–9.

Parr, J.F., Papendick, R.I., Hornick, S.B. and Meyer, R.E. (1992) Soil quality: attributes and relationship to alternative and sustainable agriculture. *American Journal of Alternative Agriculture* 7, 5–11.

Pimental, D., Harvey, C., Resosuda, P., Sinclair, K., Kurz, D., McNair, M., Crist, S., Shpritz, L., Fitton, L. and Saffouri, R. (1995) Environmental and economic costs of soil erosion and conservation benefits. *Science* 267, 1117–1122.

Power, J.F. and Broadbent, F.E. (1989) Proper accounting for N in cropping systems. In: Follett, R.F. (ed.) *Nitrogen Management and Ground Water Protection*. Elsevier, Amsterdam, pp. 159–161.

Powlson, D.S. and Johnston, A.E. (1994) Long-term field experiments: their importance in understanding sustainable land use. In: Greenland, D.J. and Szabolcs, I. (eds) *Soil Resilience and Sustainable Land Use*. CAB International, Wallingford, UK, pp. 367–394.

Rapport, D.J. (1989) What constitutes ecosystem health? *Perspectives in Biology and Medicine* 33, 120–132.

Rapport, D.J. (1995a) Ecosystem health: an emerging integrative science. In: Rapport, D.J., Gaudet, C. and Calow, P. (eds) *Evaluating and Monitoring the Health of Large-scale Ecosystems*. Springer-Verlag, Heidelberg, pp. 5–31.

Rapport, D.J. (1995b) Ecosystem health: exploring the territory. *Ecosystem Health* 1, 5–13.

Rapport, D.J. (1995c) Ecosystem services and management options as blanket indicators of ecosystem health. *Journal of Aquatic Ecosystem Health* 4, 97–105.

Rapport, D.J., Regier, H.A. and Hutchinson, T.A. (1985) Ecosystem behaviour under stress. *American Naturalist* 125, 617–640.

Schepers, J.S., Moravek, M.G., Alberts, E.E. and Frank, K.D. (1991) Maize production impacts on groundwater quality. *Journal of Environmental Quality* 20, 12–16.

Smol, J.P. (1995) Paleolimnological approaches to the evaluation and monitoring of ecosystem health: providing a history for environmental damage and recovery. In: Rapport, D.J., Gaudet, C. and Calow, P. (eds) *Evaluating and Monitoring the Health of Large-scale Ecosystems*. Springer-Verlag, Heidelberg, pp. 301–318.

Sprague, M.A. (1986) Overview. In: Sprague, M.A. and Triplett, G.B. (eds) *No-tillage and Surface-Tillage Agriculture: the Tillage Revolution*. John Wiley and Sons, Toronto, pp. 1–18.

Stork, N.E. and Eggleton, P. (1992) Invertebrates as determinants and indicators of soil quality. *American Journal of Alternative Agriculture* 7, 38–47.

Swift, M.J. (1994) Maintaining the biological status of soil: a key to sustainable land management. In: Greenland, D.J. and Szabolcs, I. (eds) *Soil Resilience and Sustainable Land Use,* CAB International, Wallingford, UK, pp. 235– 247.

Swift, M.J. and Anderson, J.M. (1993) Biodiversity and ecosystem function in agricultural systems. In: Schulze, E.D. and Mooney, H.A. (eds) *Biodiversity and Ecosystem Function*. Springer-Verlag, Berlin, pp. 15–41.

Turco, R.F., Kennedy, A.C. and Jawson, M.D. (1994) Microbial indicators of soil quality. In: Doran, J.W., Coleman, D.C., Bezdicek, D.F. and Stewart, B.A. (eds)

Defining Soil Quality for a Sustainable Environment. SSSA Special Publication No. 35. Soil Science Society of America, Madison, Wisconsin, pp. 73–90.

UNSO (1991) *Concepts and Methods of Environment Statistics: Statistics of the Natural Environment.* United Nations Statistical Office, New York. 148 pp.

Whitford, W.G. (1995) Desertification: implications and limitations of the ecosystem health metaphor. In: Rapport, D.J., Gaudet, C. and Calow, P. (eds) *Evaluating and Monitoring the Health of Large-scale Ecosystems.* Springer-Verlag, Heidelberg, pp. 273–294.

Woodward, F.I. (1993) How many species are required for a functional ecosystem. In: Schulze, E.D. and Mooney, H.A. (eds) *Biodiversity and Ecosystem Function.* Springer-Verlag, Berlin, pp. 271–291.

3

Rationale for Developing Bioindicators of Soil Health

E.T. Elliott

Natural Resource Ecology Laboratory, Colorado State University, Fort Collins, Colorado 80524, USA

Introduction

The ideal bioindicator of soil health would be simply measured, work equally well in all environments and reliably reveal what problems existed where. For the reasons outlined below, we are more likely to develop bioindicators for which information is laboriously obtained, that are specific to a given ecosystem or environmental problem and only tell us that there is a problem and not show us what the problem is. Even with this gloomy forecast, it is unquestionable that we will continue to seek ways by which we can determine the health of ecosystems, or their components, such as soil (Doran and Parkin, 1994).

The focus of this book is soil health. Soil is a component of the whole ecosystem and inextricably aligned with the health of that ecosystem. Soil may be healthy but functioning as a component of an unhealthy ecosystem. In most cases, the proximate causes of an unhealthy soil come from outside of the soil itself, for example, compaction caused by ungulate hoofs or addition of heavy metals contained in sewage sludge. However, there may be some internal proximate causes, such as loss of functional groups of organisms. If the soil is not healthy, the ecosystem will not be healthy.

One must have a clear definition of bioindicators of soil health to develop a rationale for their use (Doran and Parkin, 1994; Turco *et al.*, 1994). Soil bioindicators are biological properties or processes within the soil component of an ecosystem that indicate some state of the ecosystem. To indicate is to show or point out as a cause. Health is freedom from disease and normality of function. Therefore, one can never prove health, only the lack of measurable disease. To know for certain whether a soil is healthy, one must observe the system retrospectively. That is, after a reasonable period of time, if the system is functioning normally, we may presume that the system was healthy. However, if we observe a malfunction, we know that system was diseased. Given enough testing and analysis of various indicators, one may construct a set of probabilities that a system may be healthy based on these indicators, but absolute health can never be proven. What we desire as properties of bioindicators of soil health are measures that have the potential to show the soil to be unhealthy; that is, bioindicators of soil disease.

It is unlikely that ecosystem health (Costanza *et al.*, 1992; Rapport, 1989, 1995) can be defined with a single measure because of the multitude of components in a system which could be diseased. Rather, we must target what we believe to be key processes/components and make measures of them which are indicative of disease. For example, numerous recent studies have shown the impact of heavy metal contamination on various measures of microbial biomass and activity (Angle *et al.*, 1993; Brendecke *et al.*, 1993; Chander and Brookes, 1993; Yeates *et al.*, 1994; Valsecchi *et al.*, 1995; Frostegaård *et al.*, 1996). Theory to describe these responses to metals is becoming well developed (Speir *et al.*, 1995). Likewise, pesticide responses have been detected with bioindicators (Harden *et al.*, 1993) as have industrially contaminated soils (Rowell and Florence, 1993). In each case the measure is for disease, not health.

Some measures of soil health (lack of disease) will be more integrative and indicative than others. We can then determine health through experimentation where we alter the health of the system and determine how our measures change. Only after having done this repeatedly can we hope to discover whether our indicators are useful. These measures can then be extrapolated to known dysfunctioning systems to determine if they give an effective measure of disease. Given these considerations, it is clear that we must make a suite of measurements.

Indicators

Direct measurement of soil health is not possible, for the reasons given above. Therefore, we must make measurements of the components or processes of the system that indicate soil health, or more accurately, those that might indicate soil disease. For the purposes of this chapter, soil disease will be defined as a particular destructive process in the soil; one with a specified cause and characteristic symptoms. A soil may appear diseased, but the symptoms are derived from outside the soil system. For example, lack of residue input eventually results in a loss of fertility. It is a challenging problem to separate soil health from ecosystem health. A more general description might be given, such as soil sickness (presence of disease), but consideration of soil sickness (as was done early in this century) may be inapplicable as it is derived by analogy to human health.

Direct measurement of soil disease may be, likewise, difficult. Our key concern is for the healthy functioning of ecosystems as they exist in the field.

Usually we find it necessary to sample parts of the system for analysis, often bringing them into the laboratory. Therefore, the functions we are measuring may not be the same as those occurring in the field, but we must attempt to make them indicative of field level functions. We would like to make measurements of ecosystem structure or function and have them indicate the level of health of the system.

The Ecosystem Context

Why do we care about soil health? In a general context, we humans would like to use the natural resources to support us on into the indefinite future. To meet this goal, we must find ways to manage them wisely. We must also find reasons and approaches to further motivate ourselves and others to care about the health of the soil. As illustrated in Fig. 3.1, this can be fostered through the application of psychological principles (Geller, 1994). The active individuals are those who feel they can make a valuable difference.

Since the natural resources are the components of the ecosystem and they are interacting with each other in complex ways, it is necessary to manage the ecosystem as a whole, and not just the individual components. Alterations in one component may result in effects on other components, some of which may be unexpected. Therefore, if we are to manage soils well, we must do this within the context of managing the whole ecosystem. A framework for the analysis of the structure and function of ecosystems will be described below.

The approach described here has been referred to as the 'driving variable-process-property' (DVPP) paradigm (Elliott *et al.*, 1994). It is an attempt to explain the fundamental properties underlying the function of ecosystems. Within this framework, driving variables regulate rates of processes, which determine the flow patterns of material and energy among components. These flow patterns, in turn, determine ecosystem structure and function. At the level of resolution of ecosystem development, the set of driving variables often used are climate, organisms, relief, and parent material. These all act over time to determine ecosystem development and state. This is similar to the general theory of soil development derived by Hans Jenny (1941), except that the idea of processes has been added.

All ecosystems have the same basic set of processes, (e.g. production, decomposition, secondary consumption) even in those from extreme environments. Variations in driving variables differentially control the rates of these processes, thus generating the unique character of different ecosystems. However, as even a single driving variable varies, the whole ecosystem may be altered. As driving variables gradually change geographically, they result in definable regions between ecosystems called ecotones, e.g. the grassland–forest boundary, for natural systems, or for agriculture, the bread basket wheat–corn belt boundary. To address the difficulties of sorting out cause and effect within

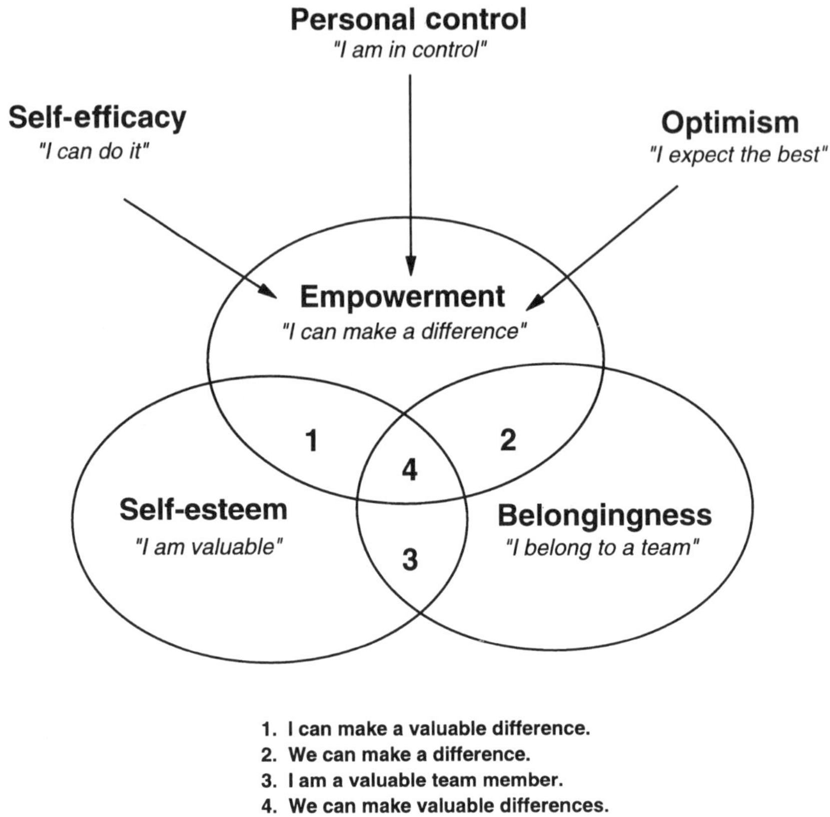

Fig. 3.1. Psychological principles that can be applied to promote 'active caring' about the health of our ecosystems (from Geller, 1994).

the mosaic of terrestrial ecosystems that spread over the face of the continents, scientists often resort to experiments, where components of the ecosystem are manipulated in ways to isolate single variables and their interactions.

Perhaps one of the most intriguing ideas generated from this set of logic is the complex role of organisms as driving variables, as differentiated from the role of organisms as response variables. Jenny (1961) explained that, as a driving variable, the organism should be conceived of as a 'dissemenule' i.e. the disseminating propagule, germplasm or biotic potential of a particular place in space and time. Since a key theme in contemporary ecosystem science is the role of organisms in determining the structure and function of ecosystems, and it is the purpose of this book to find useful biological indicators, it is worth examining in what ways changes in the biotic factor alter ecosystems. In particular, the concept of organisms as the driving variable is appropriate for agricultural systems.

The properties of the whole system may change, including soil processes, as we alter our choice of crop plant at a particular site. However, at a larger scale, cropping systems vary geographically, much like natural ecosystems, as land managers choose crops which best suit the given climate–soil–relief combination. Another driving variable of critical importance to agricultural systems is the cultural/socio-political factor, which has strong controls on the choices of land users because of market values, laws and decisions based upon cultural considerations (Cairns and Niederlehner, 1995). This group of factors might be considered an extension of the organism factor, since humans are responsible for making these decisions. Of all factors, these may have the greatest impact on soil health.

The general concepts described above may be applied to ecosystems at several spatial and temporal scales (Munkittrick and McCarty, 1995) and may be especially important when considering 'what indicates what?' For example, within the time frame of ecosystem development, soil organic matter is a response variable of the system. However, over the course of a growing season, in a particular field or forest, soil organic matter may be relatively unchanging, and may be a driving variable for smaller and faster processes, such as microbial biomass production and consumption or mineral N turnover. Therefore, what constitutes a driving variable and what constitutes a response variable (ecosystem component) depends upon the level of resolution at which one is working. Generally, driving variables are independent from each other, at a given scale, while response variables are capable of changing significantly, and usually are linked through feedbacks within the system. Even the driving variables of Jenny interact if long enough time scales are considered. For example, climate and parent material may interact over geological time to determine relief.

The ecosystem approach integrates detailed process level mechanisms into a description that operates over large areas and long time scales (Elliott and Cole, 1989). These systems may exhibit characteristics that are the result of feedbacks among components and interactions among controlling variables. These attributes can be embodied in conceptual or mathematical models. One such model which illustrates the many factors mediated by the soil biota which affects nutrient supply to plants is shown in Fig. 3.2. Here, driving variables such as climate, directly affect the soil biota by determining the soil conditions. Indirectly, the driving variables impinge on the decisions of land managers, consequently determining the management practices, which eventually influences the soil biota as well. Capturing of these processes allows us to investigate different configurations of system components while we develop new farming systems or examine the behaviour of existing systems. It is becoming increasingly important to test the projected response of new or existing systems to changes in environmental variables as concerns of global climate change

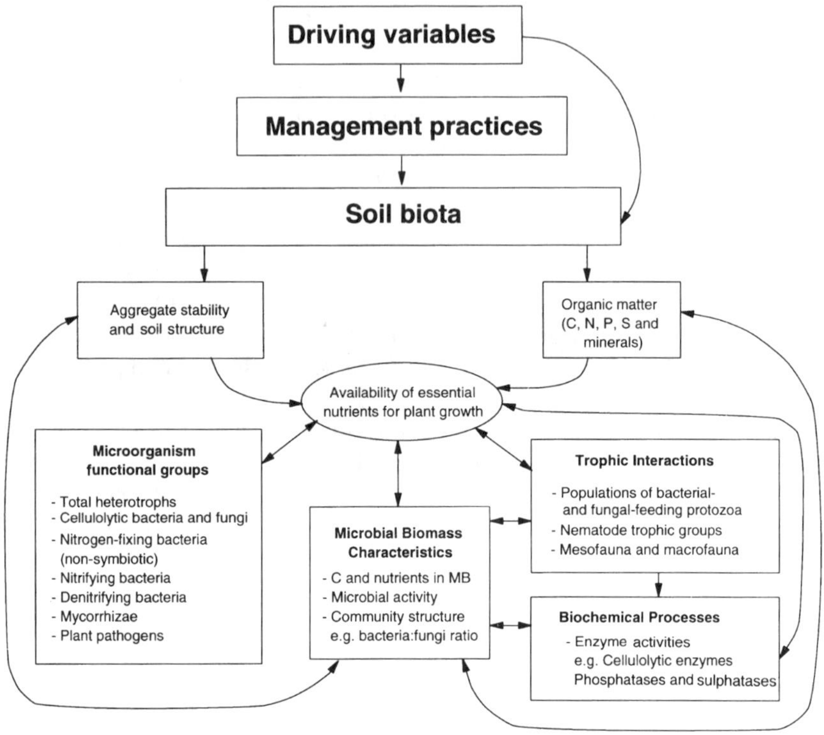

Fig. 3.2. Conceptual model showing the various factors mediated by the soil biota that affect the supply of essential nutrients to plants (modified from Roper and Gupta, 1995).

increase. Where spatially distributed environmental variables can be represented with geographic information systems, projections of response patterns can be investigated spatially (Stewart and Elliott, 1991).

Problems related to ecosystem health occur at different levels of resolution in time and space. Processes that occur at more detailed levels of resolution require explanation based on observations made at coarser levels of resolution. At the same time, finer level processes must be embedded within a general framework to evaluate the system level effects. Thus ecosystem properties and processes are linked in a hierarchical way (Fig. 3.3) (Niles and Freckman, 1996). The response variable at one level in the hierarchy may be the driving variable for the next level.

With the increasing awareness of the human dimension of ecosystems, ways must be found to explicitly include them as ecosystem components (Turner *et al.*, 1990; Cairns, 1994). They have historically been considered outside of the

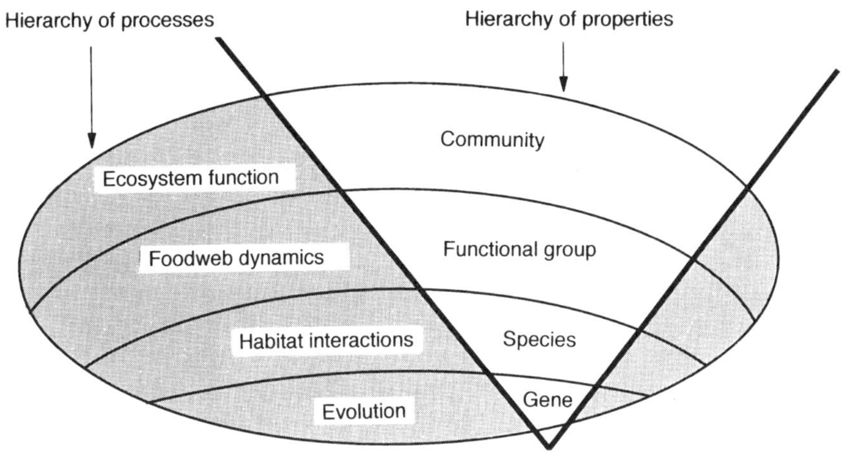

Fig. 3.3. Hierarchy of properties and processes in ecosystems (from Niles and Freckman, 1996).

ecosystem. Natural resources, such as soil, air, water, plants and animals are components of the ecosystem and compose the human environment. We degrade our environment by exploiting natural resources. The degradation of our environment has often had severe and unpredictable secondary effects which could be avoided if management decisions are made within a whole ecosystem context. We seek to maintain the biological integrity necessary to deliver the ecological services necessary to support our society (Cairns and Niederlehner, 1995).

Sustainability

Ecosystem or soil health indicators are difficult to identify because individually measured components represent the sum response of the whole system (Ikerd, 1993) in unknown ways (Elliott *et al.*, 1994). The term sustainable, used in the most general sense in this chapter, will be defined before the idea of indicators is evaluated. Any definition of sustainability (Schaller, 1993) must be defined for the whole system because it seems unlikely that a component can be sustainable without the whole being sustainable.

Constraints on inputs, outputs, structure and attainment of a steady-state have been used to define a sustainable system (Elliott *et al.*, 1993, 1994). Input constraints are controlled by physical, chemical and biological processes as they determine the potential productivity of an agroecosystem. This suite of processes is also influenced by human originated factors which are socio-economic, political and cultural. These factors determine which management practices exist in the field because of subsidies, price supports and life style inheritance.

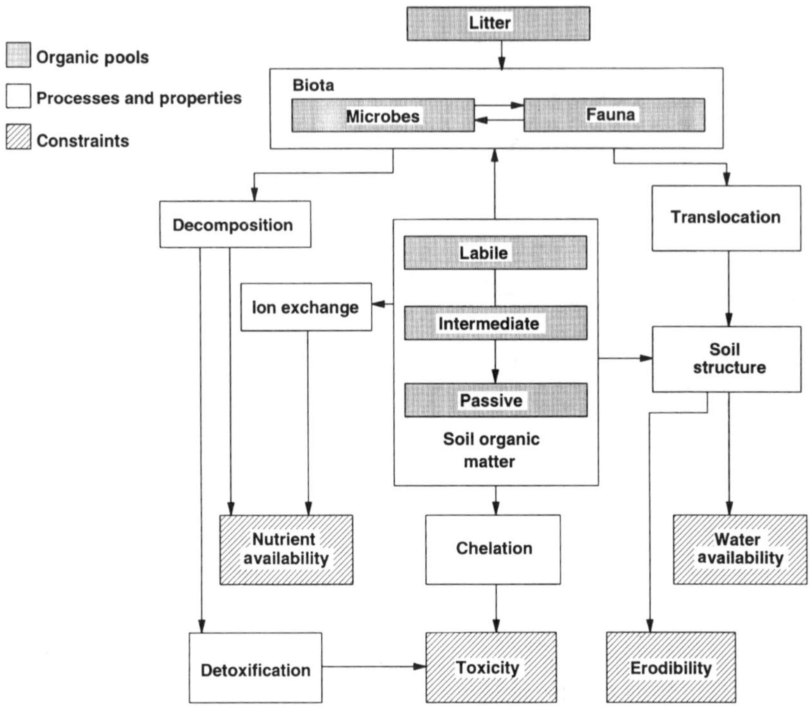

Fig. 3.4. The interrelationship between soil organic matter and biota to produce a complex array of interactions and feedbacks within the soil portion of the ecosystem. These processes and properties determine the intensity of the factors which constrain the functioning of the belowground system (from Swift and Woomer, 1993).

Output constraints can be either positive, such as acceptable limits on crop yield, negative, such as pesticide groundwater pollution, or they may be neutral. If yields do not meet farmers' requirements, or if pollution production exceeds acceptable levels set by law, it is likely that the system from which these outputs came would not exist within the landscape.

Ecosystem structure may determine sustainability because of concerns for biodiversity or species extinction. Systems may be abolished or preserved to prevent loss of diversity. Alternatively, systems not containing the desired component, cattle for example in a pastoral society, will not be sustained because it does not satisfy societal needs. Systems not attaining at least a quasi-steady-state will not exist because they would otherwise continually increase (an impossibility) or decline to an unacceptable level or disappear.

The success of this approach to sustainability is contingent upon simultaneously optimizing the constraints, not merely maximizing or minimizing them

Table 3.1. Mycorrhizal colonization was considerably reduced in sunflower with a long, compared to short, fallow period. There were significant reductions in P and Zn uptake and plant dry weight (adapted from Thompson, 1987).

Measured variable	Fallow (14 months)	Fallow (6 months)
Percentage root length colonized	3	32
Dry weight (g per plant)	12	658
Phosphorus uptake (μg per plant)	18	109
Zinc uptake (μg per plant)	160	1374

individually. Each constraint is necessary and none alone are sufficient. All must be considered together within a systems framework (Fig. 3.4, Swift and Woomer, 1993). Simulation and risk assessment (Power *et al.*, 1994) models can be used to facilitate this system optimization.

It may be difficult to find biotic indicators that meet these broad requirements, even given this very general definition of sustainability. A more reasonable approach may be to focus on specific properties or processes of sustainable systems, such as soil fertility, plant production and associated microbe assemblages (Table 3.1, Thompson, 1987; Jasper, 1994) and identify soil health indicators specifically for them.

Steady States and Trajectories

It is essential to understand the steady-state and trajectory nature of components of ecosystems if one is to apply the use of soil health bioindicators. It is inescapable that when inputs equal outputs, the system is in steady-state and the size of the component is constant (Greenland and Nye, 1959; Olson, 1963). When inputs and outputs vary over time but generally balance, the value of the component will vary around some constant value. This situation is referred to as a quasi-steady-state. This idea is useful when applied to the dynamics of soil organic matter and is applicable to other system properties as well, such as population dynamics. For a given set of driving variables, ecosystems approach a quasi-steady-state, with a climax biotic community and representative level of soil development. Ecosystems and their components vary, as driving variables change within or across seasons, usually around some mean steady-state value (Fig. 3.5, Cairns *et al.*, 1993).

An alternative to the quasi-steady-state is the 'non-equilibrium' conceptual model, which may be more applicable where the variations (possibly caused by perturbations) are so great that a steady-state value is not discernable, as Ellis and Swift (1988) suggest for the semi-arid and arid tropics. Even if it may never

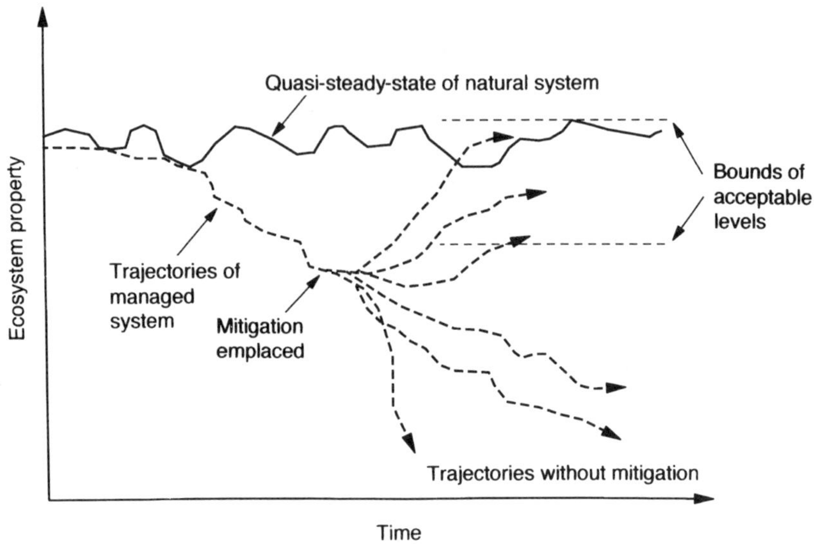

Fig. 3.5. Differences in the management of ecosystems produces differences in their trajectories. One must know the steady-state 'endpoint' solution and the current state to determine the likely future trajectory and where one is placed along it. Levels for the chosen property may be set to indicate what is an acceptable value and the level of health under the current situation or in the future (Adapted from Cairns *et al.*, 1993).

be achieved in nature, the concept of steady-state is useful as a destination which the system, or components of the system, are moving toward.

For a known set of driving variables and ecosystem structure, a component of the system, such as soil organic matter, approaches a new steady-state level based upon the balance of inputs (primary production and other organic source addition) and outputs (decomposition, erosion and leaching). The incremental set of values attained by the component as it moves from the previous state to the new state, is its trajectory. Trajectories are altered as driving variables (e.g. management or climate) change. The final steady-state may never be reached if drivers change frequently. Knowledge of where a system or component lies along a trajectory is necessary for predicting where it is moving, given that the steady-state value is known.

Processes and Properties as Indicators

Managing soil biota and using them as indicators is a useful concept, at one level of abstraction, and a more general perspective is also useful for understand-

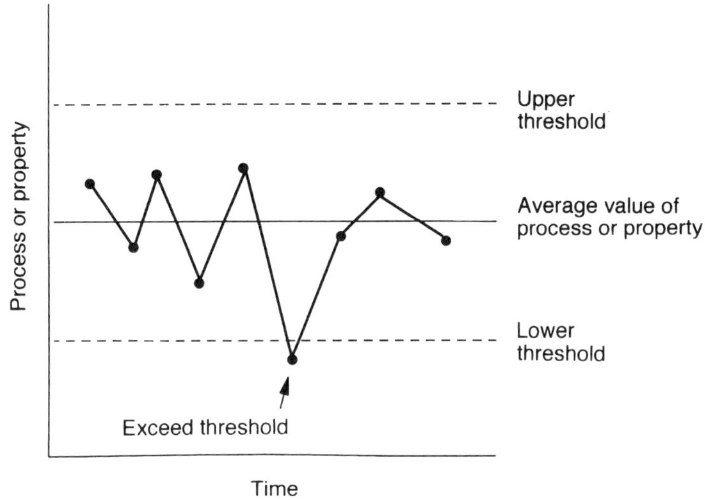

Fig. 3.6. When the values of properties or processes exceed preassigned thresholds, as shown in this generalized example of a control chart, it can be taken as an indication of poor soil quality or health (adapted from Larson and Pierce, 1994).

ing soil health bioindicators. One may measure either processes (e.g. mineralization) or properties of the soil to determine the presence of disease. Properties are expressed as an amount per unit area and can be any component of the system (Fig. 3.6) (Larson and Pierce, 1994). Processes are expressed as a change in the amount per area per unit time.

The amount of material within a component at any point in time is the net result of the sum of all inputs to the component (e.g. bacterial consumption by soil nematodes) minus the sum of all processes leading to outputs from the component (e.g. respiration, excretion, reproductive losses due to egg laying). The inputs and outputs from the component are controlled by the processes. As a result, one may measure a process (e.g. bacterial consumption by nematodes) as an indicator or one may measure the property itself (i.e. bacterial feeding nematode biomass). Because of the dynamic nature of the biological components of ecosystems (Gupta, 1994), there is a constant turnover of material within them determined by the relative rates of inputs and outputs. The amount of material within the component may remain constant but the turnover may vary considerably. Therefore, if one uses a component as an indicator, some information may be lost. Alternatively, components are the result of multiple processes and integrate many kinds of information and represent well the state of the whole system. Biotic indicators that are generally useful should behave as response variables, occur over a variety of ecosystem types and embody as much as possible of the component dynamics or, alternatively, the system as a whole.

Table 3.2. Soils obtained from farms using organic farming practices showed higher microbial activity with subsequent reductions in corky root rot severity and an increase in suppression compared with soils from conventionally managed farms ($n = 3$) (modified from Workneh and van Bruggen, 1994).

Farm type	Microbial activity[a]	Corky root severity[b]	% Suppression of corky root[c]
Organic	1.07	13.1	67.7
Conventional	0.20	20.8	41.4

[a] Fluorescein diacetate hydrolysis ($\mu g\ g^{-1}\ min^{-1}$);
[b] % diseased;
[c] % reduction (non-irradiated/radiated).

Rationale for Bioindicators

What are the properties of a good bioindicator of soil health? The presence, absence or abundance of any organism indicates something (Kremen, 1992), but what, and is it of value? The following section discusses indicators of soil health.

Whole soils, and their constituent parts, reflect the influence of other properties of ecosystems (Elliott *et al.*, 1994). As systems vary geographically, soils are affected directly, and also indirectly reflect changes in the rest of the system. For example, transition from grassland to forest creates large differences in many soil properties including total soil organic matter content, the distribution of organic matter within horizons and the composition of the soil fauna. Yet, the shift from grasslands to forest is most often in response to changes in the climate driving variables. Therefore, soils and plants change together, as do many other components of the system, including the soil biota. Soil bioindicators should reflect many of these changes and, as discussed earlier, may variously represent different aspects in different ecosystems. Therefore, if one is to use a soil component as an indicator across systems it must reflect the responses of the same set of components across all ecosystems of interest. This is unlikely to occur but the use of functional biodiversity based on substrate use (Zak *et al.*, 1994) is a promising direction for research.

Generic components, such as total organic matter content, may be more useful in this regard than more specific components, such as collembolan species composition. In contrast, more specific indicators may be of greater value if the ecosystems to be evaluated are more narrowly bound by sets of driving variables. Under local conditions, management may be the key driving variable and knowledge of soil processes, such as biocontrol, may be of particular importance (Table 3.2, Workneh and van Bruggen, 1994). Biocon-

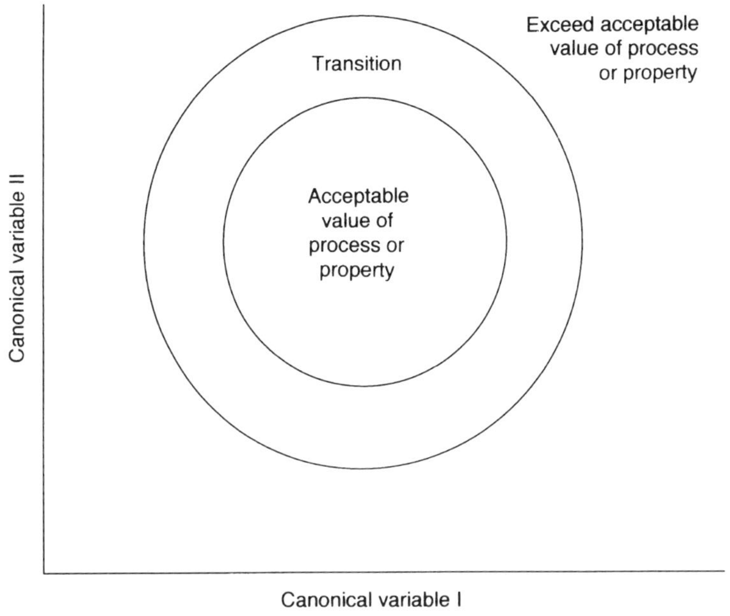

Fig. 3.7. Multiple biotic and abiotic factors, either derived from analytical measurements or models, may be statistically and/or mathematically combined into sets of corresponding canonical factors which represent state–space. These regions may be used to indicate soil health that is acceptable, in transition or unacceptable (modified from Moore and de Ruiter, 1993).

trol of disease (Mannion, 1995) has been recently demonstrated for the introduction of microflora (Dewan *et al.*, 1994; Ousley *et al.*, 1994; Pfender *et al.*, 1996), nematodes (Yeates and Wardle, 1996), collembola (Lartey *et al.*, 1994; Lussenhop, 1996) and indirectly with earthworms (Doube *et al.*, 1994). The effectiveness of biocontrol agents could be used as an indication of the rehabilitive possibilities of diseased soil. The choice of an indicator should be determined by the specific local situation.

Empirical relationships can be developed between two components or more of a system or a component and some set of properties (Fig. 3.7) (Moore and de Ruiter, 1993). For example, butterfly assemblages were used as indicators of plant species diversity by Kremen (1992). Generally, one would like to choose an indicator which is easiest to measure and most integrative (Smyth and Dumanski, 1995). This approach may provide useful indicators within the bounds of the conditions under which the relationships were developed. However, if an underlying mechanistic relationship can be determined, the indicator can be more easily extrapolated to other conditions. At least the basis for further testing can be readily identified.

Sampling in Time and Space

Useful indicators respond significantly to components one wishes to monitor. The sensitivity of an indicator to changes in other components of interest may sometimes be large, thereby creating a dynamic response of the indicator. In this case, choosing the appropriate point in time to sample becomes important. One may wish to sample frequently to obtain an understanding of the integrated response. However, it is often impractical to use indicators which must be sampled frequently.

Interestingly, indicators that respond slowly have restricted value because it may take an inordinate amount of time to obtain a measurable response. However, the slow to respond indicators are good for integrating responses and, where changes are large enough to observe, they can be confidently used. When effects accumulate and their signature is captured in a slowly responding indicator, the signal may be available for later analysis of a past environmental problem. Usually, we would like indicators that respond in an intermediate time frame so that we both: (i) capture responses that occur over usefully short time periods; and (ii) adequately integrate the response so that a reasonable sampling frequency is attainable.

An analogous discussion of the spatial distribution of indicators and sampling protocols can be made as was discussed above for time. Indicators which are highly patchy in their distribution (e.g. populations of microanthropods) will require more effort to enumerate than more uniformly distributed variables (e.g. organic carbon). Limitations set for researchers will be quite different from those set for land managers because of the differential availability of technology and resources.

Biological Considerations

Soil ecology has many facets which must be considered before they can act as measurements which indicate soil health. One can consider numbers of organisms at many taxonomic levels, both within phyletic lines and within functional groups. One must consider a range in generation times from hours to years. Requirements of indicator response times are critical. One must also consider that the density may vary from one or less per metre square to billions per gram, making sampling difficult. From the knowledge of the weight of an individual (which is commonly unavailable), one can convert numbers to biomass. Weights can vary depending upon age and life stage. Knowledge of weight may be as great a limitation to estimating biomass as estimates of numbers.

Measurement of the age distribution of the population (demography), could be used to build life tables from which calculations of growth rates could be made. However, one must know the relationship between some measurable characteristic such as size and age. In addition, the transition time for life stages to pass from one stage to the next and other life history information are needed. Clearly, this requires a great deal of knowledge about the organism. If this information is obtained, one can calculate the gross growth rate of the population, perhaps a good indicator of growth conditions in the soil. Other approaches for determining growth rates in soil, such as difference in numbers over a known time period (net growth), will generally be an underestimate of the growth potential of the organism. This type of measure does not account for predation, parasitism or other mortality factors other than old age, which may be independent from the environmental carrying capacity which one is interested in determining.

Chloroform fumigation incubation or extraction methods are direct measurements of total soil biomass as are a number of other procedures (Sparling and Ross, 1993). These have the advantage of not requiring direct counts and size conversions. Drawbacks include different extraction efficiencies for different soils, especially for microbial biomass P, and difficulties in separating root from microbial biomass. Metabolism and microbial biomass formation from added substrates may be a sensitive indicator of soil health as shown for soil contaminated with different levels of heavy metals (Fig. 3.8) (Bardgett and Sagger, 1994). With the application of image analysis, direct counting of soil microorganisms is becoming more accessible. Sparling (1991) suggested use of the ratio of microbial biomass to total soil organic matter as an indicator of biological activity of the soil and its relationship to accumulation of total soil organic matter (Swift and Woomer, 1993; Park and Cousins, 1995).

An indicator with considerable promise is the community structure of soil organisms (Moore and de Ruiter, 1993). If considered within function groups, community structure may be useful for distinguishing environmental conditions in the soil. Success in this approach requires the existence of adequate knowledge about a particular organismal group. Soil microflora are so diverse and representative isolation is so unreliable that this group may be applied as indicators only with difficulty (Roper and Ophel-Keller, Chapter 7, this volume), although Beare *et al.* (1993) were able to make distinctions for fungal species between conventional and no-till treatments.

Despite their dominant role as consumers of microflora, protozoa are poorly known taxonomically and isolation and identification are exceptionally tedious (Gupta and Yeates, Chapter 8, this volume). While microarthropod taxa are well known, sampling and extraction procedures are time consuming. They are often patchily distributed so one must take many cores in order to achieve a realistic result. Because of the necessary heat extraction procedures, cores must be maintained intact so bulking, mixing and subsampling is not an option unless one uses heptane float methods, which have their own problems. Hówever, microarthropods may be easily manipulated in laboratory experiments (Table 3.3) (Siepel and Maaskamp, 1994) and are large enough that body burdens of pollutants can be determined at the species level (van Wensem *et al.,* 1994). Soil macrofauna are often not ubiquitous (Doube and Schmidt, Chapter 11, this

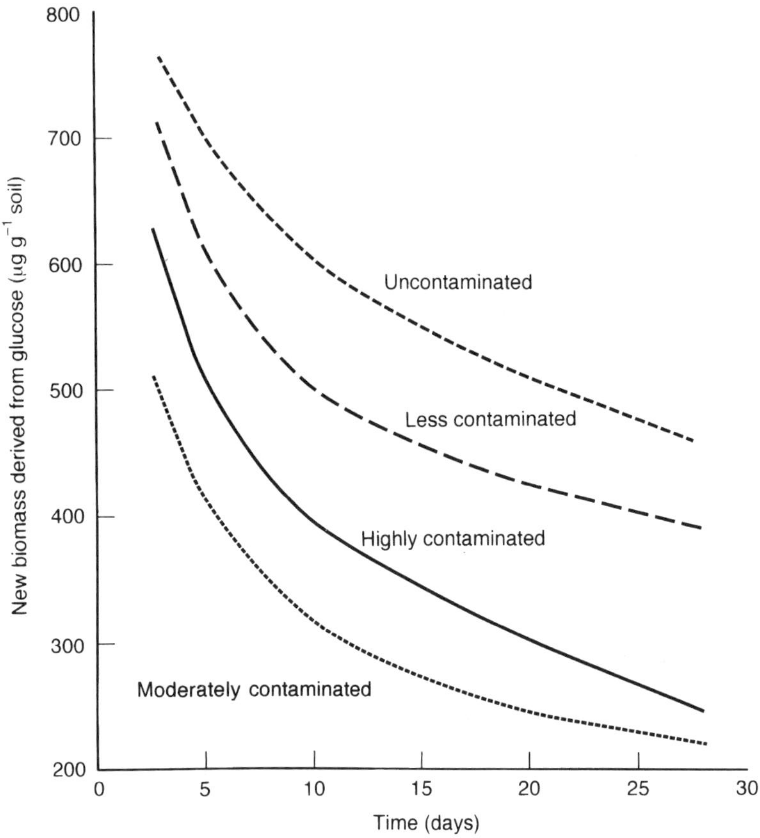

Fig. 3.8. The amount of microbial biomass developing and subsequently declining after addition of radioisotopically labelled glucose is a sensitive bioindicator of the levels of contaminating metals in soil (modified from Bardgett and Saggar, 1994).

volume), due to vagaries in biogeography and contagion and are therefore problematic for use as general community structure indicators across a wide variety of ecosystems. Soil nematodes may be the best organismal group to use for community indicator analysis because a great deal of knowledge exists on their taxonomy and feeding roles (Gupta and Yeates, Chapter 8, this volume). Also, their distribution in the soil is such that adequate samples can be taken with reasonable effort.

Aspects of the soil foodweb structure may be used as indicators but the effort required to obtain such information precludes it from common application. Moore and de Ruiter (1991) have demonstrated its usefulness for understanding energy flow and nutrient cycling in several ecosystems around the world, many

Table 3.3. Members of mite guilds that contained chitinase activity in their guts stimulated decomposition rates when added to mixed fungal cultures grown on the leaves of ground *Avenella* (modified from Siepel and Maaskamp, 1994).

Mite feeding guild	Enzyme activity: Cellulase	Trehalase	Chitinase	Effect on decomposition rate
Herbivorous grazer	+	0	0	0
Herbofungivorous grazer	+	+	+	+
Fungivorous grazer	0	+	+	+
Opportunistic herbofungivore	+	+	0	−
Fungivorous browser	0	+	0	−

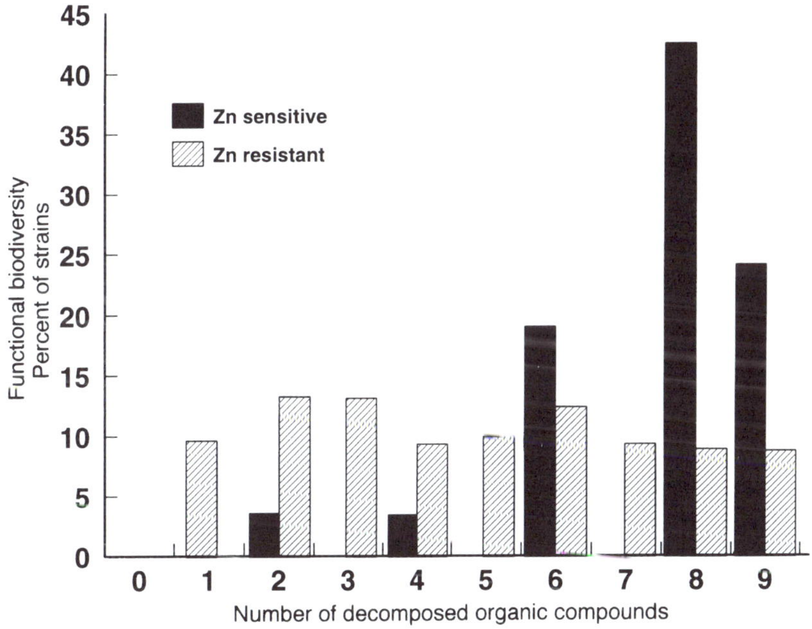

Fig. 3.9. Bacterial strains that were isolated from Zn contaminated soils were categorized as resistant or sensitive to Zn in culture. These strains were then tested for their ability to utilize a series of aromatic organic compounds. Clearly, a higher proportion of the sensitive strains were able to use the organic compounds than those resistant to Zn. These interactive effects have important implications for the health of soils contaminated both with heavy metals and toxic aromatic pollutants (modified from Doelman *et al.*, 1994).

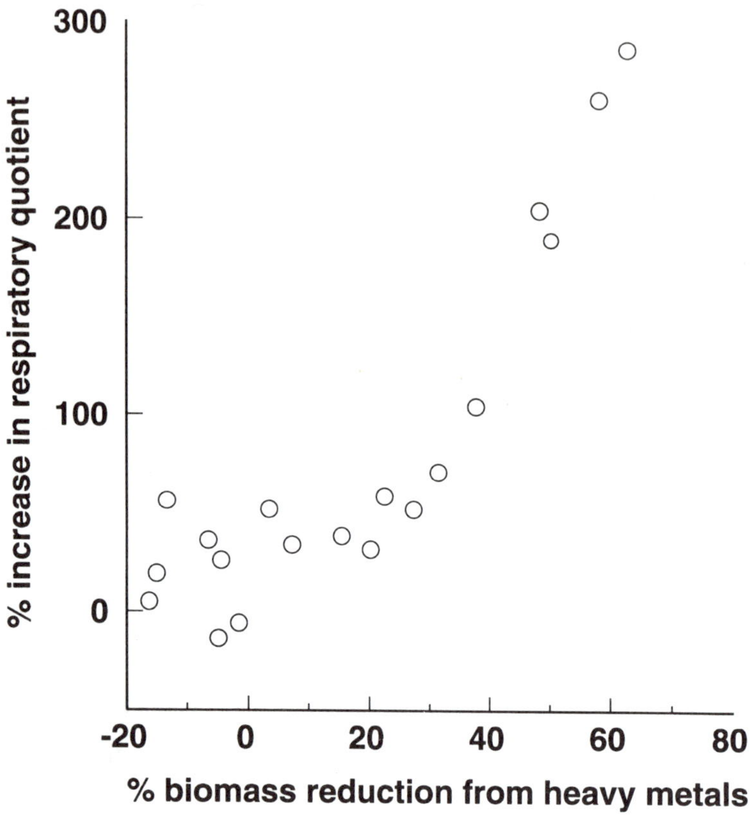

Fig. 3.10. Low levels of metal contamination added with sewage sludge initially stimulated the production of microbial biomass. However, as metal levels increased there was a concomitant decrease in biomass that was associated with increasingly less efficient biomass synthesis (modified from Fliebbach *et al.*, 1994).

of them agricultural. Energy flow paths were used to distinguish the importance of different groups. Beare *et al.* (1992) found that more energy flowed through the fungal pathway in a no-till than a conventionally tilled system. While such examination is one of the few approaches that can integrate a whole system perspective, thus making it a good candidate as an indicator of soil health, it takes several to many human year equivalents of work time to obtain such information for most ecosystems.

Biotic activity measures rely heavily upon quantification of substrate use (Fig. 3.9) (Doelman *et al.*, 1994) or mineralization (Bardgett and Sagger, 1994; Franzluebbers *et al.*, 1995; Groot and Houba, 1995). Many of these measures, such as substrate induced respiration (Anderson and Domsch, 1978) or ATP

(Jenkinson and Oades, 1979), are calibrated for use in determining biomass rather than activity, such as enzyme activity, for example (Ohtonen *et al.*, 1994). Such measurements as potentially mineralizable C or N (Groot and Houba, 1995) may serve as indicators of natural soil fertility. The specific rate of respiration (respiratory quotient) may be particularly sensitive to metal contamination (Fig. 3.10) (Fliebbach *et al.*, 1994) or pH (Anderson and Domsch, 1993). Mass loss from litter bags, and attendant analyses of the litter, are also in this category of measurements, and along with the above described mineralization measurements, are well described in Anderson and Ingram (1993).

Abiological Measures of Soil Health

Indirect indicators of soil health may be derived from the measurement of the influence of soil biota on other parts of the system (Karlan and Stott, 1994). Soil organic matter and soil structure may serve particularly well for this function (Park and Cousins, 1995). Minimum datasets have been developed as indicators of soil quality (Table 3.4, Gregorich *et al.*, 1994). Measures of porosity or aggregation are used to define soil structure. These two approaches essentially represent the study of soil holes or particles. From a systems perspective, these two approaches are not necessarily independent if one assumes that the walls of the pore space are the surfaces of the aggregates (Elliott and Coleman, 1988). Production of soil organic matter, including extracellular polysaccharides and other cellular debris, increases the capacity of the soil to maintain structure once it is formed (Fig. 3.11, Roldán *et al.*, 1994). High clay soils have greater capacity than sandy soils.

Table 3.4. Minimum datasets required for estimating different aspects of soil quality (Gregorich *et al.*, 1994).

Processes	Minimum data requirements
Soil structure development	Total organic C and N
	Microbial biomass
	Carbohydrate
	Fungal length
Nutrient storage	Total organic C and N
	Microbial biomass
	Mineralized C and N
	Light fraction and macroorganic matter
Biological activity	Microbial biomass
	Enzymes
	Mineralized C and N

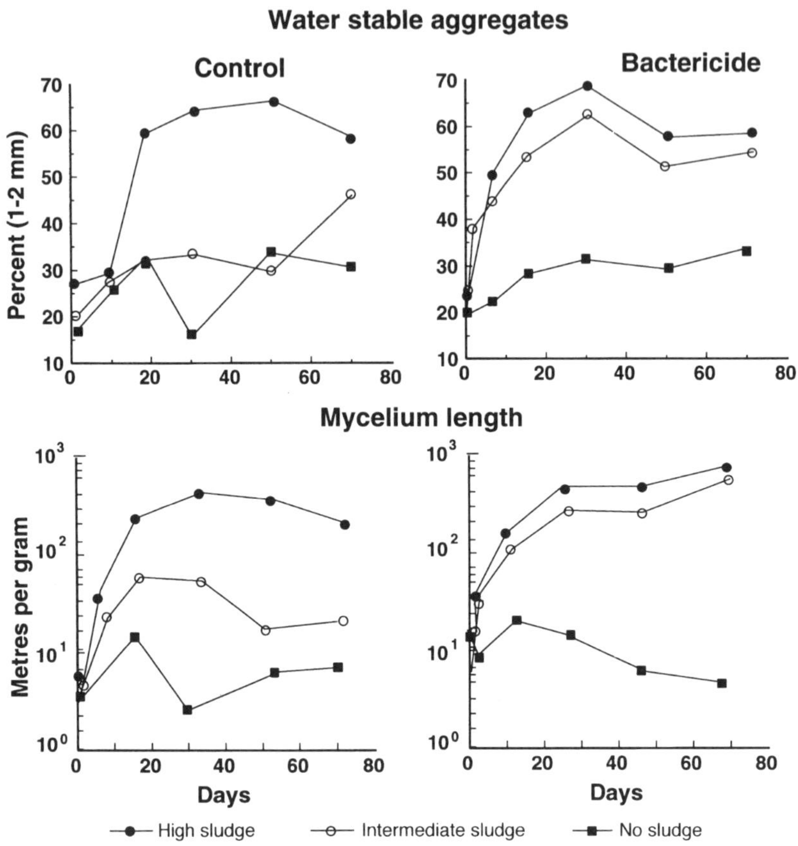

Fig. 3.11. Raw sewage sludge stimulated the growth of fungal mycelia at high levels of addition with the subsequent formation of water stable aggregates. When a bacterial inhibitor was added, fungi also proliferated at an intermediate level of sludge additions, and again, water stable aggregates were formed. Water stable aggregates may be a good indirect soil health indicator of perturbed soil (modified from Roldán *et al.*, 1994).

Soil particles may be moved into various positions by earthworms, roots, wet–dry and freeze–thaw cycles and other forces thus forming soil structure. Production of faeces by mesofauna also constitutes formation of soil structure since both particles and the spaces between them are formed. The life span of newly formed soil structure depends upon the capacity of the soil to maintain the structure and the frequency of disturbance. The capacity will change as organic materials are either formed or destroyed through microbial activity.

Nematode Community Structure as an Indicator

Productivity of the belowground system as a whole may be indicated by the number of bacteriophagic nematodes present (Freckman, 1988) as compared with direct measures of bacteria, which do not necessarily reflect total production if they are heavily predated. Standing stocks of bacteria may be reduced under predation but the turnover rate may be increased; yield may be the same in both cases. In a similar fashion, other functional groups of nematodes may represent the production of the system component upon which they feed. However, categorization of stylet-bearing forms and omnivores requires more taxonomic and biological information to distinguish feeding preferences than is required for bacterial feeders. Wasilewska (1979) suggested the use of plant feeders and predators as bioindicators of human impact. Bacterial and fungal feeding forms increased as plant feeding and predacious forms decreased in ecosystems representing increasing human intervention. Parmelee and Alston (1986) found that during the summer there is a relatively greater production of fungi in soil from a no-till treatment and bacteria in conventional till, based upon the numbers of bacterial vs fungal feeding nematodes.

Functional groupings of nematodes or use of indices based upon species level comparisons may be useful as a qualitative indicator of the status of soil health (Niles and Freckman, 1996). Sohlenius *et al.* (1987) found that total numbers, numbers in various functional groups and relative numbers in various functional groups differed among cropping treatments. Species level indices 'were little affected in soils subjected to different management practices'. He suggested that the short duration of the experiment (5 years) may have been the reason. Dmowska and Kozlowska (1988) found no effect of low levels of liquid manure application on nematode community structure, but at levels higher than what would normally be used for fertilizer application, significant changes in numbers and composition occurred. Degree of similarity of nematode communities in conventional and no-till management across seven crop rotation systems that graded from continuous corn to continuous soybean were evaluated by Niles (1991). He found that the similarity of communities varied in proportion to the representation of corn in the rotation except for the no-till system in the spring where the similarity varied in proportion to the representation of soybean in the rotation. He suggested that the more labile soybean residue may still be the predominant substrate in the spring following the previous fall harvest for the no-till system. It is encouraging that such clear patterns can distinguish among treatments differing in tillage and crop residue inputs.

Bongers *et al.* (1991) suggested a Maturity Index (MI) for marine nematodes, which was subsequently tested for terrestrial soils (de Goede and Bongers, 1994), based upon relative selection of fast (colonizers, c) versus slow (persisters, p) growing forms of nematodes in different environments.

Each species is assigned a c-p value based upon knowledge of life history traits. This approach is similar to separation of r- and K-selected forms suggested by MacArthur and Wilson (1967). For colonizers these traits may be high colonization ability, a short generation time, large population fluctuations, form of dauer larvae that bridge unfavourable conditions, large gonads, production of many small eggs, and viviparity. The opposite traits are exemplary for persisters. Samples are analysed, c-p values attributed and the weighted mean of taxon scores determined, thus yielding the MI. MI is reduced by pollution but increases with colonization processes for marine systems. As restoration occurs, MI increases. However, if high organic inputs are constant over time, MI remains low because fast growing forms predominate even if the environment is stable. A spatial equivalent of this example is where the input of new substrates into the marine benthos consistently yields new habitats for colonization thus maintaining a low MI even with low and stable input.

The above approach may have useful applications in terrestrial ecosystems (Bongers, 1990) and has been modified to include plant feeders (Yeates, 1994). For any particular location, other factors, such as texture, must be also considered. Even the best indicators may be difficult to compare across different ecosystems. A bioassay where a nematode fauna is inoculated into a set of soils and their fate followed, may give better results across different systems (Bongers, 1990). Experimental manipulations of nematode colonization and succession showed the expected trends in nematode community development moving from colonizers to persisters (Ettema and Bongers, 1993). Freckman and Ettema (1993) compared several diverse management systems at the Kellogg Biological Station, Michigan, USA. Significant differences among treatments were shown consistently only for the predator trophic group, the MI and a plant parasite index, even though many different indices were calculated. A more objective approach using multivariate canonical analysis separated the eight treatments into five groups.

Soil Organic Matter as an Indicator

Soil organic matter is a key abiological indicator of soil health (Swift and Woomer, 1993; Park and Cousins, 1995). It is a direct product of the combined biological activity of plants, microorganisms and animals plus the myriad of abiological factors. It is responsible for crucial aspects of soil function such as aeration and fertility (Elliott and Coleman, 1988). Conceptually, SOM has been divided into pools of differing turnover times and research is ongoing to better define these pools analytically (Christensen, 1996; Elliott *et al.,* 1996) and mathematically (Powlson *et al.,* 1996).

Particulate organic matter (POM) is a fraction of soil organic matter (SOM) that has the potential to serve as an indirect measure of soil health.

The half-life of POM was found to be ~10–20 years at Sidney, Nebraska, USA and may be an important contributor to soil aggregation (Cambardella and Elliott, 1993), either directly or indirectly by serving as a substrate for microbial activity. POM is found in soil and outside aggregates and the C/N of this POM may vary with tillage. As plant material is decomposed, part is used by microbes, and eventually becomes microbial derived soil organic matter [hypothetically isolated as an enriched labile fraction (ELF) by Cambardella and Elliott (1994)], and the other part, that which is not decomposed, becomes plant derived soil organic matter (POM) (Elliott *et al.*, 1996). POM and ELF were found to represent a significant part of the total SOM (about half in some soils) and have higher concentrations under reduced tillage. POM can be a substantial contributor to total SOM (40–50% in native grassland soils) but may account for considerably less of the total SOM in soils cultivated for many years.

Soil health may be indicated by the measurement of the POM fraction because the turnover time allows enough sensitivity to detect changes within a few years. POM dynamics are not large over the growing season, except perhaps in highly degraded soils, so the method is relatively insensitive to time of sampling. It may also integrate responses over a number of years. Analysis of data from 20 agricultural experiments in the Central US revealed that POM-N and POM-N/total N were the best indices for comparing treatment effects (E.T. Elliott, K. Paustian, E.A. Paul and H.P. Collins unpublished). Normalizing POM-N to total N allows one to account for variations in organic matter content, which may be affected by texture, climate or other driving variables. Before such measures can be used confidently as indicators of soil health, they must be tested over a wide range of driving variables and treatments. POM changes in response to these factors could be used to calibrate the method for general use as an indicator.

Conclusions

Developing rationale for the use of biological components of soil as indicators of soil health necessarily spans a wide range of considerations. One can never prove soil health, only the lack of measurable disease. Therefore, we must target what we believe to be key processes or components which indicate disease and measure them. Given enough testing and analysis of various indicators, one may construct a set of probabilities that a system may be healthy based on these indicators, but absolute health can never be proven. Considerable work is needed to reach this level of understanding of soil health indicators.

Ecosystem or soil health indicators are difficult to identify because individually measured components represent the summed response of the whole system. Biological indicators that are generally useful should behave as response variables, occur over a variety of ecosystem types and embody as much as possible of the component dynamics or, alternatively, the dynamics of the system as a whole. If an underlying mechanistic relationship can be determined, the indicator can be more easily extrapolated to other conditions. Indirect indicators of soil health may be derived from the measurement of the influence of soil biota on other parts of the system. The sensitivity of an indicator to changes in other components of interest may sometimes be large, thereby creating a dynamic response of the indicator and thus making it less suitable for long-term monitoring.

An indicator with considerable promise for distinguishing environmental conditions is the community structure of selected groups of soil organisms. Soil nematodes may be the best group to use for community indicator analysis because a great deal of knowledge exists on their taxonomy and feeding roles. Also, their distribution in the soil is such that adequate samples can be taken with reasonable effort. Functional groupings of nematodes or use of indices based upon species level comparisons may be useful as a qualitative indicator of the status of soil health.

Soil organic matter is a key indirect indicator of soil health. It is a product of the combined biological activity of plants, microorganisms and animals plus the myriad of abiological factors. It is responsible for crucial aspects of soil function such as aeration and fertility. Particulate organic matter (POM) is a fraction of soil organic matter that has potential to serve as a sensitive measure of soil health. POM may be an important contributor to soil aggregation, either directly or indirectly by serving as a substrate for microbial activity and the turnover time allows enough sensitivity to detect changes within a reasonable period of time.

References

Anderson, J.M. and Ingram, J.S.I. (1993) *Tropical Soil Biology and Fertility: A Handbook of Methods.* CAB International, Wallingford, UK, 221 pp.

Anderson, J.P.E. and Domsch, K.H. (1978) Mineralization of bacteria and fungi in chloroform-fumigated soils. *Soil Biology and Biochemistry* 10, 207–213.

Anderson, T.H. and Domsch, K.H. (1993) The metabolic quotient for CO_2 (qCO_2) as a specific activity parameter to assess the effects of environmental conditions, such as pH, on the microbial biomass of forest soils. *Soil Biology and Biochemistry* 25, 393–395.

Angle, J.S., Chaney, R.L. and Rhee, D. (1993) Bacterial resistance to heavy metals related to extractable and total metal concentrations in soil and media. *Soil Biology and Biochemistry* 25, 1443–1446.

Bardgett, R.D. and Saggar, S. (1994) Effects of heavy metal contamination on the short-term decomposition of labelled [^{14}C] glucose in a pasture soil. *Soil Biology and Biochemistry.* 26, 727–733.

Beare, M.H., Parmelee, R.W., Hendrix, P.F. and Cheng, W. (1992) Microbial and

faunal interactions and effects on litter nitrogen and decomposition in agroecosystems. *Ecological Monographs* 62, 569–591.
Beare, M.H., Pohlad, B.R., Wright, D.H. and Coleman, D.C. (1993) Residue placement and fungicide effects on fungal communities in conventional and no-tillage soils. *Soil Science Society of America Journal* 57, 392–399.
Bongers, T. (1990) The maturity index: An ecological measure of environmental disturbance based on nematode species composition. *Oecologia* 83, 14–19.
Bongers, T., Alkemade, R. and Yeates, G.W. (1991) Interpretation of disturbance-induced maturity decrease in marine nematode assemblages by means of the Maturity Index. *Marine Ecology Progress* 76, 135–142.
Brendecke, J.W., Axelson, R.D. and Pepper, I.L. (1993) Soil microbial activity as an indicator of soil fertility: long-term effects of municipal sewage sludge on an arid soil. *Soil Biology and Biochemistry* 25, 751–758.
Cairns, J. Jr (1994) Ecosystem health through ecological restoration: barriers and opportunities. *Journal of Aquatic Ecosystem Health* 3, 5–14.
Cairns, J. Jr and Niederlehner, B.R. (1995) Ecosystem health concepts as a management tool. *Journal of Aquatic Ecosystem Health* 4, 91–95.
Cairns, J. Jr McCormick, P.V. and Niederlehner, B.R. (1993) A proposed framework for developing indicators of ecosystem health. *Hydrobiologia* 263, 1–44.
Cambardella, C.A. and Elliott, E.T. (1993) Distribution of organic C and N within aggregate size fractions in cultivated and native grassland soils. *Soil Science Society of America Journal* 57, 1071–1076.
Cambardella, C.A. and Elliott, E.T. (1994) Carbon and nitrogen dynamics of soil organic matter size/density fractions under different management practices. *Soil Science Society of America Journal* 58, 123–130.
Chander, K. and Brookes, P.C. (1993) Residual effects of zinc, copper, and nickel in sewage sludge on microbial biomass in a sandy loam. *Soil Biology and Biochemistry* 25, 1231–1239.
Christensen, B.T. (1996) Matching measurable soil organic matter fractions with conceptual pools in simulation models of carbon turnover: revision of model structure. In: Powlson, D.S., Smith, P. and Smith, J.U. (eds) *Evaluation of Soil Organic Matter Models Using Existing Long-Term Datasets.* Springer-Verlag, Berlin, pp. 143–159.
Costanza, R., Norton, B.G. and Haskell, B.D. (eds) (1992) *Ecosystem Health: New Goals for Environmental Management.* Island Press, Washington DC.
de Goede, R.G.M. and Bongers, T. (1994) Nematode community structure in relation to soil and vegetation characteristics. *Applied Soil Ecology* 1, 29–44.
Dewan, M.M., Ghisalberti, E.L., Rowland, C. and Sivasithamparam, K. (1994) Reduction of symptoms of take-all of wheat and rye-grass seedlings by the soil-borne fungus *Sordaria fimicola. Applied Soil Ecology* 1, 45–51.
Dmowska, E. and Kozlowska, J. (1988) Communities of nematodes in soil treated with semi-liquid manure. *Pedobiologia* 32, 323–330.
Doelman, P., Jansen, E., Michels, M. and van Til, M. (1994) Effects of heavy metals in soil on microbial diversity and activity as shown by the sensitivity-resistance index, an ecologically relevant parameter. *Biology and Fertility of Soils* 17, 177–184.
Doran, J.W. and Parkin, T.B. (1994) Defining and assessing soil quality. In: Doran, J.W., Coleman, D.C., Bezdicek, D.F. and Stewart, B.A. (eds) *Defining Soil*

Quality for a Sustainable Environment. Soil Science Society of America, Special Publication. No. 35, Madison, pp. 3–21.

Doube, B.M., Stephens, P.M., Davoren, C.W. and Ryder, M.H. (1994) Interactions between earthworms, beneficial soil microorganisms and root pathogens. *Applied Soil Ecology* 1, 3–10.

Elliott, E.T. and Cole, C.V. (1989) A perspective on agroecosystem science. *Ecology* 70, 1597–1602.

Elliott, E.T. and Coleman, D.C. (1988) Let the soil work for us. *Ecological Bulletin* 39, 23–32.

Elliott, E.T., Cambardella, C.A. and Cole, C.V. (1993) Modification of ecosystem processes by management and the mediation of soil organic matter dynamics. In: Mulongoy, K. and Merckx, R. (eds) *Soil Organic Matter Dynamics and Sustainability of Tropical Agriculture.* Wiley-Sayce Publishing Company, New York & Chichester, pp. 257–267.

Elliott, E.T., Janzen, H.H., Campbell, C.A., Cole, C.V. and Meyer, R.J.K. (1994) An ecosystem approach to integrated nutrient management for sustainable land use. In: Dumanski, J. (ed.) *Sustainable Land Management in the 21st Century. Proceedings of an International Workshop,* Lethbridge, Saskatchewan, Canada, June 20–26, 1993.

Elliott, E.T., Paustian, K. and Frey, S.D. (1996) Modeling the measurable or measuring the modelable: A hierarchical approach to isolating meaningful soil organic matter fractionations. In: Powlson, D.S., Smith, P. and Smith, J.U. (eds) *Evaluation of Soil Organic Matter Models Using Existing Long-Term Datasets.* Springer–Verlag, Berlin, pp. 160–179.

Ellis, J.E. and Swift, D.M. (1988) Stability of African pastoral ecosystems: alternate paradigms and implications for development. *Journal of Range Management* 41, 450–459.

Ettema, C.H. and Bongers, T. (1993) Characterization of nematode colonization and succession in disturbed soil using the Maturity Index. *Biology and Fertility of Soils* 16, 79–85.

Fliebbach, A., Martens, R. and Reber, H.H. (1994) Soil microbial biomass and microbial activity in soils treated with heavy metal contaminated sewage sludge. *Soil Biology and Biochemistry* 26, 1201–1205.

Franzluebbers, A.J., Zuberer, D.A. and Hons, F.M. (1995) Comparison of microbiological methods for evaluating quality and fertility of soil. *Biology and Fertility of Soils* 19, 135–140.

Freckman, D.W. (1988) Bacterivorous nematodes and organic matter decomposition. *Agriculture, Ecosystems and Environment* 24, 195–217.

Freckman, D.W. and Ettema, C.H. (1993) Assessing nematode communities in agroecosystems of varying human intervention. *Agriculture, Ecosystems and Environment* 45, 239–261.

Frostegård, Å., Tunlid, A. and Bååth, E. (1996) Changes in microbial community structure during long-term incubation in two soils experimentally contaminated with metals. *Soil Biology and Biochemistry* 28, 55–63.

Geller, E.S. (1994) Psychological perspectives of ecosystem health. *Journal of Aquatic Ecosystem Health* 3, 59–62.

Greenland, D.J. and Nye, P.H. (1959) Increases in the carbon and nitrogen contents of tropical soils under natural fallows. *Journal of Soil Science* 10, 284–299.

Gregorich, E.G., Carter, M.R., Angers, D.A., Monreal, C.M. and Ellert, B.H. (1994) Towards a minimum data set to assess soil organic matter quality in agricultural soils. *Canadian Journal of Soil Science* 74, 367–385.

Groot, J.J.R. and Houba, V.J.G. (1995) A comparison of different indices for nitrogen mineralization. *Biology and Fertility of Soils* 19, 1–9.

Gupta, V.V.S.R. (1994) The impact of soil and crop management practices on the dynamics of soil microfauna and mesofauna. In: Pankhurst, C.E., Doube, B.M., Gupta, V.V.S.R. and Grace, P.R. (eds) *Soil Biota: Management in Sustainable Farming Systems.* CSIRO Press, Melbourne, Australia, pp. 107–124.

Harden, T., Joergensen, R.G., Meyer, B. and Wolters, V. (1993) Soil microbial biomass estimated by fumigation-extraction and substrate-induced respiration in two pesticide-treated soils. *Soil Biology and Biochemistry* 25, 679–683.

Ikerd, J.E. (1993) The need for a systems approach to sustainable agriculture. *Agriculture, Ecosystems and Environment* 46, 147–160.

Jasper, D.A. (1994) Bioremediation of agricultural and forestry soils with symbiotic micro-organisms. *Australian Journal of Soil Research* 32, 1301–1319.

Jenkinson, D.S. and Oades, J.M. (1979) A method for measuring adenosine triphosphate in soil. *Soil Biology and Biochemistry* 11, 193–199.

Jenny, H. (1941) *Factors of Soil Formation.* McGraw-Hill, New York.

Jenny, H. (1961) Derivation of state factor equations of soils and ecosystems. *Soil Science Society of America Proceedings* 25, 385–388.

Karlen, D.L. and Stott, D.E. (1994) A framework for evaluating physical and chemical indicators of soil quality. In: Doran, J.W., Coleman, D.C., Bezdicek, D.F. and Stewart, B.A. (eds) *Defining Soil Quality for a Sustainable Environment.* Soil Science Society of America, Special Publication No. 35, Madison, Wisconsin, pp. 53–72.

Kremen, C. (1992) Assessing the indicator properties of species assemblages for natural areas monitoring. *Ecological Applications* 2, 203–217.

Larson, W.E. and Pierce, F.J. (1994) The dynamics of soil quality as a measure of sustainable management. In: Doran, J.W., Coleman, D.C., Bezdicek, D.F. and Stewart, B.A. (eds) *Defining Soil Quality for a Sustainable Environment.* Soil Science Society of America, Special Publication No. 35, Madison, Wisconsin, pp. 37–51.

Lartey, R.T., Curl, E.A. and Peterson, C.M. (1994) Interactions of mycophagous collembola and biological control fungi in the suppression of *Rhizoctonia solani.* *Soil Biology and Biochemistry* 26, 81–88.

Lussenhop, J. (1996) Collembola as mediators of microbial symbiont effects upon soybean. *Soil Biology and Biochemistry* 28, 363–369.

MacArthur, R.H. and Wilson, E.O. (1967) *The Theory of Island Biogeography.* Princeton University Press, Princeton.

Mannion, A.M. (1995) Agriculture, environment and biotechnology. *Agriculture, Ecosystems and Environment* 53, 31–45.

Moore, J.C. and de Ruiter, P.C. (1991) Temporal and spatial heterogeneity of trophic interactions within below-ground food webs. *Agricultural Ecosystems and Environment* 34, 371–397.

Moore, J.C. and de Ruiter, P.C. (1993) Assessment of disturbance on soil ecosystems. *Veterinary Parasitology* 48, 75–85.

Munkittrick, K.R. and McCarty, L.S. (1995) An integrated approach to aquatic ecosystem health: top-down, bottom-up or middle-out? *Journal of Aquatic Ecosystem Health* 4, 77–90.

Niles, R.K. (1991) Relationship between agronomic management and nematode community structure. PhD Thesis, Purdue University, West Lafayette, Indiana.

Niles, R.K. and Freckman, D.W. (1996) From the ground up: nematode ecology in bioassessment and ecosystem health. In: Barker, K.R., Pederson, G.A. and Windham, G.L. (eds) *Plant–Nematode Interactions.* Agronomy Monograph. American Society of Agronomy, Crop Science Society of America, and Soil Science Society of America, Madison, Wisconsin (in press).

Ohtonen, R., Lähdesmäki, P. and Markkola, A.M. (1994) Cellulase activity in forest humus along an industrial pollution gradient in Oulu, Northern Finland. *Soil Biology and Biochemistry* 26, 97–101.

Olson J.S. (1963) Energy storage and the balance of producers and decomposers in ecological systems. *Ecology* 44, 322–331.

Ousley, M.A., Lynch, J.M. and Whipps, J.M. (1994) Potential of *Trichoderma* spp. as consistent plant growth stimulators. *Biology and Fertility of Soils* 17, 85–90.

Park, J. and Cousins, S.H. (1995) Soil biological health and agro-ecological change. *Agriculture, Ecosystems and Environment* 56, 137–148.

Parmelee, R.W. and Alston, D.G. (1986) Nematode trophic structure in conventional and no-tillage agroecosystems. *Journal of Nematology* 18, 403–408.

Pfender, W.F., Fieland, V.P., Ganio, L.M. and Seidler, R.J. (1996) Microbial community structure and activity in wheat straw after inoculation with biological control organisms. *Applied Soil Ecology* 3, 69–78.

Power, M., Dixon, D.G. and Power, G. (1994) Perspectives on environmental risk assessment. *Journal of Aquatic Ecosystem Health* 3, 69–79.

Powlson, D.S., Smith, P. and Smith, J.U. (1996) *Evaluation of Soil Organic Matter Models Using Existing Long-Term Datasets.* NATO ASI Series I: Global Environmental Change, Vol. 38. Springer-Verlag, Berlin, Heidelberg, New York.

Rapport, D.J. (1989) What constitutes ecosystem health? *Perspectives in Biology and Medicine* 33, 120–132.

Rapport, D.J. (1995) Ecosystem services and management options as blanket indicators of ecosystem health. *Journal of Aquatic Ecosystem Health* 4, 97–105.

Roldán, A., Garcia-Orenes, F. and Lax, A. (1994) An incubation experiment to determine factors involving aggregation changes in an arid soil receiving urban refuse. *Soil Biology and Biochemistry* 26, 1699–1707.

Roper, M.M. and Gupta, V.V.S.R. (1995) Management practices and soil biota. *Australian Journal of Soil Research* 33, 321–339.

Rowell, M.J. and Florence, L.Z. (1993) Characteristics associated with differences between undisturbed and industrially-disturbed soils. *Soil Biology and Biochemistry* 25, 1499–1511.

Schaller, N. (1993) Sustainable agriculture and the environment: The concept of agricultural sustainability. *Agriculture, Ecosystems and Environment* 46, 89–97.

Siepel, H. and Maaskamp, F. (1994) Mites of different feeding guilds affect decomposition of organic matter. *Soil Biology and Biochemistry* 26, 1389–1394.

Smyth, A.J. and Dumanski, J. (1995) A framework for evaluating sustainable land management. *Canadian Journal of Soil Science* 75, 401–406.

Sohlenius, B., Bostrom, S. and Sandor, A. (1987) Long-term dynamics of nematode communities in arable soil under four cropping systems. *Journal of Applied Ecology* 24, 131–144.

Sparling, G.P. (1991) Organic matter carbon and microbial biomass carbon as indicators of sustainable land use. In: Elliot, C.T., Latham, M. and Dumanski, J. (eds) *Evaluation for Sustainable Land Management in the Developing World*, Vol. 2; Technical Papers. IBSRAM Proceedings No. 12. Bangkok, Thailand: IBSRAM.

Sparling, G.P. and Ross, D.J. (1993) Biochemical methods to estimate soil microbial biomass: Current developments and applications. In: Mulongoy, K. and Merckx, R. (eds) *Soil Organic Matter Dynamics and Sustainability of Tropical Agriculture*. Wiley-Sayce Publishing Company, New York and Chichester, pp. 21–37.

Speir, T.W., Kettles, H.A., Parshotam, A., Searle, P.L. and Vlaar, L.N.C. (1995) A simple kinetic approach to derive the ecological dose value, ED_{50}, for the assessment of Cr (VI) toxicity to soil biological properties. *Soil Biology and Biochemistry* 27, 801–810.

Stewart, J.W.B. and Elliott, E.T. (1991) A conceptual framework for regional analysis for semiarid to subhumid agroecosystems. In: Hanson, J.D., Shaffer, M.J., Ball, D.A. and Cole, C.V. (eds) *Sustainable Agriculture for the Great Plains*. Symposium Proceedings USDA-ARS, ARS-89, pp. 17–30.

Swift, M.J. and Woomer, P. (1993) Organic matter and the sustainability of agricultural systems: definition and measurement. In: Mulongoy, K. and Merckx, R. (eds) *Soil Organic Matter Dynamics and Sustainability of Tropical Agriculture*. Wiley-Sayce Publishing Company, New York and Chichester, pp. 3–18.

Thompson, J.P. (1987) Decline of vesicular-arbuscular mycorrhizae in long fallow disorder of field crops and its expression in phosphorus deficiency of sunflower. *Australian Journal of Agriculture Research* 38, 847–867.

Turco, R.F., Kennedy, A.C. and Jawson, M.D. (1994) Microbial indicators of soil quality. In: Doran, J.W. Coleman, D.C., Bezdicek, D.F. and Stewart, B.A. (eds) *Defining Soil Quality for a Sustainable Environment*. Soil Science Society of America, Special Publication No. 35, Madison, Wisconsin, pp. 73–90.

Turner, II, B.L., Clark, W.C., Kates, R.W., Richards, J.F., Mathews, J.T. and Meyer, W.B. (1990) *The Earth as Transformed by Human Action*. Cambridge University Press, Cambridge.

Valsecchi, G., Gigliotti, C., and Farini, A. (1995) Microbial biomass, activity, and organic matter accumulation in soils contaminated with heavy metals. *Biology and Fertility of Soils* 20, 253–259.

Van Wensem, J., Vegter, J.J. and Van Straalen, N.M. (1994) Soil quality criteria derived from critical body concentrations of metals in soil invertebrates. *Applied Soil Ecology* 1, 185–191.

Wasilewska, L. (1979) The structure and function of soil nematode communities in natural ecosystems and agrocenoses. *Ekologia Polska* 27, 97–146.

Workneh, F. and van Bruggen, A.H.C. (1994) Microbial density, composition, and diversity in organically and conventionally managed rhizosphere soil in relation to suppression of corky root of tomatoes. *Applied Soil Ecology* 1, 219–230.

Yeates, G.W. (1994) Modification and qualification of the nematode maturity index. *Pedobiologia* 38, 97–101.

Yeates, G.W. and Wardle, D.A. (1996) Nematodes as predators and prey: relationships to biological control and soil processes. *Pedobiologia* 40, 43–50.

Yeates, G.W., Orchard, V.A., Speir, T.W., Hunt, J.L. and Hermans, M.C.C. (1994) Impact of pasture contamination by copper, chromium, arsenic timber preservative on soil biological activity. *Biology and Fertility of Soils* 18, 200–208.

Zak, J.C., Willig, M.R., Moorhead, D.L. and Wildman, H.G. (1994) Functional diversity of microbial communities: a quantitative approach. *Soil Biology and Biochemistry* 26, 1101–1108.

4

Bioindicators: Perspectives and Potential Value for Landusers, Researchers and Policy Makers

J.M. Lynch[1] and L.F. Elliott[2]

[1]*School of Biological Sciences, University of Surrey, Guildford, Surrey GU2 5XH, UK;* [2]*National Forage Seed Production Research Centre, United States Department of Agriculture, Agricultural Research Service, Oregon State University, Corvallis, Oregon 97331, USA*

Introduction – the Need for Bioindicators

For many years the need for bioindicators was primarily identified by scientists for ecological studies in the quantification of natural processes. Their requirements ranged from the needs to study global nutrient budgets through ecosystems to the assessment of the impact of pollutants on the environment. In recent years, however, there has been increased national and international concern that the resolution of problems arising from mistreatment of the environment should be subject to regulatory control. Commonly, an international agency has assembled a team of experts to identify such problems and then presented the evidence to an assembly of competent authorities from each country of the organization. One of the most influential agencies in this respect is the Organization for Economic Co-operation and Development (OECD) which is based in Paris and provides advice to the 24 subscribing member countries covering Europe, North America, Japan and Australasia. It has 12 directorates and the pertinent ones in the present context are Agriculture, Environment, and Science and Technology. Therein lie potential complications in that it is essential that not only is there interdisciplinary cooperation among practising professionals across international boundaries, but that there is also good communication in such large international agencies across its various divisions. It must be emphasized that international agencies such as OECD, and for example the North Atlantic Treaty Organization (NATO) and the World Health Organization (WHO), are not regulators as such, but the advice they provide is commonly incorporated into national regulations. In the case of the European Union (EU), the incorporation of OECD guidelines into EU regulations means that member governments are compelled to have national regulatory control within time limits set in Brussels. It is increasingly common that EU regulations so formulated have parallels in countries outside Europe, such as the USA, and the OECD is probably a major influence in this, leading to international standardization.

National and international policies need usually to be enforced by regulatory control to be effective. On issues concerning agriculture and the environment, regulations can only be addressed satisfactorily if reliable methods are available for assessment. Traditionally, this has been based on chemical or physical, and to a lesser degree, biological measurements on materials within or taken from the natural environment. Commonly, however, the concern is not the physical presence of a substance in the environment, but more the biotic impact of a substance or an environmental management practice. That biotic impact can be on environmental processes, the users of that environment or both. Usually, biotic measurements are preferred to chemical and physical measurements for studies of population dynamics but the problem is that the biotic measurements can be relatively variable. In these terms we can think of plants, animals and man as the users of the environment and ultimately their responses have to be regarded as the most meaningful bioindicators of change. In respect to man, the science of toxicology is well developed in terms of assessing the toxicity of chemicals by using animal or cell models, but increasingly, nucleic acid probes and predictive methods are used. The field of ecotoxicology (such as the impact of pesticides) has often been focused on invertebrates, and less attention has been paid to the impact on vertebrates and plants. The invertebrates, however, are often used as bioindicators because of their low cost, the ease of measurement and lack of restrictions. The problem in using the crop or the animal as a bioindicator is that any detrimental effects are likely to be irreversible and loss will have occurred. Additionally, these measurements are difficult to trace because complicated ecosystems are highly integrated. This therefore sets a challenge for the generation of bioindicators which can be used in assessing potential environmental impact.

Several microbiological parameters are potential bioindicators. These include: (i) a microorganism or groups of microorganisms; (ii) size of the microbial biomass; (iii) enzyme levels; (iv) enzyme end-products; and (v) gene presence and gene regulation. The microorganism is, of course, a potential bioindicator. It can be a specific indicator microorganism which has the potential to cause harm. For example, the ratio of faecal *Streptococcus* spp. which are products of man, to the faecal coliforms which are products of both man and animals, is a good indicator of whether water courses have become polluted from domestic effluent. It should also be emphasized that this source of pollution arises potentially from surface water leaching through soil macropores to groundwater. It is a rather different proposition to assess impact where, for example, a toxicant enters a soil and it may cause damage to various components of the soil biota; and one or more components of the affected biomass may affect plant and crop productivity. The toxicant can also be degraded to products more toxic than the original compound with consequent effects. The questions that then need to be asked are:

1. Which components of the biomass should be determined? It should be recognized that, although estimates vary, it is generally considered that only about 10% of the soil population is normally isolated in conventional plating procedures (see Roper and Ophel-Keller, Chapter 7, this volume). This can be improved by modifying culture media but it does appear that some organisms are inherently unculturable.
2. What fraction of the total biomass can be cultivated, and is this fraction representative of the changes occurring ?
3. What is the efficiency and precision of counting the isolated organisms and therefore what is the change in the population which would be necessary in order to detect a significant change? In this respect it is worth noting that it is rarely possible to count soil organisms, even on selective media, to give an error less than 0.5 log units, which many would regard to be a catastrophic change on an ecosystem. One improvement in the assessment of populations is the measurement of colony development on isolation plates with time (De Leij *et al.*, 1994). Counts are made each day for ten days and the number of colonies in each daily group assessed as a population and the numbers fitted to an ecophysiological index, which is mathematically the same as the Shannon index of biodiversity. This index is a more sensitive indicator of perturbation, and has more ecological significance, than routine colony counting at one moment of time.

It is clear, therefore, that microorganisms as bioindicators may have a value only in certain circumstances. As it is microbial function which is likely to have impact on ecosystems, the question arises as to whether functional indicators could be deployed. In classical biochemical terms, enzyme measurements are useful in indicating the potential activity of a biological system, but they do not measure the actual activity in the system or substrate availability. One of the major advantages of enzyme measurements, however, is that they can be made with high precision; it is common for the determinations to be made with an error within 10%. This is considerably more accurate than estimates of microbial populations using conventional counting procedures and, whereas the information provided is different, it is not necessarily any less useful. It may be possible to measure differences in systems more precisely by measuring enzyme activity in the rhizosphere (Naseby and Lynch, 1997). This is the area of greatest enzyme activity in soil and it should reflect the greatest change, especially when a suite of enzymes is measured. Another problem with enzyme measurements is that the source of the enzyme is usually not distinguished.

The downside of enzyme measurements giving only potential activity can be overcome by measuring the enzyme product, namely the accumulated pool of a biological reaction in soil. This then deploys standard chemical methods of determination, but it is often forgotten that these are biological products. A major advantage is that, with modern methods of instrumental analysis, the types and precision of information provided can be very great. For example, stable isotope analyses employing a mass spectrometer can be used to measure the actual uptake of nitrogen by a plant with a high degree of precision, usually within a few per cent, and any perturbation of that uptake pattern will be of direct significance in plant productivity (F.A.A.M. De Leij and J.M. Lynch, unpublished results). These measurements can also be used to assess potential leaching to groundwaters and any consequent environmental hazards.

The major difference between population and enzyme/product measurements is that the former attempts to consider the size of a biological pool, whereas the latter is concerned with its activity. Both types of information are non-exclusive and useful in their own ways. An alternative strategy to approach size is to measure biomass. The various indirect measurements of this, usually involving the fumigation/respiration or ATP techniques (see Sparling, Chapter 5, this volume), provide an integrated measure of the total amount of living tissue within a system, and for many purposes, this can be more useful than measuring individual components of that total tissue with all the associated problems in measurement. It must be recognized, however, that no biomass measure provides an absolute measure of biological tissue and the real value of these determinations is in the comparative mode. In this respect the precision can be quite good, usually intermediate between population and enzyme/product measurements.

At this stage nothing has been mentioned about the use of the techniques in molecular biology, which many authors in this volume (see Roper and Ophel-Keller, Chapter 7; Pankhurst, Chapter 12; White and Macnaughton, Chapter 15, this volume) see as the major opportunity for the development of bioindicators. It must be recognized, however, that these are not usually alternatives to the techniques already mentioned as they provide different but often complementary information. The simplest and most direct methods involve the use of marker or reporter genes inserted on a plasmid, or more usefully on the chromosome, which facilitates precision tracking of an organism in the environment. Such techniques, however, provide no information about resident populations and their activity, for which nucleic acid probing methods are necessary. Such probes usually involve DNA or mRNA and if suitable constructs are available they can be targeted at the gene sequences present in natural populations. Providing there is homology of the gene sequences, it will not distinguish the various components of the population carrying that gene sequence. The information thus provided is, therefore, leading to population genetics rather than population biology, and it does not necessarily provide any information about the function of the ecosystem. In order for function to be expressed it is essential that the appropriate primer or promoter sequences are associated with the gene of interest. Even then, when the full gene complex is present to allow a gene to be expressed, function which involves enzymic reactions will only take place if the necessary substrates are present in the appropriate physiological conditions. In a study of dinitrogen-fixing bacteria in soil isolated under facultatively anaerobic conditions, it was found that the *nif*-gene appeared to be present in about 50% of the bacteria isolated on an nitrogen-free medium (Kimura *et al.*, 1992); if the numbers present all actually fixed dinitrogen there would certainly be no necessity to apply nitrogen fertilizers! This then points to the fallacy of extrapolating too widely from new methodology and the need to combine the new techniques with the classical, especially those which determine the actual function of the system. It is possible, however, that some of these difficulties will be overcome with the use of mRNA probing, as this is a potential indicator of genetic activity as opposed to genetic presence.

As a final perspective to provide on the use of bioindicators, it should be considered which experimental system the bioindicator is being deployed in, and particularly whether field or laboratory/glasshouse/microcosm experiments are being carried out. Nearly always the populations and activities of soil organisms are different in the field and in any experimental system outside of the natural environment. Again, the potential problem in this respect is likely to be greater in the determination of populations rather than activities, because they are more likely to be subject to small changes in environmental conditions than the availability of substrate. It is, therefore, quite likely that different populations could utilize the substrates to yield similar products and, therefore, the product pool in the system would be more damped against oscillation than the populations. The availability of substrates can be dramatic on a system. Elliott and Lynch (1984) showed that, under nitrogen-limiting conditions, straw decomposition proceeded with the production of polysaccharides with soil stabilizing capabilities. With some conditions, polysaccharides could provide a useful soil quality indicator.

Specific experimental approaches to bioindicators as a component of soil quality/health have been covered elsewhere in this volume and in a previous volume (Doran *et al.*, 1994). We will now consider in outline some specific situations where bioindicators have been deployed, and discuss how the social sciences have an important role to play in the interpretation and use of the results of such studies. These indicators will be useful only if they are accepted by the scientists, society and policy makers. Usually, society will drive the policy makers.

Energy Budgets – Macroecology

Traditionally, the ecologist has been interested in the global cycling of nutrients through ecosystems. The carbon cycle has received the most attention because the dominant provision of primary productivity is through this cycle. Furthermore, microbiotic activity, which is probably the dominant component of the biosphere function, is generally considered to be carbon-limited (Lynch, 1988). This is, of course, not always the case and the cycles involving nitrogen, phosphorus and sulphur can be equally important. In the rhizosphere, for example, depending on the fertilizer regime, it is likely that nitrogen will become the growth limiting nutrient. In crop residues, such as straw, the ratio of carbon to nitrogen is extremely high (C80:1) and this, therefore, provides a substrate where nitrogen is certainly the growth-limiting nutrient. The essential factor to consider then, in the need for biomonitors of nutrient and energy cycles, is that all the substrate energy entering the ecosystem is passing through 'the eye of the needle', namely the soil biomass. Of particular concern in recent years has been the impact of climate change on these processes. The importance of a warming trend is recognized by the fact that boreal regions, which contain the sub-arctic zone, represent less than 10% of the world's land area but contain about 25% of the earth's total soil carbon (Hobbie and Melillo, 1984).

Initially, most of the studies on energy cycling focused only on biomass and substrate/product relationships. In full global budgets the micro- and macrofaunal relative contributions to the budgets have been targets for research. The larger the scale of the organisms, the easier it is to count by eye, and there is less need for a biomonitor as such. For example, no special gear is needed for going out to count elephants in an ecosystem! All the indications are that the animals contribute least in terms of the ecosystem biomass and activity (Lynch, 1988).

In recent years, the activity of the various specific components of the biomass has been of interest particularly because of their metabolic functions, such as the methanogenic bacteria, which can produce a product which might contribute to global change along with industrial emissions and other abiotic sources. It seems useful to have bioindicators at the physiological, biochemical and genetic levels to measure such important changes. Another area of interest in ecological terms is the recognition that organisms interact to bring about function. For example, it now seems clear that protozoan predators have important roles to play in the mineralization of bacterial cells in the rhizosphere and that this contributes to the nitrogen economy of plants (Kuikman and Van Veen, 1989). It is therefore necessary, in adopting more reductionist approaches to ecology, that bioindicators are available to measure the energetic balances and to determine the sensitivity to change brought about by any perturbations.

The Problems of Pesticides and Recalcitrance

The agrochemical industry has always been sensitive about any potential non-target effects that its pesticides might generate. Perhaps the biggest scare was in the use of DDT and its damaging effects on wildlife. International groups have taken leads in the development of bioindicator systems to test products, and produced catalogues of methods (e.g. Sommerville and Greaves, 1987).

Generally, the regulatory position is that the onus has been placed on the manufacturer/supplier to provide evidence of safety before consent is given for field trials, and then registration is allowed for commercial use.

An interesting situation has arisen in The Netherlands in recent years. In 1990, the Dutch government, wary of the potential hazards of chemical loads on the environment, announced statutory regulations that demanded a reduction of chemical emissions (including agrochemicals) into the environment of 50% over a ten year period. The Dutch have always been at the forefront in the bioindicator field and have accumulated much evidence on environmental and toxicological effects of chemicals. However, they would admit that the targets in chemical load reduction are arbitrary and merely set a potentially achievable target. The need for bioindicators to assess pesticide toxicities and effects is critical. Public perception is that all pesticides pose serious threats to human and animal health. This includes pesticides and their degradation products in drinking water and the food chain. The fear of this contamination has, and will, continue to drive policy makers. Perception that damage is imminent is just as real as if the damage is occurring. The development and availability of bioindicators to define pesticide hazards will not be adequate. Society and policy makers will have to accept the utility of the bioindicators or they will not be useful.

What is a recalcitrant chemical? A recalcitrant (or refractory) molecule is one that persists for extended periods in biologically active environments, regardless of whether the compound is, or is not, inherently non-biodegradable (Alexander, 1973). Any compound which shows resistance to degradation can be classified as recalcitrant. In this respect, natural chemicals such as lignin are in this group and, whereas they do no direct harm to the ecosystem, their tight chemical bonding with other natural polymers can provide a barrier to decomposition of what would otherwise be a substrate which can be decomposed to useful metabolites. For example, there have therefore been many attempts to promote decomposition of lignin, which is recalcitrant and can prevent the decomposition of cellulose to which it is bound, and such studies have been carried out at the ecophysiological and molecular ecological levels. These have provided a very useful set of bioindicators for decomposition, including gene probes which provide the potential to determine how the catabolism is regulated (Gaskell *et al.*, 1995). As indicated before, such modern techniques need to be coupled with the classical. In the case of lignocellulose decomposition, the chemical fractionation can be assessed by proximate analysis, and various components of the ligninase and cellulase complexes can be assayed. The problem with the latter enzymes is that they have multiple components and it can be difficult to determine which are the most relevant components to assay. Therein identifies another problem for molecular methodologies, because genes can only be characterized based on the enzyme components that they code, and it can be very easy to spend time in generating gene probes to genes coding irrelevant components of enzymes. Recently, Horwath and Elliott (1996) showed that elemental analysis was much more accurate for defining the rate of lignin degradation than proximate analysis.

The substances that would more likely be referred to as recalcitrant are the synthetic chemicals intended for industrial applications which were never intended to enter the environment as such. A good example is the polychlorinated biphenyls. They can enter soil through spillage or leakage. For example, they can be components of the insulating oils of power cables at transformer stations. Should they leak from such situations, it is essential that the leakage be identified quickly, and because of the complexity of the chemical assay it could be that a bioindicator would be useful. The catabolic pathways are relatively well characterized and this opens the scope for gene probes to be developed. The added advantage of the bioindicator in this respect is that it can also give an indication of how quickly the catabolic genes get switched on. Therefore, the gene probe provides the potential to study natural and induced bioremediation. Of course, it is important to avoid the need for bioremediation, and many industries suggest that they do not have a problem. The reality is that accidents do happen. A good recent example was the oil spillage generated by the *Exxon Valdez* oil tanker off Alaska. It provided scope to treat the crude oil in the marine/sediment situation but it took several weeks before a suitable treatment strategy was developed by enhancing the natural populations present. Fortunately an environmental catastrophe was avoided. However, it would appear to be sensible to learn lessons here and minimize such risks by having bioindicators and treatments ready to use in any comparable situations.

Crop Residue Management

In modern farming systems, particularly with the introduction of minimal tillage systems, it was evident that under wet soil conditions crop residues, normally straw, needed to be removed or the yields of a succeeding crop could be reduced (Lynch and Elliott, 1983). It was identified that many of the problems could be caused by the activity of fermentative bacteria yielding volatile fatty acids, particularly acetic, from cellulose. A bioindicator for the metabolite proved difficult to exploit because the activity only occurred at localized microsites on the surface of residues and, therefore, it was heavily diluted when extracting soil. This is a very important consideration when deploying bioindicators, especially as the soil is so heterogeneous. It can be expected that the heterogeneity issue will be somewhat easier to resolve in the future because soil has been analysed as a fractal system and this provides some semblance of order to the heterogeneity (Crawford *et al.*, 1993). Another problem in the analysis of the residue problem was that a second biotic factor was operative, the presence of deleterious rhizobacteria which colonized the straw and affected the succeeding crop (Elliott and Lynch, 1984). Subsequently, a bioassay was developed for these toxin-producing bacteria (Frederickson and Elliott, 1985) and a useful spin-off from this research is that the toxin-producing bacteria have been patented and have potential as herbicides (Elliott and Kennedy, 1991; Kennedy *et al.*, 1991). This then provides a useful secondary goal for bioindicator studies, namely that the research will lead to new opportunities in environmental biotechnology.

It should also be recognized that crop residues provide substrates for beneficial microorganisms as well as those which are deleterious. In the context of straw with its high cellulosic content, saprophytic fungi include the pathogenic *Fusarium* spp. as well as the biocontrol fungus *Trichoderma* spp. (Lynch, 1987). Given a population balance in favour of either fungus, plate studies (Lynch, 1987) and continuous-flow (thermodynamically open) column studies (Cheetham *et al.*, 1995) have shown that the fungus with the increased initial inoculum will become dominant. This, therefore, establishes a key principle of biocontrol and also identifies the need for suitable bioindicators of the pathogen and antagonist in the field situation in the context of soil suppression of disease. Selective media are available but this will only give viable propagule numbers. Other studies have shown that this can be quite misleading in the *Trichoderma*/*Pythium* interaction because the activity comes from mycelial biomass and this is inversely related to the propagule numbers. In an attempt to overcome this problem, Cheetham *et al.* (1995) developed a technique of generating hyphal fragments and counting small distinct colonies on agar by introducing a paramorphogen, sodium deoxycholate, into the growth medium. Such a procedure, which worked well under laboratory conditions, needs to be evaluated in field soil. Again, it should be mentioned that another spin-off of the *Trichoderma* studies was that some strains of the fungus appeared to be stimulating plant growth directly as well as controlling disease (Lynch *et al.*, 1991). This has now been studied in many experiments (Ousley *et al.*, 1993) to demonstrate consistency, and recently we have identified a growth-stimulatory metabolite from the fungus (J.M. Lynch, T.N. Danks and L. Pasquali, unpublished results).

The studies on crop residue decomposition have applications in another relevant area, that of composting. Here, the main target is to decompose the residues to provide benefit to the soil-plant ecosystem while minimizing any phytotoxic effects and energy inputs. Recent encouraging beneficial results have been obtained on the composting of grass residues in the Willamette Valley of Oregon (Churchill *et al.*, 1995). Potentially, one of the most beneficial actions of compost is likely to be its soil conditioning action. Bacteria and fungi which utilize straw can produce polysaccharide gums which aggregate soil particles and prevent water erosion. Whatever the net beneficial activity of the compost, it is important to have bioindicators for the components of the activity to be able to study it in scientific depth and to enhance the beneficial factors.

Cations

One of the major complications of urban and industrial waste entering the terrestrial environment is that they can contain heavy metals which can be very damaging to all kinds of biological systems. Aluminium should also be added to this group of toxicants because it can have adverse effects, which range from the toxicity to root-nodulating rhizobia to toxicity to plants, and at the top of the food chain it can be a causative agent of Alzheimer's disease in man (Crapper *et al.*, 1973). The active speciation of the metal only occurs in strongly acidic conditions. Therefore, whatever the indicator used for the metal, it must be recognized that any information obtained needs to be considered in conjunction with measurement of the soil physical conditions. For example, the aluminium toxicity is a particular problem for the acidic soils of Queensland in Australia where legumes are major crops (Date *et al.*, 1995). Similar considerations can be applied to the other metal ions but of course the impact of pH and other environmental conditions such as redox potential will vary among the different metal ions.

Not all cations or even metals will be toxic. For example, there has been much study in recent years of iron and its sequestration by siderophores. The principle siderophore studied has been pseudobactin produced by *Pseudomonas* spp. (Lynch, 1990). It was recognized that all living systems needed iron, and the production of pseudobactin in the rhizosphere by plant growth promoting rhizobacteria made iron unavailable to pathogens and, therefore, provided a route to biological control. Many assumptions were made in those deductions, not the least of which was the lack of consideration of whether the plant would have siderophores with high enough affinity to compete with the bacterial siderophores. Notwithstanding this, several academic and commercial research groups embarked on studies to develop gene probes and genetic manipulation systems for the siderophore. Generally, these studies have been successful scientifically. Unfortunately, it now seems that siderophore production is not the critical factor in the action of the plant growth promoting bacteria (Lynch, 1990). Obviously, it now seems that it might have been useful to have placed more emphasis on physiological studies in tandem with the molecular biological studies. This lesson is applicable to the studies on the other cations. At present we know rather little about the mechanisms for their transport through the soil/microorganism/plant ecosystem. Now that techniques such as X-ray probe microanalysis are available for these studies, it might be useful to establish these physiological concepts and indicators of nutrient transport before rushing into the gene probe techniques.

Risks and Benefits from the Release of Genetically Engineered and Other Microorganisms

With the concepts of environmental biotechnology developing rapidly, there has been regulatory pressure to develop biomonitors to ensure that any biotechnology products released are safe. Although the genetically modified organisms have been the trigger for this interest, many of the considerations are equally relevant to the exploitation of wild-type organisms. As already indicated, OECD has been a lead agency in the formulation of guidelines for such studies. The first 'Blue Book' was concerned with safety in biotechnology and focused on laboratory investigations (OECD, 1986) but the second concentrated on means of assessing risk following the release of genetically modified plants and microorganisms (OECD, 1992). These were followed with volumes which specifically addressed the scale-up of applications of genetically modified crop plants (OECD, 1993), and then another which concerned the scale-up of microorganisms as biofertilizers (OECD, 1995). The latter is particularly relevant in the present context as it discusses the concept of familiarity. It argues that experience with the use of wild-type organisms which are similar to those which have been genetically modified, particularly in similar environments, is relevant to release applications. The argument indirectly is that bioindicators developed for the studies of wild-type organisms in ecosystems are generally applicable. The point particularly illustrates the concept of familiarity with the long history of studies on the *Rhizobium*–legume associations, which have never created any significant adverse effects extensively using commercial inoculants. The major problem with this view is that techniques of study in microbial ecology have not been totally satisfactory. With the identification of the need to study biosafety in biotechnological applications, governments and international agencies have invested money in developing new techniques for study. The result is that the field of molecular ecology has emerged rapidly. Indeed, in other chapters of this volume it will be seen that many of the new and exciting techniques have emerged from this route (Roper and Ophel-Keller, Chapter 7; Pankhurst, Chapter 12; White and Macnaughton, Chapter 15 and Paton *et al.*, Chapter 16, this volume). Volumes which have appeared from the OECD Co-operative Research Programme on Biological Resource Management for Sustainable Agricultural Systems have addressed the specific issues on the mathematical interpretation and prediction of risk of release of organisms into the environment (Bazin and Lynch, 1994), on the release of organisms for the biological control of pests and diseases (Hokkanen and Lynch, 1995) and on the development of soil inoculants (Elliott and Lynch, 1995). This has been coupled with an extensive fellowship programme which was initially funded 1990–1994 for binational cooperation between OECD member countries, and has recently been extended to 1999. Other international and national bodies have also been very active, and particularly noteworthy is a volume which has just been produced by the secretariat of the United Nations Industrial Development Organization (UNIDO) in cooperation with the International Centre for Genetic Engineering and Biotechnology (ICGEB) for the UNIDO/UNEP/WHO/FAO Informal Working Group on Biosafety (Tzotzos, 1995).

It would not be appropriate to discuss the output from these various initiatives in any detail but a few observations are pertinent. The use of genetic constructs as reporter systems has greatly facilitated the development of microbial population dynamic studies. For example, in field release studies of a genetically modified *Pseudomonas fluorescens* (De Leij *et al.*, 1995) it was possible to use a most probable number technique (De Leij *et al.*, 1993) which was sufficiently sensitive to the recovery of the *lac ZY* gene that a single bacterial cell could be recovered from 1 kg soil. In the field, the bacterium could be recovered 2 m laterally and at least 0.5 m deep from the point of seed application. In other work linked to the same studies funded by the Department of the Environment in the UK, the *lux* marker system was developed in soil bacteria, and with the use of a charge coupled device in association with a boroscope, bacteria could be monitored *in situ* in soils and on roots (Rattray *et al.*, 1995).

The above studies demonstrated the value of using large multi-disciplinary teams and this has been developed as a concept by the European Community. In the IMPACT Programme (Interactions between Microbial inoculants and resident Populations in the rhizosphere of Agronomically important Crops in Typical soils), 17 partners from eight different countries have been funded to work together on a single project using a variety of inoculants, crops and soil types relevant to the European farmers. For example, a variety of enzyme biomonitors have been developed at Surrey and partners send soil samples for analysis to that centre, whereas at the University of Bielefeld and TUV Sudwest in Germany, the groups have received the other partners' samples for genetic fingerprinting. A problem with some of the earlier studies in this field is that it is not necessarily expected that marker genes will have any impact, although they were necessary to establish principles and concepts. In the current wave of investigations, the functional gene release is of paramount consideration. In the IMPACT project the effect of antibiotic genes on non-target organisms, especially mycorrhiza and rhizobia, is being addressed; fortunately no adverse effects are yet apparent.

Whether it is a biofertilizer or a biocontrol agent as the introduced agent, a major ecological and commercial question is whether the agent being deployed is effective in the situations to which it is applied. This highlights a potential complication but also an opportunity for the use of biomonitors. It is clear from the extensive experience with rhizobial inoculants that, where the legume crop has previously been grown in the same soil, it is seldom necessary to inoculate the crop. The reason is that sufficient populations of the relevant rhizobial populations will have already been established during the previous cropping. Thus the application of inoculant would be without effect. Similar considerations might be relevant to other inoculants and the necessity and effectiveness would depend on the natural microbial populations present in the soil. This, therefore, identifies the need for bioindicators to establish need for the inoculant to be deployed in a specific situation. Similarly, if the target is biocontrol, it would be useful to have bioindicators for the pest or pathogen. Of course, it can be argued that many agrochemicals are applied prophylactically, but that in itself is a potential waste and misuse of resource. It is clear, therefore, that the bioindicator could be marketed alongside the crop protection or enhancement agent.

Societal Values and Perception

With emotive terms such as 'rivers of death', Rachel Carson, in her book *Silent Spring*, captured the public's imagination as to the potential risks of agrochemicals. Even though the book carried much emotional terminology and inaccuracies, there was much truth and this almost certainly led to the elimination of DDT applications in many countries because of its damaging effects on wildlife and its persistence in nature.

On the issue of release of genetically engineered organisms, Jeremy Rifkin used the same kind of emotional terminology which could equally well have been extracted from John Wyndam's book *The Chrysalids* where 'a world paralysed by genetic mutation' was described. Under such pressures the US government responded by placing many restrictions and requirements on the experiments to be carried out. For example, all the scientists actively engaged in the field had to wear 'moon suits', which probably added to the public's fear and perception that experiments only previously conceived in science fiction stories were being carried out, although such precautions were probably sensible for the initial experiments. In some ways the scientists themselves were not well prepared either because they were frequently courted by the media and provoked into exaggerating the claims of what their experiments might show, or making statements which further aroused public fear. The issue developed a few years later in Europe and the response was different depending on which country was involved. In Germany, the situation became analogous to that in the US, with strong pressures coming from the Green political party. Indeed, this latter pressure made it difficult for any biological releases to occur, even of wild-types. In Britain there was much less public concern, but all potential interest groups were consulted and represented on regulatory committees where appropriate. Possibly, the media took the view that the issue was second-hand news following the US stories. Whatever the reason, public perception was much more accommodating to the experiments in the UK than in the USA. These scenarios do show the importance of being able to assess risk accurately and to provide these data to the public and policy makers. It also points out that scientists are probably not the people to provide the information; those trained in the social sciences may be better equipped for the task. Some survey work has already been undertaken (Leopold, 1995).

There are two particular issues of relevance to discuss in relation to this. Firstly, did the experimenters play excessively to the media attention, thereby making themselves more of a target for criticism than was really necessary? Secondly, did they have all the necessary background experiments completed which justified the field tests? It is on this second issue that the need for bioindicators is pertinent. For example, in the Monsanto field releases at Clemson University, South Carolina (Drahos *et al.*, 1992), a vast amount of monitoring information was collected on a bacterium with a non-functional gene inserted on a plasmid, but there was little attention to the impact of the organism on the resident microbial populations. In terms of bioindication, the question must be asked as to what the regulators and the public expected to learn from these experiments. Certainly within the experimental accuracy, information was provided on spread and survival of the introduced organism, and this should have been sufficient to allay fears that the bacterium would spread over great distances. It was evident that none of the participants in the experiment mutated or became ill. It was also clear that in terms of the plant as the ultimate agronomic bioindicator, there were also no adverse effects. To an extent much of the output from these experiments would appear to have quelled some of the public fear, and it is also clear that regulators are able to take a more lenient view on some of the demands and preconditions for future field releases. Ultimately, the story which needs to be conveyed to all interested parties is the benefits and risks. The place of the biomonitor is, therefore, not only to provide the scientific evidence of benefit and risk, but to provide the evidence in such a form that this can be understood by the public at large. This strategy is not just important to convince the public of safety, but it is equally valuable to funding agencies, commercial organizations and regulatory agencies that will provide the necessary support for the development and exploitation of biotechnological products.

The Profit Motive, Taxation and Penalties – is there a Conflict between the Needs of Industry and Society?

The initial groundwork studies by Monsanto on microbial gene release proved to be extremely expensive and did not result in useful industrial products. This probably was a major factor in their decision to withdraw completely from the field shortly after completing the studies. Therein lies a major complication for the development of many biotechnologically based products, namely excessive regulation will be a disincentive to industrial development yet there is an obligation to protect the public against unnecessary risks. The question then arises, however, as to what is risk. Many scientists would argue that gene transfer is a risk to be avoided. But who or what will come to harm if a gene exchanges, especially when it is considered that genes are constantly exchanging in the environment? A serious risk would be if somebody died or was adversely affected as a direct or indirect result of a genetically-engineered organism being released, yet nobody has ever really suggested that this could happen. By contrast there have been many suggestions that pesticides could cause cancer and therefore death. A recent article discusses the conflicting evidence on the risk ratios of contracting various forms of cancer from activities ranging from smoking to nutritional factors, but it is clearly an area where much more investigation is required (Taubes, 1995). With such confusion on this important area of epidemiology, the prospects for understanding some of the issues discussed in this article would not appear to be good! Whereas it seems unlikely that biological products would give many problems, it must be recognized that some metabolites of microorganisms can be carcinogenic and this needs to be evaluated in risk assessment. Such problems can hopefully be identified well before serious consideration is given to a release. Perhaps a rather more serious threat is that an organism being released has only a small genetic difference from a plant pathogen because most people are aware of the catastrophic effects plant pathogens can have on human populations, such as occurred in the Irish potato famine. Thus the release of an organism which would have the potential to become a pathogen would need to be considered seriously in risk assessment procedures.

The balance of risk factors of different forms of agrochemicals would certainly appear to favour biologicals against chemicals. There is also increased concern for the environment by the public at large in most countries. Under such circumstances and under a climate where the world must be set in a mode of sustainable development, much more information is necessary of the type that can be provided by bioindicators in natural environments. This needs to be set alongside the energy economic budgets that can be obtained for different methods of land management. Ultimately there does not need to be a conflict between the needs of industry and society because governments have the power to regulate the economics. The price support system of agriculture has been made primarily by politicians and it would seem to be appropriate that environmental economics should have a part to play in this support. Two alternative strategies seem possible. Either the principles of the Dutch ten year plan referred to above could be adopted or an ecotax could be introduced to penalize chemical usage which is environmentally damaging. The problem in implementation at present is that this could only be achieved on a perceived scale rather than actual damage that is created.

Conclusion – the Place of Strategic Research in the Natural and Social Sciences

It must be clear from all that has been said so far that there is need for natural and social scientists to collaborate closely with regulators and politicians in formulating environmental policy. One attempt at this was in the UN conference on the Environment held in Rio de Janeiro in 1992. One of the major foci of this debate was the identified need to protect the biodiversity of the planet. There is some evidence that increased plant and faunal biodiversity can improve the function and stability of ecosystems (Naeem *et al.*, 1994; Tilman and Downing, 1994). The implicit assumption then is that a high microbial biodiversity is good and this should, therefore, be a good target for bioindicator studies. This argument unfortunately is potentially fallacious and has been for a large part promoted by those interested in biosystematics. There is no evidence that increased microbial biodiversity increases the stability of an ecosystem, nor is any other beneficial effect obvious.

The public become alarmed if they are told that any compound of the biodiversity of an ecosystem is being reduced. Why should they be so alarmed? The social scientists should be able to establish the bases for this alarm. Meanwhile, the natural scientists should be combining in the largely reductionist approaches which are studied in their respective sciences into using more holistic approaches to their work. The bioindicator has a potentially pivotal role in this merger between reductionist and holistic approaches to issues concerning the environment.

Acknowledgements

We are grateful to the European Union; the Ministry of Agriculture, Fisheries and Food and the Department of the Environment in the UK; the Organization for Economic Cooperation and Development; and the United States Department of Agriculture for financial support of our experimental studies and the preparation of this paper.

References

Alexander, M. (1973) Nonbiodegradable and other recalcitrant molecules. *Biotechnology and Bioengineering* 15, 611–647.

Bazin, M.J. and Lynch, J.M. (eds) (1994) *Environmental Gene Release: Models, Experiments and Risk Assessment.* Chapman & Hall, London, 166 pp.

Cheetham, J.L., Bazin, M.J., Markham, P. and Lynch, J.M. (1995) A method of utilising mycelial fragments to estimate relative biomass densities of fungal species in mixed culture. *Journal of Microbiological Methods* 21, 113–122.

Churchill, D.B., Bilsland, D.M. and Elliott, L.F. (1995) Method for composting grass seed straw residue. *Journal of Applied Engineering in Agriculture* 11, 275–279.

Crapper, D.R., Krishnan, S.S. and Dalton, A.J. (1973) Brain aluminium distribution in Alzheimer's disease and experimental neurofibrillary degeneration. *Science* 180, 511–513.

Crawford, J.W., Ritz, K. and Young, I.M. (1993) Quantification of fungal morphology, gaseous transport and microbial dynamics in soil: an integrated framework utilising fractal geometry. *Geoderma* 56, 157–172.

Date, R.A., Grunden, N.J., Rayment, G.E. and Probert, G.E. (eds) (1995) *Plant–Soil Interactions at Low pH: Principles and Management.* Kluwer, Dordecht, 822 pp.

De Leij, F.A.A.M., Bailey, M.J., Whipps, J.M. and Lynch, J.M. (1993) A simple most probable number technique for the sensitive recovery of genetically modified *Pseudomonas aureofaciens* from soil. *Letters in Applied Microbiology* 16, 307–310.

De Leij, F.A.A.M., Whipps, J.M. and Lynch, J.M. (1994) The use of colony development for the characterisation of bacterial communities in soil and on roots. *Microbial Ecology* 27, 81–97.

De Leij, F.A.A.M., Sutton, E.J., Whipps, J.M., Fenlon, J.S. and Lynch, J.M. (1995) Impact of field release of genetically modified *Pseudomonas fluorescens* on indigenous microbial populations on wheat. *Applied and Environmental Microbiology* 61, 3443–3453.

Doran, J.W., Coleman, D.C., Bezdicek, D.F. and Stewart, B.A. (eds) (1994) *Defining Soil Quality for a Sustainable Environment.* Soil Science Society of America, Madison, Wisconsin, Special Publication no. 35, 244 pp.

Drahos, D.J., Barry, G.F., Hemming, B.C., Brandt, E.J., Kline, E.L., Skipper, H.D. and Kluepfel, D.A. (1992) Use of *lacZY* chromosomal marker system to study spread and survival of genetically engineered bacteria in soil. In: Day, M.J. and Fry, J.C. (eds) *Environmental Release of Genetically Engineered Microorganisms.* Academic Press, London, pp. 147–159.

Elliott, L.F. and Lynch, J.M. (1984) The effect of available carbon and nitrogen in straw on soil and ash aggregation and acetic acid production. *Plant and Soil* 78, 335–343.

Elliott, L.F. and Lynch, J.M. (1995) The international workshop on establishment of microbial inocula in soils: Cooperative research project on biological resource management of the Organization for Economic Cooperation and Development (OECD). *American Journal of Alternative Agriculture* 10, 50–73.

Elliott, L.F. and Kennedy, A.C. (1991) Method for screening bacteria and application thereof for field control of the weed downy brome. Patent No. 5, 030, 562.

Fredrickson, J.K. and Elliott, L.F. (1985) Effects on winter wheat seedling growth by toxin-producing rhizobacteria. *Plant and Soil* 83, 399–409.

Gaskell, J., Wymelenberg, A.V. and Cullen, D. (1995) Structure inheritance and transcriptional effects of Pcel, an insertional element within *Phanerochaete chrysosporium* lignin peroxidase gene lipl. *Proceedings of the National Academy of Sciences of the USA* 92, 7465–7469.

Hobbie, J.E. and Melillo, J.M. (1984) Role of microbes in global carbon cycling. In: Klug, M.J. and Reddy, C.A. (eds) *Current Perspectives in Microbial Ecology.* American Society for Microbiology, Washington, DC, pp. 389–393.

Hokkanen, H.M.T. and Lynch, J.M. (eds) (1995) *Biocontrol Agents: Benefits and Risks.* Cambridge University Press, Cambridge, 304 pp.

Horwath, W.R. and Elliott, L.F. (1996) The degradation of lignin and ryegrass straw components during mesophilic and thermophilic decomposition. *Biology and Fertility of Soils* 21, 227–232.

Kennedy, A.C., Elliott, L.F., Young, F.L. and Douglans, C.L. (1991) Rhizobacteria suppressive to the weed downy brome. *Soil Science Society of America Journal* 55, 722–727.

Kimura, R., Lynch, J.M., Katoh, K. and Miyashita, K. (1992) Enumeration of specifically functional soil bacteria by the DNA probe method. In: Guerrero, R. and Pedros-Alio, C. (eds) *Trends in Microbial Ecology.* Spanish Society for Microbiology, Barcelona, pp. 655–658.

Kuikman, P.J. and Van Veen, J.A. (1989) The impact of protozoa on the availability of bacterial nitrogen to plants. *Biology and Fertility of Soils* 8, 13–18.

Leopold, M. (1995) Public perception of biotechnology. In: Tzotzos, G.T. (ed.)

Genetically Modified Organisms: A Guide to Biosafety. CAB International, Wallingford, UK, pp. 8–16.

Lynch, J.M. (1987) *In vitro* identification of *Trichoderma harzianum* as a potential antagonist of plant pathogens. *Current Microbiology* 16, 49–55.

Lynch, J.M. (1988) The terrestrial environment. In: Lynch, J.M. and Hobbie, J.E. (eds) *Micro-organisms in Action: Concepts and Applications in Microbial Ecology*. Blackwell, Oxford, pp. 103–131.

Lynch, J.M. (1990) Microbial metabolites. In: Lynch, J.M. (ed.) *The Rhizosphere*. John Wiley, Chichester, pp. 177–206.

Lynch, J.M. and Elliott, L.F. (1983) Minimizing the potential phytotoxicity of wheat straw by microbial degradation. *Soil Biology and Biochemistry* 15, 221–222.

Lynch, J.M., Wilson, K.L., Ousley, M.A. and Whipps, J.M. (1991) Response of lettuce to *Trichoderma* treatment. *Letters in Applied Microbiology* 12, 59–61.

Naeem, S., Thompson, L.J., Lawlor, S.P., Lawton, J.H. and Woodfin, R.M. (1994) Declining biodiversity can alter the stability of ecosystems. *Nature* 368, 734–737.

Naseby, D.C. and Lynch, J.M. (1997) Rhizosphere soil enzymes as indicators of perturbations caused by inoculation of a genetically modified strain of *Pseudomonas fluorescens* on wheat seed. *Soil Biology and Biochemistry* (in press).

OECD (1986) *Recombinant DNA Safety Considerations*. Organization for Economic Co-operation and Development, Paris.

OECD (1992) *Safety Considerations for Biotechnology*. Organization for Economic Co-operation and Development, Paris.

OECD (1993) *Safety Considerations for Biotechnology: Scale-up of Crop Plants*. Organization for Economic Co-operation and Development, Paris.

OECD (1995) *Safety Considerations for Biotechnology: Scale-up of Micro-organisms as Biofertilizers*. Organization for Economic Co-operation and Development, Paris.

Ousley, M.A., Lynch, J.M. and Whipps, J.M. (1993) Potential of *Trichoderma* spp. as consistent plant growth stimulators. *Biology and Fertility of Soils* 17, 85–90.

Rattray, E.A.S., Prosser, J.L., Glover, L.A. and Killham, K. (1995) Characterisation of rhizosphere colonisation by luminescent *Enterobacter cloacae* at the population and single cell level. *Applied and Environmental Microbiology* 61, 2950–2957.

Sommerville, L. and Greaves, M.P. (eds) (1987) *Pesticide Effects on Soil Microflora*. Taylor and Francis, London, 240 pp.

Taubes, G. (1995) Epidemiology faces its limits. *Science* 269, 164–169.

Tilman, D. and Downing, J.A. (1994) Biodiversity and stability in grasslands. *Nature* 367, 363–368.

Tzotzos, G.T. (ed.) (1995) *Genetically Modified Organisms: A Guide to Biosafety*. CAB International, Wallingford, UK, 213 pp.

5

Soil Microbial Biomass, Activity and Nutrient Cycling as Indicators of Soil Health

G.P. Sparling
Manaaki Whenua – Landcare Research, Private Bag 3127, Hamilton, New Zealand

Introduction

Vast numbers of microorganisms naturally reside in soil and perform a wide range of functions which are essential for a normal and healthy soil. Soil microbes decompose organic matter, release nutrients into plant-available forms, and degrade toxic residues. They also form symbiotic associations with roots, act as antagonists to pathogens, influence the weathering and solubilization of minerals, and contribute to soil structure and aggregation. The role of the microbial fraction in mediating soil processes, and their relatively high rate of turnover, logically suggest that the microbial fraction could be a sensitive indicator and early predictor of other changing soil organic matter processes (Powlson and Jenkinson, 1981; Powlson *et al.*, 1987). The time-frame of microbial metabolism is also a meaningful one for humans with rates of turnover in soil typically being 0.2–6 years for the soil microbial biomass compared to >20 years for the bulk of the organic matter (Jenkinson and Ladd, 1981; Stout *et al.*, 1981; O'Brien, 1984; Jenkinson, 1990).

Microbial indices have been suggested as indicators of soil quality (Carter and Rennie, 1982; Doran and Parkin, 1994), but in my opinion it is important to differentiate quality from health which are sometimes used interchangeably (Harris and Bezdicek, 1994; Garlynd *et al.*, 1994). A short, precise definition of soil quality and health has proved elusive. The Soil Science Society of America (committee chaired by D.L. Karlen) defined soil quality in the following way:

> The capacity of a specific kind of soil to function, within natural or managed ecosystem boundaries, to sustain plant and animal productivity, maintain or enhance water and air quality, and support human health and habitation.

(eds C.E. Pankhurst, B.M. Doube and V.V.S.R. Gupta)

A major problem with definitions such as the above is that soil quality, when defined in terms of natural ecosystems (such as sustainability and biodiversity), differ very greatly from soil quality when defined in terms of plant and animal productivity (such as profitability of production, low biodiversity). Quality has been defined as 'fit for purpose' (Larson and Pierce, 1994; Doran and Parkin, 1994); if we accept this definition, quality becomes dependent on the intended land use (purpose). I feel this is a useful point at which to separate quality from health. The SSSA committee had this to say regarding quality and health, also making the point that the term 'health' was favoured by farmers, and the term 'quality' favoured by scientists:

> Characterisation of soil quality by scientists focuses on analytical/quantitative properties of soil with a separately defined quantitative link to the functions of soil quality. Characterisation of soil health by farmers focuses on descriptive/quantitative properties of soil with a direct value judgement (unhealthy to healthy) integrated into the options for a given property; in addition, interwoven into the properties of soil per se are value-based descriptive properties of plant, water, air and animal/human systems considered by farmers to be an integral part of soil health characterisation.

I wish to argue that some soils of low quality by the criteria suggested above should still be regarded as healthy. I have in mind soils at early successional stages and in extreme environments (e.g. sand-dunes, deserts, alpine and polar regions) which have low biodiversity and low productive potential, but this is a natural stage of development and should not be regarded as unhealthy. Another group of soils are from an ecosystem climax, with potentially high biodiversity but typically of low fertility (e.g. tropical rainforest soils). I suggest all these soils should be regarded as healthy, although their quality status will vary depending on the actual or intended land use. A possible analogy with human health may be to consider a baby, an elderly person and an athlete. The baby and the elderly person may be frail and vulnerable, and of poor quality in terms of athletic ability, but their condition is natural and 'healthy'. The athlete would get a high ranking for athletic ability, but some athletes (e.g. sumo wrestlers) would not automatically be rated as 'healthy' and may be maintaining their athletic 'quality' by unhealthy practices.

This review examines the value of the microbial biomass content of soils, the activity of the biomass, some derived relationships and selected microbial nutrient processes, as indicators of soil health.

Microbial Indices of Soil Health

Two microbial indices that have been suggested to monitor soil health are: (i) the microbial biomass, usually determined by biochemical methods (Jenkinson and Ladd, 1981; Sparling and Ross, 1993; Martens, 1995); and (ii) soil microbial respiration (Anderson, 1982; Insam, 1990). Two derived indices will also be examined in this review: (i) the microbial quotient, which is the proportion of the total soil organic C represented by the microbial C; and (ii) the respiratory quotient (qCO_2) which is the specific rate of respiration (CO_2-C h^{-1}) per unit of microbial biomass C, usually expressed as μg CO_2-C h^{-1} mg^{-1} microbial C. The microbial mineralization of soil organic nitrogen is considered as an example of a nutrient transformation process being used as an index of soil health.

Microbial Biomass

The microbial biomass is defined as the living component of soil organic matter (Jenkinson and Ladd, 1981) but excludes macrofauna and plant roots. Typical methods of determination are by fumigation–incubation, fumigation–extraction, or by substrate-induced respiration methods (Sparling and Ross, 1993). Microbial and organic matter indices are particularly useful to compare the health of soils of inherently low soil chemical fertility such as long-term native forests. The amounts of available nutrients in such soils are low, with a very high proportion of the nutrients immobilized in the living biomass (Jordan, 1985), so that standard soil fertility tests are of little value. It makes much more sense to measure the living soil microbial biomass and microbial activity, and obtain an estimate of pool size, potentially-available nutrients and the rate of cycling. The microbial biomass contents of soils can be useful measures to reveal effects of forest management (Diaz-Ravina *et al.*, 1995; Hossain *et al.*, 1995), and permit comparisons between soils of natural ecosystems with those that have been modified under agricultural use.

Amounts of microbial biomass in soils

As a very broad generalization, the amount of microbial biomass in a soil reflects the total organic matter content, with the living microbial component forming a low proportion of the total (Tables 5.1, 5.2). The proportion present as living microbial cells (microbial biomass C μg^{-1} g^{-1} soil) typically comprises 1–5% (w/w) of total soil organic C, and microbial N forms 1–6% (w/w) of total soil organic N (Jenkinson and Ladd, 1981; Sparling, 1985; Wardle, 1992). The total amounts of microbial C and N in soil consequently tend to reflect the total organic matter contents of soils, which are generally (but not invariably) greater in soils from cooler, wetter regions compared to warmer, drier regions (Allison, 1973; Spain *et al.*, 1983). This generalization is greatly modified by soil texture, mineralogy and land use. Extreme environments generally have low organic matter and very low microbial biomass contents (Table 5.1).

Table 5.1. Organic matter, microbial biomass and the microbial quotient in a range of soils of different class, mineralogy, climate and land use.

Soils and environment	USDA soil taxonomy	Land use	Total C (%)	Micro-bial-C μg g^{-1}	Quotient (%)	Source
Soil types						
Sand dune primary	Entisol	Pine forest	0.53	121	2.33	1
Tropical clay soil	Oxisol	Pasture	2.53	659	2.60	2
Sand dune, stable	Entisol	Pine forest	2.70	351	1.30	1
Clay loam	Inceptisol	Pasture	5.00	1139	2.27	4
Vocanic ash loam	Andosol	Pasture	6.20	1046	1.69	3
Volcanic ash loam	Andosol	Pasture	10.7	2088	1.95	4
Organic peat soil	Histosol	Scrub	39.2	5236	1.34	4
Extreme climates						
Antarctic maritime	NA	barren <2 mm	0.64	58	0.90	5
Saline desert	NA	dwarf scrub	0.47	50–225		6

1. Ross and Sparling, 1993; 2. Fiegl *et al.*, 1995; 3. Sparling and West, 1988; 4. Sparling and Searle, 1993; 5. Roser *et al.* 1993; 6. Sarig and Steinberger, 1994.
NA, Not available.

Soil texture

Within a given climatic region, finer-textured clay soils generally have greater organic matter contents than coarser-textured sandy soils under a similar land use (Spain *et al.*, 1983) and so, overall, the finer-textured clay soils tend to have more microbial biomass. Additionally, the microbial fraction in the clay soils tends to comprise a greater proportion of the total organic C than it does on sandy soils (Wardle, 1992). This is apparently because of the protective effect of the clays on the microbial biomass (van Veen *et al.*, 1985; Merckx *et al.*, 1985). Consequently, some soil types such as heavily-weathered tropical clay soils (Oxisols) support a relatively high microbial biomass in relation to their modest organic matter content (Luizao *et al.*, 1992). Some examples of the range of microbial biomass contents in different soil types and the relationship to the total organic matter content are shown in Table 5.1. There is a wide range in organic matter and microbial biomass values but all these soils would be considered healthy.

Volcanic ash soils containing allophanic clays (Andosols) tend to have high microbial biomass contents (Table 5.1), but in this case, the proportion of microbial biomass is often lower than expected in relation to the very high organic matter contents of these soils (Sparling, 1991). The explanation seems to be that the allophanic clays, as well as stabilizing the microbial cells, are also able to stabilize a large proportion of inert organic matter (Ross *et al.*, 1982). Organic

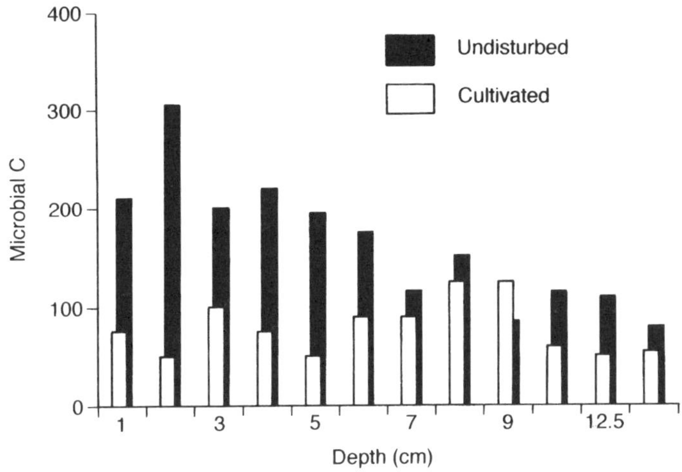

Fig. 5.1. Distribution of microbial C (μgC g^{-1} soil) over 0–15 cm depth in an undisturbed and a cultivated (tilled) soil (Compiled from data of Woods, 1989).

soils form another special case, where, despite the very high organic matter content, the microbial biomass is relatively low.

Microbial biomass in the soil profile

The microbial biomass is concentrated in the top few centimetres of soil and in non-tilled soils declines rapidly with depth (Fig. 5.1). Therefore, when measuring the microbial biomass content of soils it is important to define the sampling depth, particularly with till and no-till comparisons. Adjustment for any changes in bulk density under different land uses should also be made, and the strategy of 'equivalent sampling depth' should be adopted (Powlson and Jenkinson, 1981). There is good justification for expressing results on a volumetric rather than a weight basis as this compensates for changes in bulk density.

Seasonal variability

There is still contention about the degree of seasonal variation in the microbial biomass. Indications are that under most climatic regimes, the biomass shows some seasonal variation. Figure 5.2 shows relative fluctuations in microbial C under different climatic regimes. Fluctuations appear to be greater in the dry tropics and mediterranean climates with larger seasonal changes in soil temperature and moisture (Granatstein *et al.*, 1987; Srivastava, 1992); but can

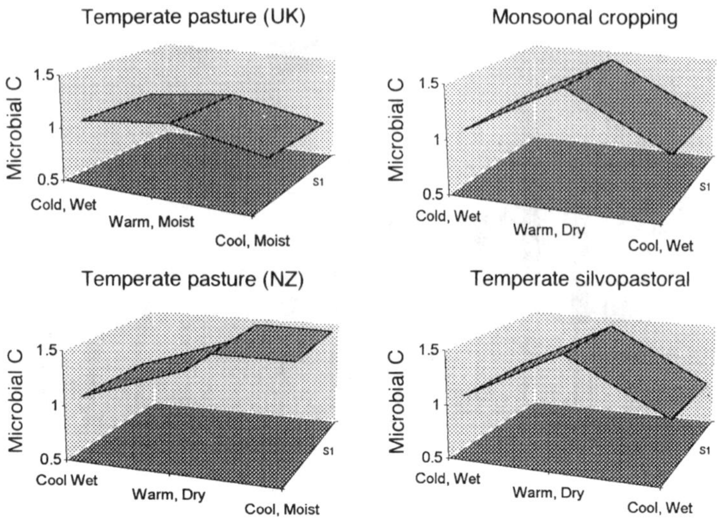

Fig. 5.2. Relative changes in microbial biomass reported under four different climatic regimes. (Compiled from Patra *et al.*, 1990; Srivastava, 1992; Tate *et al.*, 1991 and Campbell *et al.*, 1995.)

still be substantial in more temperate regions (Ritz and Robinson, 1988; Tate *et al.*, 1991). Clearly, if absolute microbial biomass is used as an indicator of health in these soils, then the time of sampling needs to be taken into consideration. Patra *et al.* (1990) found little change in microbial biomass C of pasture in Britain, and argued that seasonal fluctuations would be relatively small and the time of sampling was not crucial. However, I concur with the opinion of Granatstein *et al.* (1987) who stated 'numerous factors affect soil microbial biomass and that cropping history and seasonal changes must be taken into account when microbial biomass data are compared'.

The range in microbial biomass values in Table 5.1 should indicate that the absolute microbial C content of a soil is of limited value as an indicator of health because the microbial C content of different soil types differs, even when they are both under a similar land use and in a similar climatic region. For the microbial biomass content of a partiular soil to be used as an indicator of soil health it is necessary to have some soil-specific baseline for comparison. This baseline or 'target' value is normally derived from the same soil type under alternative land management. This approach is necessary because we do not know how to define the 'ideal' microbial biomass content for a healthy soil. Insam *et al.* (1989) proposed a function to predict the microbial biomass C content using precipitation and evaporation parameters. The approach gave a reasonable prediction of microbial biomass in some North American soils, but the function was a poor predictor of microbial biomass in New Zealand soils (Sparling, 1991), and does not appear to have been tested elsewhere.

Extreme environments

Care is needed when examining soils with low contents of microbial biomass. A low microbial biomass will usually suggest low organic matter contents and poor soil quality. However, low microbial biomass contents are normal in early stages of soil development (Insam and Haselwandter, 1989) and reflects the sparse vegetation cover and low organic matter content. I consider such soils to be healthy but very likely of low quality. A case for alarm would be where a soil had no detectable biomass at all. This would mean the soil was biologically dead, the ultimate case in ill-health! In practice, some microbial population has always been detectable in all but the most extreme of natural ecosystems (such as boiling mud pools in volcanic areas). Microbial numbers and mass may be naturally very low in extreme terrestrial environments such as alpine, polar (Insam and Haselwandter, 1989; Cochran *et al.*, 1989; Roser *et al.*, 1993) and arid or semi-arid regions (Sarig and Steinberger, 1994; Smith *et al.*, 1994; Sparling *et al.*, 1994a). It is important to differentiate these soils with naturally low populations from those that have been diminished by soil degradation or contamination.

Microbial Biomass as an Early Indicator of Changing Soil Processes

The microbial biomass can be a sensitive indicator of changes in soil processes because it has a much faster rate of turnover than the total soil organic matter (Jenkinson and Ladd, 1981; Paul, 1984) and several authors have suggested that trends in the microbial biomass content of soils will predict longer term trends in total organic matter contents (Powlson and Jenkson, 1981; Powlson *et al.*, 1987; Sparling, 1992). This is consistent with the approach of Larson and Pierce (1994) who suggest that the rates of change in soil parameters, rather than the absolute values, can provide an assessment of longer term soil quality and health.

There are numerous examples of organic matter and microbial biomass decline under agricultural or land disturbance (Table 5.2), indicating exploitation of the organic resource and the modifying influences of differing tillage systems, fertilizers and crop rotation (Carter and Rennie, 1982; Dalal and Mayer, 1987; Granatstein *et al.*, 1987; Srivastava and Singh, 1989; Luizao *et al.*, 1992; Sparling *et al.*, 1992; Weigand *et al.*, 1995). Relative to some reference site, those land management techniques that have least impact on microbial biomass and organic matter can be identified. Such studies generally assume that those land

Table 5.2. Comparison of organic matter, microbial C and the microbial quotient in topsoils (0–10 cm depth) of paired sites under alternative land uses.

Soil	Land use	Total C (%)	Mic-C μg g^{-1}	Quotient (%)
1. Kairanga (NZ)	Long-term pasture	5.24	1557	2.97
silty clay loam	Maize cropping	3.59	540	1.50
2. Judgeford (NZ)	Native forest	9.9	1295	1.30
silt loam	Long-term pasture	8.7	1905	2.19
3. Taita (NZ)	Native forest	4.90	622	1.27
clay loam	Pine forest	4.15	538	1.30
	Long-term pasture	4.40	690	1.56
4. Tammin (WA)	Native woodland	1.19	151	1.26
sandy loam	Long-term pasture	0.59	139	2.34
	Reforested pasture	0.39	171	4.34
5. Banaras (India)	Cropping with no fertilizer	0.72	202	2.81
sandy loam	Cropping with manure	1.12	300	2.67
	Cropping with NPK fertilizer	0.82	240	2.93

1. Sparling *et al.*, 1992; 2. Yeates *et al.*, 1991; 3. Sparling *et al.*, 1994b; 4. Sparling *et al.*, 1994a; 5. Ghoshal and Singh, 1995.

management methods causing greater rates of decline in organic matter will become less sustainable in the longer term because of the deteriorating physical, chemical and biological fertility of the soil. However, from the examples given in Table 5.2, it is clear that change in the microbial fraction is not always greater than that in the total C content, and can even show conflicting trends (e.g. the Tammin and Judgeford soils).

Microbial biomass contents of degraded soils can provide a useful index to monitor soil restoration techniques. A non-degraded example of the soil provides a target value for the microbial biomass content, and the soil being restored can be monitored over time to provide an index of recovery. Microbial indices have been used to monitor the recovery of soil after topsoil stripping and lignite mining in New Zealand (Ross *et al.*, 1984, 1992) and in Germany (Insam and Domsch, 1988), and to follow changes in stockpiled soils (Williamson and Johnson, 1994).

Microbial Biomass and Soil Fertility

Some authors (Hart *et al.*, 1986; Williams and Sparling, 1988; Insam *et al.*, 1991; Srivastava and Singh, 1991) have suggested a strong link between soil microbial biomass, soil fertility and soil health. A positive relationship between the microbial biomass content of soils and the amount of potentially available

N as determined by anaerobic incubation or other tests has been shown several times (Hart *et al.*, 1986; Myrold, 1987; Williams and Sparling, 1988; Stockdale and Rees, 1994). In contrast, a higher microbial biomass did not necessarily relate to greater soil fertility as assessed by the amounts of 'plant-available' phosphorus extracted from soil, or plant yield (Sorn-srivichai *et al.*, 1988). More biomass C and N was found under wheat with no fertilizer than under continuous wheat receiving N fertilizer (Biederbeck *et al.*, 1984); soil respiration was greater from less productive than more productive soils (Dinwoodie and Juma, 1988). Tate *et al.* (1991) found more organic matter and a greater microbial biomass C on a low fertility pasture site compared to a high fertility site. Organic matter and microbial biomass in an uncleared forest soil of very low fertility were similar to those under a productive pasture fertilized with lime and superphosphate for some 30 years (Sparling *et al.*, 1994b).

There are several possible reasons for these trends. Biederbeck *et al.* (1984) suggested that a larger but less active biomass was present in the poorly fertilized system compared to a low fertility soil. Crops grown on soils with greater quantities of available nutrients may need less extensive root systems, less below-ground plant biomass and consequently a decreased microbial biomass. Long-term fertilizer use can improve soil nutrient status but also cause soil acidification and perhaps decrease the microbial biomass while still improving crop yield (Ladd *et al.*, 1994). High levels of inorganic nutrients and lower pH on fertilized plots may interfere with biochemical assays to estimate microbial biomass (Widmer *et al.*, 1989; Amato and Ladd, 1994).

Microbial Biomass and Soil Pollution

Soil microbial biomass measures have been used to assess the impacts of fungicides (Anderson *et al.*, 1981; Heilmann *et al.*, 1995), herbicides (Wardle and Parkinson, 1990) and heavy metal contaminants (Brookes *et al.*, 1986; Dumontet and Mathur, 1989; Fliebbach *et al.*, 1994; Leita *et al.*, 1995; Brookes, 1995) on soil health. In general, the soil microbial biomass has not been a sensitive indicator of the effects of herbicides except when added at more than the recommended dosage, when the microbial biomass and soil respiration was decreased. Usually, when used at the recommended dosages, only small effects of pesticides on the soil microbial biomass have been reported. In contrast, heavy metals can have marked effects on the microbial biomass. Brookes and McGrath (1984) reported that the microbial biomass in soils amended with contaminated sludge over 30 years previously were still lower than uncontaminated soil. Organic matter contents and soil respiration of sludge-amended soils were not decreased and so were not good indicators of soil health. At current permissible levels in Europe, heavy metal contamination has little effect on soil respiration (Brookes, 1995), but Brookes and McGrath (1984) suggested that the respiratory quotient (see later), could be a more sensitive index of soil pollution than either the soil biomass or respiration considered alone.

Microbial Quotient

Several authors have suggested that the microbial quotient (microbial-C/total-C), indicates changing soil processes and soil health, and is a more useful measure than either microbial C or total C considered individually (Anderson and Domsch, 1989; Insam and Domsch, 1988; Sparling, 1992). Because the quotient is a ratio, the approach avoids the problems of working with absolute values and comparing across soils with different organic matter contents. Anderson and Domsch (1989) showed differences in the microbial biomass quotients of a range of soils, with soils under monocropping having lower quotients than those under multicropping. Sparling *et al.* (1992) found the approach useful to 'normalize' data from a chronosequence of cropping sites, and to obtain a much clearer indication of changing organic C contents with time. Generally, if a soil is being used exploitively, then the microbial C pools will decline at a faster rate than the total organic matter, and the microbial quotient will decrease (Table 5.2).

The reverse process is true for soils in early stages of ecosystem succession where the proportion of the total organic matter as microbial biomass tends to be greater than in established soils (Insam and Haselwandter, 1989). Insam and Domsch (1988) showed a marked decrease in the microbial quotient over a 20 year period as forest re-established on soils after disturbance for lignite mining.

Care is necessary when using the microbial quotient to compare soils of differing mineralogy, with soils such as Andosols having generally lower microbial quotients than Oxisols (Table 5.1) It is again necessary to select a 'target' value from a reference soil of the same type. The microbial quotients are useful to determine trends with time, and to compare soils, but our current knowledge is such that there is no particular value that can be regarded as 'healthy'.

Microbial Respiration

Soil respiration is a well established parameter to monitor decomposition (Anderson, 1982), but soil respiration is also highly variable and can show wide natural fluctuation depending on substrate availability, moisture and temperature (Orchard and Cook, 1983; Alvarez *et al.*, 1995; Brookes, 1995). Soil organisms can respond very rapidly to a change in soil conditions even after long periods of inactivity. Only minutes after the rewetting of a seasonally dry soil there were increases in respiration and mineralization of C and N from soil organic matter (Kieft *et al.*, 1987; West *et al.*, 1988a,b; van Gestel *et al.*, 1992). The great variability in respiration means that this measure taken alone is very difficult to interpret in terms of soil health (Brookes, 1995). For valid comparisons between soils, respiration measurements must normally be made under controlled laboratory conditions (Anderson, 1982). The rate of CO_2 efflux, under conditions where moisture and temperature are not limiting, can provide an indication of organic matter quality and whether the soil environment is conducive to the decomposition process. Again, interpretation of respiration in terms of soil health or quality is problematic. Rapid decomposition of organic matter is not necessarily desirable because stable organic matter has an important role in soil physical and chemical characteristics. On the other hand, the decomposition of organic residues to release nutrients at time of plant demand, is a desirable characteristic.

Respiratory Quotients (qCO_2)

The respiratory quotient is obtained by expressing the rate of soil respiration in terms of the microbial biomass. This measure is also termed the specific respiration rate and is usually expressed as μg CO_2-C h^{-1} mg^{-1} microbial biomass C, or some equivalent unit. Anderson and Domsch (1990) and Insam and Haselwandter (1989) suggested that the respiratory quotient of soil organisms had an analogy to plant ecosystems and could be used to investigate both soil development, substrate quality, ecosystem development and response to stress. The theory states that in a young developing ecosystem there is less competition for energy, and less incentive for efficient use. As the ecosystem gains maturity and diversity, there is greater competition for available energy, and a stronger selective pressure towards efficient use of available resources (Insam and Haselwandter, 1989). In soil biomass terms, this may be translated into a greater biomass and lower specific respiration rates in the established ecosystem. Such a trend was found by Insam and Haselwandter (1989) on a chronosequence of sites, and Santrucková and Straskraba (1991) noted a tendency for respiratory quotients to be greater on those soils with low microbial biomass contents. Gupta *et al.* (1994) presented data showing that the qCO_2 responses of soils changed after incorporation of different crop residues or after burning. However, there appeared to be no consistent trend. Even so, they advocated that 'a measurement of microbial activity (qCO_2) must be taken into consideration when the changes in the quality of OM and the health of a soil are being considered', but gave no guidance as to how the qCO_2 value should be interpreted.

Respiratory quotients were used by Brookes and McGrath (1984) to assess the effects of applications to soils of heavy metal contaminated sewage sludge. The respiratory quotient (qCO_2), was substantially greater in the sludge-amended soils. In this case the higher respiratory quotients suggest a stress response and 'poor health'.

Care needs to be taken in using respiratory quotients too literally. If the quality of the carbon substrates differs greatly, or soils are of different soil types, then interpretation becomes very difficult. Does a high respiratory quotient infer

Table 5.3. Influence of land use on the organic matter content (%), C:N ratio, microbial C content (μg g^{-1} soil), microbial quotient (%) and respiratory quotients (qCO_2) of a Western Australian Oxisol and a New Zealand Dystrochrept soil under long-term native woodland/forest or introduced pastures.

Land use	Total C (%)	C:N ratio	Microbial C	Microbial quotient (%)	Respiratory quotient qCO_2
Australia					
Woodland	0.94	25.7	604	6.4	3.51±0.62
Pasture	1.05	13.2	665	6.3	2.45±0.20
New Zealand					
Woodland	6.00	28.6	819	1.37	0.46±0.07
Pasture	6.80	13.0	1055	1.55	1.00±0.15

stress, an immature ecosystem or a more respirable substrate? Conflicting trends have sometimes been obtained when comparing 'established, natural' ecosystems with 'unstable' agricultural ecosytems. Table 5.3 shows directly opposite trends in qCO_2 values of a Western Australian *Banksia* woodland soil compared with adjacent pasture, and a New Zealand *Nothofagus* woodland compared with adjacent ryegrass/clover pasture. In the Western Australian example, the qCO_2 was greater in the woodland compared to the pasture; in the New Zealand soil the qCO_2 of the pasture was greater than the woodland.

Such opposing trends suggest that the transfer of plant ecological principles directly to microbial biomass and respiration relationships may need to be reconsidered. An obvious difference for the soil ecosystem is that the major source of energy (organic matter) is not of uniform quality. The quality and amount of respirable C-substrate is likely to influence qCO_2, but the results in Table 5.3 show that the relationships are more complex than can be explained by organic matter contents and C:N ratio.

Mineralization of Soil Organic Nitrogen

The mineralization and cycling of nutrients by soil microorganisms is an essential function for a healthy soil. The microbial mineralization of organic nitrogen has been suggested as a useful index of soil quality and health because both the accumulation and mineralization of N in soil are predominantly biological processes. The mineralization of soil organic N through to nitrate reflects the quality and quantity of soil organic nitrogen and links the substrate with the functioning and activity of a range of soil organisms. A large range of soil organisms are able to decompose organic N and release ammonium N, but the subsequent oxidations of ammonium to nitrite and nitrate are predominantly performed by small specialist groups of chemotrophic bacteria which are more fastidious in their growth requirements (Birch, 1960; Shen *et al.*, 1984; Grant, 1994). The mineralization of organic N therefore reflects previous biological activity in the soil, general decomposer activity (ammonification) and the activity of specialist groups (ammonium and nitrite oxidizers), and has potential as an indicator of soil health on several levels.

Mineralization

The amount of N mineralized in a soil is a good integrative measurement of climatic factors with the chemical, physical and biological fertility of a soil. Greater amounts of N are mineralized from those soils with readily decomposable organic N sources, which reflects previous accumulations of organic N (Allison, 1973; Dalal and Mayer, 1987; Cabrera and Kissel, 1988). Factors such as the presence and abundance of legumes, efficient N-fixation, N fertilizer applications, moderate or neutral soil pH, low bulk density, adequate moisture retention and residue incorporation favour accumulation of N in plants and organic matter (Allison, 1973). These are also all factors that would generally be regarded as beneficial for a healthy soil. Mineralization is also strongly influenced by physical and chemical conditions in the soil (Grant, 1994). Strongly acidic soils, cold temperatures, very dry conditions and poorly aerated or waterlogged soils inhibit mineralization. Mineralization is enhanced by neutral pH, well aerated soil and warm, moist conditions (MacDuff and White, 1985; Pilbeam *et al.*, 1993). While soils can readily be ranked in terms of N mineralization, it is very difficult to define a particular value that can be regarded as healthy, and perceptions of soil health and N mineralization differ depending upon the land use (see 'Interpretation of Microbial Indices of Soil Health').

In the presence of an available carbon source any nitrogen mineralized may be immediately immobilized by soil microbes, so that net mineralization can be zero or negative (Shen *et al.*, 1984; Barraclough, 1988). For this reason anaerobic (waterlogged) mineralization of N has sometimes been preferred to aerobic mineralization, because there is less microbial immobilization of N under waterlogged conditions (Gianello and Bremner, 1986a,b). A close relationship has been obtained between the amount of N mineralized under anaerobic (waterlogged) incubation and the amount of microbial biomass (Hart *et al.*, 1986; Myrold, 1987; Stockdale and Rees, 1994), suggesting that at least some of the N mineralized during incubation was derived from microbial cells killed by the waterlogging. Such techniques allow some comparisons of the 'health' of soils of widely differing characteristics.

An alternative method to assess N mineralization activity is to measure gross rather than net mineralization, using the isotope dilution technique (Kirkham and Bartholomew, 1954; Barraclough, 1991). The isotope dilution

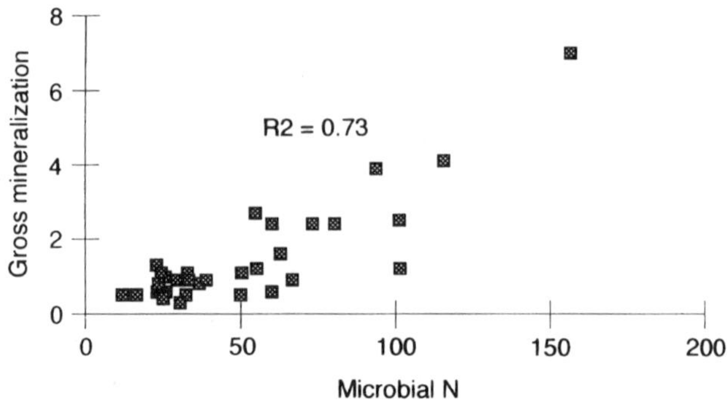

Fig. 5.3. Relationship between gross mineralization of soil organic N and soil microbial biomass N in a range of soils. (Data from Murphy, Sparling and Fillery, unpublished results.)

method allows the comparison of N mineralization in a wide range of soils, even those that are net immobilizers of N. Short-term gross ammonification rates showed a reasonable linear relationship with the microbial biomass content in undisturbed cores from a range of soils in Western Australia (Fig. 5.3 from Murphy, Sparling and Fillery, unpublished data) and demonstrated a link between activity (gross mineralization) and pool size or mass (microbial biomass N).

Nitrification

After ammonification, the ammonium is oxidized to nitrite and subsequently to nitrate. These steps are predominantly carried out by specialist chemoautotrophic bacteria which are active over a much narrower range of soil conditions and much less abundant than the general ammonifying population (MacDuff and White, 1985). Nitrification is inhibited by poor aeration and by both high and low pH, at which point ammonification may still be occurring, so that ammonium accumulates in soil (Grant, 1994). There is also evidence that root exudates of some plants can suppress the activity of the nitrifier populations (Dyck *et al.*, 1985). The rapid accumulation of nitrate in soil indicates the presence of mineralizable organic N in addition to favourable soil conditions for the functioning of ammonifier organisms and the nitrite and nitrate oxidizing bacteria. Nitrogen mineralization is therefore a very useful indicator of both organic matter quality and microbial processes in soil. High N mineralization rates indicate a rapid decomposition of soil organic N and an active microbial population. Land users interested in high levels of available nutrients and utilization of soil organic reserves for plant production would generally regard these characteristics as positive and indicating a healthy soil of good quality. However, a healthy soil does not always require high levels of available nutrients (see 'Interpretation of Microbial Indices of Soil Health').

Interpretation of Microbial Indices of Soil Health

Microbial indices can be sensitive measures of changing soil processes, but a major limitation to their application to assess soil health is that we do not know what are the 'ideal' values for a healthy soil, and we generally acknowledge they will differ for different soils and land uses. A farmer interested in production will generally approve of rapid N mineralization at times of plant demand, indicating active soil populations, organic matter of high N status and conditions favouring mineralization, so that the crop may benefit from the mineralized N. However, nutrient losses from the system could be of lesser concern. Conversely, a hydrologist interested in water quality would prefer consistently low N status and immobilization to decrease the risk of nitrates leaching into ground water. Other concerns may be how soil health relates to the stability and sustainability of a particular ecosystem, and maintenance of biodiversity, rather than production-oriented or pollution issues.

Microbial indices have most value where a 'target' value can be justifiably defined, and examples have been given of practical uses to monitor soil recovery during rehabilitation, or to show the adverse effects of toxic residues on soil processes. Microbial biomass and activity measures have also given early indications of longer term trends resulting from different cropping, rotation and cultivation practices. However, in my opinion, we now need to be more specific about what these trends mean, particularly in the context of cropping and sustainable land use. It is useful to observe that microbial biomass and activity are lower under a particular land use rather than another, but at what point do we move from observation to concern? What is to be used as the baseline to assess the degree of change? If we monitor the soil itself over time how much can organic matter and microbial biomass change before we should be alarmed?

To answer these questions we need to establish justifiable 'baseline' or target values of microbial biomass and activity and to define acceptable rates of change. Such values will be soil specific and will show regional differences. The sensitivity of the indices needs much more stringent testing, as well as analyses of the cost effectiveness of various indices. Measuring rates of change rather than absolute values of microbial indices may be a productive route, but trigger points still need to be defined for any management actions to be justified. To set trigger points, the consequences of a change in the microbial indices need to be identified. Direct linkages between soil health and human health and prosperity will be essential to justify changes in management practices, particularly those that involve social and economic costs. Our actions will be largely defined by the prevailing social, economic and political attitudes.

Conclusions

There are few instances where the absolute microbial biomass content of a soil can provide a definitive measure of soil health. Interpretation is hampered by the natural range in microbial biomass contents in different soil types and ecosystems, seasonal fluctuations and inconsistent trends in relation to soil fertility and plant production. Further, our criteria of 'soil health' change depending on how the soil is being used. Current knowledge is such that there are no accepted baseline or reference values. Similarly, there are currently no defined trigger points above or below which microbial biomass indices could be justifiably regarded as healthy or unhealthy.

The soil microbial biomass can provide a more sensitive measure of change than total organic matter contents, and can usefully reveal trends within 1–5 years rather than decades. By comparing the microbial biomass and respiration under a particular land management technique with a reference soil of similar type, an estimate of the rate of change under that land management is obtained.

Derived indices such as the microbial quotient and qCO_2 can provide additional information about the microbial biomass, but currently there are no absolute values above or below which a soil can be defined as healthy or unhealthy.

Mineralization of N from the soil organic matter can provide a useful integration of chemical, physical and biological aspects of soil health because it combines both the accumulation of N through previous biological activity, the present organic matter status of the soil, and the current N mineralization activity of the soil microorganisms. N mineralization reflects both the decomposer activity of the general soil population (ammonification) and the activity of the more fastidious nitrifier population (nitrification).

Soils with low natural contents of microbial biomass and low activity, such as occur in harsh natural environments, should not be regarded as unhealthy, and need to be differentiated from degraded and contaminated soils.

Acknowledgement

This review was commenced while I was Senior Wheatgrower Research Fellow at the Department of Soil Science and Plant Nutrition, The University of Western Australia, with funding provided by the Grains Research and Development Corporation, and completed at Manaaki Whenua – Landcare Research, Hamilton, New Zealand.

References

Allison, F.E. (1973) Soil organic matter and its role in crop production. In: *Developments in Soil Science.* No. 3. Elsevier, Amsterdam.

Alvarez, R., Santanatoglia, O.J. and Garcia, R. (1995) Effect of temperature on soil microbial biomass and its metabolic quotient *in situ* under different tillage systems. *Biology and Fertility of Soils* 19, 227–230.

Amato, M. and Ladd, J.N. (1994) Application of the ninhydrin-reactive N assay for microbial biomass in acid soils. *Soil Biology and Biochemistry* 26, 1109–1115.

Anderson, J.P.E. (1982) Soil Respiration. In: *Methods of Soil Analysis, Part 2, Chemical and Microbiological Properties.* American Society of Agronomy, Madison, Wisconsin, pp. 831–871.

Anderson, J.P.E., Armstrong, R.A. and Smith, S.N. (1981) Methods to evaluate pesticide damage to the biomass of the soil microflora. *Soil Biology and Biochemistry* 13, 149–153.

Anderson, T.-H. and Domsch, K.H. (1989) Ratios of microbial biomass carbon to total organic carbon in arable soils. *Soil Biology and Biochemistry* 21, 471–479.

Anderson, T.-H. and Domsch, K.H. (1990) Application of eco-physiological quotients (qCO_2 and qD) on microbial biomasses from soils of different cropping histories. *Soil Biology and Biochemistry* 22, 251–255.

Barraclough, D. (1988) Studying mineralization/immobilization turnover in field experiments: The use of nitrogen-15 and simple mathematical analysis. In: Jenkinson, D.S. and Smith, K.A. (eds) *Nitrogen Efficiency in Agricultural Soils.* Elsevier Applied Science, London, pp. 409–417.

Barraclough, D. (1991) The use of mean pool abundances to interpret ^{15}N tracer experiments I. Theory. *Plant and Soil* 131, 89–96.

Biederbeck, V.O., Campbell, C.A. and Zentner, R.P. (1984) Effect of crop rotation and fertilization on some biological properties of a loam in southwestern Saskatchewan. *Canadian Journal of Soil Science* 64, 355–367.

Birch, H.F. (1960) Nitrification in soils after different periods of dryness. *Plant and Soil* 12, 81–96.

Brookes, P.C. (1995) The use of microbial parameters in monitoring soil pollution by heavy metals. *Biology and Fertility of Soils* 19, 269–279.

Brookes, P.C. and McGrath, S.P. (1984) Effects of metal toxicity on the size of the soil microbial biomass. *Journal of Soil Science* 35, 341–346.

Brookes, P.C., Heijen, C.E., McGrath, S.P. and Vance, E.D. (1986) Soil microbial biomass estimates in soil contaminated with metals. *Soil Biology and Biochemistry* 18, 383–388.

Cabrera, M.L. and Kissel, D.E. (1988) Evaluation of a method to predict nitrogen mineralized from soil organic matter under field conditions. *Soil Science Society of America Journal* 52, 1027–1031.

Campbell, C.D., Chapman, S.J. and Urquhart, F. (1995) Effect of nitrogen fertiliser on temporal and spatial variation of mineral nitrogen and microbial biomass in a silvopastoral system. *Biology and Fertility of Soils* 19, 177–185.

Carter, M.R. and Rennie, D.A. (1982) Changes in soil quality under zero tillage farming systems: Distribution of microbial biomass and mineralizable C and N potentials. *Canadian Journal of Soil Science* 62, 587–597.

Cochran, V.L., Elliott, L.F. and Lewis, C.E. (1989) Soil microbial biomass and enzyme activity in subarctic agricultural and forest soils. *Biology and Fertility of Soils* 7, 283–288.

Dalal, R.C. and Mayer, R.J. (1987) Long term trends in fertility of soils under continuous cultivation and cereal cropping in southern Queensland. VII. Dynamics of nitrogen mineralization potentials and microbial biomass. *Australian Journal of Soil Research* 25, 461–472.

Diaz-Ravina, M., Acea, M.J. and Carballas, T. (1995) Seasonal changes in microbial biomass and nutrient flush in forest soils. *Biology and Fertility of Soils* 19, 220–226.

Dinwoodie, G.D. and Juma, N.G. (1988) Allocation and microbial utilisation of C in two soils cropped to barley. *Canadian Journal of Soil Science* 68, 495–505.

Doran, J.W. and Parkin, T.B. (1994) Defining and assessing soil quality. In: Doran, J.W., Coleman, D.C., Bezidicek, D.F. and Stewart, B.A. (eds) *Defining Soil Quality for a Sustainable Environment.* Soil Science Society of America Special Publication No. 35, Madison, Wisconsin, pp. 3–21.

Dumontet, S. and Mathur, S.P. (1989) Evaluation of respiration-based methods for measuring microbial biomass on metal-contaminated acidic mineral and organic soils. *Soil Biology and Biochemistry* 21, 431–436.

Dyck, W.J., Will, G.M. and Messina, M.G. (1985) The influence of radiata pine forestry on soil. In: *Proceedings of the Soil Dynamics and Land-Use Seminar,* Blenheim, May 1985. New Zealand Society of Soil Science, Lower Hutt, and New Zealand Soil Conservators Society, pp. 263–287.

Feigl, B.J., Sparling, G.P., Ross, D.J. and Cerri, C.C. (1995) Microbial biomass in Amazonian soils: evaluation of methods and estimates of pool sizes. *Soil Biology and Biochemistry* 27, 1467–1472.

Fliebbach, A., Martens, R. and Reber, H.H. (1994) Soil microbial biomass and microbial activity in soils treated with heavy metal contaminated sewage sludge. *Soil Biology and Biochemistry* 26, 1201–1205.

Garlynd, M.J., Roming, D.E., Harris, R.F. and Kurakov, A.V. (1994) Descriptive and analytical characterization of soil quality/health. In: Doran, J.W., Coleman, D.C., Bezidicek, D.F. and Stewart, B.A. (eds) *Defining Soil Quality for a Sustainable Environment.* Soil Science Society of America Special Publication No. 35, Madison, Wisconsin, pp. 159–169.

Gianello, G. and Bremner, J.M. (1986a) Comparison of chemical methods of assessing potentially available organic nitrogen in soil. *Communications in Soil Science and Plant Analysis* 17, 215–236.

Gianello, G. and Bremner, J.M. (1986b) A simple chemical method of assessing potentially available organic nitrogen in soil. *Communications in Soil Science and Plant Analysis.* 17, 195–214.

Ghoshal, N. and Singh, K.P. (1995) Effects of farmyard manure and inorganic fertilizer on the dynamics of soil microbial biomass in a tropical dryland agroecosystem. *Biology and Fertility of Soils* 19, 231–238.

Granatstein, D.M., Bezdicek, D.F., Cochran, V.L., Elliott, L.F. and Hammel, J. (1987) Long-term tillage and rotation effects on soil microbial biomass, carbon and nitrogen. *Biology and Fertility of Soils* 5, 265–270.

Grant, R.F. (1994) Simulation of ecological controls on nitrification. *Soil Biology and Biochemistry* 26, 305–315.

Gupta, V.V.S.R., Grace, P.R. and Roper, M.M. (1994) Carbon and nitrogen mineralization as influenced by long-term soil and crop residue management systems in Australia. In: Doran, J.W., Coleman, D.C., Bezidicek, D.F. and Stewart,

B.A. (eds) *Defining Soil Quality for a Sustainable Environment.* Soil Science Society of America Special Publication No. 35, Madison, Wisconsin, pp. 193–200.

Harris, R.F. and Bezdicek, D.F. (1994) Descriptive aspects of soil quality/health. In: Doran, J.W., Coleman, D.C., Bezidicek, D.F. and Stewart, B.A. (eds) *Defining Soil Quality for a Sustainable Environment.* Soil Science Society of America Special Publication No. 35, Madison, Wisconsin, pp. 23–35.

Hart, P.B.S., Sparling, G.P. and Kings, J.A. (1986) Relationship between mineralisable nitrogen and microbial biomass in a range of plant litters, peats, and soils of moderate to low pH. *New Zealand Journal of Agricultural Research* 29, 681–686.

Heilmann, B., Lebuhn, M. and Beese, F. (1995) Methods for the investigation of metabolic activities and shifts in the microbial community in a soil treated with a fungicide. *Biology and Fertility of Soils* 19, 186–192.

Hossain, A.K.M.A., Raison, R.J. and Khanna, P.K. (1995) Effects of fertilizer application and fire regime on soil microbial biomass carbon and nitrogen, and nitrogen mineralization in an Australian subalpine eucalypt forest. *Biology and Fertility of Soils* 19, 246–252.

Insam, H. (1990) Are the soil microbial biomass and basal respiration governed by climatic regime? *Soil Biology and Biochemistry* 22, 525–532.

Insam, H. and Domsch, K.H. (1988) Relationship between soil organic carbon and microbial biomass on chronosequences of reclamation sites. *Microbial Ecology* 15, 177–188.

Insam, H. and Haselwandter, K. (1989) Metabolic quotient of the soil microflora in relation to plant succession. *Oecologia* 79, 174–178.

Insam, H., Michell, C.C. and Dormaar, J.F. (1991) Relationship of soil microbial biomass and activity with fertilization and crop yield of three Ultisols. *Soil Biology and Biochemistry* 23, 459–464.

Insam, H., Parkinson, D. and Domsch, K.H. (1989) Influence of macroclimate on soil microbial biomass. *Soil Biology and Biochemistry* 21, 211–221.

Jenkinson, D.S. (1990) The turnover of organic carbon and nitrogen in soil. *Philosophical Transactions of the Royal Society (London) Series B* 329, 361–368.

Jenkinson, D.S. and Ladd, J.N. (1981) Microbial biomass in soil: measurement and turnover. In: Paul, E.A. and Ladd, J.N. (eds) *Soil Biochemistry,* Volume 5 Marcel Dekker, Inc, New York and Basel, pp. 415–471.

Jordan, C.F. (1985) *Nutrient Cycling in Tropical Forest Ecosystems.* John Wiley, Chichester.

Kieft, T.L., Soroker, E. and Firestone, M.K. (1987) Microbial biomass response to a rapid increase in water potential when dry soil is wetted. *Soil Biology and Biochemistry* 19, 119–126.

Kirkham, D. and Bartholomew, W.V. (1954) Equations for following nutrient transformations in soil utilizing tracer data. *Soil Science Society of America Proceedings* 18, 33–34.

Ladd, J.N., Amato, M., Li-Kai, Zhou and Schultz, J.E. (1994) Differential effects of rotation, plant residue and nitrogen fertilizer on microbial biomass and organic matter in an Australian Alfisol. *Soil Biology and Biochemistry* 26, 821–831.

Larson, W.E. and Pierce, F.J. (1994) The dynamics of soil quality as a measure of sustainable management. In: Doran, J.W., Coleman, D.C., Bezidicek, D.F. and

Stewart, B.A. (eds) *Defining Soil Quality for a Sustainable Environment.* Soil Science Society of America Special Publication No. 35, Madison, Wisconsin, pp. 37–51.

Leita, L., Nobili, M. De, Muhlbachova, G., Mondini, C., Marchiol, N. and Zerbi, G. (1995) Bioavailability and effects of heavy metals on soil microbial biomass survival during laboratory incubation. *Biology and Fertility of Soils* 19, 103–108.

Luizao, R.C.C., Bonde, T.A. and Rosswall, T. (1992) Seasonal variations of soil microbial biomass – the effects of clearfelling a tropical rainforest and establishment of pasture in the central Amazon. *Soil Biology and Biochemistry* 24, 805–813.

Macduff, J.H. and White, R.E. (1985) Net mineralization and nitrification rates in a clay soil measured and predicted in permanent grassland from soil temperature and moisture content. *Plant and Soil* 86, 151–172.

Martens, R. (1995) Current methods for measuring microbial biomass C in soil: potentials and limitations. *Biology and Fertility of Soils* 19, 87–99.

Merckx, R., Den Hartog, A. and Van Veen, J.A. (1985) Turnover of root-derived material and related microbial biomass formation in soils of different texture. *Soil Biology and Biochemistry* 17, 565–569.

Myrold, D.D. (1987) Relationship between microbial biomass nitrogen and a nitrogen availability index. *Soil Science Society of America Journal* 51, 1047–1049.

O'Brien, B.J. (1984) Soil organic carbon fluxes and turnover rates estimated from radiocarbon enrichments. *Soil Biology and Biochemistry* 16, 115–120.

Orchard, V.A. and Cook, F.J. (1983) Relationship between soil respiration and soil moisture. *Soil Biology and Biochemistry* 15, 447–453.

Patra, D.S., Brookes, P.C., Coleman, K. and Jenkinson, D.S. (1990) Seasonal changes of soil microbial biomass in an arable and a grassland soil which have been under uniform management for many years. *Soil Biology and Biochemistry* 22, 739–742.

Paul, E.A. (1984) Dynamics of organic matter in soils. *Plant and Soil* 76, 275–285.

Pilbeam, C.J., Mahapatra, B.S. and Wood, M. (1993) Soil matric potential effects on gross rates of nitrogen mineralization in an Orthic Ferralsol from Kenya. *Soil Biology and Biochemistry* 25, 1409–1413.

Powlson, D.S., Brookes, P.C. and Christensen, B.T. (1987) Measurement of soil microbial biomass provides an early indication of changes in total soil organic matter due to straw incorporation. *Soil Biology and Biochemistry* 19, 159–164.

Powlson, D.S. and Jenkinson, D.S. (1981) A comparison of the organic matter, biomass, adenosine triphosphate and mineralizable nitrogen contents of ploughed and direct-drilled soils. *Journal of Agricultural Science* 97, 713–721.

Ritz, K. and Robinson, D. (1988) Temporal variations in soil microbial biomass C and N under a spring barley crop. *Soil Biology and Biochemistry* 20, 625–630.

Roser, D.J., Seppelt, R.D. and Ashbolt, N. (1993) Microbiology of ornithogenic soils from the Windmill Islands, Budd Coast, continental Antarctica: some observations on methods for measuring soil biomass in ornithogenic soils. *Soil Biology and Biochemistry* 25, 177–183.

Ross, D.J. and Sparling, G.P. (1993) Comparison of methods to estimate microbial C and N in litter and soil under *Pinus radiata* on a coastal sand. *Soil Biology and Biochemistry* 25, 1591–1599.

Ross, D.J., Tate, K.R. and Cairns, A. (1982) Biochemical changes in a yellow-brown loam and a central gley soil converted from pasture to maize in the Waikato area. *New Zealand Journal of Agricultural Research* 25, 35–42.

Ross, D.J., Speir, T.W., Tate, K.R., Cowling, J.C. and Watts, H.M. (1984) Restoration of pasture after topsoil removal: changes in soil biochemical properties over a 5-year period – a note. *New Zealand Journal of Science* 27, 419–422.

Ross, D.J., Speir, T.W., Cowling, J.C. and Feltham, C.W. (1992) Soil restoration under pasture after lignite mining: management effects on soil biochemical properties and their relationships with herbage yields. *Plant and Soil* 140, 85–97.

Santrucková, H. and Straskraba, M. (1991) On the relationship between specific respiration activity and microbial biomass in soils. *Soil Biology and Biochemistry* 23, 525–532.

Sarig, S. and Steinberger, Y. (1994) Microbial biomass response to seasonal fluctuation in soil salinity under the canopy of desert halophytes. *Soil Biology and Biochemistry* 26, 1405–1408.

Shen, S.M., Pruden, G. and Jenkinson, D.S. (1984) Mineralization and immobilization of nitrogen in fumigated soil and the measurement of microbial biomass nitrogen. *Soil Biology and Biochemistry* 16, 437–444.

Smith, J.L., Halvorson, J.J. and Bolton, H., Jr (1994) Spatial relationships of soil microbial biomass and C and N mineralization in a semi-arid shrub-steppe ecosystem. *Soil Biology and Biochemistry* 26, 1151–1159.

Sorn-srivichai, P., Syers, J.K., Tillman, R.W. and Cornforth, I.S. (1988) An evaluation of water extraction as a soil-testing procedure for phosphorus II. Factors affecting the amounts of water-extractable phosphorus in field soils. *Fertilizer Research* 15, 225–236.

Spain, A.V., Isbell, R.F. and Probert, M.E. (1983) Soil organic matter. In: *Soils: an Australian Viewpoint.* CSIRO, Melbourne/Academic Press, London, pp. 551–563.

Sparling, G.P. (1985) The soil biomass. In: Vaughan, D. and Malcolm, R.E. (eds) *Soil Organic Matter and Biological Activity.* Nijhoff/Junk, Dordrecht, The Netherlands.

Sparling, G.P. (1991) Organic-matter C and soil microbial biomass C as indicators of sustainable land use. In: *Evaluation for Sustainable Land Management in the Developing World.* Volume 2: Technical Papers. Bangkok, Thailand: International Board for Soil Research and Managment, IBSRAM Proceedings No. 12, 563–580.

Sparling, G.P. (1992) Ratio of microbial biomass carbon to soil organic carbon as a sensitive indicator of changes in soil organic matter. *Australian Journal of Soil Research* 30, 195–207.

Sparling, G.P. and Ross, D.J. (1993) Biochemical methods to estimate soil microbial biomass: Current developments and applications. In: Mulangoy, K. and Merckx, R. (eds) *Soil Organic Matter Dynamics and Sustainability of Tropical Agriculture.* Wiley, Chichester, pp. 21–37.

Sparling, G.P. and Searle, P.L. (1993) Dimethyl sulphoxide reduction as a sensitive indicator of microbial activity in soil: relationship with microbial biomass and mineralisation of nitrogen and sulphur. *Soil Biology and Biochemistry* 25, 251–256.

Sparling, G.P. and West, A.W. (1988) A direct extraction method to estimate soil microbial C: calibration in situ using microbial respiration and 14C-labelled cells. *Soil Biology and Biochemistry* 20, 337–343.

Sparling, G.P., Shepherd, T.G. and Kettles, H.A. (1992) Changes in soil organic C, microbial C and aggregate stability under continuous maize and cereal cropping, and after restoration to pasture in soil from the Manawatu region. *New Zealand Soil and Tillage* 24, 225–241.

Sparling, G.P., Brandenburg, A. and Zhu, C. (1994a) Microbial C and N in revegetated wheatbelt soils in Western Australia: Estimation in soil, humus and leaf-litter using the ninhydrin method. *Soil Biology and Biochemistry* 26, 1179–1184.

Sparling, G.P., Hart, P.B.S., August, J.A. and Leslie, D.M. (1994b) A comparison of soil and microbial carbon, nitrogen, and phosphorus contents, and macro-aggregate stability of a soil under native forest and after clearance for pastures and plantation forest. *Biology and Fertility of Soils* 17, 91–100.

Srivastava, S.C. (1992) Microbial C, N and P in dry tropical soils: seasonal changes and influence of soil moisture. *Soil Biology and Biochemistry* 24, 711–714.

Srivastava, S.C. and Singh, J.S. (1989) Effect of cultivation on microbial carbon and nitrogen in dry tropical forest soil. *Biology and Fertility of Soils* 8, 343–348.

Srivastava, S.C. and Singh, J.S. (1991) Microbial C, N and P in dry tropical forest soils: Effects of alternate land-uses and nutrient flux. *Soil Biology and Biochemistry* 23, 117–124.

Stockdale, E.A. and Rees, R.M. (1994) Relationship between biomass nitrogen and nitrogen extracted by other nitrogen availability methods. *Soil Biology and Biochemistry* 26, 1213–1220.

Stout, J.D., Goh, K.M. and Rafter, T.A. (1981) Chemistry and turnover of naturally occurring resistant organic compounds in soil. In: Paul, E.A. and Ladd, J.N. (eds) *Soil Biochemistry.* Volume 5, Marcel Dekker, New York, pp. 1–73.

Tate, K.R., Speir, T.W., Ross, D.J., Parfitt, R.L., Whale, K.N. and Cowling, J.C. (1991) Temporal variations in some plant and soil P pools in two pasture soils of widely different P fertility status. *Plant and Soil* 132, 219–232.

Van Gestel, M., Ladd, J.N. and Amato, M. (1992) Microbial biomass responses to seasonal change and imposed drying regimes at increasing depths of undisturbed topsoil profiles. *Soil Biology and Biochemistry* 24, 103–111.

van Veen, J.A., Ladd, J.N. and Amato, M. (1985) Turnover of carbon and nitrogen through the microbial biomass in a sandy loam and a clay soil incubated with [^{14}C(U)] glucose and [^{15}N]$(NH_4)_2SO_4$ under different moisture regimes. *Soil Biology and Biochemistry* 17, 747–756.

Wardle, D.A. (1992) A comparative assessment of factors which influence microbial biomass carbon and nitrogen levels in soil. *Biological Reviews* 67, 321–358.

Wardle, D.A. and Parkinson, D. (1990) Effects of three herbicides on soil microbial biomass and activity. *Plant and Soil* 122, 21–28.

Weigand, S., Auerswald, K. and Beck, T. (1995) Microbial biomass in agricultural topsoils after 6 years of bare fallow. *Biology and Fertility of Soils* 19, 129–134.

West, A.W., Sparling, G.P., Speir, T.W. and Wood, J.M. (1988a) Comparison of microbial biomass C, N-flush and ATP and certain enzyme activities of different textured soils subjected to gradual drying. *Australian Journal of Soil Research* 26, 217–219.

West, A.W., Sparling, G.P., Speir, T.W. and Wood, J.M. (1988b) Dynamics of microbial C, N-flush and ATP and enzyme activities of gradually dried soils from a climosequence. *Australian Journal of Soil Research* 26, 519–530.

Widmer P., Brookes P.C. and Parry L.C. (1989) Microbial biomass nitrogen measurements in soils containing large amounts of inorganic nitrogen. *Soil Biology and Biochemistry* 21, 865–867.

Williams, B.L. and Sparling, G.P. (1988) Microbial biomass C and readily mineralized nitrogen in peat and forest humus. *Soil Biology and Biochemistry* 20, 579–581.

Williamson, J.C. and Johnson, D.B. (1994) Conservation of mineral nitrogen in restored soils at opencast coalmine sites: II. The effects of inhibition of nitrification and organic amendments on nitrogen losses and soil microbial biomass. *European Journal of Soil Science* 45, 319–326.

Woods, L.E. (1989) Active organic matter distribution in the surface 15 cm of undisturbed and cultivated soil. *Biology and Fertility of Soils* 8, 271–278.

Yeates, G.W., Bamforth, S.S., Ross, D.J., Tate, K.R. and Sparling, G.P. (1991) Recolonization of methyl bromide sterilized soils under four different field conditions. *Biology and Fertility of Soils* 11, 181–189.

6

Soil Enzyme Activities as Integrative Indicators of Soil Health

R.P. Dick
Department of Crop and Soil Science, Oregon State University, Corvallis, Oregon 97331, USA

Introduction

Environmental degradation poses a significant threat to the global ecosystem due to rising world populations, acceleration of industrial activities, and technological developments over the last 100 years. Soils play a central role in the function and long-term sustainability of ecosystems, but because changes can be slow and difficult to quantify, measuring its 'health' or developing 'soil quality standards' has been largely ignored compared to air or water quality standards. Microorganisms and fauna are the primary agents for the innumerable processes that occur in soils. Soil enzymes are the mediators and catalysts of important soil functions that include: (i) decomposition of organic inputs; (ii) transformation of native soil organic matter; (iii) release of inorganic nutrients for plant growth; (iv) N_2 fixation; (v) detoxification of xenobiotics; (vi) nitrification; and (viii) denitrification.

Soil health can be viewed as the ability of a soil to perform functions that are required for the biological components of an ecosystem within the constraints of local environmental factors (i.e. water and temperature regimes, landscape position). In this respect, assessment of enzymes in soils offers potential as an integrative index of the soil biological status or the ability of a soil to carry out discreet enzyme-catalysed processes.

The focus of this chapter is to provide an overview of soil enzyme activity as a biological, process-level indicator for impacts of natural processes and anthropogenic activities on soils. In the following section, background information is provided as a framework for discussing soil enzymes and soil health. For more in-depth discussions of soil enzyme methodology and biochemistry, the reader is referred to Skujins (1967, 1976, 1978), Burns (1978), Kiss *et al.* (1975),

(eds C.E. Pankhurst, B.M. Doube and V.V.S.R. Gupta)

Ladd (1985) and Tabatabai (1994). Other useful information on soil enzymes related to soil quality and applications of soil enzymes in the plant-soil environment is presented by Dick (1994), and Dick and Tababatai (1993), respectively.

Soil Enzyme Properties

Enzymes are present in two general locations in soils; those associated with viable cells (intracellular enzymes) or as extracellular enzymes. The function of intracellular enzymes is obvious with their central role in the innumerable life processes of cells. Extracellular or abiontic enzymes as defined by Skujins (1976) are those outside living cells.

An inherent difficulty of studying soil enzymes is that only small amounts of the total enzymes found in the soil can be extracted from soils. Stronger extractants generally denature proteins such as enzymes by disrupting the stereospecific structure of enzymes that is necessary for biochemical reactions. Consequently, most investigations on soil enzymes are done by measuring their activity directly in the soil. This has a number of implications for interpreting and understanding the role of enzymes in soils.

The first consideration is the assay itself. Enzyme assays are carried out by adding a substrate solution of known concentration to a known amount of soil and measuring the rate of conversion of substrate to product. The assay is done under a strict set of conditions that includes temperature, buffer pH, and ionic strength. Thus the results are operationally defined and any change in these conditions will change the measured activity. This means that enzyme assays measure the potential activity under optimal conditions and not the *in situ* activity because assay conditions are quite different from those in the original soil, particularly substrate concentration which saturates the system.

A second implication of soil enzyme assays is that enzymatic activity of the abiontic or extracellular enzymes cannot be separated from that of living cells. Enzyme assays, particularly those with long incubations, will use antiseptics such as toluene to inhibit microbial growth and metabolism during the assay (Tabatabai, 1994). This does not inhibit the activity of preexisting enzymes in viable cells. Other procedures using antibiotics or high energy irradiation, although insightful, have not been successful in clearly differentiating between the enzyme activities of viable cells and abiontic enzymes. These and other studies provide indirect evidence that enzymes outside viable cells are stabilized in the soil matrix (Skujins, 1978). Furthermore, estimates of the amount of microbial biomass based on the soil enzyme activity far exceed the amount of bacterial numbers or fungal mycelia that is possible in soil (Ramirez-Martinez and McLaren, 1966).

Except for a few enzymes like dehydrogenase, most enzymes studied in soils have a significant portion of the enzymatic activity associated with abiontic enzymes. These enzymes enter the soil matrix as extracellular enzymes excreted into the soil solution, or are released upon cell lysis or as a part of cell debris. It seems likely that the vast majority of extracellular enzymes released into soil survive for only short periods and are readily decomposed or denatured. However, some of these enzymes are stabilized in the soil matrix and remain catalytic (Skujins, 1978). Abiontic enzymes can exist as stabilized enzymes in two locations: (i) adsorbed to internal or external clay surfaces; and (ii) complexed with humic colloids through adsorption, entrapment, or co-polymerization during humic matter genesis (Boyd and Mortland, 1990). Enzymes can be linked to soil humic substances or mineral surfaces by ionic or hydrogen bonding. But only a small amount of the total activity is associated with enzymes in such a linkage (McLaren, 1975). A more important form of enzyme bonding in soils is by covalent bonding to humic components (Gosewinkel and Broadbent, 1986).

Enzymatic activity associated with cytoplasmic functions plays a critical role in the life processes of soil organisms. Enzymes excreted by microorganisms into the soil solution could be important: (i) in hydrolysing substrates that are too large or insoluble to be taken up directly by cells; (ii) in detoxifying the surrounding environment; or (iii) in creating, in some other way, a favourable environment for the survival of the organism. The ecological function of abiontic enzymes that are stabilized in the soil matrix or with cell debris is less clear. Burns (1982) hypothesized that humic-enzyme complexes may be important for substrate catalysis for some organisms. The product released by the complexed soil enzyme could be taken up by microorganisms or be broken down further by other soil enzymes. It may be advantageous for a microbial cell to be located on the surface of a humic colloid containing a number of enzyme molecules because it may not have the ability to induce or produce certain enzymes and obtain the benefits of organisms that can excrete extracellular enzymes.

As stated above, the enzyme assay cannot distinguish between activity originating from living cells or abiontic enzymes or identify the exact sources of enzymes in soil. Soil enzymes may originate from plants, animals, fungi and bacteria. Although it is generally agreed that the microbial component is the main source of enzymes in soils, the specific properties of isoenzymes (enzymes that catalyse the same reactions but may differ in origin and have slight differences in kinetic properties or amino acid sequencing) from each source cannot be determined. Thus, not all isoenzymes are assayed under their optimal conditions and a soil enzyme assay then measures the activity of the sum of all the isoenzymes.

Soil Enzyme Activity as a Measure of Soil Health

Defining soil health

The discussion above illustrates the complexity of both the function of enzymes in soil and the limitations of research methodology for studying soil enzymes.

Nevertheless, the specificity and integrative nature of soil enzyme activity provide a potential basis for the use of soil enzyme activity as an indicator of certain functions in soils. Table 6.1 presents a brief description of the ecological function of enzymes that commonly have been studied in soils.

It is important to distinguish soil quality from soil health, which often are used interchangeably (see Doran and Safley, Chapter 1, this volume). Soil quality is defined largely by soil function or use and represents a composite of its physical, chemical, and biological properties that provide various functions to an ecosystem (e.g. plant growth medium, partitioning water, environmental buffer) (Doran *et al.*, 1996). Soil health can be thought of in terms of its condition. Doran *et al.* (1996) defined soil health as 'the continued capacity of soil to function as a vital living system, within ecosystem and land-use boundaries, to sustain biological productivity, maintain the quality of air and water environments, and promote plant, animal, and human health'. In this respect, soil enzymes have a unique and appropriate role in assessing soil health because soil as a whole (not only the biological component), can be thought of as a living biological entity (Quastel, 1946) capable of carrying out certain biochemical reactions without viable cells. In particular, soil enzyme activity should relate to: (i) plant productivity; (ii) microbial biomass; (iii) biogeochemical cycling potential; (iv) impacts of pollutants on soils; or (v) status of remediated soils.

Soil enzymes and plant productivity

Relating soil enzyme activity to plant productivity has produced mixed results. Early work showed no close relationship of enzyme activity to crop yields or soil nutrient status (Koepf, 1954; Drobnik, 1957; Galstyan, 1960; Haban, 1967). Conversely, Verstraete and Voets (1977) showed that the activity of selected soil enzymes (phosphatase, invertase, β-glucosidase, and urease) were positively correlated with crop yields and that these measurements were superior to measurement of microbial abundance in correlating with soil fertility or crop yield. In managed systems, other factors may confound or override the relationship between soil enzyme activity and plant productivity. This is likely to be true for agroecosystems where external inputs of nutrients and water can greatly stimulate plant growth without a corresponding response by soil microorganisms. Yaroschevich (1966) showed that manure-amended soil increased soil respiration and enzyme activity but inorganic fertilizer-amended soils showed depressed enzyme activity. Crop yields, however, were the same when adequate nutrients were supplied from either inorganic or organic sources. There is some evidence that the activity of certain soil enzymes is better related to plant productivity under native conditions and in highly disturbed landscapes (Pancholy and Rice, 1973a,b; Pancholy *et al.*, 1975; Kiss *et al.*, 1993).

Table 6.1. Major groupings of commonly assayed soil enzymes and their ecological function.

Enzyme Commission group and subgroup number	Enzyme	Ecological or soil health function
Oxidoreductases		
1.1	Dehydrogenase	Exists as integral part of intact cell and reflects total oxidative activities of soil microflora/important in oxidizing soil organic matter
1.1	Glucose oxidase	Oxidizes glucose
1.11	Catalase, peroxidase	Release oxygen from hydrogen peroxide
1.10	Polyphenol oxidases	Oxidize phenolic compounds and are involved in humification of soils
Hydrolases		
3.1	Phosphatase (mono- and diester)	Releases plant available PO_4 from organic matter
	Sulphatase	Releases plant available SO_4 from organic matter
3.2	Amylase	Hydrolyses starch into maltose
	Cellulase	Endohydrolysis of 1,4-β-D glucosidic linkage in cellulose, a major component of wood and plant fibres
	Xylanase	Cleaves 1,3-β-D-xylosidic linkages of xylan, a polysaccharide found with cellulose
	β- and α-glucosidases	Release glucose, an important energy source for microbial activity
	β- and α-galactosidase	Hydrolysis of melibiose and lactose, respectively
	Invertases, saccharase, sucrase	Hydrolyses sucrose to glucose and fructose, providing energy for microbial activity
3.4	Proteinases	Hydrolyses proteins, releasing amino compounds/important in N cycling and N mineralization
	Peptidase	Hydrolyses dipeptides, releasing 2 amino acids/important in N cycle and N mineralization

Table 6.1. Continued

Enzyme Commission group and subgroup number	Enzyme	Ecological or soil health function
3.5	Asparginase, glutaminase	Act on C–N bonds (other than peptide bonds) on respective amino acids releasing NH_3/ important in N mineralization to provide plant available N
	Amidase	Hydrolyses C–N bonds of amides releasing NH_3/important for N mineralization to provide plant available N form
	Urease	Belongs to group of enzymes acting on C–N bonds of urea, a fertilizer source and a major constituent in urine of grazing animals
3.5	Arylacylamidase	Hydrolyses propanil, which is used as a herbicide
Transferases		
2.4	Dextransucrase	Hydrolyses sucrose, releasing glucose and fructose
2.8	Thiosulphate *S*-transferase (rhodanese)	Performs intermediate step in oxidation of elemental S which is found in small amounts in soils or is added as a S fertilizer
Lyases		
4.1	Glutamate decarboxylase	Hydrolyses aspartic acid
4.1	Tyrosine decarboxylase	Hydrolyses tyrosine, a product of proteinase activities and involved in N mineralization
4.3	L-Histidine ammonia lyase	Deaminates histidine and involved in N mineralization
Broad spectrum enzymes assay	Fluorescein diacetate hydrolysis	Provides general indicator of soil hydrolytic activity by assaying 3′,3′-diacetylfluorescein hydrolysis which is carried out by proteases, lipases, and esterases

Relationship of enzymes to biological properties in soils

Some studies have shown various microbial indicators to be correlated with the activity of dehydrogenase, protease, cellulase, phosphatase and urease (Laugesen, 1972; Laugesen and Mikkelsen, 1973; Nannipieri *et al.,* 1978; Ross and Cairns, 1982; Tiwari *et al.,* 1989). Alef *et al.* (1988) have proposed arginine ammonification as a biological index. In a study based on 22 soils from Germany and Austria, they reported this to be a relatively simple assay that showed high correlations with biomass C, heat output, soil ATP content, and soil protease activity ($r>0.90$) but not with viable microbial counts. In an earlier study, Frankenberger and Dick (1983) evaluated 11 enzymes in ten diverse soils for their potential relationships with microbial respiration, biomass, viable plate counts and other soil properties. They found that alkaline phosphatase, amidase and catalase activities were correlated with both microbial respiration and total biomass but not to microbial plate counts in glucose-amended soils. The lack of correlation with microbial plate counts suggested that enzyme activities were associated with microorganisms which were active in soil but whose numbers were not accurately reflected by the viable plate counts or whose activity was unrelated to the plate counts.

There are many exceptions where microbial biomass or activity was not closely related to enzyme activities. In some cases, this may be due to the choice of enzyme assay or the inadequacy of soil biological measurements. Soil enzymes that have a significant abiontic activity may confound attempts to correlate enzyme activity with measurement of the microbial biomass because the enzyme assay can not distinguish activity from the viable cells and abiontic enzymes. A more appropriate enzyme is one directly associated with microbial activity and is rapidly inactivated with cell death. Furthermore, most enzymes perform a very specific reaction, and therefore only a small fraction of the population may possess that enzyme at any given time.

The most widely studied enzyme indicator of soil biological activity is dehydrogenase because it should exist only in viable cells. Dehydrogenases use oxygen directly as a hydrogen acceptor (aerobic dehydrogenases) or to operate through other hydrogen acceptors (anaerobic dehydrogenases). However, it has not correlated consistently with microbial activity (Skujins, 1978; Frankenberger and Dick, 1983). Howard (1972) presented data that provided a potential explanation of these results. Using oxygen uptake to calculate theoretical dehydrogenase activity, he showed that the observed dehydrogenase activity was substantially less than the theoretical dehydrogenase activity. He hypothesized that extracellular phenol oxidases, which are known to exist in soil, also may act as alternative hydrogen acceptors. Similarly, Bremner and Tabatabai (1973) found that some common anions in soil, such as nitrate, reduced measured dehydrogenase activity, suggesting these were also acting as alternative hydrogen acceptors. With alternative hydrogen acceptors in the soil, the terminal hydrogen acceptor for dehydrogenase systems in the assay, 2,3,5-triphenyltetrazolium chloride (TTC) (produces coloured triphenylformazan together with hydrogen chloride) is not the sole hydrogen acceptor in soil. In this case, dehydrogenase activity is underestimated. Another problem with this assay is that Cu can interfere with the analytical procedure so that soils high in soil solution Cu or that received Cu-contaminated amendments will show artificially low dehydrogenase activity levels (Chander and Brookes, 1991).

Protease is often well correlated with microbial biomass (Nannipieri *et al.*, 1978; Asmar *et al.*, 1992) because it apparently exists only in microbial cells or is excreted into soil solution as an extracellular enzyme. Extracellular protease does not survive longer than six to seven days in soils (Nannipieri *et al.*, 1978; Asmar *et al.*, 1992) and apparently does not stabilize in the abiontic form. Extracellular protease activity in soil solution is highly correlated with ATP content ($r = 0.74$) and total bacteria counts ($r = 0.94$) (Asmar *et al.*, 1992). Besides showing a relationship between enzyme activity and soil biology, it also provided evidence for a functional role of extracellular enzymes as hypothesized by Burns (1982), i.e. extracellular enzymes begin hydrolysing compounds outside the cell that may be too large or insoluble for microorganisms to use directly.

Enzymes and ecological or inhibition dose models

Soil enzyme activity has potential to assess the impacts of pollutants on specific processes in soils and the ecological dose value (ED_{50}) provides a mechanism for quantifying these impacts. The ED_{50} is analogous to LD_{50} (lethal dose at 50% kill rate) used for assessing the toxicity of substances on animal and human life. Applying this to enzyme activity would mean that an ED_{50} value would be the pollutant (inhibitor) concentration required to cause a 50% inhibition of enzyme activity in soil.

This approach, first applied to soils by Babich *et al.* (1983), uses mathematical models to assess the impact of heavy metals on microbially mediated processes in soils. Two general models have been used: the sigmoidal dose-response curve model as defined by Haanstra *et al.* (1985); and the Michaelis-Menten kinetic model as defined by Babich *et al.* (1983) (see Appendix and Fig. 6.5 for details of these two models).

The sigmoidal dose-response model has been used to assess heavy metal impacts on urease (Doelman and Haanstra, 1986), phosphatase (Doelman and Haanstra, 1989), and arylsulphatase activities (Haanstra and Doelman, 1991). For example, Haanstra and Doelman (1991) found that the ED_{50} (i.e. the toxicity) of heavy metals was highest in sand and sandy loam and lowest in sandy peat soils. They were also able to discriminate toxicities among various heavy metals added to soils and changes in ED_{50} over time. They went on to suggest that this model be used to revise critical heavy metal concentration standards for soils.

Speir *et al.* (1995) argued that the Michaelis-Menten kinetic model is more appropriate for assessing toxicities than the sigmoidal dose-response model because an enzyme kinetic model has physical interpretation whereas the sigmoidal model has no physical interpretation. Secondly, they pointed out that by plotting concentrations on a logarithmic scale (as is done in the sigmoidal model), the initial section of the curve becomes greatly elongated which can mean that the lowest concentration tested may be well beyond this section of the curve. In this case, a simple plot of activity against metal concentration might have yielded a rectangular hyperbola, which would be predicted by simple kinetic models of inhibition. They were able to rank the sensitivity of soil biological properties to Cr(VI) using an ED_{50} with a Michaelis-Menten kinetic model in the following order: denitrification > dimethyl sulphoxide-reducing activity > sulphatase activity ≅ biomass C > phosphatase activity > urease activity > respiration. They were also able to identify that mineral surface areas, organic matter content, and CEC were sensitive soil properties that affected the ED_{50}.

So far, this approach has been used only to assess heavy metal pollution of soils. Presumably it could be used, on a relative basis, to quantify the impact of other soil pollutants or soil perturbations (e.g. mining) on soil enzyme activities.

Enzyme Activity as a Biological Indicator of Soil Management, Pollution or Perturbations

There is growing evidence that soil biological properties are sensitive to soil management and environmental stresses (Dick, 1992). Serious attention to the role of soils in environmental issues may change previously-held environmental standards. A good example is heavy metal standards developed for municipal waste applications to soils which are based on plant uptake and animal health. As Brookes (1995) pointed out, applications of European Community standards for heavy metal concentrations of soils results in significant negative impacts on microbial biomass and activity, indicating the greater sensitivity of the soil to these impacts than of plants or animals.

The soil microbial component and soil enzyme activities are attractive as indicators for monitoring various impacts on soils because of their central role in the soil environment. An important question is what constitutes a 'significant' impact on a soil biological parameter because these properties typically vary widely under natural conditions. For example, Domsch *et al.* (1983) in their survey of reports on bacterial fluctuations under field conditions showed that reductions of 90% are common. These extremes in variation occur naturally due to fluctuations in water potential, gas exchange, and temperature. They recommended that a stress that results in full recovery in <30 days is normal but >60 days would indicate a severe impact that merits investigation. This is an important concept that needs further consideration and should be part of the criteria for developing soil health standards.

This section will discuss the soil health relative to the impact of soil pollution (xenobiotics), land management and physical perturbation on soil enzyme activities.

Xenobiotics

Xenobiotics are, by definition, unnatural compounds (e.g. pesticides, industrial wastes) but the wider definition includes naturally occurring compounds (e.g. heavy metals) that are synthesized or are present in unnaturally high concentrations in the environment (Skladany and Metting, 1993). Such compounds are of crucial concern in the soil environment as they could affect many biological and biochemical reactions in soils.

Pesticides

Pesticides including herbicides, fungicides, etc. introduced into the soil environment have potential to affect non-target organisms and soil biochemical processes. Pesticide applications have been shown to have little or no effect on enzyme activity in soils (Burns, 1978; Ladd, 1985).

When pesticides enter the soil environment, they are subject to four microbial processes (Bollag and Liu, 1990): (i) biodegradation (pesticide serves as a substrate for growth); (ii) cometabolism (pesticide transformed by metabolic reactions but does not serve as an energy source for microbes); (iii) polymerization or conjugation (pesticide reacts with other pesticides or soil organic compounds); and (iv) accumulation in microorganisms. Additionally, pesticides can have secondary effects on microorganisms by inducing changes in the soil environment such as changes of pH, or redox potential. Biodegradation is considered to be the primary mechanism for transforming pesticides in soils and is largely facilitated by heterotrophic bacteria and actinomycetes, some autotrophic bacteria, fungi, and specific protozoa (Cheng, 1990). Indeed, phenoloxidases found in soils appear to protect soil 'health' by detoxifying pesticides by the mechanisms listed above (Filip and Preusse, 1985).

Pesticides also are subject to abiotic reactions (Pierzynski *et al.*, 1994). In soils, pesticides or their degradation products can be sorbed by clay and metal oxide surfaces, and by humic substances. In turn, sorbed compounds can affect bioactivity and bioavailability of pesticides or degradation products or later be desorbed. Pesticides can be transformed by abiotic reactions in the liquid or at the liquid–solid interface, most commonly by hydrolysis and redox reactions. This is the likely basis for the varied responses of pesticides on soil biology among different soil types. Chemical reactions may affect the availability and/or toxicity of pesticides to biological systems or processes.

When pesticides are applied at recommended field rates, short-term studies often show an initial stimulatory, but small, effect on dehydrogenase activity: this may or may not occur with other enzymes. An example of such a response

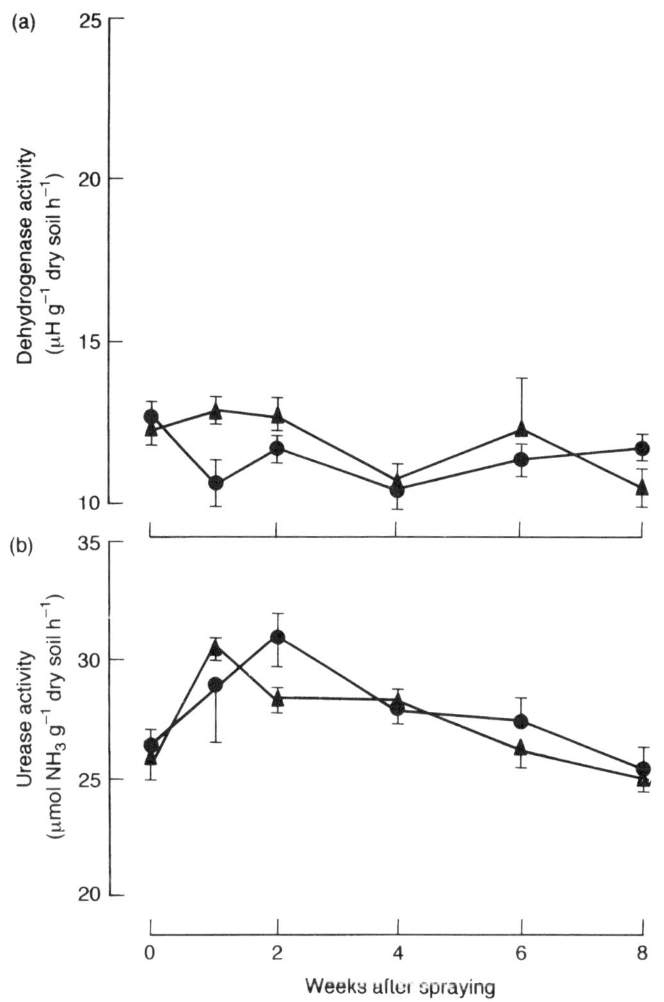

Fig. 6.1. The effect of 21.6 kg^{-1} ha^{-1} glyphosate on (a) dehydrogenase and (b) urease activities in soil. (●) Control; (▲) glyphosate. Bars show standard errors. (Adapted from Davies and Greaves, 1981.)

to glyphosate [*N*-(phosphonomethyl)glycine] is shown in Fig. 6.1. Tu (1981), in a test of 32 pesticides, found that except for fumigants which depressed dehydrogenase activity, the other insecticides, fungicides, and herbicides stimulated soil dehydrogenase activity after two weeks incubation. Similar results were found by Baruah and Mishra (1986) on a rice field soil for three herbicides (2, 4-dichlorophenoxy acetic acid, 2,4-D; 2-chloro 2′6′ diethyl-n-butoxy-methyl acetanilide, butachlor; and 2-chloro-1-3 ethoxy-4-nitrophenoxy-4 benzene, oxyfluorfen). A stimulatory effect suggests that pesticides can be a substrate for metabolism and stimulation of microbial activity. This is corroborated by studies on various pesticides that have shown concurrent increases in other measures of biological activity such as CO_2 evolution (Baruah and Mishra, 1986; Davies and Marsh, 1977), and O_2 consumption (Tu, 1992). Soil-free extracts of soil have been shown to degrade chloroanilines (Bartha and Bordeleau, 1969) and malathion (Getzin and Rosefield, 1968).

In relation to the effects of pesticides in soils, the two most widely-studied enzymes other than deydrogenase are phosphatase and urease. Again, short-term studies involving applications of pesticides to soils at recommended dosages for periods ranging from a few days up to 8 weeks have shown slight increases or no significant effect on activity of these two enzymes (Davies and Greaves, 1981; Baruah and Mishra, 1986; Tu, 1992, 1993). These results might be expected because these enzymes are known to exist as abiontic enzymes and thus, unless there was a direct effect on the enzyme reaction, there should be little effect on an abiontic enzyme.

To overcome the confounding effect of the multiple sources of isoenzymes in soils, studies have been conducted on the effect of pesticides on pure enzymes. Gianfreda *et al.* (1993) studied the effect of three herbicides and one insecticide on pure enzymes in free solution and found that responses could not be generalized and were enzyme and pesticide specific. For example, glyphosate and paraquat showed a marked activation of invertase activity but urease and phosphatase activities were unaffected by these pesticides. Carbaryl inhibited urease and invertase activities but had no effect on the activity or kinetics of acid phosphatase. Atrazine did not affect the kinetics of urease, phosphatase or invertase except at very high concentrations. Further work showed that the 'state of an enzyme' can affect its activity response to the presence of a pesticide. Purified urease was unaffected by paraquat or glyphosate, but urease activity complexed on montmorillonite increased in the presence of paraquat or glyphosate (Gianfreda *et al.* 1994). It was hypothesized that the pesticide displaced some inactive/immobilized urease from the clay surface which regained catalytic capabilities upon release into solution. These simplified systems show the complexity of the mechanisms involved in pesticide–enzyme interactions.

When pesticides are applied to soils at very high concentrations such as when there is an accidental spill, enzyme activities have been significantly affected. Alachlor (10,000 mg kg^{-1}) alone or in mixture with atrazine and metolachor severely depressed dehydrogenase activity for 125 days whereas esterase was only affected by the herbicide mixture (Dzantor and Felsot, 1991). Although bacterial numbers recovered, fungal numbers were still inhibited 90 days after the pesticide was applied. Adding the herbicide imaxethapyr at 100× the recommended rate to soil showed decreases in microbial biomass-C and dehydrogenase activity whereas hydrolytic enzyme activities (protease and 3,6′-diacetylfluorescein hydrolysis, FDA; a broad spectrum enzyme assay, see Table

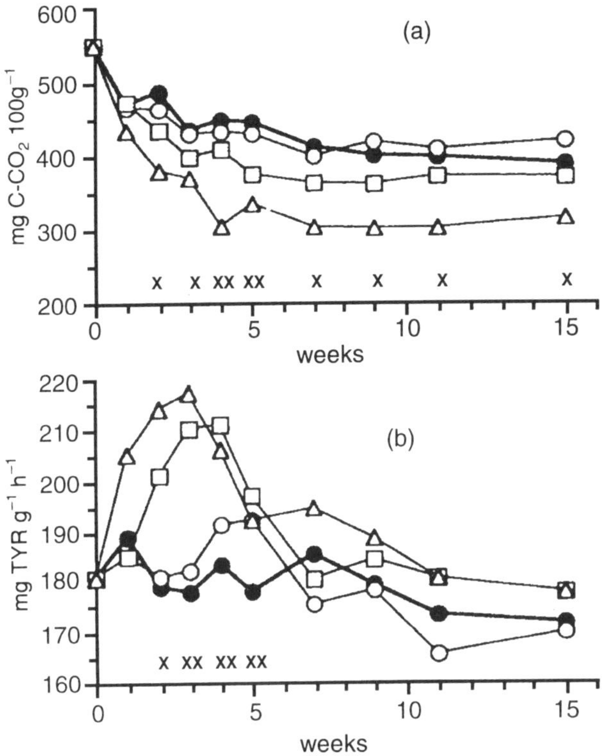

Fig. 6.2. Biomass-C content (a), and protease activity (b), in control soil (●) and in soil treated with imazethapyr at the recommended field rate (○), and at 10 (□) and 100 (△) times the field rate in a laboratory experiment lasting 15 weeks. Each value is the mean of three replicates. X and XX indicates means that at 100-fold and at both 10- and 100-fold application rates are significantly different from control ($P<0.05$, Duncan's multiple range test), respectively. (Adapted from Perucci and Scarponi, 1994.)

6.1) showed corresponding increases up to 15 weeks after application of the herbicide (Perucci and Scarponi, 1994) (Fig. 6.2). In this case, Perucci and Scarponi (1994) hypothesized that the hydrolytic enzymes were released during lysis of microbial cells killed by imaxethapyr.

The difference between the enzyme response at low and high pesticide concentration is due to the persistence of the pesticide which results in the inability of microbial populations to degrade or flourish in its presence (Davidson *et al.*, 1980; Junk *et al.*, 1984; Schoen and Winterlin, 1987). For example, Junk *et al.* (1984) found that alachlor and atrazine applied alone or in combination at rates of 15 or 300 mg kg^{-1} soil showed no degradation after 68 weeks. Dzantor and

Felsot (1991) found that an emulsifiable concentrate of alachlor (10,000 mg a.i. kg soil^{-1}) was stable for 337 days which caused depressed bioactivity (dehydrogenase and esterase activity) and reduced bacterial and fungal numbers. Felsot and Dzantor (1995) suggested that high pesticide concentrations cause biotoxicity and that addition of nutrients to pesticide-contaminated soils appeared to stimulate pesticide degradation. This further indicated that some specific microbial species could survive these high concentrations and that readily-available nutrients may be limiting in pesticide-contaminated soil.

Although single applications of pesticides have shown minimal effects on soil biological properties, it may be more important to consider the effects of repeated applications over many years. Voets *et al.* (1974) showed that long-term atrazine applications significantly reduced the activity of phosphatase, invertase, β-glucosidase, and urease in soils. They hypothesized that this was due to reduction of biological activity rather than a direct effect on the catalytic behaviour of these enzymes. Rai (1992) found that the effect of long-term (15 years) applications of 2,4-D on dehydrogenase and urease activity depended on application formulation. The water-based dimethylamine salt formulation plus 2,4-D showed little effect on the activity of these enzymes over the control whereas the 2,4-D oil-based isoctyl ester formulation significantly depressed activity of these enzymes. This was thought to be due to toxic metabolite(s) formed during degradation of the ester (Rai, 1992). Alternatively, the amine may stimulate microbial degradation of 2,4-D as evidenced by increased soil respiration (Rai, 1992).

Heavy metals

Environmental pollution of soils with heavy metals and trace elements has been reported to have toxic effects on soil biology and to affect soil biochemical processes. Sources of these contaminants can come from repeated applications of sewage sludge, municipal wastes, smelting wastes, electroplating industry wastes, impurities in fertilizers, and deposition from air pollutants such as burning of fossil fuels and various industrial activities.

Enzyme reactions are inhibited by metals: (i) through complexation of the substrate; (ii) by combining with the protein-active groups of the enzymes; or (iii) by reacting with the enzyme-substrate complex. The mode of action of metals varies with enzyme and little is known about the exact mechanisms of interactions of metals and the multitude of enzymes that can exist in soils. Some ions can act as co-factors or activators and at certain concentrations can increase the activity of some enzymes (e.g. Mg, Ca, Ba, Co, Ni, Zn and Mn for pyrophosphatase; Tena *et al.*, 1981; Dick and Tabatabai, 1983). Sulphydral groups of enzymes serve as catalytic sites or as groups involved in maintaining the correct conformation of the protein. Metals can react with sulphydral groups causing inactivation or inhibition of enzyme activity (Shaw and Raval, 1961).

In studies where a wide range of trace elements have been tested, Hg, Ag, Cr and Cd have generally caused the greatest inhibition of sulphatase, L-glutaminase, cellulase, L-asparginase, and β-glucosidase (Frankenberger and Tabatabai, 1981, 1991a,b; Eivazi and Tabatabai, 1990; Deng and Tabatabai, 1995). Another important factor associated with the toxicity of heavy metals is the oxidation state of an element. For example, Cr(III) is significantly less toxic to soil biological properties than Cr(VI) because of Cr(III)'s tendency to undergo precipitation, complexation or fixation reactions in soils (Ross *et al.*, 1981). Cr(VI) is a powerful oxidizing agent which can cause enzyme degradation by oxidation of structural linkages and has been shown to cause irreversible inhibition of urease in soils (Speir *et al.*, 1995).

Enzymes vary in their degree of inhibition by trace elements. Of the soil enzymes tested so far, arylsulphatase appears to be the most sensitive to trace elements whereas acid phosphatase, urease and invertase are less affected by metals (Al-Khafaji and Tabatabai, 1979; Bardgett *et al.*, 1994; Yeates *et al.*, 1994). Thus, arylsulphatase activity could be used as a sensitive indicator of soil pollution by heavy metals and other trace elements.

The nature and degree of inhibition of soil enzymes by heavy metals is strongly related to soil type (Speir *et al.*, 1992). Greater inhibition of arylsulphatase and phosphatase has been shown in soil with low particle surface area, CEC and organic matter content which is likely to reduce the potential of the soil to inactivate metals via complexation or sorption reactions and increases the availability of metals to affect enzyme processes (Speir *et al.*, 1995).

Animal and municipal waste

Long-term addition of farmyard manures to soil stimulates enzyme activities (Yaroschevich, 1966; Khan, 1970; Verstraete and Voets, 1977; Dick *et al.*, 1988a; Goyal *et al.*, 1993; Kandeler and Eder, 1993). These studies have often shown that manure-amended soils have higher microbial biomass, N mineralization potential, and microbial activities that are related to some enzyme activities. This would be expected as elevated levels of C inputs and other nutrients would stimulate biological activity and stabilization of abiontic enzymes. Similar results have been found for municipal refuse (Werner *et al.*, 1988; Perucci, 1992). Although these types of soil amendments can contain enzymes, the stimulation of enzyme activity is likely to be related to increased microbial growth (Martens *et al.*, 1992) because introduced enzymes are rapidly degraded or denatured in soils (Dick *et al.*, 1983).

Fertilizers

A priori it would be expected that increased plant biomass resulting from fertilizer additions to soils should stimulate soil enzyme activities. Limited research at long-term sites has not shown this to be the case.

Khan (1970) showed that 40 years of annual applications of NPKS fertilizers increased soil enzyme activities but this increase was not statistically significant. Repeated inorganic N applications in the greenhouse over a period of 305 days showed no significant effects on β-glucosidase and protease activities in soil whereas organic amendments stimulated enzyme activities (Fauci and Dick, 1994).

Dick *et al.* (1988a), on soil samples from plots with treatments that had been in place since 1931, showed an effect of increasing rates of ammonia-based N fertilizer decreasing the activity of amidase and urease (both enzymes are involved in the N cycle). Conversely, activities of other enzymes not directly involved in the N cycle (arylsulphatase and β-glucosidase) did not correlate with N fertilizer input levels. A feedback mechanism was hypothesized as suppressing production of enzymes whose reaction product (NH_4) was continually added by inorganic fertilizers. This was later confirmed by McCarty *et al.* (1992) who showed that NH_4 repression of urease activity was real, but that the NH_4 effect was indirect and due to byproducts of NH_4 assimilation.

Phosphorus fertilization under field conditions has been shown to depress phosphatase activity in agricultural systems (Mathur and Rayment, 1977; Spiers and McGill, 1979) and forest systems (Clarholm, 1993). But this is likely to be a function of soil type as shown by Mathur and Rayment (1977) where a low organic matter soil increased phosphatase activity with P fertilization, but a soil with a higher organic matter content amended with P fertilizer showed no change in phosphatase activity. In cases of phosphatase inhibition, this has been correlated with high levels of PO_4 in soil solution (Chunderova and Zubets, 1969). Phosphate may inhibit the synthesis of microbial phosphatases and also it is known that orthophosphate is a competitive inhibitor of acid and alkaline phosphatase activity (Juma and Tabatabai, 1978).

Industrial amendments or contaminants

Large amounts of combustion by-products (fly ash) are produced by coal-fired power plants and other industries. There is growing interest in applying these to soils. These materials generally have a high pH and can contain high levels of trace elements. Soil enzyme activities have been used to determine whether there are positive or negative effects on soils. In an acid soil (pH 4.2), McCarty *et al.* (1994) found that fly ash had a positive effect on hydrolytic enzyme activities (urease and arylsulphatase) which closely correlated with the effects of $CaCO_3$. This suggested the positive effect was due to the liming effect (increase in pH) on this acid soil. Conversely, fly ash had a negative effect on dehydrogenase activity which would indicate that the overall microbial population was being suppressed. This might suggest there was either a selective stimulation by a portion of the remaining population that produced hydrolytic enzymes or there was induction of these hydrolytic enzymes. Alternatively, the dehydrogenase assay may have been affected by Cu in the ash. Chander and Brookes (1991) found there was a chemical reaction between triphenylformazan (the end product measured in dehydrogenase activity) and Cu, meaning that soils amended with materials containing Cu severely underestimate dehydrogenase activity. Although McCarty *et al.* (1994) do not report the Cu content of the fly ash, high amounts can be present in this material (Pichtel and Hayes, 1990).

In contrast to the findings of McCarty *et al.* (1994), Pichtel and Hayes (1990) found that inhibition of phosphatase, sulphatase, dehydrogenase and invertase activities increased with increasing ash rates (0 to 20%, w/w) but that catalase activity was unaffected in a 28-day incubation. There was a corresponding decrease in respiration and total bacterial, actinomycete and fungal counts. These results are likely to be due to the high concentration of some heavy metals found in the ash (e.g. Cu, Cd, Al, Zn, Ni). The pH of this soil was 6.2 and the liming benefits of ash would not be nearly as important as in the soil used by McCarty *et al.* (1994) (their soil had a pH of 4.2 so that liming would cause significant precipitation of metals). This shows that soil type is important when determining the impact of a given soil amendment and that soil enzyme activities are generally useful in differentiating the effects of these amendments on soil health.

Disposal of pulp and paper mill effluent on agricultural land is increasing as a way to recycle these materials in an ecological manner. Kannan and Oblisami (1990) investigated the effect of pulp and paper mill effluent on an alkaline soil (pH 7.5) that was cropped to sugarcane and received paper mill effluents as an irrigation application for periods ranging from 0 to 15 years. After two years, there was > threefold increase in total C suggesting that the applied material was recalcitrant, and only invertase and dehydrogenase activities of the enzymes tested were significantly higher than the control. Phosphatase and amylase after three years and cellulase after 15 years were significantly different than the control. Of the enzymes tested, invertase was the most sensitive early indicator of changes with a 3.3-fold increase over the control after two years of paper mill effluent applications. After 15 years amylase was the highest with an 8.6-fold increase compared to the next highest activity of invertase which had a 6.6-fold increase over the control. Apparently there was a liming effect by this effluent on the soil, as soil pH increased to 8.3 after three years of effluent irrigations.

Hydrocarbons

Accidental spills or leakage from storage facilities of hydrocarbons is a common and widespread problem. Soil enzyme activities hold potential for assessing the impact of hydrocarbons on soils and the effectiveness of remediation of soils contaminated with hydrocarbons. However, there is little information on the interaction of hydrocarbons and soil enzyme activities (Song and Bartha, 1990). In one example, Song and Bartha (1990) showed that in the absence of bioremediation (addition of fertilizer, tillage, and liming), jet fuel at 50 and 135 mg g^{-1} soil, inhibited FDA hydrolysis during an 18-week incubation period (Fig. 6.3). With bioremediation, during the first two weeks, FDA hydrolysis was initially inhibited but was then stimulated reaching a peak after most of the jet fuel had been mineralized (4 and 15 weeks for 50 and 135 mg g^{-1} soil, respectively). These results support the potential for enzyme assays to monitor changes in soil health after an environmental accident. In this case, there appeared to be a

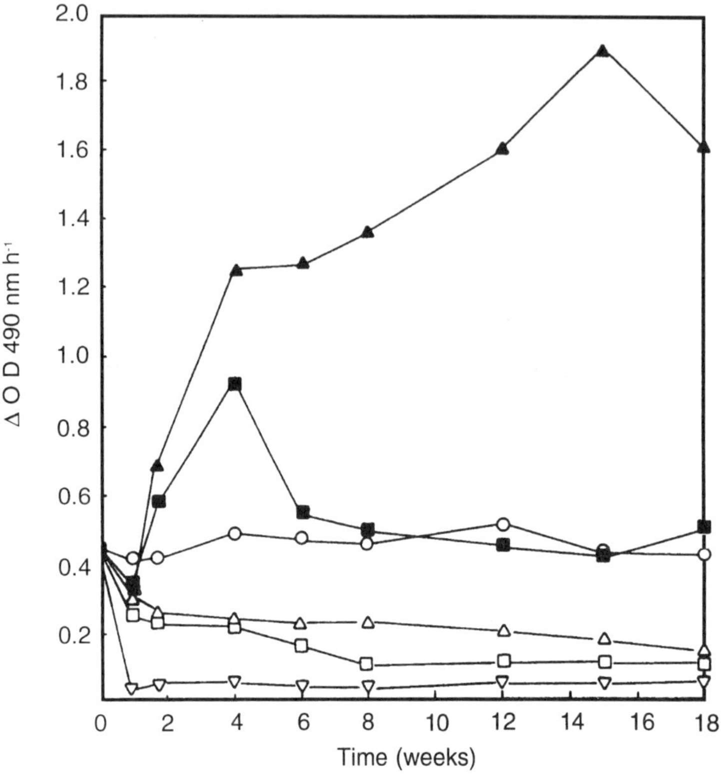

Fig. 6.3. Changes in FDA hydrolysis activity with time in surface soil contaminated by jet fuel. Symbols: ○, no jet fuel, no bioremediation; □, 50 mg of jet fuel g of soil^{-1}; ▽ 135 mg of jet fuel g of soil^{-1}; ■, 50 mg of jet fuel g of soil^{-1} plus bioremediation; ▲, 135 mg of jet fuel g of soil^{-1} plus bioremediation; △, poisoned control (1% $HgCl_2$). (Adapted from Song and Bartha, 1990.)

long-lasting effect of jet fuel on enzyme activity but with bioremediation this inhibition was short-lived with a corresponding increase in FDA hydrolysis, inferring a restoration of soil health. However, more studies on a variety of soil enzymes, soil types and types of hydrocarbons are needed to fully determine the potential of soil enzymes to serve as indicators of the impacts of hydrocarbons on soils.

Atmospheric pollution of soils

Simulated acid precipitation with extreme soil pH values (3.0–3.7) and prolonged (692 days) conditions in laboratory incubations can affect soil enzyme activities (Bitton and Boylan, 1985), but under more realistic field conditions no effect on dehydrogenase, phosphatase or urease activities was observed (except for nitrification which was suppressed) (Bitton *et al.*, 1985).

Studies along industrial air pollution gradients (N and S gases) in Finland showed that cellulase activity was better correlated with the pollution gradient than soil respiration (Ohtonen *et al.*, 1994). Comparison of soil adjacent (15 m) to a major motorway in India showed that enzyme activities (cellulase, amylase) of forest litter were significantly inhibited compared to a site 500 m from the roadway. This was partially attributed to elevated levels of heavy metals in the soil adjacent to the roadway (Joshi *et al.*, 1993).

Plant residues and vegetation management

Cropping systems that return elevated levels of green manures/crop residues significantly increased the activity of a wide range of soil enzymes over unamended soil (Verstraete and Voets, 1977; Dick *et al.*, 1988a; Martens *et al.*, 1992; Jordan *et al.*, 1995). Although these plant amendments contain enzymes, the increase in activity in soils amended with organic residues is likely to be due to stimulation of microbial activity rather than direct addition of enzymes from the organic sources (Martens *et al.*, 1992). Crop rotations which, over time, provide greater plant diversity than monoculture systems generally have a positive effect on soil enzyme activities (Khan, 1970; Blagoveshchenskaya and Danchenko, 1974; Dick, 1984; Bolton *et al.*, 1985). This effect may be due to stimulation of microorganisms in the rhizosphere and improved physical conditions of soils in crop rotations, particularly when rotations contain legume species (Miller and Dick, 1995a,b). Plant roots stimulate enzyme activity (Castellano and Dick, 1991; Tabatabai and Bremner, 1970) by creating favourable microhabitats (e.g. increased compounds, water, porosity) for microbial activity.

Altering the native vegetation through cropping or overgrazing can result in degraded soils. Garcia *et al.* (1994) found that soil enzyme activities were lower in an arid region of Spain than those reported in more humid regions and that activities of dehydrogenase, protease (*N*-a-benzoil-L-argininamide substrate), phosphatase, and β-glucosidase were better indicators of soil degradation than total organic C. In northeastern India, soils degraded by slash-and-burn agriculture had consistently lower levels of dehydrogenase, phosphatase and urease activity than soils in natural forest (Jha *et al.*, 1992). In both of these studies, enzyme activities were closely related to other microbiological measurements such as microbial biomass C or microbial counts.

Physical perturbations

Soil enzyme activities have been responsive to tillage treatments. Gupta and Germida (1988) compared Canadian soils cultivated for 69 years with adjacent

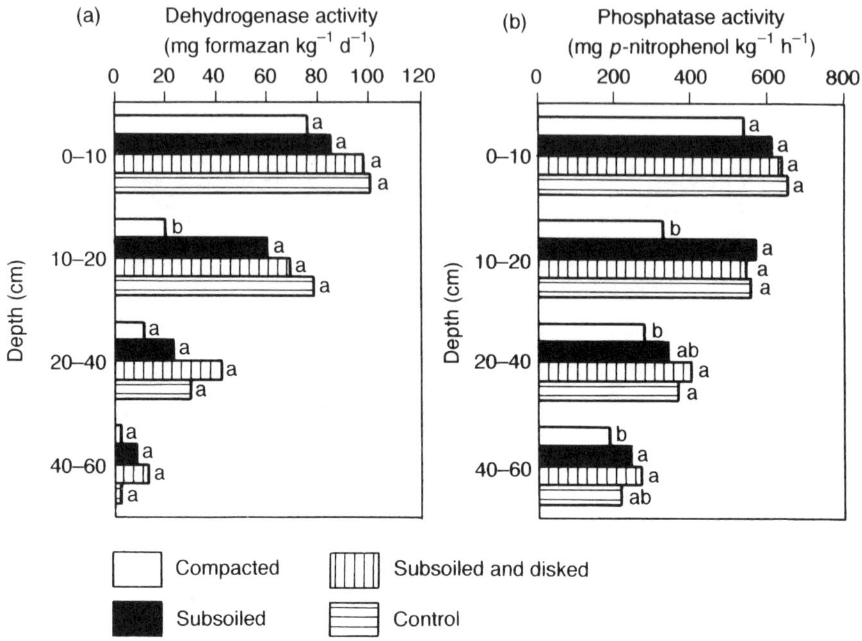

Fig. 6.4. Distribution of phosphatase (a) and dehydrogenase (b) activities in the soil profile as affected by soil compaction and tillage. Bar graphs within each soil depth followed by the same letter are not significantly different at the 0.05 probability level according to Duncan's new multiple range test. (Adapted from Dick *et al.*, 1988b.)

grassland and found that cultivation depressed phosphatase (49%) and arylsulphatase (65%) activity. Further investigation of enzyme kinetics on these soils indicated that cultivation changes the origin of arylsulphatase activity as evidenced by significant differences in K_m values (Farrell *et al.*, 1994). Effects of cultivation on soil biology can be modified by the type of tillage. Conservation tillage practices produce less soil disturbance than conventional tillage systems and have higher levels of enzyme activities in the surface soil (<10-cm depth) (Klein and Koths, 1980; Doran, 1980; Dick, 1984; Angers *et al.*, 1993).

Compacted soils would be expected to have low O_2 concentration and low air-filled porosity which decreases microbial activity (Parr and Reuszer, 1962; Linn and Doran, 1984; Bridge and Rixon, 1976). Dick *et al.* (1988b) compared the effect of various soil tillage treatments applied to compacted soils in a clear-cut forest in western Oregon, USA. Four years after the treatments were initiated, phosphatase activity was significantly lower in the compacted soil compared to the rehabilitated soil (Fig. 6.4) and all enzymes tested (phosphatase, amidase, dehydrogenase, and arylsulphatase) were depressed from 41 to 75% in

Table 6.2. Influence of revegetation on some parameters of high-elevation sites disturbed by pipeline construction as compared to native sites in the Piceance Basin, Colorado. (From Redente and Cook, 1986.)

	Years since disturbance			
Parameters	2	4	27	Native
pH	8.36	7.72	7.35	6.70
Organic matter (%)	0.95	1.49	1.04	2.66
Phosphatase activity (mg of *p*-nitrophenol g soil^{-1} h^{-1})	193	467	690	781
Mineralizable N (mg of NH_4^+-N g soil^{-1})	47.4	63.7	62.7	91.9
Dehydrogenase activity (mg of triphenylformazan g soil^{-1} 24 h^{-1})				
Actual (autochthonous)	37.1	17.7	11.5	13.6
Potential (zymogenous)	25.3	25.9	9.8	15.4
Litter cover (%)	2	0	47	63
Total plant cover (%)	19	20	60	67

soil 10–20 cm deep. Enzyme activities were negatively correlated with soil bulk density ($P<0.05$). Biological parameters such as soil enzyme activities could provide forest managers with a way to identify management practices that are improving soil properties long before such benefits could be measured by tree growth.

Soil enzyme activities have been used extensively to assess the effectiveness of reclamation of soils in drastically disturbed landscapes. Since this has been fully reviewed by Klein *et al.* (1985) and Kiss *et al.* (1993), only an overview and selected examples will be presented. Through production of mining wastes and landscape perturbations from open-pit or strip mining, large areas of the world's land surface are degraded each year and it is estimated that mining affects a larger area than severe water erosion (Kiss *et al.*, 1993). Early work by Pancholy *et al.* (1975) in Oklahoma, USA found that urease and dehydrogenase activity were good indicators of the potential for monitoring the revegetation of a landscape near an abandoned zinc smelter. In Romania, nine years of various soil organic and fertilizer treatments on iron mine spoils showed that there was a close correlation between yield of corn (*Zea mays*) and invertase activity (Dragan-Bularda *et al.*, 1987). Table 6.2 presents the time course for revegetation of disturbed pipeline soil where phosphatase activity was rapidly increasing between two and four years but still had not quite reached the same level as the surrounding soil after 27 years. This same site showed that disturbance caused elevated levels of dehydrogenase activity which suggests there may have been high rates of organic matter oxidation due to exposure of protected organic matter during the perturbation. Revegetation of soils after removal of top soil (Ross *et al.*, 1982) or at a reclaimed coal strip mining site (Ross *et al.*, 1992) in New Zealand showed that invertase and then arylsulphatase were best correlated with phytomass yields within two to three years after establishing vegetation and reclamation treatments. Activities of phosphatase, amylase, and pectinolase have successfully tracked reclamation of coal mine spoils (Hersman and Temple, 1979; Stroo and Jencks, 1982).

The conceptual role of soil enzymes for monitoring the revegetation of disturbed landscapes is substantiated by studies of plant community succession (Pancholy and Rice, 1973a,b; Rice and Mallik, 1977). The succession from abandoned farmland to climax vegetation showed that in the early successional stages, carbohydrases were highest (invertase, amylase and cellulase) which corresponded to the pioneer weed species that dominated at this stage (these species having high cellulose and polysaccharide content). In later successional stages carbohydrases declined when perennial grasses and woody species dominated, which was related to their higher lignin contents. Dehydrogenase and urease increased with plant community succession which probably reflected the general build-up of the microbial biomass. These results were independent of soil organic matter which was unaffected by successional stage showing the importance of the type of plant residues that are incorporated (substrate) and illustrating the fact that soil enzyme activities are sensitive to changes in vegetation. These findings were strikingly consistent across three different plant successions (i.e. different mixes of plant communities specific to three environments) which represented different climate regimes (ranging from 84 to 178 cm annual rainfall) and different soil orders (Pancholy and Rice, 1973a,b; Rice and Mallik, 1977).

Salinity

Large areas of the world's terrestrial ecosystems have arid or semiarid climates. Through mismanagement of crop land (particularly irrigation) or natural processes, soils in these regions can accumulate soluble salts, causing nutrient imbalances and lower osmotic potential of soil solution which can have significant negative impacts on plant productivity. Excess salinity can affect soil enzyme activities. For example, Garcia *et al.* (1994) found a negative correlation between soil electrical conductivity and dehydrogenase, phosphatase, urease, and β-glucosidase activities among 36 soils in an arid region of Spain. This reduction in enzyme activity can be due to a lower microbial biomass and to direct effects of osmotic potential and specific ions on enzyme activity. In a short-term incubation (one week), Frankenberger and Bingham (1982) ruled out that changing the soil solution pH, with various salts at rates that have negative impacts on agronomic productivity, was not a factor in inhibiting enzyme activities. From this work, they suggested that enzyme proteins are 'salted out' and although they could not conclusively isolate the osmotic effect, there was evidence that salts containing Cl showed more inhibition of enzyme activities than salts containing SO_4. Undoubtedly, both ionic and osmotic effects are important in affecting the proliferation of microorganisms, the synthesis of enzymes, and in changing the ionic conformation of enzymes.

Thermal impacts

Long-term effects of crop residue burning (55 years) have shown no significant effect on the activity of a range of different enzymes (Dick *et al.*, 1988a) which is consistent with the minimal or short-lived effects on microbial populations in the top 2.5 cm of soil after crop residue burning (Biederbeck *et al.*, 1980). From the work of Saa *et al.* (1993), the impact of thermal activity on phosphatase activity was determined by the soil temperature during the burn. In controlled forest burns with temperatures <50°C in the top 5 cm of soil, there was no effect on phosphatase activity, but wild fires which caused much higher soil temperature significantly decreased enzyme activities in soil to a depth of 10 cm. These results show the sensitivity of enzyme activities in assessing the degree of thermal stress which could have practical applications after burning events in determining thermal impacts on soil.

Potential and Limitations of Enzyme Activity as an Indicator of Soil Health

Over the past 20 to 30 years, considerable progress has been made in developing methods for measuring the activity of well over 50 enzymes found in soil. Because many enzymes are substrate-specific and can be chosen from different functional groupings (oxidoreductases, transferases, hydrolases, etc.), there is an opportunity to determine the potential of a soil to carry out a whole range of reactions that may be critical for the functioning of an ecosystem. Alternatively, enzyme activities hold the potential to determine whether a 'stressed or degraded' soil is 'impaired' to carry out specific biochemical processes.

An inherent difficulty for interpreting soil enzyme activities is the lack of understanding of the state and sources of enzymes; biotic vs. abiontic. On the other hand, these characteristics provide for a unique soil assessment that could be an advantage for assessing soil health because it can function as an overall integrative biological index. Secondly, the abiontic component which exists for many enzymes in soils provides an indication of the potential of the soil to complex and stabilize enzymes. This latter ability is likely to be a precursor of organic matter accumulation as it seems that a soil environment favourable for enzyme stabilization would also be conducive for organic matter accumulation.

Since enzyme activity is a much more sensitive measurement than organic C measurements, it could be used as an early indicator of changes in soil organic matter and soil health.

One of the most promising applications of soil enzyme activities is to determine the impact of pollution or severe perturbations on soil health or to assess the success of remediation activities. Development of the Ecological Dose Model (Babich *et al.*, 1983; Speir *et al.*, 1995) provides a means to determine a site specific/relative measure of soil health. Of course, this is possible because in these situations it is possible to have a reference value from adjacent uncontaminated/undisturbed soil. Although many studies have shown that the more subtle effects due to changes in vegetation management, tillage and soil amendments can be discriminated by soil enzyme activities in relatively short periods, this was possible because there was a control or some other treatment (e.g. native soil) for comparison on the same soil type. For many practical applications that involve land managers, there may not be a 'control' to provide a relative comparison. Critical thresholds or ranges of activity values need to be developed to aid interpretation of enzyme assays in determining soil health. The limitation for doing this is that soil enzyme activities naturally, vary widely as a function of soil type (as do most soil properties). To overcome this, internal or relative soil enzyme indexes are needed that would be independent of soil type. Examples might include ratios of enzyme activity to other enzymes, total C or microbial biomass C.

Another area that requires more research is the degree to which spatial and temporal variability affect soil enzyme arrays. To some extent, the effects of spatial variability can be removed by using sampling strategies developed for soil fertility testing (e.g. composite and stratified sampling of subareas at a sampling site). However, enzyme assays that show wide seasonal or year-to-year fluctuations would be undesirable because it would be difficult to measure changes in activity due to soil management. Preliminary reports suggest some enzymes are more stable than others (see review by Dick, 1994); however, more studies on a range of enzymes in diverse ecosystems are needed to establish the temporal variability of soil enzyme activities.

It is important to choose the appropriate soil enzyme for assessing soil health for a given situation and to know the limitations or confounding factors that might affect interpretation of the results. As pointed out in the above discussion, other factors such as fertilizers or heavy metals can stimulate or depress enzyme activity in a way that is independent of soil health. Additionally, one must always interpret results within the context of soil type and local environment.

A major advantage of enzyme assays over most soil biological measurements is that many assays are relatively straightforward and do not require sophisticated instrumentation. Relatively large numbers of assays can be completed on a daily basis. Another advantage is that for many enzymes it is possible to run the assay on air-dried samples while retaining its potential to discriminate between soil management effects (R.P. Dick, Oregon, personal communication). This greatly eases handling/storage problems and provides for a longer post-sampling time period for completion of the assay. These characteristics would allow soil enzyme assays to be run on a commercial basis with suitable market demands. In contrast, most other soil biological measurements are tedious, labour intensive, may require sophisticated equipment (e.g. some biodiversity procedures), and must be run on field-moist samples as soon as possible after soil sampling.

Conclusions

Soil enzyme activities have potential to provide a unique integrative biological assessment of soils and the possibility for assessing the health of the 'living' component of soil. Certain enzymes can exist only in viable cells providing assessments of the activity of the biological component of the soil. Many other enzymes can exist in both viable cells and as extracellular enzymes (abiontic) in soil solution or complexed in the soil matrix. This latter characteristic provides the opportunity to incorporate a 'historical' component into enzyme assays that reflects the cumulative effect of genesis or soil management on soil biology.

Soil enzyme activities have not correlated consistently with crop productivity in agroecosystems. However, under native conditions, enzymes have shown ecological relevance to plant succession and to plant productivity. In general, soil enzyme activities show relationships to biological properties in soils but not always. In this case, it is important to choose the correct enzyme assay that is best related to the soil biological component of interest. Soil enzyme assays can give information on the potential of soils to carry out certain biochemical processes.

Soil enzyme activities have successfully discriminated between a wide range of soil management practices under controlled conditions that include inorganic and organic soil amendments, applications of xenobiotics to soils, and highly disturbed landscapes. The most promising applications for soil enzymes are in assessing pollution impacts on soil, and ecological dose models can be used to quantify this effect. Conversely, in situations where there is no 'control' or native soil comparison (as is often the case for land managers) interpretation of soil enzyme activities relative to soil health remains a challenge. More research is needed across a range of soil types, ecosystems, and soil management practices to calibrate soil enzyme activities or to develop relative soil enzyme indexes that are interpretable, independent of soil type or environment.

Soil enzyme activities are attractive as one measure of soil health because they can provide unique biological information about soils with procedures that are relatively rapid and low cost. Nonetheless, soil enzyme activities should be run in conjunction with other microbial, chemical and physical measurements to fully assess and improve chances of correctly diagnosing soil health.

Appendix

Ecological dose models

Logistic response model

This model, adapted from Haanstra *et al.* (1985) and shown in Fig. 6.5a is described by the relationship expressed as:

$$Y = \frac{c}{1+e^{b(X-a)}} + E \tag{6.1}$$

where Y is the observed enzyme activity at the logarithm of the inhibitor concentration X and c is the calculated initial (uninhibited) level of enzyme activity, determined by the activity of the control sample and of samples with low concentrations. The parameter b is a slope parameter indicating the inhibition rate, and is equal to $4.39/(0.1\ c - 0.9\ c)$. The parameter a is the logarithm of the concentration at which the enzyme activity is half the uninhibited level ($a = 0.5\ c$). E is the stochastic error term, describing the deviations of the observations around the model response curve. For clarity, the logistic curve with the parameters a, b and c is shown in Fig. 6.5a. As the logarithm of the inhibitor concentration of the blanks cannot be calculated ($\ln 0 = -\sim$), a small logarithm value is substituted (e.g. 10^{-3} mmol kg^{-1}, essential for the curve but not for the inhibitor content), where the curve reaches its asymptotic value. The use of the total contaminant content of both blanks and samples eliminates this substitution but obscures the calculation of the effect of the contaminant addition, especially at low concentrations. For computations, the iterative least squares procedure can be used with a statistical program package.

The ED_{50} as discussed by Babich *et al.* (1983) is calculated from the parameter a by taking its inverse logarithm. The ED_{50} is defined as the dose range in which activity decreases from 90 to 10% of the calculated uninhibited enzyme activity.

Michaelis-Menten kinetic model

This model adapted from Speir *et al.* (1995) is a general algebraic expression for the common inhibition kinetics given by Dixon and Webb (1979) of the form $v = c/(1 + bi)$ (Model 1) or $v = c(1 + ai)/(1 + bi)$ (Model 2), when substrate concentration is assumed to be fixed. The constants a, b and c are a combined term and their values depend on the substrate concentration (s), the inhibition constant K_i (a measure of the affinity of the enzyme-inhibitor system), the maximum reaction rate V, the apparent and effective Michaelis constants K_m and, with some types of kinetics, various other constant terms. The terms a, b and c must always be positive with $b > a$, and have physical interpretations which depend upon the type of inhibition involved, e.g. fully competitive, fully non-

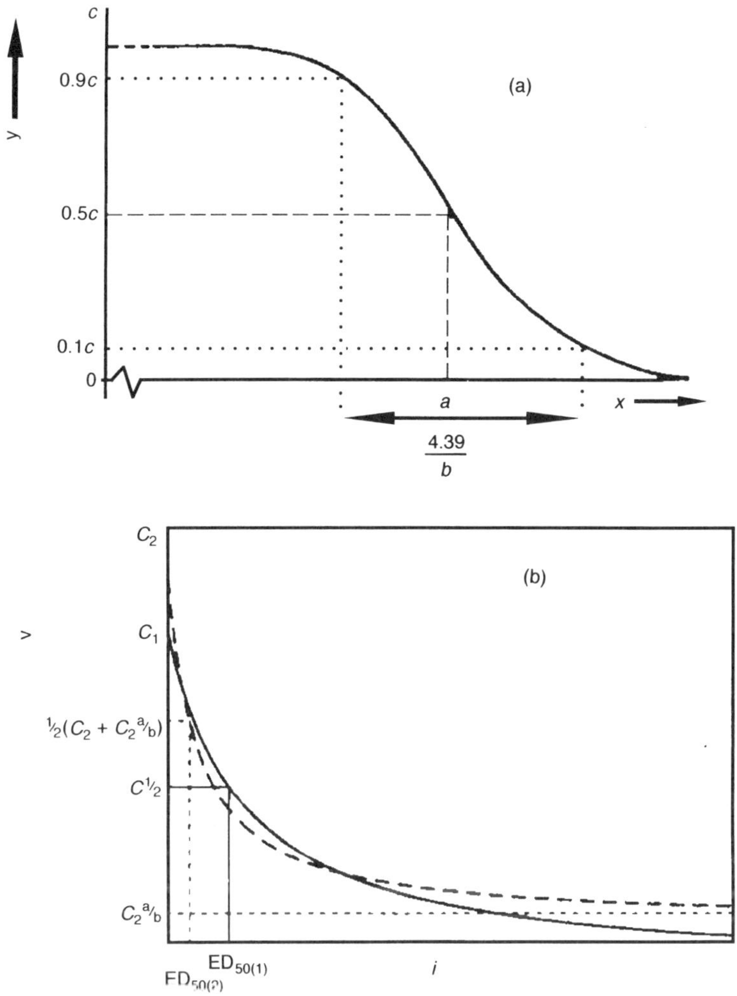

Fig. 6.5. (a) The logistic response curve with parameters *a, b* and *c*. (Adapted from Haanstra *et al.*, 1985.) (b) Michaelis–Menten kinetic models showing relationship between reaction rate (*v*) and inhibitor concentration (*i*), for a typical data set, as described by Models 1 (—) and 2 (--------). The parameters C_1 and C_2, $ED_{50(1)}$ and $ED_{50(2)}$ represent maximum (uninhibited) rates and ED_{50} values for Models 1 and 2, respectively; c_2a/b is the minimum rate (the asymptote) for Model 2. (Adapted from Speir *et al.*, 1995.)

competitive, etc. The first equation (Model 1), when written as a Michaelis equation and transformed as described in Dixon and Webb (1979), prescribes a linear relationship between the reciprocal of reaction rate (*v*) and inhibitor concentration (*I*) at constant substrate concentration. Model 1 describes full inhibition (fully competitive/fully non-competitive). Model 2 describes partial inhibition (partially competitive/partially non-competitive).

In each of the biological and biochemical analyses described above, a single substrate concentration is used, or, where no substrate was added (respiration and denitrification), is assumed. Where data fit Model 1, it is possible to simply derive an ED_{50} (inhibitor concentration or 'dose' resulting in 50% inhibition of activity) (Babich *et al.*, 1983) and, under some inhibition kinetics, to approximate K_i. Where data fit Model 2, K_i cannot be determined, but a derivation of ED_{50} is still possible. For Model 1, ED_{50} is the inhibitor concentration which reduces activity by 50%, and is defined as $ED_{50} = 1/b$. For Model 2, the equation describes a concave rectangular hyperbolic relationship (typical of enzyme kinetics) between v and i, with an asymptote ca/b, parallel to, but above the x-axis. This asymptote represents the minimum possible activity but is never zero. ED_{50} represents the inhibitor concentration which reduces activity from the maximum (uninhibited) value (c) to half of the difference between c and the minimum possible activity, and is $ED_{50} = (1 - a/b)/(b - a)$. An illustration of the relationships between v and i described by Models 1 and 2 is presented in Fig. 6.5b.

References

Alef, K., Beck, Th., Zelles, L. and Kleiner, D. (1988) A comparison of methods to estimate microbial biomass and N-mineralization in agricultural and grassland soils. *Soil Biology and Biochemistry* 20, 561–565.

Al-Khafaji, A.A. and Tabatabai, M.A. (1979) Effects of trace elements on arylsulfatase activity in soils. *Soil Science* 127, 129–133.

Angers, D.A., Bissonnette, N., Légère, A. and Samson, N. (1993) Microbial and biochemical changes induced by rotation and tillage in a soil under barley production. *Canadian Journal of Soil Science* 73, 39–50.

Asmar, F., Eiland, F. and Nielsen, N.E. (1992) Interrelationship between extracellular enzyme activity, ATP content, total counts of bacteria and CO_2 evolution. *Soil Biology and Fertility* 14, 288–292.

Babich, H., Bewley, R.J.E. and Stotzky, G. (1983) Application of 'Ecological Dose' concept to the impact of heavy metals in some microbe-mediated ecological processes in soil. *Archives of Environmental Contaminates and Toxicology* 12, 421–426.

Bardgett, R.D., Speir, T.W., Ross, D.J., Yeates, G.W. and Kettles, H.A. (1994) Impact of pasture contamination by copper, chromium, and arsenic timber preservative on soil microbial properties and nematodes. *Biology and Fertility of Soils* 18, 71–79.

Bartha, R. and Bordeleau, L. (1969) Cell-free peroxidases in soil. *Soil Biology and Biochemistry* 16, 423–424.

Baruah, M. and Mishra, R.R. (1986) Effect of herbicides butachlor, 2,4-D and oxyfluorfen on enzyme activities and CO_2 evolution in submerged paddy field soil. *Plant and Soil* 96, 287–291.

Biederbeck, V.O., Campbell, C.A., Bowren, K.E., Schnitzer, M. and McIver, R.N. (1980) Effect of burning cereal straw on soil properties and grain yields in Saskatchewan. *Soil Science Society of America Journal* 44, 103–111.

Bitton, G. and Boylan, R.A. (1985) Effect of acid precipitation on soil microbial activity: I. Soil core studies. *Journal of Environmental Quality* 14, 66–69.

Bitton, G., Volk, B.G., Graetz, D.A., Bossart, J.M., Boylan, R.A. and Byers, G.E. (1985) Effect of acid precipitation on soil microbial activity: II. Field studies. *Journal of Environmental Quality* 14, 69–71.

Blagoveshchenskaya, Z.K. and Danchenko, N.A. (1974) Activity of soil enzymes after prolonged application of fertilizers to a corn monoculture and crops in rotation. *Soviet Soil Science* 5, 569–575.

Bollag, J.-M. and Liu, S.-Y. (1990) Biological transformation processes of pesticides. In: *Pesticides in the Soil Environment.* Soil Science Society of America, Book Series no. 2.

Bolton, H., Elliot, L.F., Papendick, R.I. and Bezdicek, D.F. (1985) Soil microbial biomass and selected soil enzyme activities: Effect of fertilization and cropping practices. *Soil Biology and Biochemistry* 17, 297–302.

Boyd, S.A. and Mortland, M.M. (1990) Enzyme interactions with clays and clay-organic matter complexes. In: Bollag, J.M. and Stotzky, G. (eds) *Soil Biochemistry,* Vol. 6. Marcel Dekker, Inc., New York. pp. 1–28.

Bremner, J.M. and Tabatabai, M.A. (1973) Effect of some inorganic substances on TTC assay of dehydrogenase activity in soils. *Soil Biology and Biochemistry* 5, 385–386.

Bridge, B.J. and Rixon, A.J. (1976) Oxygen uptake and respiratory quotient of field soil cores in relation to their air-filled pore space. *Journal of Soil Science* 27, 279–286.

Brookes, P.C. (1995) The use of microbial parameters in monitoring soil pollution by heavy metals. *Biology and Fertility of Soils* 19, 269–279.

Burns, R.G. (ed.) (1978) *Soil Enzymes.* Academic Press, London, New York, and San Francisco.

Burns, R.G. (1982) Enzyme activity in soil: location and a possible role in microbial activity. *Soil Biology and Biochemistry* 14, 423–427.

Castellano, S.D. and Dick, R.P. (1991) Cropping and sulfur fertilization influence on sulfur transformations in soil. *Soil Science Society of America Journal* 54, 114–121.

Chander, K. and Brookes, P.C. (1991) Is the dehydrogenase assay invalid as a method to estimate microbial activity in Cu-contaminated and non-contaminated soils? *Soil Biology and Biochemistry* 23, 901–915.

Cheng, H.H. (ed.) (1990) *Pesticides in the Soil Environment: Processes, Impacts, and Modeling.* Soil Science Society of America, Madison, Wisconsin.

Chunderova, A.I. and Zubets, T. (1969) Phosphatase activity in dernopodzolic soils. *Pochvovedeniye* 11, 47–53.

Clarholm, M. (1993) Microbial biomass P, labile P, and acid phosphatase activity in the humus layer of a spruce forest, after repeated additions of fertilizers. *Biology and Fertility of Soils* 16, 287–292.

Davidson, J.M., Rao, P.S., Ou, L.T., Wheeler, W.B. and Rothwell, D.F. (1980) Adsorption, movement and biological degradation of large concentrations of

selected pesticides in soil. EPA-600/2-80-124, Municipal Environmental Research Laboratory, Cincinnati, Ohio.
Davies, H.A. and Greaves, M.P. (1981) Effects of some herbicides on soil enzyme activities. *Weed Research* 21, 205–209.
Davies, H.A. and Marsh, J.A.P. (1977) The effect of herbicides on respiration and transformation of nitrogen in two soils. II. Dalapon, pyrazone and trifluralin. *Weed Research* 17, 373–378.
Deng, S.P. and Tabatabai, M.A. (1995) Cellulase activity of soils: effect of trace elements. *Soil Biology and Biochemistry* 27, 977–979.
Dick, R.P. (1992) A review: long-term effects of agricultural systems on soil biochemical and microbial parameters. *Agriculture, Ecosystems and Environment* 40, 25–36.
Dick, R.P. (1994) Soil enzyme activities as indicators of soil quality. In: Doran, J.W., Coleman, D.C., Bezdicek, D.F. and Stewart, B.A. (eds) *Defining Soil Quality for a Sustainable Environment.* Soil Science Society of America Special Publication No. 35, Madison, Wisconsin, pp. 107–124.
Dick, R.P., Rasmussen, P.E. and Kerle, E.A. (1988a) Influence of long-term residue management on soil enzyme activity in relation to soil chemical properties of a wheat-fallow system. *Biology and Fertility of Soils* 6, 159–164.
Dick, R.P., Myrold, D.D. and Kerle, E.A. (1988b) Microbial biomass and soil enzyme activities in compacted and rehabilitated skid trail soils. *Soil Science Society of America Journal* 52, 512–516.
Dick, W.A. (1984) Influence of long-term tillage and crop rotation combinations on soil enzyme activities. *Soil Science Society of America Journal* 48, 569–574.
Dick, W.A. and Tabatabai, M.A. (1983) Activation of soil pyrophosphatase by metal ions. *Soil Biology and Biochemistry* 15, 59–363.
Dick, W.A. and Tabatabai, M.A. (1993) Potential uses of soil enzymes. In: Metting, F.B., Jr (ed.) *Soil Microbial Ecology: Applications in Agricultural and Environmental Management.* Marcel Dekker, Inc., New York, pp. 95–127.
Dick, W.A., Juma, N.G. and Tabatabai, M.A. (1983) Effects of soils on acid phosphatase and inorganic pyrophosphatase of corn roots. *Soil Science* 136, 19–25.
Dixon, M. and Webb, E.C. (1979) *Enzymes.* 3rd edn. Longman, London.
Doelman, P. and Haanstra, L. (1986) Short- and long-term effects of heavy metals on urease activity in soils: an ecological dose-response model approach. *Biology and Fertility of Soils* 2, 213–218.
Doelman, P. and Haanstra, L. (1989) Short- and long-term effects of heavy metals on phosphatase activity in soils: An ecological dose-response model approach. *Biology and Fertility of Soils* 8, 235–241.
Domsch, K.H., Jagnow, G. and Anderson, T.H. (1983) An ecological concept for the assessment of side-effects of agrochemicals on soil microorganisms. *Residue Review* 86, 65–105.
Doran, J.W. (1980) Soil microbial and biochemical changes associated with reduced tillage. *Soil Science Society of America Journal* 44, 765–771.
Doran, J.W., Sarrantonio, M. and Liebig, M.A. (1996) Soil health and sustainability. In: Sparks, D.L. (ed.) *Advances in Agronomy.* Academic Press, New York, 56 (in press).
Dragan-Bularda, M., Blaga, G., Kiss, S., Pasca, D., Gherasim, V. and Vulcan, R.

(1987) Effect of long-term fertilization on the enzyme activities in a technogenic soil resulting from the recultivation of iron strip mine spoils. *Studia Universitatis Babes-Bolyai, Biolgia* 32, 47–52.

Drobnik, J. (1957) Biological transformations of organic substances in the soil. *Pochvovedeniye* 12, 62–71.

Dzantor, E.K. and Felsot, A.S. (1991) Microbial responses to large concentrations of herbicides in soil. *Environmental Toxicology and Chemistry* 10, 649–655.

Eivazi, F. and Tabatabai, M.A. (1990) Factors affecting glucosidase and galactosidase activities in soils. *Soil Biology and Biochemistry* 22, 891–897.

Farrell, R.E., Gupta, V.V.S.R. and Germida, J.J. (1994) Effects of cultivation on the activity and kinetics of arylsulfatase in Saskatchewan soils. *Soil Biology and Biochemistry* 26, 1033–1040.

Fauci, M.F. and Dick, R.P. (1994) Soil microbial dynamics: short- and long-term effects of inorganic and organic nitrogen. *Soil Science Society of America Journal* 58, 801–806.

Felsot, A.S. and Dzantor, E.K. (1995) Effect of alachlor concentration and an organic amendment on soil dehydrogenase activity and pesticide degradation rate. *Environmental Toxicology and Chemistry* 14, 23–28.

Filip, Z. and Preusse, T. (1985) Phenoloxidases – their properties and effects in soil. *Pedobiologia* 28, 133–142.

Frankenberger, W.T., Jr and Bingham, F.T. (1982) Influence of salinity on soil enzyme activities. *Soil Science Society of America Journal* 46, 1173–1177.

Frankenberger, W.T., Jr and Dick, W.A. (1983) Relationships between enzyme activities and microbial growth and activity indices in soil. *Soil Science Society of America Journal* 47, 945–951.

Frankenberger, W.T., Jr and Tabatabai, M.A. (1981) Amidase activity in soils: IV. Effects of trace elements and pesticides. *Soil Science Society of America Journal* 45, 1120–1124.

Frankenberger, W.T., Jr and Tabatabai, M.A. (1991a) Factors affecting L-asparaginase activity in soils. *Biology and Fertility of Soils* 11, 1–5.

Frankenberger, W.T., Jr and Tabatabai, M.A. (1991b) Factors affecting L-glutaminase activity in soils. *Soil Biology and Biochemistry* 23, 875–879.

Galstyan, A.Sh. (1960) Enzyme activities in solonchaks. *Doklady Akademii Nauk Armyanskoi SSR* 30, 61–64.

Garcia, C., Hernandez, T. and Costa, F. (1994) Microbial activity in soils under Mediterranean environmental conditions. *Soil Biology and Biochemistry* 26, 1185–1191.

Getzin, L.W. and Rosefield, I. (1968) Organophosphorus insecticide degradation by heat-labile substances in soil. *Journal of Agricultural Food Chemistry* 16, 598–601.

Gianfreda, L., Sannino, F., Filazzola, M.T. and Violante, A. (1993) Influence of pesticides on the activity and kinetics of invertase, urease and acid phosphatase enzymes. *Pesticide Science* 39, 237–244.

Gianfreda, L., Sannino, F., Ortega, N. and Nannipieri, P. (1994) Activity of free and immobilized urease in soil: effects of pesticides. *Soil Biology and Biochemistry* 26, 777–784.

Gosewinkel, U. and Broadbent, F.E. (1986) Decomposition of phosphatase from extracted soil humic substances with electron donating reagents. *Soil Science* 141, 261–267.

Goyal, S., Mishra, M.M., Dhankar, S.S., Kapoor, K.K. and Batra, R. (1993) Microbial biomass turnover and enzyme activities following the application of farmyard manure to field soils with and without previous long-term applications. *Biology and Fertility of Soils* 15, 60–64.

Gupta, V.V.S.R. and Germida, J.J. (1988) Distribution of microbial biomass and its activity in different soil aggregate size classes as affected by cultivation. *Soil Biology and Biochemistry* 20, 777–786.

Haanstra, L. and Doelman, P. (1991) An ecological dose-response model approach to short- and long-term effects of heavy metals on arylsulphatase activity in soil. *Biology and Fertility of Soils* 11, 18–23.

Haanstra, L., Doelman, P. and Voshaar, J.H.O. (1985) The use of sigmoidal dose response curves in soil ecotoxicological research. *Plant and Soil* 84, 293–297.

Haban, L. (1967) Effect of ploughing depth and cultivated crops on the soil microflora and enzyme activity of the soil. *V d. Pr. Výsk. Úst. Rastl. Výrobý Pie tanoch* 5, 159–169.

Hersman, L.E. and Temple, K.L. (1979) Comparison of ATP, phosphatase, pectinolase, and respiration as indicators of microbial activity in reclaimed coal strip mine spoils. *Soil Science* 127, 70–73.

Howard, P.J.A. (1972) Problems in the estimation of biological activity in soil. *Oikos* 23, 235–240.

Jha, D.K., Sharma, G.D. and Mishra, R.R. (1992) Soil microbial population numbers and enzyme activities in relation to altitude and forest degradation. *Soil Biology and Biochemistry* 24, 761–767.

Jordan, D., Kremer, R.J., Bergfield, W.A., Kim, K.Y. and Cacnio, V.N. (1995) Evaluation of microbial methods as potential indicators of soil quality in historical agricultural fields. *Biology and Fertility of Soils* 19, 297–302.

Joshi, S.R., Sharma, G.D. and Mishra, R.R. (1993) Microbial enzyme activities related to litter decomposition near a highway in a sub-tropical forest of north east India. *Soil Biology and Biochemistry* 25, 1763–1770.

Juma, N.G. and Tabatabai, M.A. (1978) Distribution of phosphomonoesterases in soils. *Soil Science* 126, 101–108.

Junk, G.A., Richard, J.J. and Dahm, P.A. (1984) Degradation of pesticides in controlled water-soil systems. In: Krueger, R.F. and Seiber, J.N. (eds) *Treatment and Disposal of Pesticide Wastes.* ACS Symposium Series 259, American Chemical Society, Washington, DC, pp. 37–67.

Kandeler, E. and Eder, G. (1993) Effect of cattle slurry in grassland on microbial biomass and on activities of various enzymes. *Biology and Fertility of Soils* 16, 249–254.

Kannan, K. and Oblisami, G. (1990) Influence of paper mill effluent irrigation on soil enzyme activities. *Soil Biology and Biochemistry* 22, 923–926.

Khan, S.W. (1970) Enzymatic activity in a gray wooded soil as influenced by cropping systems and fertilizers. *Soil Biology and Biochemistry* 2, 137–139.

Kiss, S., Dragan-Bularda, M. and Radulescu, D. (1975) Biological significance of enzymes in soil. *Advances in Agronomy* 27, 25–87.

Kiss, S., Dragan-Bularda, M. and Pasca, D. (1993) *Enzymology of Technogenic Soils.* Casa Cartii de Stiinta, Cluj, Romania.

Klein, T.M and Koths, J.S. (1980) Urease, protease, and phosphatase in soil continuously cropped to corn by conventional or no-tillage methods. *Soil Biology and Biochemistry* 12, 293–294.

Klein, D.A., Sorensen, D.L. and Redente, E.F. (1985) Soil enzymes: a predictor of reclamation potential and progress. In: Tate, R.L. and Klein, D.A. (eds) *Soil Reclamation Processes.* Marcel Dekker, New York, pp.141–171.

Koepf, H. (1954) Investigations on the biological activity in soil. I. Respiration curves of the soil and enzyme activity under the influence of fertilizing and plant growth. *Zeitschrift für Acker-und Pflanzenbau* 98, 289–312.

Ladd, J.N. (1985) Soil enzymes. In: Vaughan, D. and Malcolm, R.E. (eds) *Soil Organic Matter and Biological Activity.* Martinus Nijhoff, Boston, pp. 175–221.

Laugesen, K. (1972) Urease activity in Danish soils. *Danish Journal of Plant and Soil Science* 76, 221–229.

Laugesen, K. and Mikkelsen, J.P. (1973) Phosphatase activity in Danish soils. *Danish Journal of Plant and Soil Science* 77, 252–257.

Linn, D.M. and Doran, J.W. (1984) Effect of water-filled pore space on carbon dioxide and nitrous oxide production in tilled and non-tilled soils. *Soil Science Society of America Journal* 48, 1267–1272.

Martens, D.A., Johanson, J.B. and Frankenberger, W.T., Jr (1992) Production and persistence of soil enzymes with repeated additions of organic residues. *Soil Science* 153, 53–61.

Mathur, S.P. and Rayment, A.F. (1977) Influence of trace element fertilization on the decomposition rate and phosphatase activity of a mesic fibrisol. *Canadian Journal of Soil Science* 57, 397–408.

McCarty, G.W., Shogren, D.R. and Bremner, J.M. (1992) Regulation of urease production in soil by microbial assimilation of nitrogen. *Biology and Fertility of Soils* 12, 261–264.

McCarty, G.W., Siddaramappa, R., Wright, R.J., Codling, E.E. and Gao, G. (1994) Evaluation of coal combustion byproducts as soil liming materials: their influence on soil pH and enzyme activities. *Biology and Fertility of Soils* 17, 167–172.

McLaren, A.D. (1975) Soil as a system of humus and clay immobilized enzymes. *Chemica Scripta* 8, 97–99.

Miller, M. and Dick, R.P. (1995a) Dynamics of soil C and microbial biomass on whole soil and aggregates in two cropping systems. *Applied Soil Ecology* 2, 253–261.

Miller, M. and Dick, R.P. (1995b) Thermal stability and activities of soil C enzymes as influenced by crop rotations. *Soil Biology and Biochemistry* 27, 1161–1166.

Nannipieri, P., Johnson, R.L. and Paul, E.A. (1978) Criteria for measurement of microbial growth and activity in soil. *Soil Biology and Biochemistry* 10, 223–227.

Ohtonen, R., Lähdesmäki, P. and Markkola, A.M. (1994) Cellulase activity in forest humus along an industrial pollution gradient in Oulu, Northern Finland. *Soil Biology and Biochemistry* 26, 97–101.

Pancholy, S.K. and Rice, E.L. (1973a) Soil enzymes in relation to old field succession: amylase, cellulase, invertase, dehydrogenase, and urease. *Soil Science Society of America Proceedings* 37, 47–50.

Pancholy, S.K. and Rice, E.L. (1973b) Carbohydrases in soils as affected by successional stages of revegetation. *Soil Science Society of America Proceedings* 37, 227–229.

Pancholy, S.K., Rice, E.L. and Turner, J.A. (1975) Soil factors preventing revegetation of a denuded area near an abandoned zinc smelter in Oklahoma. *Journal of Applied Ecology* 12, 337–342.

Parr, J.F. and Reuszer, H.W. (1962) Organic matter decomposition as influenced by oxygen level and flow rate of gases in the constant aeration method. *Soil Science Society of America Proceedings* 26, 552–556.

Perucci, P. (1992) Enzyme activity and microbial biomass in a field soil amended with municipal refuse. *Biology and Fertility of Soils* 14, 54–60.

Perucci, P. and Scarponi, L. (1994) Effects of the herbicide imazethapyr on soil microbial biomass and various soil enzyme activities. *Biology and Fertility of Soils* 17, 237–240.

Pichtel, J.R. and Hayes, J.M. (1990) Influence of fly ash on soil microbial activity and populations. *Journal of Environmental Quality* 19, 593–597.

Pierzynski, G.M., Sims, J.T. and Vance, G.F. (eds) (1994) *Soils and Environmental Quality*. Lewis Publishers, Boca Raton, Ann Arbor, London, Tokyo. 313 pp.

Quastel, J.H. (1946) *Soil Metabolism*. The Royal Institute of Chemistry of Great Britain and Ireland, London.

Rai, J.P.N. (1992) Effects of long-term 2,4-D application on microbial populations and biochemical processes in cultivated soil. *Biology and Fertility of Soils* 13, 187–191.

Ramirez-Martinez, J.R. and McLaren, M.A. (1966) Some factors influencing the determination of phosphatase activity in native soils and in soils sterilized by irradiation. *Enzymologia* 31, 23–28.

Redente, E.F. and Cook, C.W. (eds) (1986) *Structural and Functional Changes in Early Successional Stages of a Semiarid Ecosystem*. Department of Range Science, Colorado State University, Fort Collins.

Rice, E.L. and Mallik, M.A.B. (1977) Causes of decreases in residual carbohydrase activity in soil during old-field succession. *Ecology* 58, 1297–1309.

Ross, D.J. and Cairns, A. (1982) Effects of earthworms and ryegrass on respiratory and enzyme activities of soil *Allolobophora caliginosa, Lolium perenne*, Judgeford silt loam, New Zealand. *Soil Biology and Biochemistry* 14, 583–587.

Ross, D.S., Sjogren, R.E. and Bartlett, R.J. (1981) Behavior of chromium in soils: IV. Toxicity to micro-organisms. *Journal of Environmental Quality* 10, 145–148.

Ross, D.J., Speir, T.W., Tate, K.R., Cairns, A., Meyrick, K.F. and Pansier, E.A. (1982) Restoration of pasture after topsoil removal: effects on soil carbon and nitrogen mineralization, microbial biomass and enzyme activities. *Soil Biology and Biochemistry* 14, 575–581.

Ross, D.J., Speir, T.W., Cowling, J.C. and Feltham, C.W. (1992) Soil restoration under pasture after lignite mining: Management effects on soil biochemical properties and their relationships with herbage yields. *Plant and Soil* 140, 85–97.

Saa, A., Trasar-Cepedia, C., Gil-Sortes, F. and Carballas, T. (1993) Changes in soil phosphorus and acid phosphatase activity immediately following forest fires. *Soil Biology and Biochemistry* 25, 1223–1230.

Schoen, S.R. and Winterlin, W.L. (1987) The effects of various soil factors and amendments on the degradation of pesticide mixtures. *Journal of Environmental Science and Health* B22, 347–377.

Shaw, W.H.R. and Raval, D.N. (1961) The inhibition of urease by metal ions at pH 8.9. *Journal of the American Chemistry Society* 83, 3184–3187.

Skladany, G.J. and Metting, F.B., Jr (1993) Bioremediation of contaminated soil. In: Metting, F.B., Jr (ed.) *Soil Microbial Ecology.* Marcel Dekker, New York, pp. 483–513.

Skujins, J. (1967) Enzymes in soil. In: McLaren, A.D. and Peterson, G.H. (eds) *Soil Biochemistry.* Marcel Dekker, New York, pp. 371–414.

Skujins, J. (1976) Extracellular enzymes in soil. *CRC Critical Reviews in Microbiology* 4, 383–421.

Skujins, J. (1978) History of abiontic soil enzyme research. In: Burns, R.G. (ed.) *Soil Enzymes.* Academic Press, London, New York, and San Francisco, pp. 1–49.

Song, H.G., and Bartha, R. (1990) Effects of jet fuel spills on the microbial community of soil. *Applied and Environmental Microbiology* 56, 646–651.

Speir, T.W., Ross, D.J., Feltham, C.W., Orchard, V.A. and Yeates, G. (1992) Assessment of the feasibility of using CCA (cooper, chromium and arsenic)-treated and boric acid-treated sawdust as soil amendments. II. Soil biochemical and biological properties. *Plant and Soil* 142, 249–258.

Speir, T.W., Kettles, H.A., Parshotam, A., Searle, P.L. and Vlaar, L.N.C. (1995) A simple kinetic approach to derive the ecological dose value, ED_{50}, for the assessment of Cr(VI) toxicity to soil biological properties. *Soil Biology and Biochemistry* 27, 801–810.

Spiers, G.A. and McGill, W.B. (1979) Effects of phosphorus addition and energy supply on acid phosphatase production and activity in soils. *Soil Biology and Biochemistry* 11, 3–8.

Stroo, H.F. and Jencks, E.M. (1982) Enzyme activity and respiration in mine soils. *Soil Science Society of America Journal* 46, 548–553.

Tabatabai, M.A. (1994) Soil enzymes. In: Weaver, R.W., Mickelson, S.H. and Bigham, J.M. (eds) *Methods of Soil Analysis. Part 2 Microbial and Biochemical Properties.* Soil Science Society of America, Madison, Wisconsin.

Tabatabai, M.A. and Bremner, J.M. (1970) Factors affecting soil arylsulfatase activity. *Soil Science Society of America Proceedings* 34, 427–429.

Tena, M., Pinilla, J.A. and Prado, R. (1981) Properties of inorganic pyrophosphatase in soil. *Anales de Edafologia y Agrobiologia.* 40, 1815–1824.

Tiwari, M.B., Tiwari, B.K. and Mishra, R.R. (1989) Enzyme activity and carbon dioxide evolution from upland and wetland rice soils under three agricultural practices in hilly regions. *Biology and Fertility of Soils* 7, 359–364.

Tu, C.M. (1981) Effects of some pesticides on enzyme activities in an organic soil. *Bulletin of Environmental Contamination* 27, 109–114.

Tu, C.M. (1992) Effect of three newer pesticides on microbial and enzymatic activities in soil. *Bulletin of Environmental Contamination and Toxicology* 49, 120–128.

Tu, C.M. (1993) Influence of ten herbicides on activities of microorganisms and enzymes in soils. *Bulletin of Environmental Contamination and Toxicology* 51, 30–39.

Verstraete, W. and Voets, J.P. (1977) Soil microbial and biochemical characteristics in relation to soil management and fertility. *Soil Biology and Biochemistry* 9, 253–258.

Voets, J.P., Meerschman, P. and Verstraete, W. (1974) Soil microbiological and biochemical effects of long-term atrazine applications. *Soil Biology and Biochemistry* 6, 149–152.

Werner, W., Scherer, H.W. and Olfs, H.W. (1988) Influence of long-term application of sewage sludge and compost from garbage with sewage sludge on soil fertility criteria. *Journal of Agronomy and Crop Science* 160, 173–179.

Yaroschevich, I.V. (1966) Effect of fifty years' application of fertilizers in a rotation on the biological activity of a chernozem. *Agrokhimiya* 6, 14–19.

Yeates, G.W., Orchard, V.A., Speir, T.W., Hunt, J.L. and Hermans, M.C.C. (1994) Reduction in soil biological activity following pasture contamination by copper, chromium, arsenic timber preservative. *Biology and Fertility of Soils* 18, 200–208.

7

Soil Microflora as Bioindicators of Soil Health

M.M. Roper[1] and K.M. Ophel-Keller[2]

[1]*Division of Plant Industry, CSIRO, Private Bag, Wembley, PO, Western Australia 6014, Australia;* [2]*Cooperative Research Centre for Soil and Land Management, Private Bag No. 2, Glen Osmond, South Australia 5064, Australia*

Introduction

Soil microflora including bacteria (eubacteria and archaebacteria), fungi and algae have the potential to be important indicators of soil quality/health (Parr *et al.*, 1992). There is a large variety of microorganisms in the soil, possibly millions of species (Microbial Resources Panel, 1981). Jong (1989) even estimated that microorganisms constitute about one quarter of the total biomass on earth! In soils, microorganisms are responsible for the decomposition/transformation of organic matter including nearly all nitrogen and carbon transformations (Alexander, 1977; Apsimon *et al.*, 1990). Decomposition of carbon compounds such as cellulose, hemicellulose, polysaccharides, hydrocarbons and lignin provide energy to heterotrophic microorganisms responsible for other nutrient transformations, e.g. asymbiotic nitrogen fixation, protein and amino acid decomposition, mineralization and immobilization of nitrogen, and mineral transformations (Alexander, 1977; Roper, 1983; Sikora and McCoy, 1990). Microorganisms are responsible for a significant number of mineral transformations such as P, S, Fe, K, Ca, Mg, Mn, Al, As, Zn and Se (Alexander, 1977; Doran, 1982). All of these processes affect nutrient availability and hence soil quality. Therefore, because the microbiological component of the soil is so important in ecosystem functioning, they are potentially one of the most sensitive biological markers available and the most useful for assessing disturbances or perturbations in the ecosystem (Turco *et al.*, 1994).

A number of authors have observed that changes in the content and composition of organic matter in soils are linked to changes in the structure and function of the microbial community (Beare *et al.*, 1992; Mele and Carter, 1993;

(eds C.E. Pankhurst, B.M. Doube and V.V.S.R. Gupta)

Mikhnovskaya *et al.*, 1993; Gupta *et al.*, 1994). However, despite their wide occurrence and extensive role in soils, microorganisms have not been used widely as indicators of soil quality or soil health (Turco *et al.*, 1994). One possible reason is that microorganisms are not visible so that land users are simply unaware of their existence and the processes they perform. However, perhaps the most significant reason is that, because of technical limitations, only about 40,000 species of all the microorganisms in the soil have been cultured or identified (Microbial Resources Panel, 1981; Hawksworth and Mound, 1991). Recent studies comparing ribosomal RNA (rRNA) methodology with culture techniques have estimated that at least 90% of microbial species have not been identified in aquatic environments (Ward *et al.*, 1990; Giovannoni *et al.*, 1990) or in soil (Liesack and Stackebrandt, 1992; Embley and Stackbrandt, 1996). Even with microbial species that can be isolated it is difficult to make decisions about soil health on the basis of their occurrence or behaviour because natural environmental factors alter population dynamics (Visser and Parkinson, 1992). Nonetheless, with the development of new molecular technologies, it may be possible to use individual species, genera or populations as indicators of soil health or change in soil quality. Another approach is to measure microbial functions such as specific carbon and nitrogen transformations. Microbial functions directly affect nutrient availability and therefore may better reflect soil health, but again, in interpreting such information, natural environmental factors need to be taken into account.

Because of the complexities of microbial populations and their processes it could be difficult to make comparisons between the health of different soil types based on population sizes or particular functions. The use of microorganisms and their processes is more likely to be of value in one particular soil or soil type for assessing changes in soil health (rehabilitation or degradation) resulting from some perturbation. Ideally, the indicator species or function would be sensitive to a wide range of disturbances and signal changes in the system's structure and function (Moore and de Ruiter, 1993). The indicator species or function should be easily and economically measured, and the response of the indicator to a perturbation should be easily distinguishable from a background or control measurement (Brookes, 1995).

With increasing research on how physical and chemical perturbations affect soil microflora and their activities, parameters which may be useful indicators of changing soil health are beginning to emerge. In a review, Domsch *et al.* (1983) concluded from a number of studies that parameters which related to soil productivity (in particular mineralization processes) deserved preference and should have the highest priority. Other parameters such as nitrogen fixation which feed into mineralization processes were also considered to have a high priority. On the basis of numerous studies on the influence of agrochemicals on soil microflora, Domsch *et al.* (1983) proposed the following sensitivity groups:

1. High sensitivity groups (nitrifying bacteria, *Rhizobium* and actinomycetes) and functions (organic matter degradation and nitrification).

2. Medium sensitivity groups (general populations of algae, bacteria or fungi) and functions (soil respiration, denitrification and ammonification).
3. Low sensitivity groups (populations of *Azotobacter*, ammonifiers and protein degraders) and functions (aerobic nitrogen-fixing activity).

Other researchers, e.g. Sims (1990), regarded carbon cycling to be of low sensitivity to perturbations in the soil. It is likely that the sensitivities of microbial populations and functions differ depending on the nature of the particular perturbation/change applied to the soil. The sensitivities of individual populations or functions being used as bioindicators should provide an indication of the extent of changes in soil health resulting from any perturbation. Highly sensitive microorganisms or functions could be useful in detecting small changes in soil health resulting from minor perturbations, whereas changes in populations or functions of low sensitivity should indicate larger perturbations with potentially longer term effects.

In this chapter, the value of individual soil microorganisms and microbial functions as indicators of soil health will be evaluated, particularly in terms of the impact of managements and environmental factors on soils.

Studies of Microbial Populations

The size and composition of microbial populations have been used to look at changes in the soil biota in response to land management and therefore as indicators of the biological status of soil. Culturable populations of bacteria, either total populations or specific groups, have been used in a number of studies for this purpose (Maltby, 1975; Kale and Raghu, 1989). However, there are problems with the use of plate counts because, as indicated previously, most types of soil microorganisms cannot be cultured. Furthermore, the choice of culture medium can have an impact not only on the size of the population estimated but also on the composition of that population. A study of the use of different culture media in the same soil has shown that the bacterial populations vary in size and in composition depending on the medium used (Sorheim *et al.*, 1989). In another study, Frankenberger and Dick (1983) found significant correlations between plate counts on different media, but correlations between microbial biomass and viable plate counts were poor. Further, they found that plate counts correlated poorly with most enzymatic measurements of microbial activity. In contrast, they found that three enzymatic measurements, alkaline phosphatase, amidase and catalase, were highly correlated with both microbial biomass and respiration. On the basis of these results, Frankenberger and Dick (1983) concluded that plate counts are not a reliable measure of microbial growth and activity in soil. Despite these drawbacks however, there is evidence that plate count techniques are useful in comparative studies of specific microbial populations (Harris and Birch, 1992).

An alternative to plate count studies is the measurement of ester-linked fatty acids in phospholipids. Phospholipids are extracted from soil and, after alkaline methanolysis, fatty acid methyl esters (FAME) are measured using gas chromatography (Tunlid and White, 1990). FAME is a promising approach because phospholipids are a relatively stable proportion of the biomass. This has made it possible to use FAME to examine community structures in different soils (Zelles *et al.*, 1992). These researchers concluded that, because phospholipids correlate well with biomass and a number of enzymic indicators of microbial activity, FAME may be a useful tool for quantifying microbial populations. In support of this, Haack *et al.* (1994) could distinguish soils with different cropping histories by their phospholipid profiles. A further strength of FAME may be in its potential to determine the composition of the community by identification of phospholipid signatures. However, signatures do not yet exist for many genera of soil microorganisms and there is some overlap between signatures for those that are known. To date, FAME has been used largely to identify bacterial signatures, although identification of fungal signatures is increasing, e.g. species of vesicular arbuscular mycorrhizae have been distinguished using FAME (Bentivenga and Morton, 1994).

Direct measurements of fungal biomass in soils can be made using ergosterol, the predominant sterol in most fungi (Bentham *et al.*, 1992). In an early study, West *et al.* (1987) concluded that there were good correlations between ergosterol content in soils and total fungal mycelial length. However, recently Bermingham *et al.* (1995) compared ergosterol measurement with fungal biomass for nine fungal species *in vitro* and concluded that ergosterol was a poor measure of fungal biomass because there was a significant interspecific variation in ergosterol content. Therefore, changes in ergosterol content in soil may reflect qualitative changes in the fungal population but not changes in the size of the fungal biomass. Microscopic studies of fungal length have been used to estimate the size of fungal populations, but such assays are very tedious and time-consuming.

Individual Microorganisms or Groups of Microorganisms as Bioindicators

There are a number of examples where individual or groups of microorganisms have been used as bioindicators in soil. Perhaps one of the best known examples are coliform bacteria which have been used extensively as microbial bioindicators of faecal pollution. However, as pointed out by Holloway and Stork (1991), because coliform bacteria are not part of the natural soil ecosystem, they cannot be considered as true ecological indicators. There have been few efforts aimed at developing indicators of soil health using components of the natural soil microflora but, using existing technologies, it should be possible to evaluate some microbial components (Turco *et al.*, 1994). Again it is important to stress that measurements of the size and behaviour of individual or groups of microorganisms are really only useful for measuring changes in soil health in a particular soil and not as a tool to compare different soil types. The reasons for this are that the survival and growth of microbial populations are greatly affected by the range of physical and chemical properties of soils (Waksman, 1952; Alexander, 1977). For example, microorganisms vary in their ability to tolerate extremes of pH and thus, organisms such as *Azotobacter* spp. are not found at pH below 6.5 (Alexander, 1977). Generally, fungi are more tolerant of acid conditions than bacteria including actinomycetes (Doran, 1980a) leading to a changing balance in the microbial composition of soils as the pH varies. Clays are another factor which alter the size and composition of microbial populations. Clay soils generally support larger numbers and varieties of microorganisms (Roper and Halsall, 1986) due in part to protection from predation and parasitism (Roper and Marshall, 1978; Heijnen *et al.*, 1988) and from desiccation (Bushby and Marshall, 1977). Clays also can extend the range of oxygen tensions in soils to accommodate a wider variety of microorganisms (Paul and Clark, 1989).

Furthermore, when a microbial group is chosen as a bioindicator of a particular perturbation, e.g. tillage, other factors associated with the perturbation should be taken into account. For example, no-tillage systems are usually accompanied by increased herbicide use. Herbicides can have a negative impact on fungi (Edwards, 1989), but no-tillage increases the proportion of fungi in soils (Allison and Killham, 1988). If the size of fungal populations is to be used as an indicator of change, the effect may be masked, and changes in soil health may not be recognized, particularly if there are changes in the composition of the fungal population. Keeping in mind these potential pitfalls, individual populations still may be useful as indicators of change of soil health.

Most studies have focused on the use of bacteria as potential indicators because of the relative ease with which these organisms can be examined. Few studies have examined fungal or actinomycete populations. However, fungal biomass appears to be more sensitive to cropping history than bacterial biomass (Jordan and Kremer, 1994). Quantitative and qualitative changes in fungal populations have been observed under different cropping systems (wheat, corn) and with manure and fertilizer applications (Martyniuk and Wagner, 1978). In a study on the effects of corn residue on microbial populations, Doran (1980a) found that fungi and actinomycetes were directly influenced by the crop residue as a nutrient source. In the same study, bacterial populations were very sensitive to soil moisture levels. When the effect of herbicides on microbial populations was investigated, Camper *et al.* (1973) showed that actinomycete populations were not significantly affected by paraquat levels as high as 112 kg ha^{-1}, whereas fungal populations were significantly reduced at medium (5.6 kg ha^{-1}) and high (112 kg ha^{-1}) levels of paraquat. Therefore, where there are a number of new managements or perturbations being superimposed on each other, it may be necessary to select a range of bioindicators, each with a strong sensitivity to a particular type of perturbation, in order to best represent the potential changes in soil quality.

Maltby (1975) used a combination of microbial groups as ecological indicators of change in soils reclaimed from moorland in England. Three soils were evaluated: (i) unreclaimed; (ii) reclaimed; and (iii) reverted. Four major groups (filamentous fungi, yeasts, actinomycetes and bacteria) were enumerated using soil-dilution plate counting, and ratios of fungi:yeasts (F:Y), actinomycetes: bacteria (A:B) and fungi:bacteria (F:B) were determined. In the reclaimed soil, actinomycetes and bacteria (A+B) exceeded fungi and yeasts (F+Y) by at least an order of magnitude, whereas there was no distinction between A+B and F+Y in the unreclaimed soils. Furthermore, in the reclaimed soils, F:Y was less than or close to 1, whereas in the unreclaimed soils, the ratio was much greater than 1. F:Y in the reverted soils was intermediate between the reclaimed and unreclaimed soils. Ratios of A:B were greater than 1 in the reverted soils, but less than 1 for both the reclaimed and unreclaimed soils. Maltby (1975) proposed that such a microbiological analysis was useful in separating the developmental stages of soils under changed conditions.

A number of examples where specific groups of microorganisms have been used as bioindicators are outlined below.

Bacteria of the genus *Metallogenium* accumulate iron, aluminium or manganese in an insoluble form, and are able to grow in environments where the soluble form of these metals is high; the mass development of the *Metallogenium* is related positively to the concentration of the heavy metals (Aristovskaya and Zykina, 1980). The mineral deposits around the bacterial cells makes the organisms easily distinguishable from the rest of the soil microflora. Collectively, the bacterial cells form iron/manganese/aluminium concretions in certain soil horizons (Aristovskaya and Zykina, 1980). In soils where soluble forms of aluminium and manganese increase significantly through acidification, there can be severe negative effects on plant gowth. Aristovskaya and Zykina (1980) suggested that the activities of species of *Metallogenium* are useful indicators of mineral transformations in such soils.

Algae, in particular diatoms, have been used as indicators of the quality of water and sediments (Reid *et al.*, 1995; Whitton and Kelly, 1995). Diatoms occur in almost all aquatic environments and are very sensitive to changes in water chemistry. Diatom valves persist in benthic environments and, therefore, are used extensively as bioindicators for palaeolimnological studies (Reid *et al.*, 1995). This is unlikely to be useful in most soils but, where benthic systems have evolved into dryland environments, diatom valves may provide information about past chemical changes which may influence current soil health.

Lichens, which are symbiotic plants in which photosynthetic algae or cyanobacteria are surrounded by fungal tissue, are known to be sensitive to air pollution. When pollutants become dissolved in water, sensitive lichens growing on soil become poisoned and community structures change. If the algal symbiont is a cyanobacterium which fixes atmospheric nitrogen, pollutants are likely to affect nitrogen fixation and hence the nitrogen status of soils. In any case, the change in the lichen growth and structure may indicate other microbial changes which are otherwise not apparent. Lichens have been used as bioindicators of contamination of soils by heavy metals, non-metals, radionucleotides and organic substances (Richardson, 1991).

Biosensors are bacteria which are especially constructed to detect pollutants in the environment. In these bacteria, reporter genes encoding easily quantifiable characteristics have been integrated into the genome. For example, Burlage *et al.* (1990) fused lux genes from the bioluminescent bacterium *Vibrio fischeri* to promoter genes from the gene operon encoding naphthalene catabolism in *Pseudomonas putida*. This allowed the monitoring of naphthalene biodegradation via quantification of bioluminescence. Such an approach may be useful in both soil and water to sense the presence of chemical agents because there is a specific, quantitative relationship between the initial substrate concentration and bioluminesence (King *et al.*, 1990; Heitzer *et al.*, 1992).

Microbiological Functions as Bioindicators

Microbiological functions are probably much easier to interpret as bioindicators of soil health because the processes being measured may have a direct impact on the productivity of the soil.

Carbon transformations

Perhaps the major activity attributed to microbial populations in soils is the decomposition of organic matter. Soil microbial communities in general are heterotrophic and rely upon the inputs of carbon from outside the microbial community for energy. Therefore, processes involving mineralization of organic carbon and nutrients are likely to have the greatest impact on the functions of microbial communities and hence soil health. Organic matter in soils is largely derived from higher plants. There is a large variety of organisms in the soil responsible for the decomposition of organic carbon and as many as 90% of soil fungi as well as numerous bacteria are capable of utilizing cellulose (Sims, 1990). Measurements of microbial biomass C and the transformations of carbon as indicators of soil health are dealt with elsewhere in this volume (see Sparling, Chapter 5).

Nitrogen transformations

Nitrogen is among the most important nutrients in living systems and therefore, the effects of land degradation on the nitrogen cycle have been investigated extensively. Measurements of ammonification, nitrification and nitrogen fixation have been suggested as important measures of the effects of soil disturbances on soil quality (Visser and Parkinson, 1992) and a few have measured denitrification (Doyle and Stotzky, 1993).

Ammonification

Ammonification is the first step in the mineralization of nitrogen from organic matter, i.e. the formation of ammonium from organic compounds. Ammonification is performed by a wide variety of soil microorganisms, including bacteria, fungi and actinomycetes. Because microorganisms respond differently to perturbations, this diversity of microorganisms gives some stability and therefore, ammonification is relatively insensitive to any disturbance (Somerville *et al.*, 1987). In any soil, as changes occur in aeration or pH, different components of the population of ammonifying microorganisms come into play, although generally the process is slowed by acidification (Alexander, 1977). Alef and Kleiner (1987) examined the applicability of using ammonification as an indicator of microbial activity in a range of soils amended with arginine. They found that liberation of ammonia from arginine correlated well with measurements of oxygen consumption and hence microbial activity. They also found that arginine ammonification correlated well with the carbon content of the soil.

Nitrification

Nitrification is the second step in the mineralization of nitrogen from organic matter, i.e. the oxidation of ammonium to nitrite and nitrate. In contrast to ammonification, nitrification is carried out by only a few bacteria. According to Watson *et al.* (1989) in *Bergey's Manual of Systematic Bacteriology*, there are only nine genera classified as nitrifying bacteria (family *Nitrobacteraceae*), with five genera including *Nitrosomonas* responsible for ammonia oxidation and four genera including *Nitrobacter* responsible for nitrite oxidation. Of these, *Nitrosomonas* and *Nitrobacter* are encountered most frequently and are considered the major nitrifying chemoautotrophs (Alexander, 1977). There are a few heterotrophic microorganisms which can oxidize ammonia, but the amounts of nitrogen transformed are small compared with that attributed to *Nitrosomonas* and *Nitrobacter* (Alexander, 1977). Because of the small number of organisms involved, nitrification is very sensitive to environmental disturbance, i.e. if one organism is affected by a disturbance this has a significant effect on the overall process (Atlas, 1984; Visser and Parkinson, 1992). Nitrification is extremely sensitive to pH and in acid environments it proceeds very slowly (Alexander, 1977; Atlas, 1984). Nitrifying organisms are obligate aerobes and where wet conditions lead to anoxia, nitrification is suppressed (Alexander, 1977; Sims, 1990). Nitrification is usually measured by the disappearance of ammonium in soils dosed with ammonium sulphate or organic forms of nitrogen, coupled with the appearance of nitrite and nitrate. Measurement of nitrification is recommended as a key test for assessing changes in soil health as a consequence of agrochemical applications (Domsch *et al.*, 1983; Somerville *et al.*, 1987) and soil degradation (Sims, 1990). As an example of the latter, Miller *et al.* (1991) enumerated populations of nitrifying bacteria in soils subjected to simulated acid rain at three different pH (5.6, 4.2 and 3.0). They found significant ($P < 0.01$) decreases in the numbers of nitrite-oxidizing bacteria following acid rain but the extent of the decrease depended on the crop and the season. They concluded that nitrite-oxidizing bacteria could be used as experimental indicators of changes in soil microbial communities subjected to acid rain.

Rates of nitrification can be rapid in some soils particularly those with a high clay content (Alexander, 1977) making measurements difficult. However, by altering the temperature, moisture and the level of ammonium fertilization, Malkomes (1992) was able to adjust the sensitivity of the assays so that nitrification could be used as an ecotoxicological indicator for agrochemicals in a range of soils.

For studies on the effects of long-term chemical pollution on soils, Remde and Hund (1994) used a combination of the activity of autotrophic nitrifying bacteria (as an indicator of the potential toxicity of the chemical on soil microorganisms) and microbial respiration (as an indicator of the bioavailability and degradability of the chemicals) and concluded that this provided a comprehensive assessment of the influence of pollutants on soil microflora.

Nitrogen fixation

Two forms of nitrogen fixation can occur in soils: (i) symbiotic nitrogen fixation; and (ii) asymbiotic nitrogen fixation. Domsch *et al.* (1983), Somerville *et al.* (1987) and Brookes (1995) recommend the measurement of symbiotic nitrogen fixation as an indicator of soil stress resulting from pesticides or pollutants. Based on numerous studies reported in a review by Domsch *et al.* (1983) it was concluded that rhizobia are highly sensitive to agrochemicals in soil. Nodulation, nitrogen fixation and growth of legumes all can be limited by pesticides (Goring and Laskowski, 1982; Edwards, 1989). Symbiotic nitrogen fixation is sensitive to heavy metal contamination (Brookes, 1995) and, in the clover–*Rhizobium leguminosarum* biovar *trifolii* association, reduced nitrogen fixation and clover yields have been associated with poorer survival of the rhizobia in contaminated soils (McGrath, 1994). Symbiotic nitrogen fixation is sensitive to soil pH and Brockwell *et al.* (1991) found that pH is a major determinant of the numbers of naturally occurring *Rhizobium meliloti* in soils.

Populations of asymbiotic nitrogen-fixing bacteria may contain organisms from a number of different families and therefore are potentially less sensitive to perturbations. They range from anaerobic clostridia to aerobic *Azotobacter*, although a large number of asymbiotic nitrogen-fixing bacteria require microaerobic conditions to fix nitrogen (Dalton, 1980). Although the activities of asymbiotic nitrogen-fixing bacteria are affected by management practices, the intensity of this response is modified by environmental factors. For example, retaining crop residues increased the numbers of nitrogen-fixing bacteria in soils by 10^2 (Gupta and Roper, 1993) and their activities by more than ten times (Roper, 1983; Roper *et al.*, 1989), but the level of increase was sensitive to changes in moisture and temperature (Roper, 1983, 1985). Roper (1985) showed that soils low in clay content required higher moisture conditions (>−10 kPa water content) to reduce oxygen tensions sufficiently to support nitrogen-fixing activity. Soils which were high in clay content supported nitrogen-fixing activity at lower moisture contents (0.5 times −10 kPa water content) because of the ability of clays to reduce oxygen tensions at microsites in the soil. Generally nitrogen-fixing activity by asymbiotic bacteria is increased in the presence of clays by mechanisms other than providing favourable oxygen tensions, such as through protection from predation and parasitism (Roper and Marshall, 1978) and from desiccation (Bushby and Marshall, 1977), and increased metabolic activity resulting from the concentration of nutrients at surfaces (Stotzky, 1972). Nitrogen-fixing activity is highly responsive to pH, with pH's of 7–7.5 providing the best conditions for activity (Roper and Smith, 1991).

Asymbiotic nitrogen fixation is sensitive to heavy metal contamination (Lorenz *et al.*, 1992; McGrath, 1994; Brookes, 1995). However, Lorenz *et al.* (1992) concluded that asymbiotic nitrogen fixation by cyanobacteria and heterotrophic nitrogen-fixing bacteria is of limited suitability for widespread use as a biological indicator of metal toxicity because of inconsistencies of nitrogen-fixing activity under different conditions.

Denitrification

Denitrification (the reduction of nitrite and nitrate to nitrogen gases), is performed by at least 23 genera of denitrifying bacteria (Firestone, 1982) and is regarded as having a moderate sensitivity to perturbations in the soil (Domsch *et al.*, 1983). Denitrification is limited by the availability of carbon in the soil (Drury *et al.*, 1991) and increases of up to 44-fold in the size of the population of denitrifying bacteria have been observed in soils amended with crop residues (Doran, 1980a). Drury *et al.* (1991) found that the background denitrification rates in soil were strongly correlated with microbial biomass carbon and proposed the latter as an indicator of denitrification. Denitrifying microorganisms are sensitive to soil pH and populations are generally low at pH below 5.5 (Alexander, 1977). Oxygen availability is a critical environmental determinant. Denitrification only proceeds under anaerobic conditions, but oxygen is required for the formation of nitrite and nitrate which are essential for denitrification (Alexander, 1977). Therefore a combination of aerobic and anaerobic sites is required in soil. When soils are undisturbed, anaerobic microsites develop where oxygen demand exceeds supply. Cultivation can destroy these microsites within the soil, and Doran (1980b) observed twice the level of denitrifying bacteria in no-till soils compared with those under conventional cultivation.

Mycorrhizae

Vesicular arbuscular (VA) mycorrhizae, which form a symbiosis with the roots of a host plant, play a significant role in the delivery of nutrients to the plant, especially phosphorus (Hayman, 1980). A mycorrhizal association may also increase the resistance of the plant partner to soil-borne pathogens and to abiotic environmental stress, such as low temperatures, drought and salinity (Schönbeck and Dehne, 1989). The dependence of plants on the VA mycorrhizal symbiosis is a function of the phosphorus concentration in the soil solution (Manjunath and Habte, 1992) and high phosphorus levels are correlated with reduced mycorrhizal development. In support of this, there is increasing evidence of higher levels of VA mycorrhizal fungi in low-input sustainable agricultural systems. For example, Ryan *et al.* (1994) found that the colonization of wheat by VA mycorrhizal fungi was two to three times higher on a farm managed 'organically', without inputs of soluble phosphorus fertilizer, than on a neighbouring farm which was 'conventionally' farmed. Therefore, the extent of mycorrhizal development may be a useful indicator of the history of soluble phosphorus applications.

Mycorrhizal development is modified by soil properties and by land management practices. For example, where there are crop rotations with plants that do not form mycorrhizas, such as lupins and canola, populations of mycorrhizae decline (Abbott *et al.*, 1995). Physical disturbance of the soil through cultivation can severely reduce mycorrhizae due to breakage of hyphal networks (Evans and Miller, 1990; Jasper *et al.*, 1991). Populations of VA mycorrhizal fungi are diverse and have contrasting life cycles. This leads to differences in the way in which various components of the population respond to pH and salinity (Abbott and Robson, 1994).

Abbott *et al.* (1995) reported a calibrated bioassay designed to predict the formation of mycorrhizal associations in field soil. The bioassay measures the percentage of root colonized by mycorrhizae in a test plant as described by Brundrett and Abbott (1994). The calibrated bioassay takes into account the effects of environmental conditions such as plant species and density, soil moisture and temperature, and nutrient supply. As part of the assay, predictions about the formation of mycorrhizal associations are tested at a range of sites with the same plant species used in the bioassay as well as different species, and a range of soil managements. The bioassay should be useful as an indicator of the potential for phosphorus nutrition to a plant/crop via a mycorrhizal symbiont in a soil, and this should assist in making decisions about the need for soluble phosphorus fertilizers.

Developments in the quantification of functional groups and processes

The limitations of culturability and biases of culture media that exist with measuring total microbial populations are also encountered when evaluating functional groups within the population. One possible way to overcome some of these problems is the use of DNA methodology. Gene probes can be useful in two ways to study functional groups within soil by: (i) developing genus- or species-level probes to groups strongly identified with a particular process in soil, e.g. the genus *Azospirillum* which is identified with nitrogen fixation; and (ii) using gene probes which encode a function, e.g. the nitrogenase enzyme to study nitrogen fixation. Both these approaches bypass the need to culture microorganisms from soil because DNA is extracted directly from the soil (Holben, 1994) followed by DNA hybridization or polymerase chain reaction (PCR).

Quantification of functional groups

Using rRNA methodology it has been possible to develop highly specific probes at the genus or species level (Pace *et al.*, 1986; Embley and Stackebrandt, 1996). However, there are problems with using this approach for functional groups because: (i) in general, soil functions are not often associated with only one genus of bacteria or fungi but rather a consortium of rather unrelated microorganisms; and (ii) the technology for using gene probes or PCR in soil in a quantitative fashion is not well advanced. In general, the use of gene probes and PCR in soil has been successful for well-characterized groups such as *Rhizobium* (Saano and Lindstrom, 1992), *Bradyrhizobium* (Holben *et al.*, 1988) and *Frankia* (Picard *et al.*, 1992). Little work has been done with fungi other than plant pathogens (Henson and French, 1993) although genus level probes have been developed for vesicular arbuscular mycorrhizae (Simon *et al.*, 1993). A number of studies have combined most-probable-number (MPN) analysis with DNA hybridization (Fredrickson *et al.*, 1988; Ryder *et al.*, 1994) or PCR (Picard *et al.*, 1992) and this should increase the quantitative potential of this new technology.

Quantification of functions

Very few researchers have used gene probes to study directly specific functions in soil. Smith and Tiedje (1992) developed a gene probe which encoded nitrate reductase (*nir*) which was hybridized to denitrifying bacteria from five separate genera; they were able to use this probe to detect denitrifiers in bioreactors and aquifer microcosms but not in soil. DNA probes encoding genes for mercury resistance (Bruce *et al.*, 1992) and 2,4-D degradation (Holben *et al.*, 1992) have been successfully hybridized in soil. In the latter example, using quantitative DNA hybridization, Holben *et al.* (1992) observed an increase in 2,4-D degrading microorganisms in soils after amendment with 2,4-D. The combination of gene probes encoding enzymes with specific functions and the use of quantitative PCR or DNA hybridization may allow a more direct quantification of function in soil.

The BIOLOG system provides information about the metabolic function of the soil as a whole rather than about specific functions. In this system, individual microorgansisms or whole soils are characterized on the basis of the utilization of a range of carbon sources (Garland and Mills, 1991; Winding and Hendrikson, 1992). Principal component analysis on the information generated from BIOLOG can be used to differentiate soil communities. This may be useful for identifying responses to changes in soils.

Conclusions

Since the discovery of complex microbial communities in soil, attempts have been made to use them as indicators of soil quality/health. In the early 1900s Waksman considered several microbial criteria as indicators of soil fertility, including numbers of microorganisms, nitrifying capacity, CO_2 evolution, cellulose decomposing and nitrogen-fixing capacities (Waksman, 1927). However, he also recognized that soil properties, climatic factors, and cropping and soil management all altered microbial behaviour in the soil making it difficult to use microbial criteria as indicators of soil fertility. Somerville (1987) listed some of the effects of environmental and management factors on microbial processes and showed that factors such as changes in temperature and water potential, physical disturbance, pH variations and changes in soil aeration and nutrient supply could produce fluctuations ranging from 50 to 100% in the size and activity of microbial populations. This makes it almost impossible to use microbiological populations and processes as direct indicators of soil health in order to compare different soils. With our current state of knowledge, microbial indicators are probably only useful for measuring changes in soil health resulting from some perturbation (degradation/rehabilitation). Even then, it is essential that adjacent to the test site there are controls which have been subject to the same environmental/management factors as the test site other than the perturbation being evaluated.

A further complication in sampling soils in the field for microbial populations and activities is spatial heterogeneity (Cook and Greaves, 1987; Robertson, 1994) and frequently within-field variability is as high as between-field variability. Cook and Greaves (1987) found that, by collecting and pooling several cores from the same treatment/site to give composite samples, the estimates they produced reduced spatial variation but still allowed monitoring of temporal changes. More recently, there have been developments in the area of geostatistics which attempt to describe spatial heterogeneity or spatial autocorrelation and the impact of managements upon that heterogeneity. Robertson (1994) demonstrated that geostatisics provides a means of defining spatial autocorrelation and of using the degree and scale of autocorrelation to interpolate distributions across an area. This new technology promises to provide information about the impact of soil properties and managements on spatial variability of microbial populations and processes in soils.

Perhaps one of the biggest limitations to the use of microbial populations and their processes as indicators of soil health is our lack of information about the impact of ecological parameters on the 90% or more of the microbial populations in soil that have not yet been cultured or identified. New technologies such as FAME and rRNA methodologies can be used to identify or at least place phylogenetically some unculturable microorganisms, but these technologies are still in their infancy when it comes to telling us about the function or response of such organisms to changes in conditions.

The development of appropriate theories and methodologies for identifying ecologically significant effects should be a high priority in the study of microbial ecology (Doyle and Stotzky, 1993). This should lead to a definition of the factors to be included in a data set of microbial indicators of soil health. Research should be directed at identifying cost effective biological parameters that accurately respond to perturbations and include both populations and functions of the microbial community which determine soil quality/health (Turco *et al.*, 1994; Brookes, 1995). In the past, the best indicators of change have been based on the measurement of multiple microbial parameters, such as the components of populations and/or particular functions (Maltby, 1975; Zwolinski *et al.*, 1987; Sims, 1990; Remde and Hund, 1994; Brookes, 1995). Consideration of these factors along with the physical and chemical characteristics of the soil is essential. As new technologies develop it is clear that there is considerable potential for increasing our understanding of the microbiological factors that determine soil health.

References

Abbott, L.K. and Robson, A.D. (1994) The impact of agricultural practices on mycorrhizal fungi. In: Pankhurst, C.E., Doube, B.M., Gupta, V.V.S.R. and Grace, P.R. (eds) *Soil Biota: Management in Sustainable Farming Systems.* CSIRO Press, Melbourne, pp. 88–95.

Abbott, L.K., Robson, A.D. and Scheltema, M.A. (1995) Managing soils to enhance mycorrhizal benefits in mediterranean agriculture. In: Varma, A. (ed.) *Fungi in Tropical and Stressed Soils for Crop Production. Critical Reviews in Biotechnology.* CRC Press, Boca Raton, Florida vol.15, pp. 213–228.

Alef, K. and Kleiner, D. (1987) Applicability of arginine ammonification as indicator of microbial activity in different soils. *Biology and Fertility of Soils* 5, 148–151.

Alexander, M. (1977) *Introduction to Soil Microbiology.* John Wiley, New York.

Allison, M.F. and Killham, K. (1988) Response of soil microbial biomass to straw incorporation. *Journal of Soil Science* 39, 237–242.

Apsimon, H., Thornton, I., Fyfe, W., Hong, Y., Leggett, J., Nriagu, J.O., Pacyna, J.N., Page, A.L., Price, R., Skinner, B., Steinnes, E. and Yim, W. (1990) Anthropogenically induced global change – report of working group three, IUGS workshop on global change past and present. *Palaeogeography Palaeoclimatology Palaeoecology* 82, 97–111.

Aristovskaya, T.V. and Zykina, L.V. (1980) Microorganisms as indicators of the accumulation of iron, aluminium, and manganese in soils. *Soviet Soil Science* 11, 36–43.

Atlas, R.M. (1984) Use of microbial diversity measurements to assess environmental stress. In: Klug, M.J. and Reddy, C.A. (eds) *Current Perspectives in Microbial Ecology*. American Society for Microbiology, Washington, DC, pp. 540–545.

Beare, M.H., Parmelee, R.W., Hendrix, P.F. and Cheng, W. (1992) Microbial and faunal interactions and effects on litter nitrogen and decomposition in agroecosystems. *Ecological Monographs* 62, 569–591.

Bentham, H., Harris, J.A., Birch, P. and Short, K.C. (1992) Habitat classification and soil restoration assessment using analysis of soil microbiological and physicochemical characteristics. *Journal of Applied Ecology* 29, 711–718.

Bentivenga, S.P. and Morton, J.B. (1994) Stability and heritability of fatty acid methyl ester profiles of glomalean endomycorrhizal fungi. *Mycological Research* 98, 1419–1426.

Bermingham, S., Maltby, L. and Cooke, R.C. (1995) A critical assessment of the validity of ergosterol as an indicator of fungal biomass. *Mycological Research* 99, 479–484.

Brockwell, J., Pilka, A. and Holliday, R.A. (1991) Soil pH is a major determinant of the numbers of naturally occurring *Rhizobium meliloti* in non-cultivated soils in central New South Wales. *Australian Journal of Experimental Agriculture* 31, 211–219.

Brookes, P.C. (1995) The use of microbial parameters in monitoring soil pollution by heavy metals. *Biology and Fertility of Soils* 19, 269–279.

Bruce, K.D., Hiorns, W.D., Hobman, J.L., Osborn, A.M., Strike, P. and Ritchie, D.A. (1992) Amplification of DNA from native populations of soil bacteria by using the polymerase chain reaction. *Applied and Environmental Microbiology* 58, 3413–3416.

Brundrett, M.C. and Abbott, L.K. (1994) Mycorrhizal fungus propagules in the jarrah forest. I. Seasonal study of inoculum levels. *New Phytologist* 127, 539–546.

Burlage, R.S., Sayler, G.S. and Larimer, F. (1990) Monitoring of naphthalene catabolism by bioluminescence with nah-lux transcriptional fusions. *Journal of Bacteriology* 172, 4749–4757.

Bushby, H.V.A. and Marshall, K.C. (1977) Some factors affecting the survival of root-nodule bacteria on desiccation. *Soil Biology and Biochemistry* 9, 143–147.

Camper, N.D., Moherek, E.A. and Huffman, J. (1973) Changes in microbial populations in paraquat-treated soil. *Weed Research* 13, 231–233.

Cook, K.A. and Greaves, M.P. (1987) Natural variability in microbial activities. In: Somerville, L. and Greaves, M.P. (eds) *Pesticide Effects on Soil Microflora*. Taylor and Francis, London, pp. 15–43.

Dalton, H. (1980) The cultivation of diazotrophic microorganisms. In: Bergersen, F.J. (ed.) *Methods for Evaluating Biological Nitrogen Fixation*. John Wiley & Sons, Chichester, pp. 13–64.

Domsch, K.H., Jagnow, G. and Anderson, T.-H. (1983) An ecological concept for the assessment of side-effects of agrochemicals on soil microorganisms. *Residue Reviews* 86, 65–105.

Doran, J.W. (1980a) Microbial changes associated with residue management with reduced tillage. *Soil Science Society of America Journal* 44, 518–524.

Doran, J.W. (1980b) Soil microbial and biochemical changes associated with reduced tillage. *Soil Science Society of America Journal* 44, 765–771.

Doran, J.W. (1982) Microorganisms and the biological cycling of selenium. *Advances in Microbial Ecology* 6, 1–32.

Doyle, J.D. and Stotzky, G. (1993) Methods for the detection of changes in the microbial ecology of soil caused by the introduction of microorganisms. *Microbial Releases* 2, 63–72.

Drury, C.F., McKenney, D.J. and Findlay, W.I. (1991) Relationships between denitrification, microbial biomass and indigenous soil properties. *Soil Biology and Biochemistry* 23, 751–755.

Edwards, C.A. (1989) Impact of herbicides on soil ecosystems. *Critical Reviews in Plant Science* 8, 221–257.

Embley, T.M. and Stackebrandt, E. (1996) The use of 16S ribosomal RNA sequences in microbial ecology. In: Pickup, R.W. and Saunders, J.R. (eds) *Ecological Approaches to Environmental Microbiology*. Prentice Hall, Oxford, pp. 39–62.

Evans, D.G. and Miller, M.H. (1990) The role of the external mycelial network in the effect of soil disturbance upon vesicular-arbuscular mycorrhizal colonization of maize. *New Phytologist* 114, 65–71.

Firestone, M.K. (1982) Biological denitrification. In: Stevenson, F.J. (ed.) *Nitrogen in Agricultural Soils*. Agronomy No. 22, American Society of Agronomy, Madison, Wisconsin, pp. 289–326.

Frankenberger, W.T. and Dick, W.A. (1983) Relationships between enzyme activities and microbial growth and activity indices in soil. *Soil Science Society of America Journal* 47, 945–951.

Frederickson, J.K., Bezdicek, D.F., Brockman, F.J. and Li, S.W. (1988) Enumeration of Tn5 mutant bacteria in soil by using a most-probable-number-DNA hybridization procedure and antibiotic resistance. *Applied and Environmental Microbiology* 54, 446–453.

Garland, J.L. and Mills, A.L. (1991) Classification and characterization of heterotrophic microbial communities on the basis of patterns of community-level sole-carbon-source utilization. *Applied and Environmental Microbiology* 57, 2351–2359.

Giovannoni, S.J., Britschgi, T.B., Moyer, C.L. and Field, K.G. (1990) Genetic diversity in Sargasso sea bacterioplankton. *Nature* 345, 60–63.

Goring, C.A.I. and Laskowski, D.A. (1982) The effects of pesticides on nitrogen transformations in soils. In: Stevenson, F.J. (ed.) *Nitrogen in Agricultural Soils*. Agronomy No. 22, American Society of Agronomy, Madison, Wisconsin, pp. 689–720.

Gupta, V.V.S.R. and Roper, M.M. (1993) Impact of different crop residue management systems on the functional groups of microflora and microfauna. *Australian Microbiologist* 14, 99.

Gupta, V.V.S.R., Roper, M.M., Kirkegaard, J.A. and Angus, J.F. (1994) Changes in microbial biomass and organic matter levels during the first year of modified tillage and stubble management practices on a red earth. *Australian Journal of Soil Research* 32, 1339–1354.

Haack, S.K., Garchow, H., Odelson, D.A., Forney, L.J. and Klug, M.J. (1994) Accuracy, reproducibility and interpretation of fatty acid methyl ester profiles of model bacterial communities. *Applied and Environmental Microbiology* 60, 2483–2493.

Harris, J.A. and Birch, P. (1992) Land reclamation and restoration. In: Fry, J.C.,

Gadd, G.M., Herbert, R.A., Jones, C.W. and Watson-Craik, I.A. (eds) *Microbial Control of Pollution.* Society of General Microbiology Symposium No. 48, Cambridge University Press, pp. 269–291.

Hawksworth, D.L. and Mound, L.A. (1991) Biodiversity databases: the crucial significance of collections. In: Hawksworth, D.L. (ed.) *The Biodiversity of Microorganisms and Invertebrates: Its Role in Sustainable Agriculture.* CAB International, Wallingford, UK, pp. 17–29.

Hayman, D.S. (1980) Mycorrhiza and crop production. *Nature* 287, 487–488.

Heijnen, C.E., Van Elsas, J.D., Kuikman, P.J. and Van Veen, J.A. (1988) Dynamics of *Rhizobium leguminosarum* biovar *trifolii* introduced in soil: the effect of bentonite clay on predation by protozoa. *Soil Biology and Biochemistry* 20, 483–488.

Heitzer, A., Webb, O.F., Thonnard, J.E. and Sayler, G.S. (1992) Specific and quantitative assessment of naphthalene and salicylate bioavailability by using a bioluminescent catabolic reporter bacterium. *Applied and Environmental Microbiology* 58, 1839–1846.

Henson, J.M. and French, R. (1993) The polymerase chain reaction and plant disease diagnosis. *Annual Review of Phytopathology* 31, 81–109.

Holben, W.E. (1994) Isolation and purification of bacterial DNA from soil. In: Weaver, R.W. *et al.* (eds) *Methods in Soil Analysis. Part 2. Microbiological and Biochemical Properties.* Soil Science Society of America, Madison, Wisconsin. pp. 727–751.

Holben, W.E., Jansson, J.K., Chelm, B.K. and Tiedje, J.M. (1988) DNA probe method for the detection of specific microorganisms in the soil bacterial community. *Applied and Environmental Microbiology* 54, 703–711.

Holben, W.E., Schroeter, B.M., Calabrese, V.G.M., Olsen, R.H., Kukor, J.K., Biederbeck, V.O., Smith, A.E. and Tiedje, J.M. (1992) Gene probe analysis of soil microbial populations selected by amendment with 2,4-dichlorophenoxyacetic acid. *Applied and Environmental Microbiology* 58, 3941–3948.

Holloway, J.D. and Stork, N.E. (1991) The dimension of biodiversity: the use of invertebrates as indicators of human impact. In: Hawksworth, D.L. (ed.) *The Biodiversity of Microorganisms and Invertebrates: Its Role in Sustainable Agriculture.* CAB International, Wallingford, UK, pp. 37–63.

Jasper, D.A., Abbott, L.K. and Robson, A.D. (1991) The effect of soil disturbance on vesicular-arbuscular mycorrhizal fungi in soils from different vegetation types. *New Phytologist* 118, 471–476.

Jong, S.C. (1989) Microbial germplasm. In: Knutson, L. and Stone, A.K. (eds) *Biotic Diversity and Germplasm Preservation, Global Imperatives.* Kluwer Academic Publications, Dordrecht, pp. 241–273.

Jordan, D. and Kremer, R.J. (1994) Potential use of soil microbial activity as an indicator of soil quality. In: Pankhurst, C.E., Doube, B.M., Gupta, V.V.S.R. and Grace, P.R. (eds) *Soil Biota. Management in Sustainable Farming Systems.* CSIRO Press, Melbourne, Australia, pp. 245–249

Kale, S.P. and Raghu, K. (1989) Relationship between microbial numbers and other microbial indices in soil. *Bulletin of Environmental Contamination and Toxicology* 43, 941–945.

King, J.M.H., DiGrazia, P.M., Applegate, B., Burlage, R., Sanseverino, J., Dunbar, P., Larimer, F. and Sayler, G.S. (1990) Rapid sensitive bioluminescent reporter technology for naphthalene exposure and biodegradation. *Science* 249, 778–781.

Liesack, W. and Stackebrandt, E. (1992) Occurrence of novel groups of the Domain Bacteria as revealed by analysis of genetic material isolated from an Australian terrestrial environment. *Journal of Bacteriology* 174, 5072–5078.

Lorenz, S.E., McGrath, S.P. and Giller, K.E. (1992) Assessment of free-living nitrogen fixation activity as a biological indicator of heavy metal toxicity in soil. *Soil Biology and Biochemistry* 24, 601–606.

Malkomes, H.-P. (1992) Nitrification as an ecotoxicological indicator for agrochemicals in soil under different test conditions. *Zentralblatt für Mikrobiologie* 147, 250–260.

Maltby, E. (1975) Numbers of soil microorganisms as ecological indicators of changes resulting from moorland reclamation on Exmoor, UK. *Journal of Biogeography* 2, 117–136.

Manjunath, A. and Habte, M. (1992) External and internal P requirements of plant species differing in their mycorrhizal dependency. *Arid Soil Research and Rehabilitation* 6, 271–284.

Martyniuk, S. and Wagner, G.H. (1978) Quantitative and qualitative examination of soil microflora associated with different management systems. *Soil Science* 125, 343–350.

McGrath, S.P. (1994) Effects of heavy metals from sewage sludge on soil microbes in agricultural ecosystems. In: Ross, S.M. (ed.) *Toxic Metals in Soil–Plant Systems.* John Wiley and Sons, New York, pp. 247–274.

Mele, P.M. and Carter, M.R. (1993) Effect of climatic factors on the use of microbial biomass as an indicator of changes in soil organic matter. In: Mulongoy, K. and Merckx, R. (eds) *Soil Organic Matter Dynamics and Sustainability of Tropical Agriculture.* John Wiley and Sons Ltd, Chichester, pp. 57–63.

Microbial Resources Panel (1981) Proceedings of the United States Strategy Conference on Biological Diversity. Nov. 16–18, 1981.

Mikhnovskaya, A.D., Kirichenko, T.P. and Panchenko, V.P. (1993) Microbiological processes of transformation of organic matter with different cultivation practices of typical chernozem. *Eurasian Soil Science* 25, 100–109.

Miller, K.W., Cole, M.A. and Banwart, W.L. (1991) Microbial populations in an agronomically managed mollisol treated with simulated acid rain. *Journal of Environmental Quality* 20, 845–849.

Moore, J.C. and de Ruiter, P.C. (1993) Assessment of disturbance on soil ecosystems. *Veterinary Parasitology* 48, 75–85.

Pace, N.R., Stahl, D.A., Lane D.J. and Olsen, G.J. (1986) The analysis of natural microbial populations by ribosomal RNA sequences. *Advances in Microbial Ecology* 9, 1–55.

Parr, J.F., Papendick, R.I., Hornick, S.B. and Meyer, R.E. (1992) Soil quality: attributes and relationship to alternative and sustainable agriculture. *American Journal of Alternative Agriculture* 7, 5–11.

Paul, E.A. and Clark, F.E. (eds) (1989) *Soil Microbiology and Biochemistry.* Academic Press, New York.

Picard, C., Ponsonnet, C., Paget, E., Nesme, X. and Simonet, P. (1992) Detection and enumeration of bacteria in soil by direct DNA extraction and polymerase chain reaction. *Applied and Environmental Microbiology* 58, 2717–2722.

Reid, M.A., Tibby, J.C., Penny, D. and Gell, P.A. (1995) The use of diatoms to assess past and present water quality. *Australian Journal of Ecology* 20, 57–64.

Remde, A. and Hund, K. (1994) Response of soil autotrophic nitrification and soil respiration to chemical pollution in long-term experiments. *Chemosphere* 29, 391–404.

Richardson, D.H.S. (1991) Lichens as biological indicators – recent developments. In: Jeffrey, D.W. and Madden, B. (eds) *Bioindicators and Environmental Management*. Academic Press, London, pp. 263–272.

Robertson, G.P. (1994) The impact of soil and crop management practices on soil spatial heterogeneity. In: Pankhurst, C.E., Doube, B.M., Gupta, V.V.S.R. and Grace, P.R. (eds) *Soil Biota. Management in Sustainable Farming Systems*. CSIRO Press, Melbourne, Australia, pp. 156–161.

Roper, M.M. (1983) Field measurements of nitrogenase activity in soils amended with wheat straw. *Australian Journal of Agricultural Research* 34, 725–739.

Roper, M.M. (1985) Straw decomposition and nitrogenase activity (C_2H_2 reduction): effects of soil moisture and temperature. *Soil Biology and Biochemistry* 17, 65–71.

Roper, M.M. and Halsall, D.M. (1986) Use of products of straw decomposition by N_2-fixing (C_2H_2-reducing) populations of bacteria in three soils from wheat-growing areas. *Australian Journal of Agricultural Research* 37, 1–9.

Roper, M.M. and Marshall, K.C. (1978) Effects of a clay mineral on microbial predation and parasitism of *Escherichia coli*. *Microbial Ecology* 4, 279–289.

Roper, M.M. and Smith, N.A. (1991) Straw decomposition and nitrogenase activity (C_2H_2 reduction) by free-living microorganisms from soil: effects of pH and clay content. *Soil Biology and Biochemistry* 23, 275–283.

Roper, M.M., Marschke, G.W. and Smith, N.A. (1989) Nitrogenase activity (C_2H_2 reduction) in soils following wheat straw retention: effects of straw management. *Australian Journal of Agricultural Research* 40, 241–253.

Ryan, M.H., Chilvers, G.A. and Dumaresq, D.C. (1994) Colonisation of wheat by VA-mycorrhizal fungi was found to be higher on a farm managed in an organic manner than on a conventional neighbour. *Plant and Soil* 160, 33–40.

Ryder, M., Ophel-Keller, K.M., Dindsa, H. and Veal, D. (1994) Detection and characterisation of a genetically modified biocontrol bacterium in field soil. In: Ryder, M.H., Stephens, P.M. and Bowen, G.D. (eds) *Improving Plant Productivity with Rhizosphere Bacteria*. CSIRO Press, Melbourne Australia, pp. 261–266.

Saano, A. and Lindstrom, K. (1992) Detection of *Rhizobium galegae* from one-gram soil samples by non-radioactive DNA–DNA hybridization. *Soil Biology and Biochemistry* 24, 969–977.

Schönbeck, F. and Dehne, H.-W. (1989) VA-mycorrhizae and plant health. In: Vancura, V. and Kunc, F. (eds) *Interrelationships between Microorganisms and Plants in Soil*. Proceedings of an International Symposium, Liblice, Czechoslavakia. Elsevier, Amsterdam, pp. 83–91.

Sikora, L.J. and McCoy, J.L. (1990) Attempts to determine available carbon in soils. *Biology and Fertility of Soils* 9, 19–24.

Simon, L., Levesque, R.C. and Lalonde, M. (1993) Identification of endomycorrhizal fungi colonising roots by fluorescent single-strand conformation polymorphism-polymerase chain reaction. *Applied and Environmental Microbiology* 59, 4211–4215.

Sims, G.K. (1990) Biological degradation of soil. *Advances in Soil Science* 11, 289–330.

Smith, G.B. and Tiedje, J.M. (1992) Isolation and characterization of a nitrate reductase gene and its use as a probe for denitrifying bacteria. *Applied and Environmental Microbiology* 58, 376–384.

Somerville, L. (1987) Perspectives on side-effects testing. In: Somerville, L. and Greaves, M.P. (eds) *Pesticide Effects on Soil Microflora.* Taylor and Francis, London, pp. 5–13.

Somerville, L., Greaves, M.P., Domsch, K.H., Verstraete, W., Poole, N.J., van Dijk, H. and Anderson, J.P.E. (1987) Recommended laboratory tests for assessing the side effects of pesticides on soil microflora. In: Somerville, L. and Greaves, M.P. (eds) *Pesticide Effects on Soil Microflora.* Taylor and Francis, London, pp. 205–229.

Sorheim, R., Torsvik, R. and Goksoyr, J. (1989) Phenotypical divergences between populations of soil bacteria isolated on different media. *Microbial Ecology* 17, 181–192.

Stotzky, G. (1972) Activity, ecology, and population dynamics of microorganisms in soil. *CRC Critical Reviews in Microbiology* 2, 59–137.

Tunlid, A. and White, D.C. (1990) Biochemical analysis of biomass, community structure, nutritional status and metabolic activity of microbial communities in soil. In: Stotzky, G. and Bollag, J.M. (eds) *Soil Biochemistry. Vol. 7.* Marcel Dekker, New York, pp. 229–262.

Turco, R.F., Kennedy, A.C. and Jawson, M.D. (1994) Microbial indicators of soil quality. In: Doran, J.W., Coleman, D.C., Bezdicek, D.F. and Stewart, B.A. (eds) *Defining Soil Quality for a Sustainable Environment.* Soil Science Society of America Special Publication No. 35, Madison, Wisconsin, pp. 73–90.

Visser, S. and Parkinson, D. (1992) Soil biological criteria as indicators of soil quality: soil microorganisms. *American Journal of Alternative Agriculture* 7, 33–37.

Waksman, S.A. (1927) *Principles of Soil Microbiology.* Williams and Wilkins, Baltimore.

Waksman, S.A. (1952) *Soil Microbiology.* John Wiley, New York.

Ward, D.M., Weller, R. and Bateson, M.M. (1990) 16S rRNA sequences reveal numerous uncultured microorganisms in a natural community. *Nature* 345, 63–65.

Watson, S.W., Bock, E., Harms, H., Koops, H.-P. and Hooper, A.B. (1989) Aerobic chemolithotrophic bacteria and associated organisms. A. Nitrifying bacteria. In: Staley, J.T., Bryant, M.P., Pfenning, N. and Holt, J.G. (eds) *Bergey's Manual of Systematic Bacteriology, Volume 3.* Williams and Wilkins, Baltimore, pp. 1808–1834.

West, A.G., Grant, W.D. and Sparling, G.P. (1987) Use of ergosterol, diaminopimelic acid and glucosamine contents of soils to monitor changes in microbial populations. *Soil Biology and Biochemistry* 19, 607–612.

Whitton, B.A. and Kelly, M.G. (1995) Use of algae and other plants for monitoring rivers. *Australian Journal of Ecology* 20, 45–56.

Winding, A. and Hendriksen, N.B. (1992) BIOLOG microtiterplates for metabolic analysis of soil communities. *Proceedings of the International Symposium of Microbial Ecology,* Barcelona, Spain, p. 246. (Abstract).

Zelles, L., Bai, Q.Y., Beck, T. and Beese, F. (1992) Signature fatty acids in phospholipids and lipopolysaccharides as indicators of microbial biomass and community structure in agricultural soils. *Soil Biology and Biochemistry* 24, 317–323.

Zwolinski, J., Olszowska, G. and Zwolinska, B. (1987) Soil biological activity as an indicator of industrial pressure on the forest environment (microbiological and biochemical activity). *Acta Agraria et Silvestria* 26, 25–44.

8

Potential Use of Plant Root Pathogens as Bioindicators of Soil Health

D. Hornby and G.L. Bateman
IACR – Rothamsted, Harpenden, Hertfordshire AL5 2JQ, UK

Introduction

Some symbioses involving fungi are useful bioindicators. Lichens, for instance, attracted interest as indicators of air purity as long ago as last century (Roloff, 1989), because, in general, more woolly and frondose species occur in the least polluted atmospheres (Buczacki and Harris, 1981). Ectotrophic mycorrhizae may be useful indicators of forest decline, for example through direct or indirect (e.g. via host photosynthesis) effects of pollutants on root symbioses (Horak, 1991). Interest in the potential usefulness of some basidiomycete fungi (e.g. species of *Lepista, Lycoperdon* and *Mycena*) as indicators of soil pollution by heavy metals is relatively recent and currently limited by the need for more detailed information about the factors influencing uptake and translocation of heavy metals (Wondratschek and Röder, 1993). Similarly, some soil microfungi with the ability to solubilize otherwise insoluble metal compounds may provide information on fungal populations in polluted habitats and on metal tolerance (Sayer *et al.*, 1995). Inside buildings, certain fungi (particularly *Stachybotrys*) in air samples, or isolated from surfaces, are useful in indicating moisture and health problems associated with indoor mycofloras (Samson *et al.*, 1994).

A literature search indicated a big increase in publications on the subject of bioindicators in recent years, but fungal root pathogens were hardly, if at all, mentioned. This apparent absence of published information implies no formalized use of root pathogens as bioindicators and explains why instances of their use as bioindicators of soil health do not spring readily to mind. However, many of the principles of bioindication are in use in the day-to-day business of plant pathology, because a diseased plant indicates a problem involving a pathogen that, in cultivated plants at least, needs a solution. 'Biodiversity', like 'bioindicator', is a word that has come much to the fore in the last few years

 Biological Indicators of Soil Health
(eds C.E. Pankhurst, B.M. Doube and V.V.S.R. Gupta)

and describes something that may be useful as an indicator (Pankhurst, 1994). In questioning what 'biodiversity' represents, Harper amd Hawksworth (1995) listed the possibilities: a new linguistic bottle for the wine of old ideas, a changed fashion label designed to attract funding, or new and fundamental questions in science. The idea of using root pathogens as bioindicators needs to be questioned similarly, except that the last item might be phrased 'or new and useful applications for plant pathological knowledge'. This chapter opens the debate, concentrating on issues, rather than attempting a comprehensive coverage of possible uses of root pathogens as bioindicators.

Types and Properties of Environmental Indicators and Bioindicators

Bioindication has been defined as the use of an organism, a part of an organism, or a society of organisms to obtain information on the quality of all or a part of its environment; bioindicators are the organisms that give this information (Wittig, 1993). Others (Pankhurst, 1994) draw attention also to the bioindication possibilities of processes and the end products of activity. Types of bioindicators and environmental monitors have been defined and classified in a variety of ways (e.g. Chaphekar, 1978; Roloff, 1989; Wittig, 1993; MacGillivray, 1994). For the purposes of the subsequent discussion, three main types of bioindicator and some sub-categories are presented simply in Table 8.1, with notes on the type of information each might provide. The categories are not exclusive and there is some overlap between the monitoring type and the test or indicator type,

Table 8.1. Types of bioindicators.

Type	Information provided
Indicator	Presence or absence allows inferences about an environmental problem, rarely quantitative
Test	Response indicates extent of problem; tests usually highly standardized, e.g. cress to test for air pollution (Garber, 1974)
Monitor	Provides evidence of changes; quantitative conclusions may be possible with calibration
Passive	Monitoring organisms already present in an ecosystem
Active	Monitoring organisms introduced
Reactor	Response is a functional change, or reaction
Accumulator	Response is an accumulation of pollutants

After Roloff (1989).

both of which may also be passive or active and reactors or accumulators (Roloff, 1989).

Soil Health and Root Pathogens

What is soil health?

This is hard to define. One view is that a 'healthy soil' is a productive soil, characterized by the presence of organic matter and the absence of physical, chemical and biological constraints (Pankhurst, 1994). 'Soil quality' and 'fertility', terms used in discussions of biodiversity and agricultural sustainability, convey similar ideas (Kennedy and Smith, 1995). Soil quality is 'the soil's capacity to function in a desired manner such as to produce healthy crops, animals and humans, resist erosion, and minimize environmental impacts' (Parr *et al.*, 1992). It has been argued (Pankhurst, 1994) that the evolution of agroecosystems may have led to loss of species diversity and biomass. Conservation, sustainability in agriculture, and productivity and profitability with reduced input and maximal use of natural resources are now advocated widely.

In agriculture, poor soil health could mean that (i) a chosen crop plant fails to grow well, (ii) a crop does not grow well in certain environmental conditions, or (iii) no crop plant grows well. Although all these outcomes are possible in the absence of disease, disease is often a factor in the first two instances and where no obvious above-ground symptoms occur the possibility of root disease should be investigated. This possibility justifies the inclusion of root diseases as a biological constraint affecting soil health. If no, or very few, plants grow well in a plot of land, this is unlikely to be the result of a disease problem caused by a fungus, since even general pathogens, such as those that cause damping-off disease, seem not to be a scourge of all plant species. *Phytophthora cinnamomi*, for instance, is an important, soil-borne plant pathogen with a wide host range and worldwide distribution. In central Taiwan, avocado (*Persea americana*), an introduced species, suffered severe decline caused by *P. cinnamomi*, whereas the native flora in the natural forest remained healthy, despite the presence of both A1 and A2 mating types of the fungus (Ko *et al.*, 1978).

Indicators of soil health

Soil condition is an indicator of the state of the environment (MacGillivray, 1994). The soil has many vital roles in food and timber production, maintenance of biodiversity, as a reservoir for water, and a buffer and filter for pollutants, so its protection is crucial for sustainability. Agricultural intensification, afforestation with commercial timber plantations and pollution have resulted in some loss of soil function and structure in localized areas of the UK (MacGillivray, 1994). However, it is difficult to find a single emotive and meaningful factor to use as a fundamental indicator, because soils vary widely. A one-off set of data on organic matter content, for example, will be of limited value (see Sparling, Chapter 5, this volume).

Aspects of soil condition that might be measured by a bioindicator include productivity, stability, sustainability and pollution. Amongst indicators of good productivity and sustainability, Pankhurst (1994) listed low levels of soil-borne disease. Further, amongst indicators of soil pollution, he listed presence or absence of certain organisms and populations resistant to toxicants.

The scheme in Fig. 8.1 is based on the assumption that bioindicators of soil health will be biological entities, the processes they carry out, or the end product of their activity (Pankhurst, 1994). It suggests some bioindication possibilities for root pathogens (biological entities) and comments on disease (the product), but not on infection (the process leading to disease). Although it is quite possible that physiological processes (e.g. the formation of lignitubers in challenged cells) could be related to soil health, we know of no clear examples with bioindication potential.

The rationale behind using root pathogens as bioindicators for detecting and qualifying trends in soil health and sustainability needs exploring. This also applies to ideas of using root pathogens with indicators of other types and possibly in indicator packages, or using functional groups, or functional processes, that behave as bioindicators and have root pathogens as useful components.

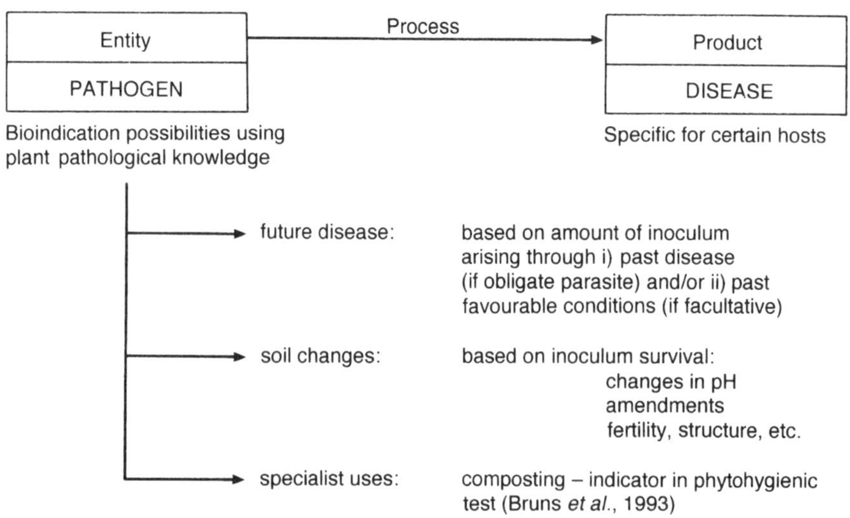

Fig. 8.1. A scheme showing some bioindicator possibilities for root pathogens and a limitation of using disease as a general indicator of soil health.

There has been much research on root pathogens: how they react to soil and crop management, their interactions with physicochemical properties of the soil, and the extent to which they are constrained by climatic and geographic zones and plant production systems. The point at issue, therefore, is whether all this information and interpretations of it can enhance the potential usefulness of root pathogens as bioindicators of soil health.

Superficially it may seem reasonable to claim that no soil is healthy if root disease is present, that consequently severity of disease is a potential health indicator and that measuring the pathogen is a simple way of quantifying the disease. This is all too simplistic: the diseased condition may be only temporary, for instance because of unfavourable weather; low levels of disease may, on occasions, actually increase crop yields (Hornby and Bateman, 1990), severe disease may be the prelude to disease decline and weather and microbial antagonism may supplant inoculum as major determinants of disease (Hornby, 1981).

Root pathogens

If high levels of soil-borne plant pathogens are currently considered an indication of poor soil health (Pankhurst, 1994), this is, presumably, because of the occurrence or threat of disease. Therefore, in contrast to recognized conventional bioindicators, a root pathogen would be the indicator of a problem of which it is the cause. The bioindicator would also be an organism that needs suppressing. Although the presence of a root pathogen in soil is an indication of disease, past, present, or future, it is not necessarily an indication of an environmental problem causing a deterioration of soil health. The severity of disease caused by soil-borne plant pathogens will not always relate directly to amounts of inoculum present in the soil, but will be the outcome of an interaction between this inoculum, the plant and the environment (the 'disease triangle', or 'disease pyramid' if time is included as a factor (Browning *et al.*, 1977)). Root pathogens could, of course, exacerbate the effects of other problems such as poor soil structure, nutrient deficiency, drought or waterlogging. Such interactions are an important consideration in any use of bioindicators. For example, to avoid misinterpreting results from plants used as bioindicators of air pollution, the effect of environmental factors on sensitivity must be taken into account (Heidt and Kehlberger, 1983).

Presence or absence of undesirable organisms (e.g. deleterious rhizobacteria and root-pathogenic bacteria), or beneficial fungi (e.g. decomposers of cellulosic and lignified plant residues and those that form mycorrhizae) may be used as indicators of soil health and productivity (Pankhurst, 1994), but the simple presence of root pathogens is not necessarily a reliable indication of poor soil health. For disease to become a soil health problem, the infection process and disease development are required (Fig. 8.1) and, as Fig. 8.2 shows, disease is not just dependent on a host and the presence of a pathogen, but is influenced by extraneous (e.g.

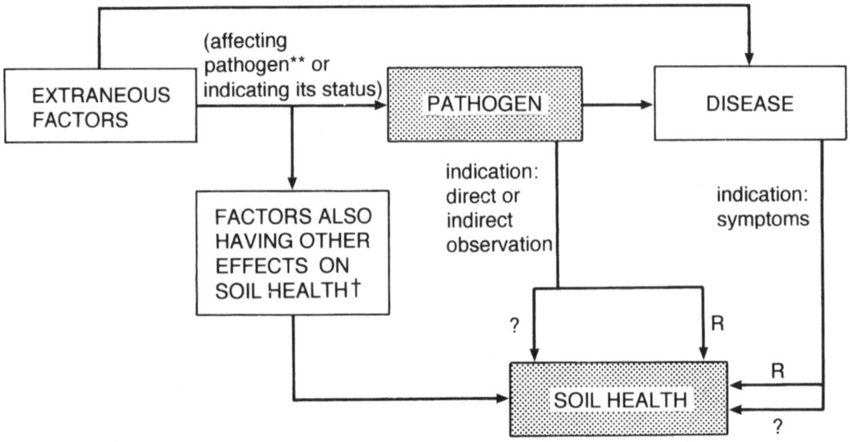

Fig. 8.2. A scheme showing considerations to be made in evaluating root pathogens as bioindicators of soil health. * e.g. environmental change affecting host response to disease; ** e.g. environmental change affecting the pathogen (leaf disease examples are rose black spot (*Diplocarpon rosae*) and acer tar spot (*Rhytisma acerinum*) being less of a problem in industrial areas where atmospheric sulphur content is high (Booth, 1974; Buczacki and Harris, 1981)); †factors that also have effects on soil health independent of those on root fungi, may be conflict, e.g. minimum tillage for conservation soil management and root disease (Pankhurst, 1994); ? possible use of root pathogens (or disease) as bioindicators of soil health problems that are not root disease; R use of root pathogens (or disease) to indicate potential root disease problems.

environmental) factors. There may be conflict with other indicators: minimum tillage, a useful conservation practice, favours growth of *Rhizoctonia solani* in many Australian soils (Pankhurst, 1994). Unless the impact of a disease is over-riding, priority is usually given to other factors in managing crops. Fig. 8.3, a scheme based on Rovira *et al.* (1990), is a simplification of how cropping systems may influence rhizosphere organisms that affect plant health. A cropping system will affect both detrimental and beneficial organisms in the rhizosphere, which in turn influence root health and crop yield. In a cropping system imposed for longer than a year, increased root disease would suggest the wrong choice of system, particularly if similar, adjacent crops under different systems were healthy. Cropping systems provide a means of managing some root pathogens and beneficial organisms, including biological control agents, but the effects of the majority of management practices on root disease remain unpredictable (Neate, 1994).

Monoculture often leads to root disease problems specific to a crop, such as take-all in cereals, where there has been no cultivar resistance to exploit (Hornby and Bateman, 1991). A classic example, where certain cultivars exacerbate the problem, is PRS (poor root syndrome) of sugarcane in northern Aus-

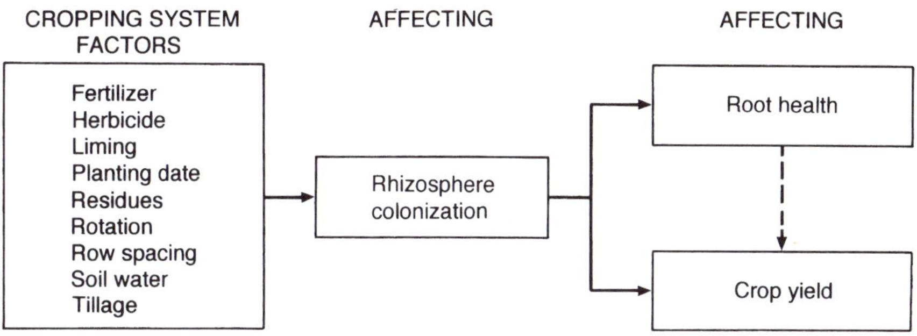

Fig. 8.3. Factors in a cropping system that may affect root health and crop yield indirectly through effects on the rhizosphere (after Rovira *et al.*, 1990).

tralia (Rovira *et al.*, 1990). The pathogens *Pythium graminicola* and *Pachymetra chaunorhiza*, which are found only in old sugarcane soils, are thought to be involved in PRS and acute yield decline, although with chronic yield decline there have been no readily identified major pathogens (Magarey, 1994). The build-up of pathogens in monoculture, the increase in root disease and the corresponding decrease in yield are what plant pathologists would expect, but is it possible to extract further information about soil health from such observations?

Although what happens in the rhizosphere affects root health, our understanding of the rhizosphere's involvements in root disease is incomplete and uneven (Hornby, 1990). Conventional wisdom is that microorganisms in the rhizosphere exist in a finely-balanced equilibrium, susceptible to a whole range of edaphic and plant-mediated modifying influences. Soil amendments can shift this equilibrium to control certain root diseases (Katznelson and Richardson, 1948; Katznelson, 1960). However, the methodology to monitor and measure rhizosphere effects under field conditions is inadequate and rarely standardized (Curl and Truelove, 1986) and most purported involvements of the rhizosphere in root disease are so correlated with other factors that cause and effect are blurred (Hornby, 1990). Since the expectations raised by insights on how the rhizosphere exerts influence on root diseases through rhizodeposition have yet to be realised in new and significant ways (Hornby, 1990), none of this augurs well for using root pathogens as simple and easily-understood bioindicators.

How Well do Root Pathogens Fit the Requirements of a Bioindicator?

Scope and opportunities for bioindication

In the scheme (Fig. 8.2) provided for discussion, either observation of the pathogen itself, or, often more easily, the symptoms of disease on the plant, are potential indicators of the soil's health. By the time above-ground symptoms have been seen, it is usually too late for successful preventive or curative action against the disease, but the information can be used to explain yield losses and allow planning for appropriate management of the soil subsequently. The possibility that a root pathogen, or the disease it causes, might usefully indicate problems other than root disease is also incorporated into Fig. 8.2.

Interactions with environmental factors

Environmental factors are an important element of the extraneous factors in Fig. 8.2, and their interactions with plant pathogens are a major consideration. The pathogen may be affected directly: survival of inoculum of the cereal take-all fungus in soil, for example, depends on temperature and water potential, with cool, dry-to-moist soil favouring survival (Hornby and Bateman, 1991) and high temperatures and moisture substantially decreasing survival of the fungus (Cotterill and Sivasithamparam, 1987). Declining levels of atmospheric sulphur (sulphur is a fungicide) have been associated with increases in some leaf diseases (two well-known examples are given in Fig. 8.2). We are not aware of similar effects on root diseases, although sulphur supplied to deficient soils can decrease take-all (Huber, 1981). Environmental factors may affect the pathogen indirectly via the plant, causing changes in host susceptibility, for instance. An example of this, also involving interactions between parasites and symbionts, which usually have opposite effects on plant health, concerns malnutrition in a declining stand of spruce and firs in the northern Black Forest (Germany) which was associated with weakening of the trees and increased susceptibility to pathogens (Haug *et al.*, 1988).

Another example of interaction is where areas of root damage attributable to the take-all fungus are seen as patches of poorly-developed, or prematurely-ripening, plants. Such patches often also indicate poor soil conditions (e.g. compaction, low pH, inadequate nutrients) likely to be detrimental to the plants even in the absence of take-all. The presence of take-all, a direct cause of poor soil health, may be obvious only where poor soil conditions, contributing to poor soil health independently of disease, are interacting with the disease to produce visible, above-ground symptoms. Such associations led to the idea that take-all could be used as an indicator of areas in fields where the soil needed attention (Hornby and Bateman, 1991).

What sorts of bioindication are possible using soil-borne plant pathogens?

Some ways in which root pathogens are indicators are given in the scheme in Fig. 8.1, but how these could be interpreted as indicating soil health is not always obvious. Should the soil health problem be clearly root disease, it could indicate:

1. Plants growing in the wrong place, e.g. wheat on sites having had successive wheat crops and therefore at risk from take-all.

2. Plants being grown in the wrong way, e.g. applying fertilizers incorrectly, or failing to irrigate potatoes during dry periods where common scab (caused by the persistent actinomycete, *Streptomyces scabies*) is known to be a problem.
3. A shift in the balance of the microflora to favour the pathogen. This may result from 1 or 2 above, or from changes in the environment, perhaps caused by weather, cultivations, other husbandry practices or a combination of factors. Plant pathologists have studied and applied this equilibrium concept for decades and, before that, growers' empirical knowledge incorporated much of it unknowingly.
4. A general environmental impact starting to affect crops. For instance, predictions of global warming and associated scenarios are well publicized, but with scarce mention of diseases of plant roots (e.g. Jäger and Ferguson, 1991; United Kingdom Climate Change Impact Review Group, 1991).

Figure 8.1 suggests that root disease is limited as an indicator of poor soil health, because different plants have different diseases. Contrary indications would arise where disease and host both appear to do well nationally, e.g. reports of more than usual take-all in wheat in high-yielding years (Hornby and Bateman, 1991). In agriculture, many root disease problems may be remedied by crop rotation, unless conditions, policies or economics severely limit alternative crops. Because of this, many disease problems in arable crops could be regarded as temporary problems of poor soil health, in contrast to longer-term and more general problems exemplified by heavy metal pollution of soils. In forests, however, changing hosts as a quick remedy for poor soil health attributed to root disease is not really an option.

Some possible areas of use

The traditional uses of bioindicators are in such areas as climate change and chemical pollution. Consequently, the potential use of root pathogens as bioindicators should include these areas as well as the obvious area of root disease, as outlined in Fig. 8.1. Unexpectedly finding a plant disease more usually associated with a different climatic zone might be evidence of climatic change. This would be especially important if the occurrence were to the detriment of the crops being presently grown. However, higher temperatures or increased CO_2 might actually benefit the crop and possibly even outweigh the effect of the disease. *Fusarium moniliforme* (*Gibberella fujikuroi*) is a warm-climate pathogen of maize that may become prevalent in higher latitudes as climate changes (Booth, 1974); monitoring crops for such fungi and the diseases they cause is routine to plant pathologists and has yet to be represented as 'bioindication'.

Reports of the potential use of soil organisms as bioindicators of toxicants in soil do not appear specifically to include root pathogens. Pesticides may cause or increase diseases that are not the targets of the pesticides (iatrogenic plant disease). In the UK, fungicide treatment for eyespot in wheat caused a second disease, sharp eyespot, to become more important. In such cases the diseases observed, or the pathogens responsible, are the indicators and the main direct cause of the poor health problem; the pesticide may be considered as another interacting factor. Review papers in the area of iatrogenic disease contain:

1. Lists of herbicides that were known to interact with disease and beneficial mycorrhizae (Altman and Campbell, 1977). Increased disease might indicate that better chemicals, formulations, doses or targeting are needed.
2. Examples of predisposition to disease and other adverse effects of fungicides, usually caused by removing antagonistic or competitive microorganisms (Papavizas and Lewis, 1979). For instance, *Pythium ultimum* and *Rhizoctonia solani* caused damping off in pine seedlings after organomercury trials. Additionally, quintozene increased strawberry wilt (*Verticillium albo-atrum*) due to interaction with a nematode (*Pratylenchus penetrans*).
3. Examples of pesticides in soil adversely affecting the microbial equilibrium (Rodriguez-Kabana and Curl, 1980). For instance, adverse conditions were created by quintozene, which selected in favour of *Pythium*, *Fusarium* and *Phytophthora* and so increased diseases caused by them.

Griffiths (1981) wrote that 'root pathogens are but a small part of the soil's microbial population whose ecology is notoriously complex. Competitive, synergistic, and antagonistic interactions abound and there are many examples of soil treatments, designed to control root pathogens, which have produced quite opposite effects'. A major caution in the use of organisms to indicate, by their suppression, the presence of a pollutant (e.g. a pesticide) is that increased tolerance of the pollutant might be induced (Chaphekar, 1978). Induced resistance to fungicides has been observed frequently in non-root pathogens.

The case for and against root pathogens as bioindicators

Meeting the requirements of good environmental indicators

Kennedy and Smith (1995) concluded that fungal populations were only moderately sensitive for indicating the fertility of a soil. Using the presence of root pathogens to indicate problems of soil health, other than root disease, is not an established activity. Therefore below, in considering root pathogens in relation to the criteria for a good environmental indicator (Roloff, 1989; MacGillivray, 1994), the soil health problem to be indicated is taken to be disease.

1. Is the bioindicator of low cost, easily handled and readily available? Generally, the answer is 'no', although molecular biological, including ELISA, methods may in the near future give quick access to individual species of soil-inhabiting or root-infecting fungi.
2. Is it homogeneous material? This is unlikely, genetic variability being commonplace amongst root pathogens. Types may be distinguished on the basis of host range or other criteria and this could complicate any use of these organisms as indicators. A fungus species name may mask much variation and an increase

in the number of formal descriptions of plant pathogens as species, even though not easily distinguished morphologically, is predicted (O'Donnell *et al.*, 1995). *Ophiostoma novo-ulmi*, the cause of the European elm disease pandemic in the mid-1970s, is an example, for it is now recognized as distinct from the almost morphologically identical *O. ulmi* (O'Donnell *et al.*, 1995).

3. Is it measurable (does it have easily-recognized and easily-evaluated symptoms with quantifiable effects)? Yes it is, but rarely simply or quickly. Direct and indirect conventional methods all have their limitations. New methods are becoming available for monitoring, and some diagnostic kits using ELISA have been widely publicized. These kits record presence or absence of a pathogen and do not usually provide quantitative information unless many samples are taken.

4. Is it meaningful? Tests will not necessarily indicate how much fungus is present. In terms of indicating a disease risk, the quantity of pathogen may not be easy to interpret, particularly where other factors put the plant under stress. The disease indicates a current problem and the presence of the pathogen, but neither necessarily indicates a persistent, serious problem of soil health.

5. Is it resonant and easy to understand? This refers to public understanding, awareness and acceptance of the indication. The answer must be 'no', because even in the scientific world the generally invisible events below ground tend to be poorly understood and often ignored.

6. Is there a short response time? Where the bioindicator (a root pathogen) is also the cause of the problem (root disease) this may not always be an appropriate question. The period between infection and symptom development varies considerably, depending on the disease and the presence or not of latent periods. The question might have more significance if root pathogens were shown to be useful indicators of soil health problems other than disease.

7. Are procedures for comparing results standardized? No, but this reflects absence of motivation and agreement, rather than absence of suitable procedures. Standardization of procedures has been introduced in other areas of applied mycology such as in monitoring resistance to fungicides in plant pathogens (e.g. Gisi, 1992).

8. Are there reaction conditions that are widely known? Metabolic changes may be early indicators in plants, but methods of detecting these present problems and most routinely-employed indicator plants produce external and well-known symptoms (Roloff, 1989). The symptoms of most diseases are well documented, but not always correctly interpreted.

9. Is there comparability between geographic areas and at an international level? Many diseases have regional responses, interacting with factors that tend to differ in different regions; therefore comparability is not usually possible. Comparability will be most likely within a climatic zone. Between zones, it may be necessary to make use of completely different pathogens that infect different crops appropriate to the regions. Some *Fusarium* spp. are considered to be cosmopolitan, with broad ecological tolerances, although it is not known whether individual clones are adapted to wide climatic ranges (Sangalang *et al.*, 1995). Broad ecological tolerance may limit an organism's usefulness as a bioindicator.

All the above properties and requirements are seldom, if ever, found in one bioindicator of any kind.

In plant pathology, the introduction of root pathogens into soil has been used for testing different types of soil suppressiveness (Hornby, 1983), but equating soil suppressiveness with soil health may mislead. On the one hand, soils with induced suppressiveness still tend to develop more disease than soils where crops are grown in good rotations. On the other hand, disease conduciveness may indicate a soil health problem only for a susceptible crop, and non-host crops would grow healthily. If there were scope for introducing a root pathogen into soil for the purposes of indicating the soil's health, experience already gained in other spheres would be valuable. Biological control agents (BCAs) for disease control may be organisms introduced into, or resident in, the soil. However, most species of microorganisms that could thrive in a particular soil are there already and soils will tend to reject exotic introductions (Garrett, 1965). Researchers still strive to overcome this barrier to introduced BCAs, by seeking to identify, for instance, the traits that contribute to the rhizosphere competence of an organism. Currently, the most exciting and potentially successful means to improve biocontrol activity and root colonization by plant growth-promoting rhizobacteria (PGPR) is recombinant DNA technology, which has already produced recombinants with enhanced production of antibiotics (Weller and Thomashow, 1994).

The idea of using root pathogens to monitor the accumulation of pesticides seems contrived. Non-persistence, at least beyond the short term, is usually a requirement for pesticides applied deliberately to soil, and so decreases in pathogens, killed by fungicides, would not usefully monitor pesticide pollution. This tends to suggest that chemical analysis is better where specific chemical compounds are involved and assessing the whole microflora is better where nonspecified pollutants are suspected.

Biodiversity

Paoletti *et al.* (1991) listed 'diversity, biological richness, and animal and plant abundance as measures of environmental quality'. Maintaining and creating biodiversity are considered important objectives in both natural and cropped environments. Although plant disease is generally considered as unwanted and something to be eliminated in cultivated plants, in natural ecosystems organisms like pollinating insects, mutualistic symbionts, population-regulating pathogens and biocontrol agents can affect the biodiversity of a site (Harper and Hawksworth, 1995). In the wild, host–pathogen associations and disease levels are held in check, at least partially, by a range of ecological and phenological phenomena that presumably have little or no importance in agricultural systems (Burdon, 1987). The losses caused by plant pathogens may represent only a harvest by the pathogens of the plant population's sustainable yield (Harper, 1990). In these circumstances, the activities of pathogens may affect the outcome of intra- and inter-specific competition, the distribution of plant species, the genetic structure of populations and possibly the evolution of sex. For instance, damping-off affects the early recruitment stages of seedlings in tropical forests and therefore has the potential to play a significant role in diversifying the forest and increasing its spatial and genetic complexity (Augspurger, 1990). Cropping, especially of single plant species, may decrease the diversity, but not necessarily the abundance, of microorganisms and favour root disease. The disease problem may be alleviated, although the diversity will not necessarily be much improved, by rotating crops that induce different host-specific pathogenic fungi. This, along with maintaining an adequate nutrient status and soil structure, are the well-known reasons for rotating crops. Total eradication of a pathogen, not usually a practical option, is on the face of it, contrary to the principle of maintaining biodiversity.

In natural soils, especially productive ones, plant pathogenic fungi form a part of the biodiversity that is generally considered to be an important requirement for healthy soils. Introduced pathogens or newly evolved races of pathogens may, on the other hand, contribute to decreased diversity in natural or partly natural systems by eliminating a species, or cause a shift in balance, as happened to a large extent in the case of Dutch elm disease caused by the (non-root-infecting) fungus *Ophiostoma novo-ulmi.*

Pankhurst (1994) considered the soil mesofauna as sensitive bioindicators of soil disturbance and earthworms as indicators of soil condition. Biodiversity of populations for monitoring soil health needs to be explored; however, this will be a difficult challenge because many soil organisms are not well known taxonomically (see Pankhurst, Chapter 12, this volume).

Kinds of pathogens and symptoms

The objectives for plant pathologists are usually to improve the productivity, edibility, or appearance of plants by controlling disease. The solution in the case of soil-borne diseases may be to avoid the disease problem by not growing a susceptible crop, to minimize it by adopting some husbandry strategy, or to attempt a cure by applying a fungicide. Changing the crop will not usually be an option in enterprises such as forestry or plantation crops, where individual plants are grown for many years. Two examples of pathogens in such circumstances are *Armillaria mellea* and *Heterobasidion annosum*, which caused serious problems in conifer plantations, whilst they were harmless in natural vegetation (Shipton, 1979). Changing the crop may also be less of an option in other situations, e.g. where alternatives are limited or where the soil harbours an unspecialized pathogen. Consideration of root pathogens and edaphic factors known to support and favour disease should be a part of any critical assessment of soil health. The methodology used in the study of phenomena such as soil sickness, soil suppressiveness and soil receptivity needs re-examination and re-evaluation for bioindication potential.

The time scale of individual crops also seems important. Specific diseases of long-lived plants (e.g. Dutch elm disease, Jarrah die-back) contrast with short-term, general diseases such as damping-off (caused by various fungi, including *Pythium*, *Rhizoctonia*, and *Phytophthora*). Plant pathology already contains a wealth of information about what these diseases indicate to growers and options available for dealing with infested soils. Consequently the information is used for management decisions and helps in determining curative or protective treatments, or in developing future strategies for minimizing disease. Its use outside the conventional bounds of plant pathology is uncommon.

Soils are rarely in irredeemably poor health because of pathogens, unless cropping options are extremely limited. The effects of crop rotation on disease development and persistence is a much-researched topic in plant pathology. In simple terms, where a particular crop is grown frequently, or continuously, there is more time and opportunity for its root pathogens to increase sufficiently and for disease to become rife. A popular view is that this, in turn, triggers a build-up of organisms that are able to antagonize the pathogens; consequently disease may decline, despite continued cropping of the host. When this occurs the soil is said to have become suppressive. How these trends are affected by rotation, or the lack of it, are summarized in Table 8.2.

Some root diseases are less easily controlled by rotation (see Table 8.3).

In some specific sites with 'resistant soils' certain diseases do not develop because the pathogen may not be present, or if present, it is suppressed. This is also discussed in references cited in Table 8.2.

Plant pathogens exploit host weaknesses and stress. Climate change could, or may already, influence diseases in the UK, as is suggested by the examples in Table 8.4. However, it is doubtful that changes in disease would be better indicators of climate change than more obvious events.

The simple proposition that the amount of a root pathogen in soil is proportional to soil health is difficult to sustain in the face of the complex interactions that affect both pathogen and disease (Fig. 8.2). The view that where several

Table 8.2. Some effects of rotation, or the lack of it, on root diseases and infectivity of soil.

Frequency of a specific crop in the rotation	Disease effects	Soils
Low	Tend to be minimized	Conducive to disease
High	Exacerbated	Conducive to disease
100% (monoculture)	May decline	Suppressive soils may develop

After Baker and Cook (1974) and Hornby (1983).

Table 8.3. Persistence of some root diseases in soil.

Disease	Persistence
Clubroot (*Plasmodiophora brassicae*)	The recommendation for cleaning fields where severe clubroot had occurred on oilseed rape in Scotland was at least 5 years without susceptible crops. Resting spores can probably survive in the soil for more than 20 years.
Onion smut (*Urocystis cepulae*)	Spores may persist in the soil for up to 15 years.
Potato wart (*Synchytrium endobioticum*)	Thick-walled resting sporangia may still release infective zoospores after 30 years in soil, but where legislation limits the planting of susceptible cultivars this disease is now of little practical significance.
Rhizomania (association between beet necrotic vein virus and *Polymyxa betae*)	This disease of sugarbeet is able to persist almost indefinitely in soil once established. The fungal resting spores are very resistant to chemicals and microbial degradation and the virus particles survive within them probably for as long as the spores remain viable.

From Buczacki and Harris (1981), Asher (1993) and Anon. (1995).

pathogens occur together, they are likely to curb each other (Zogg, 1969; Mańka, 1995), also contradicts the proposition. Thus, despite high concentrations of pathogens, the result could be disease 'decline' and a soil healthier than expected. It might be argued that the potential for severe disease is a characteristic of poor soil health, but with take-all disease, severe disease (clearly poor soil health, using the disease criterion) may be transitory, as for instance in third crops in winter wheat monocultures developing take-all decline (Hornby and Bateman, 1991). Severe disease may also be transitory for rhizoctonia root rot in sugarbeet monculture, because of a rapid decline in the pathogen's inoculum potential and an increasing suppressive effect of antagonistic microorganisms (Hyakumachi *et al.*, 1990). Consequently, in exploring the idea of pathogens as bioindicators, it emerges that their presence in soil is not a conclusive indication of poor soil health, nor does severe disease necessarily indicate future poor soil health, even if the same crop continues to be grown.

Table 8.4. How some diseases may respond in future, or already are responding, to climate change in the UK.

Disease	Response to climate change
Black dot of potatoes (*Colletotrichum coccodes*)	The nature of the disease could change and introduced crops (e.g. sunflowers) would be at risk
Eyespot of cereals (*Pseudocercosporella herpotrichoides*)	Warmer winters may favour the W-type over the R-type of the pathogen
Soil-borne, fungally-transmitted virus diseases	Winter weather was important in the occurrence in barley of BaYMV and BaMMV in the UK in 1980s Rhizomania of sugarbeet and soil-borne wheat mosaic virus are perceived as threats to UK agriculture should the weather change towards a more continental climate
Take-all of cereals (*Gaeumannomyces graminis*)	Increased disease at Woburn, Bedfordshire, in the 1980s may have been affected by a trend to smaller potential soil moisture deficits

From Leigh (1990).

What Methods are Available?

The large number of methods for identifying, isolating or enumerating root pathogens can be classified as 'traditional/conventional' (e.g. those in Tuite, 1969; Johnson and Curl, 1972; Dhingra and Sinclair, 1985) or 'molecular biological' (e.g. those in Schots *et al.*, 1994). Some of them are of general use whilst many, especially those involving molecular methods, are specific to single races, species or other restricted groupings. This existing methodology is relevant to a discussion of the potential of root pathogens as bioindicators of soil health.

Conventional methods

The simplest approach is to observe plant disease. Take-all patches (stunted or prematurely ripening areas) in a wheat crop are often an indication of soil that is in need of management, often to improve its structure to allow better root growth.

Using the pathogen rather than symptoms of disease may involve one of, or a combination of: (i) direct observation and enumeration of macroscopic or microscopic structures on or in soil or on plants; (ii) isolation on selective media; (iii) colonization of introduced substrates; or (iv) bioassays.

Bioassay is frequently the only approach where direct enumeration is not possible, as for example with obligate parasites (e.g. some fungi that are virus vectors), or ecologically-obligate parasites (e.g. the take-all fungus). Simple bait plants may be used to detect the presence of the pathogen. However, quantification of inoculum is important in many epidemiological studies involving soil-borne plant pathogens and would be so where root pathogens were used as bioindicators. Most probable number (MPN) estimations have been used to quantify inoculum and used with care this approach should enable differences of an order of magnitude to be detected, but the results should be regarded as relative, rather than absolute. An example, the soil infectivity bioassay, in which host plants are used as baits, is slow and cumbersome and quantification of inoculum using MPN estimations presents other biological and statistical difficulties (Adams and Welham, 1995). In the absence of distinctive foliar symptoms, root samples need to be examined for infection using appropriate methods such as inspection for characteristic lesions or signs, microscopy, back-inoculation to a host which develops local lesions (for viruses), or ELISA (thereby combining conventional and molecular procedures).

An example of bait plants being used as indicators of root disease, though not always reliable ones, is the soil bioassay for the cereal take-all fungus. These bioassays have long been in use in take-all research for indicating broadly whether there is much or little infectivity and whether this is likely to lead to a significant amount of disease in the subsequent crop (Hornby, 1981). They are probably not useful in indicating poor soil health that results from factors that interact with (exacerbate) take-all, and, without the complication of adding inoculum, are not especially useful in indicating when the disease has declined ('take-all decline'), so making it safer to grow cereals. This limitation, the slowness in obtaining results and the scarcity of effective measures for controlling take-all mean that such bioassays have modest value as a take-all management tool and presumably also in bioindication.

Professor K. Mańka's 'biotic series method' (Mańka, 1995) is a test of the effect of saprophytic soil fungi on pathogens and gives an indication of the probable effect of the environment on pathogenic activity. It scores interactions on agar plates between fungi isolated from the soil population and a pathogenic fungus and the total 'biotic effect' of the fungi selected is used to describe the health of the soil as it relates to the particular pathogen. The method has been applied mostly to undisturbed forest soils and the idea clearly has bioindication implications. However, it may prove difficult to adapt to easy and routine use.

Although fungicides are not used widely in the control of root pathogens on field crops, their use as tools has on occasions helped to identify, through large yield increases, situations where problems had been unrecognized. Soil fumigation also has shown up disease and pest constraints, referred to as 'soil microbiological limitations', on crop productivity (Rovira *et al.*, 1990). Usually these biological constraints affect only one, or a limited range of crops, and measuring such constraints to indicate soil health may be useful only in the context of these crops.

Molecular biological methods

Of the few such methods in use for diagnosing plant disease and for disease forecasting, most are ELISA methods based on immunological detection. New procedures are being developed continually and would undoubtedly enhance any role plant pathogens might come to have in bioindication of soil health.

Molecular techniques (DNA and ELISA) for identifying root pathogens were described in Schots *et al.* (1994). Some of the procedures were in the early stages of development and some had limitations as to their usefulness. Examples are:

1. Monoclonal antibody-based assays for *Rhizoctonia solani* that use soil extracts directly. Diagnostic kits using ELISA for soil-borne, root-infecting *Pythium* spp. and for other fungi such as *Pseudocercosporella herpotrichoides* that causes eyespot on cereal shoot bases.
2. Methods for identifying *Heterobasidion annosum*, a fungus that causes heartwood and root rots in coniferous trees, using monoclonal antibody and DNA methods.
3. Detection of the take-all fungus using DNA probes.
4. Detection of *Phytophthora fragariae* in raspberry roots using PCR.
5. Detection of *Armillaria* spp. using DNA and antibody-based methods.

Conclusions

The presence of root pathogens in soil may be useful in alerting us to past, present or potential root disease problems, but such problems tend to relate to particular hosts and are rarely a cause of a general poor soil health that adversely affects a majority of organisms. Many root disease problems are also likely to be easier to remedy or avoid than problems such as pollution. Root pathogens currently have no reputation as good, general bioindicators of soil health and, apart from a few examples, such as the use of root pathogens as indicators of the health of garden compost (Fig. 8.1), no clear and useful role has emerged for them in this context, beyond what is already deduced in relation to disease. Regarding them as bioindicators of disease is unconventional, as well as limited, because they would be both the indicator and the problem. However, disease may indicate the existence of other soil problems and therein lies scope for bioindication. Developing this role would require a re-examination of existing plant pathological knowledge and practice from a different standpoint, lateral thinking and ingenuity.

In agroecosystems, plant disease is usually a bad thing and the concept of root pathogens serving as bioindicators of this particular aspect of soil health is reasonable, but it seems to us that this is plant pathology by another name. Plant pathologists have used such information in the past to ameliorate or resolve the problem. In balanced, 'natural' ecosystems root pathogens are a part of the biodiversity and not necessarily bad, but insufficient is known of their behaviour in these systems to allow us to exploit them as bioindicators of soil health.

A practical limitation on using root pathogens as bioindicators is that they are often difficult to detect, although methods based on serology or on the recognition of specific DNA sequences are likely to become increasingly available for their diagnosis.

We conclude that the use of root pathogens for bioindication of soil health is an idea whose future depends on finding convincing answers to the following questions:

1. What could root pathogens reliably indicate (future disease, past inoculum, an environmental problem, etc.)?
2. How would this be interpreted in terms of soil health?
3. What aspects of the pathogens would need to be assessed or measured?
4. What suitable methods exist for assessment/measurement?

References

Adams, M.J. and Welham, S.J. (1995) Use of the most probable number technique to quantify soil-borne plant pathogens. *Annals of Applied Biology* 126, 181–196.

Altman, J. and Campbell, C.L. (1977) Effect of herbicides on plant diseases. *Annual Review of Phytopathology* 15, 361–385.

Anon. (1995) Scotland hit by severe clubroot in OSR crops. *Farmers Weekly*, 24 March 95, 67.

Asher, M.J.C. (1993) Rhizomania. In: Cooke, D.A. and Scott, R.K. (eds), *The Sugar Beet Crop*. Chapman and Hall, London, pp. 311–346.

Augspurger, C.K. (1990) Spatial patterns of damping-off disease during seedling recruitment in tropical forests. In: Burdon, J.J. and Leather, S.R. (eds) *Pests, Pathogens and Plant Communities*. Blackwell Scientific, Oxford, pp. 131–144.

Baker, K.F. and Cook, R.J. (1974) *Biological Control of Plant Pathogens*. W.H. Freeman, San Francisco, 433 pp.

Booth, C. (1974) The changing flora of microfungi with emphasis on the plant pathogenic species. In: Hawksworth, D.L. (ed.) *The Changing Flora and Fauna of Britain*. Academic Press, London, pp. 87–95.

Browning, J.A., Simons, M.D. and Torres, E. (1977) Managing host genes: epidemiologic and genetic concepts. In: Horsfall, J.G. and Cowling, E.B. (eds) *Plant Disease. An Advanced Treatise*. Vol. 1. Academic Press, London, pp. 191–212.

Bruns, C., Gottschall, R., Zeller, W. Schüler, C. and Vogtmann, H. (1993) Survival rates of plant pathogens during composting of biogenic wastes in commercial composting plants under different decomposition conditions. In: Paoletti, M.G., Foissner, W. and Coleman, D. (eds) *Soil Biota, Nutrient Cycling, and Farming Systems.* Lewis Publishers, Boca Raton, pp. 41–51.

Buczacki, S.T. and Harris, K.M. (1981) *Collins Guide to the Pests, Diseases and Disorders of Garden Plants.* Collins, London, 512 pp.

Burdon, J.J. (1987) *Diseases and Plant Population Biology.* Cambridge University Press, Cambridge, 208 pp.

Chaphekar, S.B. (1978) Biological indicators: the concept and new additions. *International Journal of Ecology and Environmental Sciences* 4, 45–52.

Cotterill, P.J. and Sivasithamparam, K. (1987) Intermittent wetting of soils at high temperature reduces survival of the take-all fungus. *Plant and Soil* 103, 289–291.

Curl, E.A. and Truelove, B. (1986) *The Rhizosphere.* Springer-Verlag, Berlin, 288 pp.

Dhingra, O.D. and Sinclair, J.B. (1985) *Basic Plant Pathology Methods.* CRC Press, Boca Raton, 355 pp.

Garber, K. (1974) Schwermetalle als Luftverunreinigung – Blei-Zink-Cadmium-Beeinflussung der Vegetation. *Staub-Reinhaltung der Luft* 34, 1–7.

Garrett, S.D. (1965) Toward biological control of soil-borne plant pathogens. In: Baker, K.F. and Snyder, W.C. (eds), *Ecology of Soil-borne Plant Pathogens.* John Murray, London, 4–16.

Gisi, U. (ed.) (1992) FRAC methods for monitoring the sensitivity of fungal pathogens to phenylamide fungicides. *EPPO Bulletin* 22, 298–322.

Griffiths, E. (1981) Iatrogenic plant diseases. *Annual Review of Phytopathology* 19, 69–82.

Harper, J.L. (1990) Pests, pathogens and plant communities: an introduction. In: Burdon, J.J. and Leather, S.R. (eds) *Pests, Pathogens and Plant Communities.* Blackwell Scientific, Oxford, pp. 3–14.

Harper, J.L. and Hawksworth, D.L. (1995) Preface. In: Hawksworth, D.L. (ed.), *Biodiversity. Measurement and Estimation.* Chapman and Hall, London, pp. 5–12.

Haug, I., Weber, G. and Oberwinkler, F. (1988) Intracellular infection by fungi in mycorrhizae of damaged spruce trees. *European Journal of Forest Pathology* 18, 112–120.

Heidt, V. and Kehlberger, S (1983) Möglichkeiten und Grenzen des Einsatzes von Bioindikatoren. *Verhandlungen der Gesellschaft für Ökologie* 10, 151–156.

Horak, M. (1991) Ektomykorrhiza – Einfluss von Umweltfaktoren und Methodik zur Charakterisierung. *Boletus* 15, 10–18.

Hornby, D. (1981) Inoculum. In: Asher, M.J.C. and Shipton, P.J. (eds) *Biology and Control of Take-all.* Academic Press, London, pp. 271-293.

Hornby, D. (1983) Suppressive soils. *Annual Review of Phytopathology* 21, 65–85.

Hornby, D. (1990) Root diseases. In: Lynch, J.M. (ed.) *The Rhizosphere.* Wiley-Interscience, Chichester, pp. 233–258.

Hornby, D. and Bateman, G.L. (1990) Artificial infestation of soil with *Gaeumannomyces graminis* var. *tritici* to study the relationship between take-all and wheat yields in field experiments. *Soil Use and Management* 6, 209-217.

Hornby, D. and Bateman, G.L. (compilers and eds) (1991) *Take-all Disease of Cer-*

eals. HGCA Research Review No. 20, Home-Grown Cereals Authority, London, 147 pp.

Huber, D.M. (1981) The role of nutrients and chemicals. In: Asher, M.J.C. and Shipton, P.J. (eds) *Biology and Control of Take-all*. Academic Press, London, pp. 317–341.

Hyakumachi, M., Kanzawa, K. and Ui, T. (1990) Rhizoctonia root rot decline in sugarbeet monoculture. In: Hornby, D. (ed.) *Biological Control of Soil-borne Plant Pathogens*. CAB International, Wallingford, UK, pp. 227–247.

Jäger, J. and Ferguson, H.L. (eds) (1991) *Climate Change: Science, Impacts and Policy*. Proceedings of the Second World Climate Conference. Cambridge University Press, 578 pp.

Johnson, L.F. and Curl, E.A. (1972) *Methods for Research on Ecology of Soil-borne Plant Pathogens*. Burgess Publishing Co., Minneapolis, 247 pp.

Katznelson, H. (1960) Observations on the rhizosphere effect. In: Parkinson, D. and Waid, J.S. (eds) *The Ecology of Soil Fungi*. Liverpool University Press, Liverpool, pp. 192–201.

Katznelson, H. and Richardson, L.T. (1948) Rhizosphere studies and associated microbiological phenomena in relation to strawberry root rot. *Scientific Agriculture* 28, 293–308.

Kennedy, A.C. and Smith, K.L. (1995) Soil microbial diversity and the sustainability of agricultural soils. *Plant and Soil* 170, 75–86.

Ko, W.H., Chang, H.S. and Su, H.J. (1978) Isolates of *Phytophthora cinnamomi* from Taiwan as evidence for an Asian origin of the species. *Transactions of the British Mycological Society* 71, 496–499.

Leigh, R.A. (compiler) (1990) The effect of climate change on soils, crops, pests and diseases. Research Theme. *IACR Report 1989*, 59–66.

MacGillivray, A. (ed.) (1994) *Environmental Measures. Indicators for the UK Environment*. Environmental Challenge Group, RSPB, WWF and New Economics Foundation, 107 pp.

Magarey, R.C. (1994) Yield decline of sugarcane. In: Rao, G.P., Gillaspie, A.G., Jr, Upadhyaya, P.P., Bergamin, A., Agnihotri, V.P. and Chen, C.T. (eds) *Current Trends in Sugarcane Pathology*. International Books and Periodicals Supply Service, Dehli, pp. 393–412.

Mańka, M. (1995) Non-pathogenic soil fungi reflecting soil environment. In: Mańka, M. (ed.), *Environmental Biotic Factors in Integrated Plant Disease Control*. The Polish Phytopathological Society, Poznań, pp. 27–36.

Neate, S.M. (1994) Soil and crop management practices that affect root diseases of crop plants. In: Pankhurst, C.E., Doube, B.M., Gupta, V.V.S.R. and Grace, P.R. (eds) *Soil Biota: Management in Sustainable Farming Systems*. CSIRO Press, Melbourne, pp. 96–106.

O'Donnell, A.G., Goodfellow, M. and Hawksworth, D.L. (1995) Theoretical and practical aspects of the quantification of biodiversity among microorganisms. In: Hawksworth, D.L. (ed.) *Biodiversity. Measurement and Estimation*. Chapman and Hall, London, pp. 65–73.

Pankhurst, C.E. (1994) Biological indicators of soil health and sustainable productivity. In: Greenland, D.J. and Szabolcs, I. (eds) *Soil Resilience and Sustainable Land Use*. CAB International, Wallingford, UK, pp. 331–351.

Paoletti, M.G., Favretto, M.R., Stinner, B.R., Purrington, F.F. and Bater, J.E. (1991)

Invertebrates as bioindicators of soil use. *Agriculture, Ecosystems and Environment* 34, 341–362.

Papavizas, G.C. and Lewis, J.A. (1979) Side-effects of pesticides on soil-borne plant pathogens. In: Schippers, B. and Gams, W. (eds) *Soil-borne Plant Pathogens.* Academic Press, London, pp. 483–505.

Parr, J.F., Papendick, R.I., Hornick, S.B. and Meyer, R.E. (1992) Soil quality: attributes and relationship to alternative and sustainable agriculture. *American Journal of Alternative Agriculture* 7, 2–3.

Rodriguez-Kabana, R. and Curl, E.A. (1980) Nontarget effects of pesticides on soil-borne pathogens and disease. *Annual Review of Phytopathology* 18, 311–332.

Roloff, A. (1989) Pflanzen als Bioindikatoren für Umweltbelastungen. I. Prinzipien der Bioindikation und Beispeil Waldbodenvegetation. *Forstarchiv* 60, 184–188.

Rovira, A.D., Elliot, L.F. and Cook, R.J. (1990) The impact of cropping systems on rhizosphere organisms affecting plant health. In: Lynch, J.M. (ed.), *The Rhizosphere.* Wiley-Interscience, Chichester, pp. 389–436.

Samson, R.A., Flannigan, B., Flannigan, M.E., Verhoeff, A.P., Adan, O.C.G. and Hoekstra, E.S. (eds) (1994) *Health Implications of Fungi in Indoor Environments.* Elsevier, Amsterdam, 602pp.

Sangalang, A.E., Burgess, L.W., Backhouse, D., Duff, J. and Wurst, M. (1995) Mycogeography of *Fusarium* species in soils from tropical, arid and mediterranean regions in Australia. *Mycological Research* 99, 523–528.

Sayer, J.A., Raggett, S.L. and Gadd, G.M. (1995) Solubilization of insoluble metal compounds by soil fungi: development of a screening method for solubilizing activity and metal tolerance. *Mycological Research* 99, 987–993.

Schots, A., Dewey, F.M., and Oliver R.P. (eds) (1994) *Modern Assays for Plant Pathogenic Fungi: Identification, Detection and Quantification.* CAB International, Wallingford, UK, 267 pp.

Shipton, P.J. (1979) Experimental evidence for the effect on soil-borne diseases of changes in techniques of crop and soil cultivations. In: Schippers, B. and Gams, W. (eds) *Soil-borne Plant Pathogens.* Academic Press, London, pp. 385–397.

Tuite, J. (1969) *Plant Pathological Methods.* Burgess Publishing Co., Minneapolis, 239 pp.

United Kingdom Climate Change Impact Review Group (1991) *The Potential Effects of Climate Change in the United Kingdom.* HMSO, London, 124 pp.

Weller, D.M. and Thomashow, L.S. (1994) Current challenges in introducing beneficial microorganism into the rhizosphere. In: O'Gara, F., Dowling, D.N. and Boesten, B. (eds) *Molecular Ecology of Rhizosphere Microorganisms.* VCH, Weinheim, pp. 1–18.

Wittig, R. (1993) General aspects of biomonitoring heavy metals by plants. In: Markert, B. (ed.) *Plants as Biomonitors.* VCH, Weinheim, pp. 3–27.

Wondratschek, I. and Röder, U. (1993) Monitoring of heavy metals in soils by higher fungi. In: Markert, B. (ed.) *Plants as Biomonitors.* VCH, Weinheim, pp. 345–363.

Zogg, H. (1969) Crop rotation and biological soil desinfection [*sic*]. *Qualitas Plantarum et Materiae Vegetabiles* 18, 256–273.

9

Soil Microfauna as Bioindicators of Soil Health

V.V.S.R. Gupta[1] and G.W. Yeates[2]

[1]*Cooperative Research Centre for Soil and Land Management, Private Bag No. 2, Glen Osmond, South Australia 5064, Australia;* [2]*Manaaki Whenua – Landcare Research, Private Bag 11052, Palmerston North, New Zealand*

Introduction

Soil biota regulate most of the soil processes that determine plant growth in terrestrial ecosystems and a healthy biological system is essential for sustaining productivity without damaging the environment. Extensive use of agrochemicals, e.g. fertilizers and pesticides, has become an integral part of modern-day agriculture but in recent times their negative effects on soil health (both on-site and off-site) have gained attention. Similarly, continued use of traditional agricultural practices, e.g. intensive tillage and residue removal, has caused deterioration of soil structure resulting in loss of soil through wind and water erosion. A number of alternative agricultural practices, e.g. minimum tillage with residue retention, have recently been investigated and in many instances these practices are helping to improve the capacity of soil to sustain production without causing damage to the surrounding environment. An effective monitoring system is essential to assess the health of agroecosystems and to monitor the effects of management practices.

Health of a living organism is defined as the 'condition of an organism to perform its vital functions normally and properly' (*The Oxford Dictionary*). Soil is a living and dynamic entity that requires a unique balance between its physical, chemical and biological components in order to remain productive. Proper functioning of the soil is vital for the continued functioning of terrestrial ecosystems. A number of definitions of soil health have been proposed (Doran *et al.*, 1994; Acton and Gregorich, 1995; Doran and Safley, Chapter 1, this volume). In this chapter, we discuss the role of microfauna in the functioning of soil in the ecosystem and consider whether they can be used as indicators of soil health.

 Biological Indicators of Soil Health (eds C.E. Pankhurst, B.M. Doube and V.V.S.R. Gupta)

In particular we highlight and discuss the practical and methodological issues associated with enumerating and characterizing soil microfauna and illustrate their potential as biological indicators by discussing the impact of agricultural management practices and soil pollutants on their abundance and diversity.

Role of Microfauna in Soils

Microfauna in soils consist of protozoa, nematodes and small-sized collembola and mites (Swift *et al.*, 1979). They form a critical link between the primary decomposers (i.e. the microflora) and the larger fauna in the detritus food-web in the soil. They are the primary agents for the release of nutrients immobilized by the soil microflora.

Protozoa are unicellular organisms which are classified as flagellates, ciliates, and naked or testate amoebae. These are distinguished primarily by their morphological features. Free-living protozoa in soils belong to bacterial-feeding or fungal-feeding groups based on their food source and they range in size from 2 to 1000 μm. Soil nematodes are vermiform animals ranging in size from 0.3 to 5.0 mm in length (adults). They are identified under known taxa or classified into different trophic groups based on their feeding habits viz. bacterial feeders, fungivores, plant parasites, omnivores and predators. Both protozoa and nematodes are water dependent organisms and require water-filled pores or continuity of water films for their activity and movement. They are ubiquitous inhabitants of all soils, are abundant (>million m^{-2} in most soils) and diverse (e.g. >30 taxa of nematodes).

Microfauna are involved in a variety of ecosystem processes including; decomposition/turnover of organic matter, nutrient mineralization, regulation of population densities of microflora including plant pathogenic organisms, and decomposition of agrochemicals (Stout and Heal, 1967; Yeates, 1981; Freckman and Caswell, 1985; Old, 1986; Gupta and Germida, 1989; Henkinet *et al.*, 1990; Gupta, 1993; Darbyshire, 1994). Nematodes may also directly or indirectly affect nutrient uptake by plants by their effects on plant health, e.g. plant parasitic nematodes feeding on plant roots (Norton, 1978), and free-living nematodes feeding on symbiotic microflora such as mycorrhizae (Hussey and Roncadori, 1981) and rhizobia (Westcott and Barker, 1976), and other microflora such as plant pathogenic bacteria (Chantanao and Jensen, 1969; Freckman and Caswell, 1985), and biocontrol bacteria (Bird and Ryder, 1993). Protozoa are the main consumers of microbial biomass (e.g. causing reduction in the populations of introduced bacteria) and their selective grazing pressure on microorganisms may change the composition of the microflora in the soil (Clarholm, 1981; Darbyshire, 1994). This grazing pressure helps to keep the microorganism populations physiologically young and active which in turn stimulates microbial processes in the soil (Elliott and Coleman, 1977; Gupta and Germida, 1989). Predation by microfauna (mycophagous amoebae and fungal-feeding nematodes) has also been reported to reduce the inoculum levels of plant pathogenic fungi (e.g. *Rhizoctonia solani*, *Gaeumannomyces graminis* var. *tritici*, *Verticillium* wilt) in soil (Curl, 1988; Old, 1986; Gupta *et al.*, 1996a).

Identification and Enumeration of Protozoa and Nematodes

A potential major constraint to using protozoa or nematodes as indicators of soil health is the difficult methodology associated with the identification of taxa or trophic groups of these organisms. Some of these difficulties are discussed below together with issues associated with sampling, enumeration, indices used to describe abundance and diversity, seasonal dynamics and spatial heterogeneity.

Identification of protozoa

Both 'culture techniques' and 'direct count methods', used for estimation of protozoan abundance and species composition, require prolonged periods of microscopy. Identification of protozoa is commonly based on descriptions made using light microscopy which may be unreliable especially for smaller protozoa (e.g. flagellates). Identification of protozoan genera is based on both locomotive and internal structures of the trophozite (active) form and the morphology of cysts (inactive, resting stage). Although there are reviews dealing with methods used to identify ciliates (Foissner, 1987), naked amoebae (Page, 1976, 1988), testate amoebae (Chardez and Lambert, 1981) and flagellates (Patterson and Larsen, 1991), there is currently no comprehensive review describing all types of protozoa (taxonomically). In addition, classification of protozoa into trophic groups (e.g. bacteria-feeding, fungal-feeding) is difficult at present since published information on the food preferences of protozoa is very limited and a number of protozoa feed on both bacteria and fungi.

Protozoa are traditionally classified into four morphological groups, viz. flagellates, ciliates, naked and testate amoebae. In order to relate the dynamics of protozoan populations to key soil processes they have also been grouped on the basis of their feeding habits (e.g. bacterial or fungal feeding), their ecological niche (e.g. pH preference – acidophilic or neutrophilic), their position on the *r*/*K* continuum (e.g. colonizer or persister) and their degree of soil association ('autochthonous', 'strongly edaphic' and 'also in fresh water') (Old, 1986; Foissner, 1987; Bamforth, 1995; Gupta, 1996).

Identification of nematodes

Identification of nematodes into different taxa and classification into trophic groups is based on the structures in the head and oesophageal regions. Identification keys are available for different groups of nematodes (Goodey, 1963; Maggenti, 1981; Bongers, 1988; Maggenti *et al.*, 1988; Jairajpuri and Ahmad, 1992) but such keys rapidly become out-dated because many taxa are still being described.

Nematodes show great morphological diversity in their head structures and specialized forms of oesophagus (pharynx) which are related to their feeding habits. Yeates *et al.* (1993) presented the most recent grouping of different nematodes. These groupings include: (i) plant feeders; (ii) fungal (hyphal) feeders; (iii) bacterial feeders; (iv) substrate ingestors; (v) predators of soil animals (protozoa, rotifers, enchytraeids and other nematodes); (vi) unicellular eukaryote feeders (diatoms and other algae); (vii) omnivores (generally used for dorylaimid nematodes); and (viii) animal parasites.

Sampling and analytical methods

Advantages and drawbacks of the different methods used to estimate protozoan abundance are summarized by Gupta (1996). Direct counting methods provide realistic population estimates of ciliates and testate amoebae (Foissner, 1987; Cowling, 1994), but they have not been widely tested for other protozoa. The opaque nature of soil particles creates difficulties for identifying flagellates and naked amoebae using light microscopy. Direct methods tend to underestimate populations of active protozoa if analysis is not completed before they have time to encyst, or to overestimate populations if they have time to reproduce. Culture techniques (indirect methods) which require incubation of soil suspensions, supplied with bacteria or fungi as food, may overestimate the number of active protozoa. The type of bacteria used as food is also important as this can affect estimates of protozoan abundance (Gupta and Roper, 1993). The use of a bacterial species that inhibits eukaryotes (e.g. *Pseudomonas corrugata*) would lead to the underestimation of protozoan populations (V.V.S.R. Gupta, unpublished results). The species richness of testate amoebae is estimated by inspecting 'tests' concentrated using gas bubble flotation or filtration techniques (Grospietsch, 1965; Lousier and Parkinson, 1981). Methods recommended for other groups of protozoa (viz. the non-flooded petri-dish method, the nutrient agar/liquid culture media method and the double-chamber petri-dish method) require prolonged periods (a minimum of 2–3 weeks) of microscopy (Foissner, 1987; Gupta, 1996).

A number of standard procedures are available for the collection of soil samples and extraction and handling of nematodes. Active migration techniques (e.g. the Baermann funnel technique and many modifications to it) are the most widely used techniques although these tend to be replaced by mechanical elutriators in major laboratories (Freckman and Ettema, 1993; Yeates and Bird, 1994; Yeates and Bongers, 1997). Yeates and Bongers (1997) have recently described a modified version of the 'tray method' for the extraction of nematodes.

At present, technical expertise is essential in order to identify species of protozoa and nematodes. Molecular techniques utilizing RFLP patterns, universal eukaryotic probes and primers for 18S or 5S rRNA genes, when developed sufficiently for routine use, could greatly reduce errors in species identification (Vodkin *et al.*, 1992; Badgley and Odelson, 1993; van der Knapp *et al.*, 1993; Bowers and Pratt, 1995). It would be advantageous if this technology could also differentiate between functional groups. Cellular fatty acid methyl ester (FAME) analysis which has been developed for the identification of bacteria, fungi and actinomyctes and more recently used to investigate the community structure of soil microbial communities (Cavigelli *et al.*, 1995), may have similar potential for the identification of microfauna.

Indices and other parameters used to describe abundance and diversity

A list of different parameters and indices that have been used to study the abundance and diversity of microfauna is given in Table 9.1. In general, the indices used reflect community parameters such as species richness, abundance, dominance and evenness (Norton and Niblack, 1991; Bernard, 1992; Foissner, 1994; Neher *et al.*, 1995). However, separation of microfaunal populations into functional or ecologically related groups helps to relate changes in their abundance to ecosystem function. For example, by separating protozoan populations into groups based on their food preference, e.g. bacterial-feeders or mycophagous amoebae, a link between protozoa and the ability of a particular agricultural system to suppress pathogenic fungi was established (Old, 1986; Gupta *et al.*, 1996a).

The Shannon–Weaver diversity index (H′), the maturity index (MI) and the weighted coenotic index (WCI) are three of the indices most commonly used to assess the composition of protozoan and nematode faunae.

The 'Weighted Coenotic Index' (WCI) (Wodarz *et al.*, 1992) is a diversity index that unifies the total abundance, logarithmic dominance structure, species richness and ecological weightings for different species. Ecological weightings are based on habitat preferences and positions of species in the *r*/*K*-continuum. The inclusion of a logarithmic dominance structure (since most species abundance distribution patterns fit a logarithmic series) and ecological weightings is an advantage the WCI has compared to other indices such as the Shannon-Weaver diversity index which does not include an ecological component. However, the WCI is a complex index to calculate and requires expertise in the analysis of species composition, knowledge of their ecological weightings as well as a reference (control) site for a conclusive interpretation.

Bongers (1990) proposed the maturity index (MI) in order to utilize nematode population data to explain the condition of a soil ecosystem. Different nematode taxa are given values (*c-p* values) based on their position on the colonizer-persister spectrum which is based on reproductive rate and other cor-

Table 9.1. Parameters and statistics used in the analysis of microfaunal communities*.

Parameters	Description/formula†	Comments
Total population	Abundance of all groups of protozoa or nematodes	Per gram or per square metre basis
Taxonomic groups	Abundance based on taxonomic description	e.g. ciliates, amoebae, rhabditids, dorylaimids
Ecological/functional groups		
Trophic diversity	$T = 1/\Sigma\, pi^2$ (where pi is the proportion of trophic group i)	Relative abundance of functional groups
Fungivores/bacteriovore ratio	FBR = F/B or F/(F+B), (F and B are abundances of fungal and bacterial feeding microfauna)	Describes decomposition pathway in foodweb (protozoa and nematodes)
Bacteriovores/plant parasites	Ratio of bacterial-feeding vs. plant parasitic nematodes	Decomposition relative to plant productivity
Acidiphilic/neutrophilic	Based on pH preferences (e.g. testate amoeba)	Habitat preference (protozoa)
Diversity indices		
Number of species	Total number of taxa	–
Species richness (abundance)	$SR = (s-1)/\log_e N$	Number of species/specimens identified
Species evenness	$J' = H'/H'_{max}$ where H'_{max} is $\log_e s$	Distribution of individuals within species/genera
Sorenson's similarity index	$C_s = 2j/(a+b)$	–
Simpson index for dominance	$\lambda = \Sigma\, p_i^2$	Higher index with greater disturbance
Diversity	$H_2 = -log_e\, \lambda$	–
Diversity (Shannon–Weaver index)	$H' = -\Sigma_{i=1}^{s}\, p_i \log_e p_i$	Greater weight to rare species
Simpson's diversity index	$D = 1/\Sigma(n_i/N)^2$	Greater weight to common species
Morisita–Horn index	$C_{MH} = (2\Sigma\,(a_{ni} * b_{ni}))/((da+db)*(aN*bN))$	Species and abundance similarities
Maturity index	$\Sigma\, MI = \Sigma_{i=1}^{n}\, c\text{-}p\,(i)\; p\,(i)$	Based on colonizer-persister spectrum
MI for non-parasites	similar to MI but excludes plant parasites	Lower index indicates disturbance
Plant parasite index (PPI)	similar to MI but excludes free-living taxa	Higher index = increased plant production
C/P ratio	Ratio of colpodid/polyhymenophoran ciliates (species number or abundance)	Based on position on *r/K* continuum (protozoa)
Weighted coenotic index (WCI)	Unifies total abundance, logarithmic dominance structure, species richness and ecological weightings	Requires a reference value for comparison

* Based on information given in Margalef (1968); Twinn (1974); Pielou (1977); Magurran (1988); Bongers, (1990); Freckman and Ettema (1993); Darbyshire (1994); Foissner (1994); Neher *et al.* (1995) and Yeates and Bongers (1997).

† Reference should be made to these for details of the equations and calculations; N, number of individuals identified; s, number of taxa identified; p, the proportion of individuals in the i th taxon; n, number of individuals in i th species.

related characteristics. The MI is the weighted mean of the *c-p* values for the nematode taxa in a sample. Bongers (1990) also proposed that a separate Σ MI could be calculated for plant parasites (PPI). Yeates (1994), however argued that MI should be based on the data for all nematodes in a system. Bongers *et al.* (1995) suggested that the PPI/MI ratio might be a useful index to reflect enrichment of plant parasitic nematodes in an ecosystem.

Statistical methods such as canonical discriminant analysis and principal component analysis have also been used effectively to describe the community composition of microfaunal populations, and to assess environmental impacts on community ecology (Freckman and Ettema, 1993; Foissner, 1994).

Seasonal dynamics

If observations recorded at specific times of the year are to have practical value as indicators, it is important to consider the variability that can arise due to differences in habitat characteristics and to temporal fluctuations in environmental and nutritional conditions. Seasonal fluctuations in populations of all types of protozoa and nematodes has been recorded in all climates around the world (Cutler *et al.*, 1922; Stout and Heal, 1967; Yeates, 1981; Foissner, 1987; Gupta, 1993). Soil moisture (water holding capacity and pF relationships), soil temperature, availability of food (quantity and type) and predation by larger fauna are some of the environmental and biological factors that are responsible for the seasonal fluctuations in the populations of microfauna. Although observations at specific times of the year provide information on the population structure in an ecosystem, data on seasonal fluctuations enables evaluation of the importance of particular functional groups during the critical stages of crop growth. Such information, (though time consuming and costly to obtain) would also give a better view of the long-term effects of management practices on the microfauna and their functioning (Crossley *et al.,* 1989).

In general, populations of active forms of protozoa and nematodes are lowest during the hot and dry summer periods with maximum activity following rainfall events and during the warm and moist periods of the year. For example, in a red brown earth (alfisol) in South Australia, Gupta and Roper (1996) observed a 10-fold increase in protozoan abundance following the first rainfall in May, when maximum soil temperature was <10°C, whereas during late June, when soil temperatures were ~18°C, a rainfall event caused an increase in protozoan activity of 10–100-fold (Fig. 9.1). Populations, biomass and species richness of large amoebae and ciliates were also greatest following the rainfall in June (Gupta and Roper, 1996). These results suggest that temperature and moisture are important factors affecting the abundance and activity of protozoa. There is also evidence for temperature adaptations amongst protozoa (Buitkamp, 1979; Petz *et al.*, 1985; Bamforth, 1995; Robinson *et al.*, 1996).

Generation time, fecundity, presence of predators and interspecific compet-

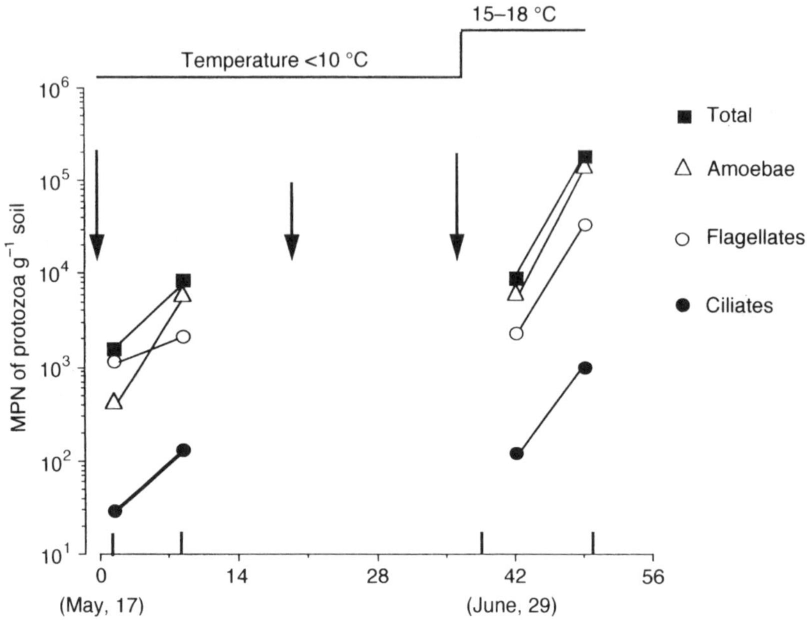

Fig. 9.1. Seasonal changes in the populations of different groups of protozoa, during 1993, in a long-term trial on an alfisol at the Waite Agricultural Research Institute, Glen Osmond, South Australia (arrows indicate rainfall events). (Adapted from Gupta and Roper, 1996.)

itive interactions are important factors which regulate the activity and species composition of soil nematode populations (Freckman and Caswell, 1985; Yeates, 1994). Yeates (1981) and Yeates and Bird (1994) reported that changes in the composition of the nematode faunae with season were dependent on the type of vegetation present in the ecosystem. In South Australian soils, Yeates and Bird (1994) observed that the fungivore to bacteriovore ratio in a soil under wheat increased (0.27 to 0.47) between September to October under a conventional tillage system whereas the trend was opposite (0.88 to 0.33) under a minimum tillage treatment. The proportion of plant parasitic nematodes decreased (35%) between September and October in agricultural systems but in shrubland soils they increased from 1.2 to 4.2%. Since climatic conditions were similar for both treatments, soil- and plant-associated differences were considered to be responsible for these effects.

Spatial heterogeneity

Populations of protozoa and nematodes are greatest in the surface soil (0–5 cm) and decline with soil depth. Species richness and taxonomic diversity has also been observed to decline with soil depth (Stout and Heal, 1967; Foissner, 1987; Schönborn *et al.*, 1987; Yeates and Bongers, 1997) reflecting changes in soil fertility, texture, moisture, availability of food and the rooting pattern of plant species. The rooting pattern of plants can have direct effects on plant-feeding nematodes. Differences in abundance and composition of microfaunal populations at different soil depths may be greater in forest soils containing thick layers of rich organic detritus than in agricultural soils under arable or grassland management (Jones *et al.*, 1969; Cowling, 1994; Yeates and Bongers, 1997).

The abundance of microfauna in the soil is patchy, due to the heterogeneous nature of the soil matrix (especially the variation in resource availability and microclimatic conditions), pore-size restrictions for microfaunal movement, and the discontinuity of water-filled pores and water-films surrounding soil particles. Microsites associated with available carbon and nutrient sources, (e.g. rhizosphere and detritus) support higher populations of microfauna than does the bulk soil (Fig. 9.2) (Zwart *et al.*, 1994; Beare *et al.*, 1995; Gupta and van Vliet, 1996). The R/S ratio (ratio of populations in rhizosphere soil versus non-rhizosphere soil) for protozoa and nematodes may range from 2:1 to 25:1 depending on the plant species. The ratio may also change with different groups of protozoa or nematodes (Darbyshire and Greaves, 1967; Boström and Sohlenius, 1986; Griffiths, 1990; Gupta, 1993; Foissner, 1994; Yeates and Bongers, 1997). For example, more flagellates were stimulated in the rhizosphere soils of white clover than in the rhizosphere soils of white mustard or perennial rye grass (Darbyshire and Greaves, 1967). Different plant species are also reported to support different populations of nematodes (Yeates *et al.*, 1983; Griffiths, 1990; Wasilewska, 1995).

Decomposing crop residues support large populations of microorganisms which serve as a food source for protozoa and nematodes. It is not surprising therefore that the abundance of different taxonomic groups of protozoa and nematodes is greater near decomposing residues than in bulk soil (Foissner, 1987; Pradhan *et al.*, 1988; Griffiths *et al.*, 1993; Gupta, 1993). Species richness and abundance of protozoa and nematodes near decomposing residues is also influenced by the placement of the residue, e.g. surface placed or buried in soil, and the quality of the residue, e.g. its state of decomposition. For example, populations of fungal-feeding nematodes were present in higher numbers during the later stages of barley straw decomposition which coincided with an increase in fungal biomass and the proportion of lignin in the substrate (Sohlenius and Boström, 1984).

From the evidence given above, it is clear that temporal and spatial variation in abundance and composition must be taken into account when considering the potential use of microfauna as indicators of soil health. Temporal variation could be minimized by collecting soil samples at a particular time in the year (when the short-term effects are at a minimum and are not masking the long-term trends) or by sampling more than once in a year/growing season. Adoption of suitable sample collection procedures (e.g. a defined number of individual soil

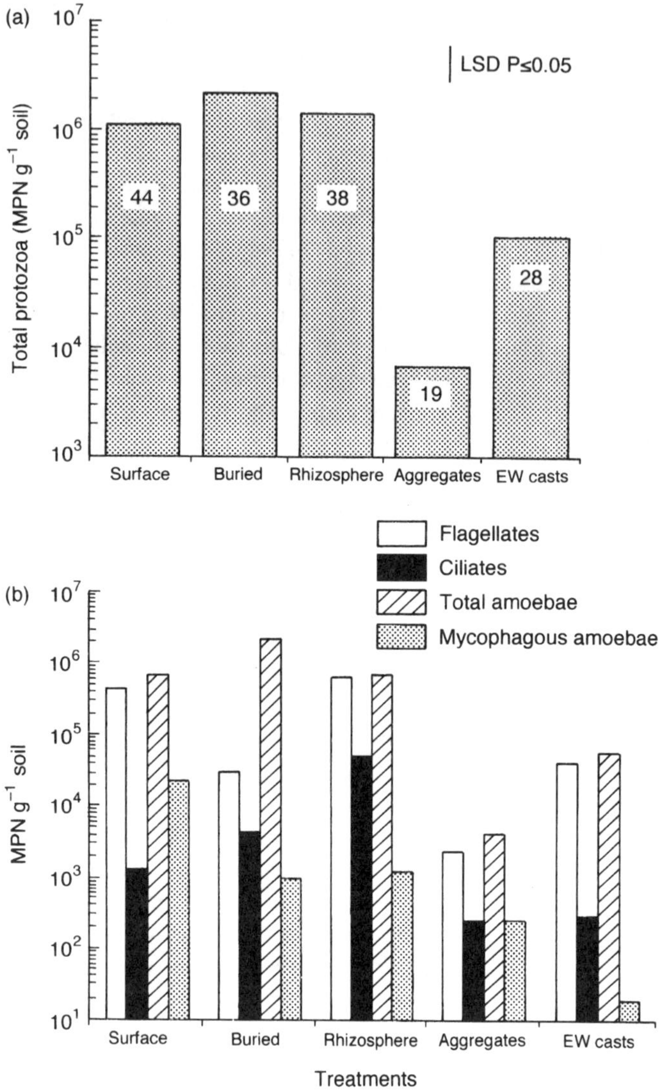

Fig. 9.2. Abundance (a) and composition of different groups (b) of protozoa at microsites differing in physical and biological properties in an agricultural soil (alfisol) in South Australia. (Numbers indicate the number of taxa observed in a 21 day bioassay; Gupta and van Vliet, 1996.)

cores from each field bulked to give a pooled sample) could help reduce the errors introduced by spatial heterogeneity.

Effects of Agricultural Practices on the Abundance and Diversity of Microfauna

Effects of tillage and plant residue management

The effects of tillage and plant residue management on the size, activity and community structure of microfaunal populations may be attributed to one or more of the following factors:

1. Food supply (e.g. higher populations of bacteria and fungi where residues are retained).
2. Changes in the populations of secondary decomposers (e.g. microarthropods as nematode predators and nematodes as protozoan predators).
3. Altered soil physical properties (e.g. improved aggregate stability in reduced tillage systems and improved moisture availability with residue retention).
4. Changes in the distribution of organic matter residues down the soil profile; e.g. more even distribution of organic matter in cultivated soil compared to that in minimum tilled soil where residues are concentrated on the soil surface.

Residue retention has been found to increase the abundance and diversity of protozoa (Gupta, 1993), whereas both stimulatory and negative effects of tillage on protozoan abundance have been reported (Foissner, 1987; Gupta, 1993). If cultivation of an agricultural soil improves soil fertility and crop production this is generally reflected in an increase in the abundance and species richness of protozoa (especially bacterial feeders) in the soil (Foissner, 1987). Similarly, reduced or minimum tillage which tends to promote a fungal-dominated microflora (Beare *et al.*, 1992; Gupta and Roper, 1993) may result in increased populations of mycophagous amoebae (Roper and Gupta, 1995), especially in soil near the decomposing residues (Gupta and van Vliet, 1996).

Both stimulatory and negative effects of cultivation on nematode abundance have been reported (Minton, 1986; Parmelee and Alston, 1986; Yeates and Hughes, 1990; Freckman and Ettema, 1993; Yeates and Bird, 1994). The type of response is specific to different trophic groups. For example the destruction of root material and residues following ploughing caused a decline in root feeding nematodes but an increase in the abundance of nematodes feeding on microorganisms (Thomas, 1978). Yeates and Bird (1994) observed higher numbers of nematodes under minimum tillage compared to conventional cultivation treatments but the effects were small and variable during the season. According to Freckman and Caswell (1985), predacious nematodes prefer the upper layers of moist soils. Since minimum tilled soils tend to have higher moisture, especially

Table 9.2. Indices of the nematode faunae under various management regimes at Avon, South Australia[1].

	CC[2]	DD[3]	Pasture	Native shrub	CF[4]
Taxa identified (s)	16.0	17.0	21.0	29.0	24.0
Specimens identified (*n*)	337.0	290.0	322.0	265.0	208.0
% bacterial feeding	37.6	37.7	27.6	63.6	44.4
% fungal feeding	7.2	4.8	8.5	12.0	11.2
% predacious	–	–	–	0.3	1.6
% plant pathogenic	49.2	45.7	46.8	2.8	10.6
% plant associated	4.1	6.4	7.7	1.4	17.2
% omnivorous	1.8	5.5	9.5	19.7	15.0
Fungivores/bacterial feeder ratio	0.19	0.13	0.31	0.19	0.25
Trophic diversity (*T*)	2.56	2.77	3.15	2.18	3.66
Species richness (*SR*)	2.58	2.82	3.46	5.02	4.31
Shannon–Weaver index (*H′*)	1.94	1.85	2.36	2.61	2.97
Evenness (*J′*)	0.70	0.65	0.78	0.78	0.94
Simpson index for dominance (λ)	0.20	0.24	0.13	0.11	0.09
Diversity (H_2)	1.63	1.42	2.03	2.21	2.46
Maturity index (Σ MI)	2.33	2.50	2.36	2.52	3.38
Maturity index for non-parasites (MI)	2.05	2.20	2.33	2.53	2.25
Plant parasite index (PPI)	2.62	2.88	2.41	2.61	2.89

[1]Adapted from Yeates and Bird (1994); [2]conventionally cultivated wheat plots; [3]direct-drilled wheat plots; [4]wheat field subject to continuous traditional farming for a century.

near the soil surface, they may favour this trophic group. The effects of minimun tillage on plant parasitic nematodes are variable (Parmelee and Alston, 1986; Thomas, 1978; Yeates and Bird, 1994).

Freckman and Ettema (1993) studied the effect of disturbance of varying intensity, including tillage, chemical inputs and successional treatments with little human impact, on soil nematode communities. They used a variety of ecological indices (similar to those reported by Yeates and Bird, 1994; Table 9.2) to describe effects on nematode community structure. Species diversity was greatest in the successional treatments, whereas nematode abundance was highest in the high input and organic systems. Canonical discriminant analysis based on the ecological indices was shown to be a useful way to demonstrate the effects of the different degrees of disturbance on nematode communities.Yeates and Bird (1994) also reported changes in nematode communities in soils exposed to different management regimes (Table 9.2). They found that ecological indices such as species richness and the Shannon index were the best of the indices in revealing trends in the trophic structure of nematode communities. However, since several ecological indices were confounded by soil textural effects, they cautioned against using faunal indices to extract generalized trends across different soils and climatic conditions.

Long-term studies of reduced tillage and conventional tillage systems have revealed distinct differences in both biological (including microfauna) and functional processes in the soil (Beare *et al.*, 1992; Groffman *et al.*, 1986; Hendrix *et al.*, 1986; Parmelee and Alston, 1986; Roper and Gupta, 1995). Tillage generally favours organisms with short generation times, high metabolic rates and rapid dispersal (Andrén and Lagerlof, 1983). This may be why residue incorporation leads to more bacterial growth and hence increased abundance of bacterial-feeding protozoa, bacterial-feeding nematodes and enchytraeids. In contrast, under no-tillage systems fungi dominate over bacteria (because of their capacity to access limiting nutrients across the soil–residue interface) resulting in increased abundance of fungal-feeding protozoa and fungal-feeding nematodes (Andrén and Lagerlof, 1983; Gupta and van Vliet, 1996). The dynamics of the detritus food web and its influence on organic matter turnover and nutrient recycling is extensively discussed by Hendrix *et al.* (1986), Hunt *et al.* (1987), Beare *et al.* (1992) and Wardle *et al.* (1995).

Effects of organic and inorganic fertilizers

The effects of organic amendments, such as farmyard manure (FYM), on soil microfaunal populations depend largely upon the effects of the amendment on soil fertility and plant growth. The total number of protozoa and the number of active amoebae and ciliates was significantly higher in soil treated with FYM or complete mineral fertilizer than in untreated soil (Singh, 1949; Detcheva, 1965). Similarly, increases in protozoan abundance in soils treated with inorganic fertilizers (e.g. nitrogenous fertilizers) have been reported (Viswanath and Pillai, 1977; Elliott and Coleman, 1977; Lehle and Funke, 1989; Griffiths, 1990). In contrast, Solhenius (1990) and Angus *et al.* (1996) observed a reduction in the biomass of protozoa as a result of high levels of inorganic fertilizers (>100 kg N ha^{-1}) and Berger *et al.* (1986) found no significant changes in the populations of ciliates and testaceae in alpine pasture soils fertilized with 600 kg NPK ha^{-1}. These differences in fertilizer response are probably due to differences in habitat type and time between fertilizer application and sampling of the protozoan communities.

Asecht and Foissner (1991, 1992) observed, in a reforested alpine soil, that fertilizer application may cause significant changes to the relative proportions of different protozoan groups (e.g. bacterial-feeding or fungal-feeding ciliates and acidophilic or neutrophilic testaceans) present in the soil (Fig. 9.3). Changes in soil pH following fertilizer application would explain the differences in the composition of testaceans whilst an increased fungal biomass in response to the fertilizers was probably the reason for changes in the composition of ciliates. In another study, Gupta and Germida (1988) reported that application of elemental

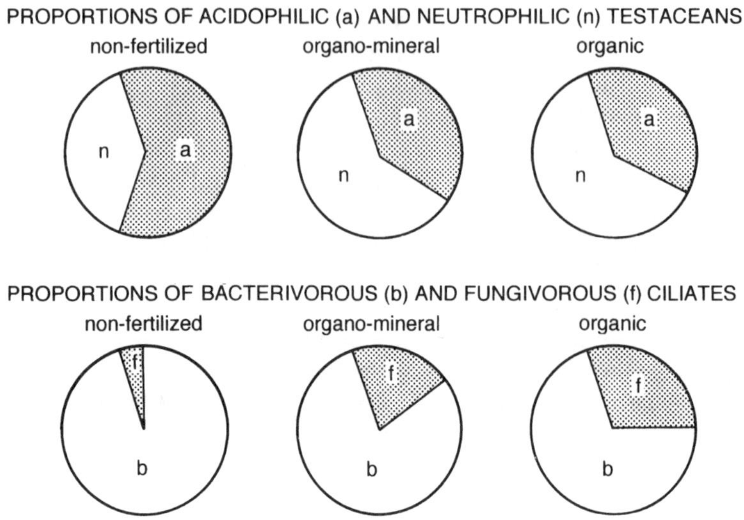

Fig. 9.3. Proportions of indicator species of protozoa (testaceans and ciliates) in a reforested and fertilized site near the alpine timberline. (Adapted from Aescht and Foissner, 1991, 1992.)

sulphur fertilizer to soil for five years resulted in a 30–71% decline in populations of bacterial feeding protozoa and a >84% decline in populations of mycophagous amoebae (MA). The changes in the populations of protozoa paralleled changes in microbial biomass and activity, especially in the case of MA. They observed a significant positive relationship ($R = 0.82$, $P < 0.01$) between fungal biovolume and populations of MA. This suggests that the amount of food available is the most important factor limiting populations of MA.

The effects of fertilizer application on nematode populations are varied and changes in both the abundance and trophic diversity of nematodes in response to fertilizers have been reported. Sohlenius and Boström (1986) and Sohlenius (1990) reported that N fertilization (120 kg N ha^{-1}) resulted in a large increase in nematode abundance compared to no N fertilization. This positive effect of N fertilization was attributed to an increase in plant production, root biomass and microbial activity in the soil. In contrast, Sohlenius and Wasilewska (1984) and Berger *et al.* (1986) observed a reduction in nematode numbers following the application of fertilizer N. Rodriguéz-Kábana (1986) reported that organic (oil cake) and inorganic amendments (300 kg N ha^{-1} as anhydrous ammonia and 130 kg N ha^{-1} as urea) were effective nematode suppressants. There was a direct relationship between the amount of protein N in the organic amendments and their effectiveness in suppressing nematode populations.

Application of both organic and inorganic fertilizers may cause significant changes in nematode trophic diversity. Generally, application of fertilizers increases the proportion of root feeders (Yeates, 1982) and bacterial feeders (Yeates, 1982; Sohlenius and Wasilewska, 1984; Sohlenius and Boström, 1986; Hyvönen and Huhta, 1989; Bohlen and Edwards, 1994) and decreases the proportion of fungal feeders and omnivores (Sohlenius and Wasilewska, 1984; Sohlenius and Boström, 1986; Sohlenius, 1990). Such trends have been found in both agricultural and forest systems. Increased plant production is considered to be the main cause for the increase in root feeding nematodes, whilst stimulation of microbial activity and bacterial populations in the rhizosphere may be responsible for the increase in bacterial feeders. Sohlenius and Boström (1986) attributed the decline in fungal-feeding nematodes to the reduction in fungal biomass in fertilized plots (Schnurer *et al.*, 1986).

De Goede *et al.* (1993) used the *c-p* triangle (which classifies nematode fauna into enrichment opportunists *c-p* 1, general opportunists *c-p* 2 and persisters *c-p* 3–5) to show changes in nematode communites following different fertilizer treatments. Ettema and Bongers (1993) also used this approach to investigate the relative abundance of nematode trophic groups in soil following manuring and/or fumigation (Fig. 9.4A). A temporary increase in food supply following manuring or fumigation caused an enrichment of opportunists (*c-p* 1) which were succeeded by *c-p* group 2 and finally supplemented by persisters (*c-p* 3–5). This type of pictorial representation of data helps to provide meaningful interpretation to large quantities of ecological data. Bongers *et al.* (1995) used a modified version of the *c-p* triangle to follow the changes in nematode diversity following eutrophication, acidification and manure application (Fig. 9.4B).

Most of the changes in protozoan and nematode abundance and trophic diversity is found to be short-term and once the effects of the C and nutrient inputs from the fertilizer application have disappeared, the microfaunal population will return to its original status, unless significant changes have occurred to the physical and chemical properties of the soil (Wodarz *et al.*, 1992; Ettema and Bongers, 1993).

Soil pH and salinity

Experimental evidence of the direct effects of pH on individual protozoan species is limited. However, many flagellates, ciliates and amoebae appear to grow over a relatively wide range of pH (pH 3.5–9.75), and protozoan populations usually respond positively and rapidly to the application of lime (Cutler and Crump, 1935; Stout and Heal, 1967; Homma *et al.*, 1983; Berger *et al.*, 1986; Foissner, 1987; Lehle and Funke, 1989). Some effects of adverse pH on protozoa include reduced activity of spore-perforating amoebae (Homma and Cook, 1985) and malformations in euglyphid testates (Rauenbusch, 1987).

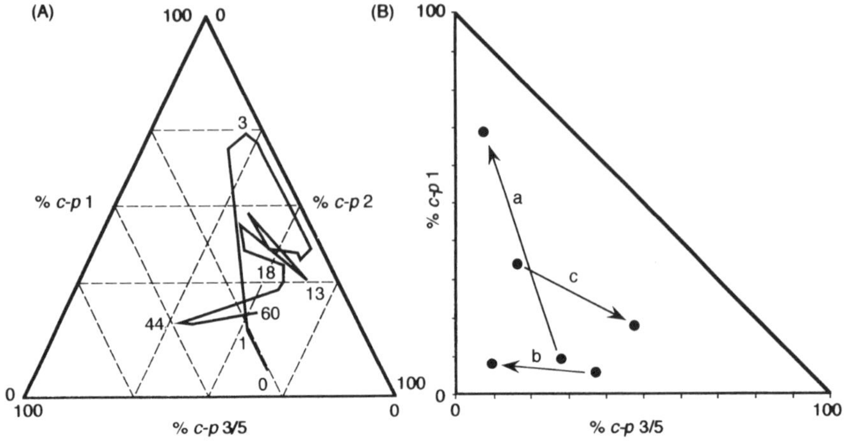

Fig. 9.4. Colonizer-persister (*c-p*) triangle showing (A) successional changes in nematode (*c-p* group) distribution in manured soil (numbers indicate weeks after manuring; Ettema and Bongers, 1993) (B) shifts during a) eutrophication, initial situation and 2 weeks after adding powdered cow-dung (Ettema and Bongers, 1990); b) artificial acidification of coniferous soil (Hyvönen and Persson, 1993) and c) recovery, 33 and 44 weeks after organic manuring (Ettema and Bongers, 1990). (Adapted from Ettema and Bongers, 1993 and Bongers *et al.*, 1995).

The available data on nematode distribution in different soils and agroecosystems indicates no direct limitation of soil pH, although liming which increased soil pH was correlated with an increase in nematode abundance (Yeates, 1981). Information of the effect of pH on the trophic diversity of nematode populations is limited. Hyvönen and Persson (1990) observed a shift in nematode composition (reduction in % abundance of *c-p* 3–5, persisters; suggesting a community under stress) following the artificial acidification of a coniferous forest soil (Fig. 9.4B).

Although most soil protozoa can adapt to a wide range of salinities (up to 45%) an increase in salinity may inhibit reproduction and cause encystment in some species (Stout, 1955). Pomp and Wilbert (1988) observed no significant correlation between the occurrence and abundance of certain ciliate species and the salt content of soils from Australia. The concentration and composition of salts in soil solution, however, has been reported to influence the trophozite form of *Naegleria gruberi* (Willmer, 1963).

Effects of pesticides

A number of studies have demonstrated negative effects of herbicides on protozoan and nematode populations (Yeates *et al.*, 1976; Popovici *et al.*, 1977; Pons and Pussard, 1980; Laminger and Maschler, 1986; Sály, 1989; Wardle *et al.*, 1995). Popovici *et al.* (1977) found a dose-dependent decline in the abundance of taxonomic groups of protozoa (e.g. flagellates, ciliates and naked amoebae) following application of the herbicide atrazine which persisted in the soils for four months; flagellates were the most sensitive group. Pons and Pussard (1980) found differences in the response of different monoxenic cultures of naked amoebae to the same herbicide, atrazine (40 mg l^{-1}). In contrast, Odeyemi *et al.* (1988) did not find any adverse effects of soil application of the herbicides Gramoxone, Dacthal, Preforan and Dual on protozoan populations. However, Pizl (1985) found that exposure of earthworms to the herbicide zeazin 50 increased their infection by monocystid gregarines (protozoa) suggesting that herbicide stress reduced the resistance of earthworms to infection.

While some direct negative effects of herbicides on nematode populations have been recorded (Wong *et al.*, 1993), many of the effects are indirect arising from the changes in the quantity and quality of plant inputs (e.g. dead organic matter from weeds) to the soil. Herbicide induced stimulation of nematode populations is attributed to the increased microbial activity from the dead weed residues (Mahn and Kastner, 1985; Edwards, 1989; Wardle *et al.*, 1995). Long-term use of herbicides, however has been shown to reduce both the abundance and diversity of nematodes in agricultural soils and vineyards (Yeates *et al.*, 1976; Ishibashi *et al.*, 1978; Wardle *et al.*, 1995).

The effects of insecticides and fungicides on protozoa (especially ciliates and testaceans) and nematodes are more pronounced than those of herbicides (Foissner, 1987; Yeates and Bongers, 1997). Petz and Foissner (1989) observed acute toxicity to the insecticide lindane in ciliates; the number and composition of the ciliate faunae remained significantly altered 90 days after its application. Negative effects of lindane on the abundance and composition of testate amoebae were also reported by Wanner (1991). Populations of fungal-feeding ciliates were significantly reduced by the application of the fungicide mancozeb (Petz and Foissner, 1989), probably through its effects on fungi. Yeates *et al.* (1991) reported that fumigation by methyl bromide eliminated nematode populations and that 166 days after fumigation nematode populations were still lower than those in untreated soils.

Effects of Sewage, Heavy Metals and Industrial Pollutants on Microfauna

Sewage effluent, sludge and heavy metals

The use of municipal sewage effluents and sludge (activated or digested) as fertilizers or soil conditioners has increased in recent years (McLaren and Smith, 1996). Soil microfauna are able to adapt to effluent irrigation and associated altered moisture regimes (Cairns *et al.*, 1978). Stout (1978) irrigated ten soil types over 18 months with either water or secondary effluent and found that protozoan populations were on average, five times higher in the effluent irrigated soils than in the water irrigated soil. Ciliates, *Colpoda steinii* and *Leptopharynx costalus* were equal in numbers in both treatments, whereas *C. infata, Uroleptus* sp. and *Trinema lineare* were less common in effluent treated soils. In contrast, Yeates (1978) found that nematode abundance in pasture soils was not affected by effluent application but found a proportional increase in Rhabditida populations in the effluent treated soils. Similarly, Gupta *et al.* (1996b) found no increase in the abundance of nematode faunae in soils under *Eucalyptus camaldulensis* and *Casuarina glauca* plantations in South Australia following application of sewage effluent. However, a proportional increase in the populations of bacterial-feeding nematodes, and an increase in the abundance and species richness of protozoa was observed in the treated soils.

Unless the effluent or sewage sludge is rich in heavy metals, the beneficial effects of their application on soil properties (e.g. increased water stable aggregates, increased microbial activity) are conducive to higher protozoan populations. The effects of heavy metals on aquatic protozoa have been studied extensively but little is known about their effects on soil protozoa (Foissner, 1987). Brookes *et al.* (1984) found no effect of heavy metals on total number of protozoa in sludge treated soils. In contrast, Forge *et al.* (1993) reported that the relative toxicity of heavy metals on the growth of *Colpoda steinii* were Ni> Cd>Cu>Zn. Chapman and Dunlop (1981) and Iftode *et al.* (1985) found that Ca^{2+} increased the heavy metal concentration necessary to reduce the growth of the protozoan species *Tetrahymena pyriformis* and *Euplotes vannus*.

Low levels of soil contamination with heavy metals that have no negative effects on plant growth have minimal effects on soil nematode populations. However, high levels of contamination that reduce plant growth have been shown to affect the relative abundance of different functional groups of nematodes in the soil. Popham and Webster (1980) reported that heavy metals such as Cd are toxic to nematodes and that Cd interferes with nutrient uptake and assimilation in *Caenorhabditis elegans*. Korthals *et al.* (1993) reported that in arable field soils in the Netherlands, application of sewage effluent contaminated with Cu (750 kg ha^{-1}) reduced the abundance and diversity (lower MI) of nematodes. Yeates *et al.* (1994) reported that contamination of pasture soils with Cu, Cr and As influenced the abundance and diversity of nematodes. Nematode abundance and diversity increased at low levels of contamination (e.g. 109–161 mg kg^{-1}) compared to control soils, whereas medium or high levels of contamination (>700 mg kg^{-1}) decreased nematode abundance and diversity at both 0–5 and 5–10 cm depths (Fig. 9.5). Reduced plant growth in the highly contaminated soils paralleled a shift in dominance from plant-feeding nematodes in control plots to bacterial-feeding nematodes in contaminated plots (Yeates *et al.*, 1994). Weiss and Larink (1991) also found an increased abundance of predatory nematodes in soils following the addition of sewage sludge and heavy

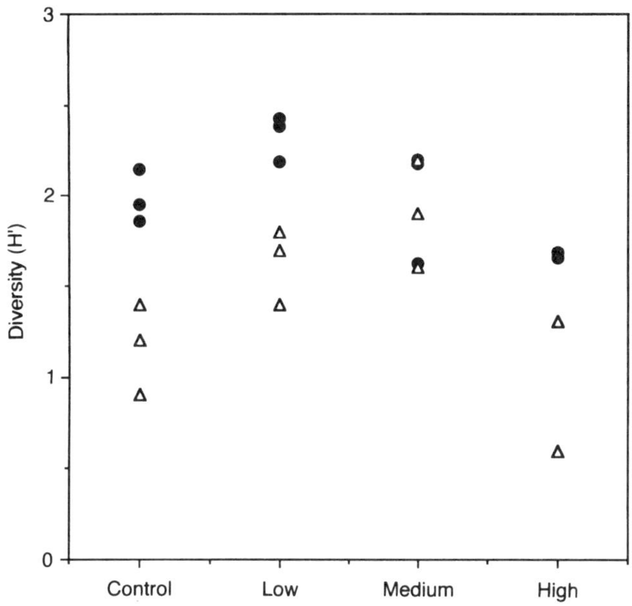

Fig. 9.5. Diversity of nematode faunae in 0–5 (●) and 5–10 cm (△) soil depth as affected by heavy metal contamination. Heavy metal (Cu, Cr, As) concentrations in soils ranged between 12–47 mg kg^{-1} for control, 109–148 mg kg^{-1} for low, 382–469 for medium and 739–835 for high level of contamination. (Adapted from Yeates *et al.*, 1994.)

metals. These studies indicate that nematode abundance (in terms of trophic groups) and diversity parameters are potentially good indicators of changes in soil health resulting from contamination of soils with heavy metals.

Industrial pollutants

Contamination of soils by airborne pollutants (e.g. industrial emissions containing heavy metals, ammonia, oxides of sulphur and nitrogen, and acid rain) is reflected in reduced abundance and diversity of protozoa in agricultural and forest soils (Tomescu, 1987; Foissner, 1994). Tomescu (1987) observed that heavily polluted soils in Romania contained only r-selected colpodids and some flagellates. Steinberg *et al.* (1990) reported that naked amoebae could not encyst in soils contaminated with pyralene (a commercial product of polychlorinated biphenyls) and thus died. Purrini (1983) observed that SO_2 pollution increased the infection of soil invertebrates with parasitic protozoa, an observation comparable to that for the effects of herbicide on the parasitic protozoa of earthworms (Pizl, 1985).

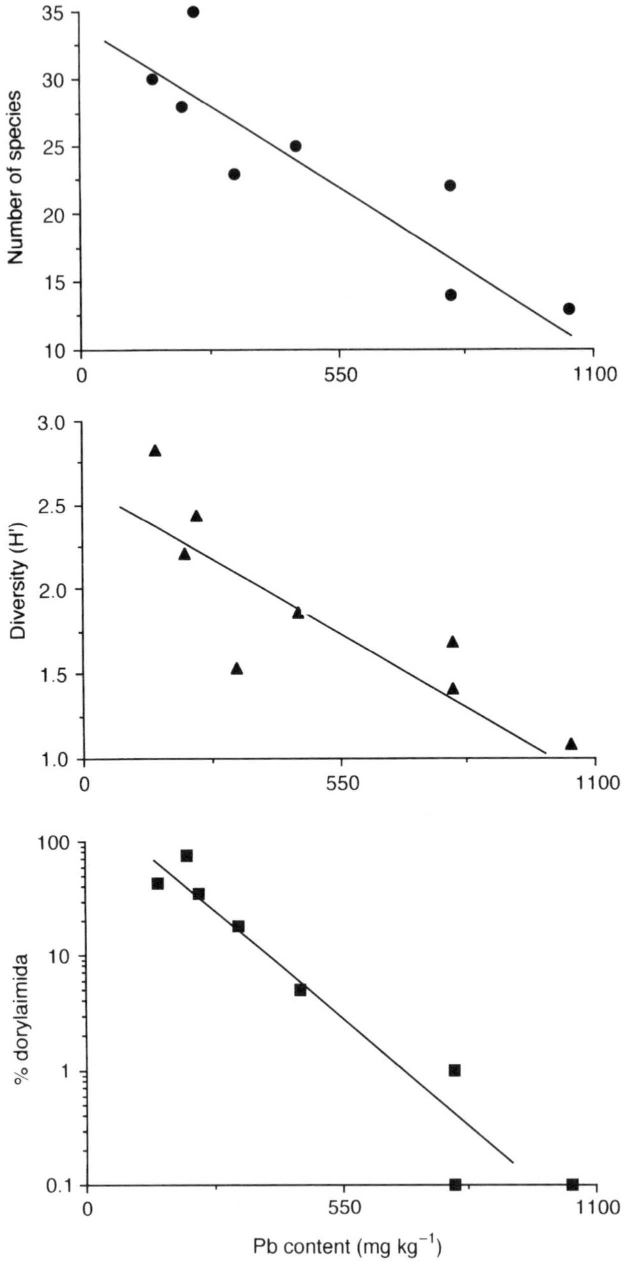

Fig. 9.6. Effect of increasing aerial pollution with lead on nematode species, diversity (H′) and proportion of Dorylaimida in the nematode fauna. (Adapted from Zullini and Peretti, 1986.)

Soil contamination by industrial pollutants has varied effects on soil nematode populations but consistently alters the relative proportions of trophic groups (Yeates and Bongers, 1997). For example, contamination of soils by Pb, Cu and Zn contained in the exhaust fumes and fine powder of metallurgical plants and smelters reduced the abundance and diversity (H′ and MI) of nematodes but increased the proportion of *r*-strategists (bacterivores) (Zullini and Peretti, 1986; Popovici, 1994; Popovici and Korthals, 1995) (Fig. 9.6). Ruess and Funke (1995) reported that pH-induced changes caused by acid rain resulted in an increase in the ratio of bacterial-feeding to fungal-feeding nematodes.

Use of protozoa in bioassays to determine the bioavailability of pollutants in soil

The use of protozoan growth and behaviour as a measure of the bioavailability of heavy metals, pesticides and other pollutants (e.g. detergents) in soils has potential as a novel bioassay (Pratt *et al.*, 1988, Schreiber and Brink, 1989; Forge *et al.*, 1993; Ekelund *et al.*, 1994).

Some of the features of protozoa that make them potentially suitable for toxicity assays are:

1. In their active form protozoa have delicate external membranes and are extremely sensitive to a wide range of toxic compounds.
2. They are relatively easy to observe and enumerate in pure culture.
3. They have short generation times (hours) compared to other soil fauna and thus respond rapidly to perturbations. A short-term bioassay is essential in order to avoid interference due to the development of tolerance to the pesticide/heavy metal.

Using aquatic protozoa as test organisms, Pratt *et al.* (1988) evaluated the potential toxicity of soil leachates containing organics and/or heavy metals. They found that soil samples from an uncontaminated agricultural field showed no toxicity in the tests whereas soils contaminated with toxic wastes (e.g. heavy metals, coal, organics etc.) showed different degrees of toxicity (Fig. 9.7). They observed a significant negative relationship between protozoan species survival and the concentration of leachates from the contaminated soils (Fig. 9.7).

Schreiber and Brink (1989) evaluated the effects of pesticides (such as chlorex, MCPA, dichlorprop, benlate and sumicidin) on different protozoan groups and recommended a laboratory-based toxicity test using an absorbance method to measure protozoan growth. When compared to a most-probable-number method (to measure protozoan populations) the absorbance method offered less variation and was labour-efficient. In this *in vitro* toxicity test the freshwater ciliate and flagellate species used showed sensitivities to pesticides at concentrations normally used in agriculture.

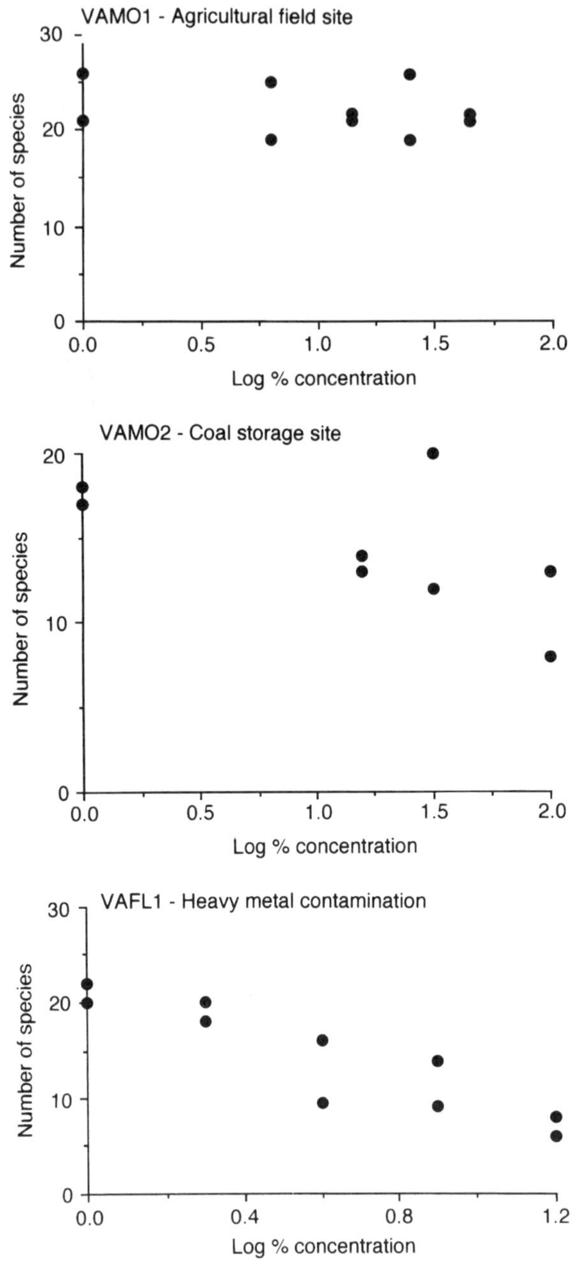

Fig. 9.7. Protozoan species survival in microcosms developed with leachates of soils from an agricultural field (VAMO1), coal storage site (VAMO2) and a site contaminated with Cu, Cd, Ni, Pb and Zn (VAFL 1). Values plotted on the Y-axis are for control samples with no leachates. (Adapted from Pratt *et al.*, 1988.)

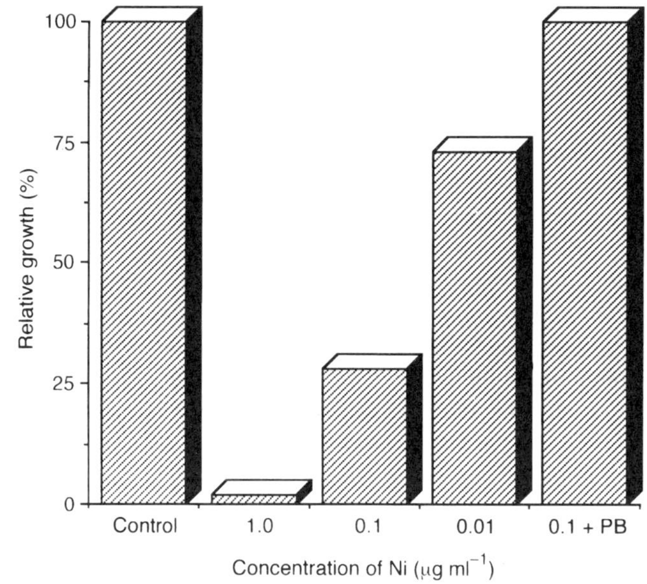

Fig. 9.8. Relative growth of the soil ciliate *Colpoda* sp. at different concentrations of Ni in a 24 h laboratory incubation assay. Addition of phosphate buffer mixture (PB) to the test solution resulted in the reduction of the 'active' (bioavailable) form of Ni in the test solution. (V.V.S.R. Gupta, unpublished.)

Laboratory bioassays using protozoans such as *Colpoda steinii*, *Acanthamoeba* sp., *Saccamoeba* sp., *Bodo* spp., *Oikomonas* sp. and *Cercomonas* sp. as test organisms have all proved to be useful in providing rapid assessment of the bioavailability of heavy metals in soils (Forge *et al.*, 1993; Ekelund *et al.*, 1994 and V.V.S.R. Gupta, unpublished). Heavy metals, such as Ni and Cd, cause growth abnormalities to both the active and cystic forms of the ciliate *Colpoda steinii* and reduce its rate of growth. Similar abnormalities were caused by pesticides in soil solution (Foissner, 1994). The magnitude of the inhibitory effect on protozoan growth is strongly related to the concentration of heavy metal in the 'active' (bioavailable) form rather than the total concentration of the metal in the sample. For example, the relative growth rate of *Colpoda* sp. at different concentrations of Ni was significantly improved in the presence of phosphate buffer which reduces the level of Ni in the 'active' form thereby reducing its bioavailability (Fig. 9.8).

Such bioassays can be used in parallel with chemical methods in routine analysis of contaminated soils to determine the bioavailability of heavy metals and other pollutants. Schreiber and Brink (1989) suggested that such toxicological evaluations could be used in remedial studies of cleaned sites to ensure the toxic materials have been effectively removed from the site.

Conclusions

Can soil microfauna be used as bioindicators of soil health?

The foregoing discussion has demonstrated that soil protozoa and nematodes are generally very responsive to changes in the condition or health of the soil. As individual groups they appear to possess the general features required of bioindicators (Elliott, Chapter 3, this volume) including: (i) having a critical role in the functioning of the soil; (ii) being present in measurable numbers and being widely distributed; (iii) having suitable techniques for their enumeration; and (iv) being sensitive enough to respond to changed soil management practices within an appropriate time frame. They are ubiquitous, abundant and diverse. They are intimately involved in regulating decomposition and plant nutrient cycling in the soil and they have various measurable attributes (abundance, species diversity, trophic diversity) which make them potentially suitable as indicators. In common with other soil organisms however, their enumeration in soils is greatly influenced by temporal and spatial variability. Care must therefore be taken to ensure that soil samples are representative of the system under study and that comparisons are made at selected (and appropriate) times throughout the year or throughout the cropping cycle.

Functional or trophic diversity of protozoa and nematodes would appear to be the most useful attribute of these organisms that could be used as a bioindicator of soil health. This diversity will largely be governed by the available food source (status of the detritus food web) in the soil, e.g. in soils subjected to minimum tillage and residue retention, the soil microflora is fungal dominated, which is reflected in a predominance of mycophagous protozoa and fungal-feeding nematodes. Another good example is the increased abundance of predatory nematodes in soils contaminated with heavy metals and the decline in diversity of protozoa in soils containing residues of some herbicides.

The major constraint to using either protozoa or nematodes as bioindicators is the fact that identification to species level requires a high level of technical expertise. At present there are no reliable biochemical analyses that can substitute for detailed microscopic examination. Diversity studies however, have shown that identification to genus or family level may be adequate as this data gives trends similar to those achieved if species level identification is carried out. Thus, indices such as the Shannon-Weaver diversity index (H′), the nematode maturity index (MI) and the weighted coenotic index (WCI) may become very useful in summarizing information on the abundance and diversity of soil microfauna in relation to soil health.

Another very promising line of current research is the development of laboratory bioassays utilizing the growth and behaviour of protozoa (e.g. *Colpoda*, *Acanthamoeba*, *Oikomonas* sp.) to measure the concentrations and bioavailability of chemical pollutants and heavy metals in soils.

Acknowledgements

The authors wish to thank the reviewers for their helpful suggestions to improve the manuscript. Financial support for VVSRG was provided by the CRC for Soil and Land Management and for GWY by Landcare Research and the New Zealand Foundation for Research, Science and Technology.

References

Acton, D.F. and Gregorich, L.J. (1995) *The Health of our Soils: Toward Sustainable Agriculture in Canada.* Agriculture, Agri-food Canada, Ottawa.

Aescht, E. and Foissner, W. (1991) Bioindikation mit mikroskopisch kleinen Bodentieren. *VDI Berichte* 901, 985–1002.

Aescht, E. and Foissner, W. (1992) Effects of mineral and organic fertilizers on the microfauna in a high-altitude reafforestation trial. *Biology and Fertility of Soils* 13, 17–24.

Andrén, O. and Lagerlof, J. (1983) Soil fauna (microarthropods, enchytraeids, nematodes) in Swedish agricultural cropping systems. *Acta Agricultarae Scandinavica* 33, 33–52.

Angus, J.F., Gupta, V.V.S.R., Good, A.J., Howe, G.N. and Pitson, G.D. (1996) Anhydrous ammonia applied as coldflo™ to wheat. *Proceedings of the Australian Agronomy Conference*, Towoomba, Australia, p. 617.

Badgley, A. and Odelson, D. (1993) Enumeration of soil protozoa by nucleic acid methodologies. *Research Findings of the Center for Microbial Ecology.* Michigan State University, East Lansing, USA, pp. 73–75.

Bamforth, S.S. (1995) Interpreting soil ciliate biodiversity. *Plant and Soil* 170, 159–164.

Beare, M.H., Parmelee, R.W., Hendrix, P.F. and Cheng, W. (1992) Microbial and faunal interactions and effects on litter nitrogen and decomposition in agroecosystems. *Ecological Monographs* 62, 569–591.

Beare, M.H., Coleman, D.C., Crossley, D.A., Jr, Hendrix, P.F. and Odum, E.P. (1995) A hierarchical approach to evaluating the significance of soil biodiversity to biogeochemical cycling. *Plant and Soil* 170, 5–22.

Berger, H., Foissner, W. and Adam, H. (1986) Field experiments on the effects of fertilizers and lime on the soil microfauna of an alpine pasture. *Pedobiologia* 29, 261–272.

Bernard, E.C. (1992) Soil nematode biodiversity. *Biology and Fertility of Soils* 14, 99–103.

Bird, A.F. and Ryder, M.H. (1993) Feeding of the nematode *Acrobeloides nanus* on bacteria. *Journal of Nematology* 25, 493–499.

Bohlen, P.J. and Edwards, C.A. (1994) The response of nematode trophic groups to organic and inorganic nutrient inputs in agroecosystems. In: Doran, J., Bezdicek, D. and Coleman, D.C. (eds) *Defining Soil Quality for a Sustainable Environment.* Soil Science Society of America Special Publication No. 35, Madison, Wisconsin, USA, pp. 193–200.

Bongers, T. (1988) *De Nematoden van Nederland.* Pirola, Schoorl, 408 pp. (reprinted without changes in 1994).

Bongers, T. (1990) The maturity index: an ecological measure of environmental disturbance based on nematode species composition. *Oecologia* 83, 14–19.

Bongers, T., de Goede, R.G.M., Korthals, G.W. and Yeates, G.W. (1995) Proposed changes of *c-p* classification for nematodes. *Russian Journal of Nematology* 3, 61–62.

Boström, S. and Sohlenius, B. (1986) Short-term dynamics of nematode communities in arable soil – influence of a perennial and an annual cropping system. *Pedobiologia* 29, 345–357.

Bowers, N.J. and Pratt, J.R. (1995) Estimation of genetic variation among soil isoloates of *Colpoda inflata* (Stokes) (Protozoa, Ciliophora) using polymerase chain reaction and restriction fragment length polymorphism analysis. *Archiv für Protistenkunde* 145, 29–36.

Brookes, P.C., McGrath, S.P., Klein, D.A. and Elliott, E.T. (1984) Effects of heavy metals on microbial activity and biomass in field soils treated with sewage sludge. In: *Environmental Contamination. International Conference,* London, July 1984, CEP Consultants Ltd, Edinburgh, pp. 574–583.

Buitkamp, U. (1979) Vergleichende Untersuchungen zur Temperaturadaptation von Bodenciliaten aus klimatisch verschiedenen Regionen. *Pedobiologia* 19, 221–236.

Cairns, A., Dutch, M.E., Guy, E.M. and Stout, J.D. (1978) Effect of irrigation with municipal water or sewage effluent on the biology of soil cores. *New Zealand Journal of Agricultural Research* 21, 1–9.

Cavigelli, M.A., Robertson, G.P. and Klug, M.J. (1995) Fatty acid methyl ester (FAME) profiles as measures of soil microbial community structure. *Plant and Soil* 170, 99–113.

Chantanao, A. and Jensen, H.J. (1969) Saprozoic nematodes as carriers and disseminators of plant pathogenic bacteria. *Journal of Nematology* 1, 216–218.

Chapman, G. and Dunlop, S. (1981) Detoxification of zinc and cadmium by the freshwater protozoan *Tetrahymena pyriformis. Water Research* 21, 81–86.

Chardez, D. and Lambert, J. (1981) Thecamoebiens indicateurs biologiques (Protozoa Rhizopoda testacea). *Bulletin des Recherches Agronomiques de Gembloux* 16, 181–203.

Clarholm, M. (1981) Protozoan grazing of bacteria in soil – impact and importance. *Microbial Ecology* 7, 343–350.

Coleman, D.C. (1994) The microbial loop concept as used in terrestrial soil ecology studies. *Microbial Ecology* 28, 245–250.

Cowling, A.J. (1994) Protozoan distribution and adaptation. In: Darbyshire, J.F. (ed.) *Soil Protozoa.* CAB International, Wallingford, UK, pp. 5–41.

Crossley, D.A., Jr, Coleman, D.C. and Hendrix, P.F. (1989) The importance of the fauna in agricultural soils: research approaches and perspectives. *Agriculture, Ecosystems and Environment* 27, 47–55.

Curl, E.A. (1988) The role of soil microfauna in plant-disease suppression. *CRC Critical Reviews in Plant Sciences* 7, 175–196.

Cutler, D.W. and Crump, L.M. (1935) *Problems in Soil Microbiology.* Longmans, Green and Company, London, 104 pp.

Cutler, D.W., Crump, L.W. and Sandon, H. (1922) A quantitative investigation of the bacterial and protozoal populations of the soil with an account of the protozoan fauna. *Philosophical Transactions of the Royal Society of London* 211, 317–350.

Darbyshire, J.F. (1994) *Soil Protozoa.* CAB International, Wallingford, UK, 209 pp.

Darbyshire, J.F. and Greaves, M.P. (1967) Protozoa and bacteria in the rhizosphere of *Sinapis alba* L., *Trifolium repens* L., and *Lolium perenne* L. *Canadian Journal of Microbiology* 13, 1057–1067.

de Goede, R.G.M., Bongers, T. and Ettema, C.H. (1993) Graphical presentation and interpretation of nematode community structure: c-p triangles. *Mededelingen Faculteit Landbouwkundige en Toegepaste Biologische Wetenschappen Universiteit Gent* 58/2b, 743–750.

Detcheva, R.B. (1965) On the study of the soil protozoa in Bulgaria. *Izvestiya na Zoologicheskiya Institut, Sofia* 18, 199–223 (in Bulgarian with English summary).

Doran, J.W., Sarrantonio, M. and Janke, R. (1994) Strategies to promote soil quality and health. In: Pankhurst, C.E. Doube, B.D., Gupta, V.V.S.R. and Grace, P.R. (eds) *Soil Biota: Management in Sustainable Farming Systems.* CSIRO Press, Melbourne, Australia, pp. 230–237.

Edwards, C.A. (1989) Impact of herbicides on soil ecosystems. *Critical Reviews in Plant Science* 8, 221–257.

Ekelund, F., Ronn, R. and Christensen, S. (1994) The effect of three different pesticides on soil protozoan activity. *Pesticide Science* 42, 71–78.

Elliott, E.T. and Coleman, D.C. (1977) Soil protozoan dynamics in a shortgrass prairie. *Soil Biology and Biochemistry* 9, 113–118.

Ettema, C.H. and Bongers, T. (1993) Characterization of nematode colonization and succession in disturbed soil using the Maturity Index. *Biology and Fertility of Soils* 16, 79–85.

Foissner, W. (1987) Soil protozoa: fundamental problems, ecological significance, adaptations in ciliates and testaceans, bioindicators and guide to the literature. *Progress in Protistology* 2, 69–212.

Foissner, W. (1994) Soil protozoa as bioindicators in ecosystems under human influence. In: Darbyshire, J.F. (ed.) *Soil Protozoa.* CAB International, Wallingford, UK, pp. 147–194.

Forge, T.A., Berrow, M.L., Darbyshire, J.F. and Warren, A. (1993) Protozoan bioassays of soil amended with sewage sludge and heavy metals, using the common soil ciliate *Coploda steinii. Biology and Fertility of Soils* 16, 282–286.

Freckman, D.W. and Caswell, E.P. (1985) The ecology of nematodes in agroecosystems. *Annual Review of Phytopathology* 23, 275–296.

Freckman, D.W. and Ettema, C.H. (1993) Assessing nematode communities in agroecosytems of varying human intervention. *Agriculture, Ecosystems and Environment* 45, 239–261.

Goodey, T. (1963) *Soil and freshwater nematodes.* 2nd edn, revised by J.B. Goodey. Methuen, London, 544 pp.

Griffiths, B.S. (1990) A comparison of microbial-feeding nematodes and protozoa in the rhizosphere of different plants. *Biology and Fertility of Soils* 9, 83–88.

Griffiths, B.S., Ékelund, F., Ronn, R. and Christensen, S. (1993) Protozoa and nematodes on decomposing barley roots. *Soil Biology and Biochemistry* 25, 1293–1295.

Groffman, P.M., House, G.J., Hendrix, P.F., Scott, D.E. and Crossley, D.A., Jr, (1986) Nitrogen cycling as affected by interactions of components in a Georgia piedmont agroecosystem. *Ecology* 67, 80–87.

Grospietsch, T. (1965) Schalenamoben im Boden. *Mikrokosmos* 54, 14–18.

Gupta, V.V.S.R. (1993) The impact of soil and crop management practices on the dynamics of soil microfauna and mesofauna. In: Pankhurst, C.E., Doube, B.M., Gupta, V.V.S.R. and Grace, P.R. (eds) *Soil Biota: Management in Sustainable Farming Systems.* CSIRO Press, Melbourne, Australia, pp. 107–124.

Gupta, V.V.S.R. (1996) Diversity of protozoa – its ecological significance in terrestrial ecosystems. *Proceedings of the 7th International Symposium on Microbial Ecology* (in press).

Gupta, V.V.S.R. and Germida, J.J. (1988) Populations of predatory protozoa in field soils after five years of elemental S fertilizer application. *Soil Biology and Biochemistry* 20, 787–791.

Gupta, V.V.S.R. and Germida, J.J. (1989) Influence of bacterial–amoebal interactions on sulfur transformations in soil. *Soil Biology and Biochemistry* 21, 787–791.

Gupta, V.V.S.R. and Roper, M.M. (1993) Impact of different crop residue management practices on microbial biomass and nutrient availability in soil. *Australian Microbiologist* 13, 121.

Gupta, V.V.S.R. and Roper, M.M. (1996) The importance of soil protozoa for the mineralization in Southern Australian soils. *Soil Science – Raising the Profile* Proceedings of the Australian and New Zealand National Soils Conference. Volume 3, 91–92.

Gupta, V.V.S.R. and van Vliet, P.C.J. (1996) Diversity and abundance of protozoa at different microsites in an agricultural soil. Paper presented at the *XII International Colloquium of Soil Zoology*, July 1996, Dublin, Ireland.

Gupta, V.V.S.R., Coles, R., McClure, S., Neate, S., Roget, D. and Pankhurst, C.E. (1996a) The role of microfauna in the survival and effectiveness of plant pathogenic fungi in soil. Paper presented at the *International meeting of the Australian and New Zealand Societies for Microbiology*, 29 September – 4 October, 1996, Christchurch, New Zealand.

Gupta, V.V.S.R., Rogers, S. and Naidu, R. (1996b) Effects of secondary treated sewage effluent application on the populations of microfauna in a forest soil. In: *Contaminants and the Soil Environment in the Australasia-Pacific Region.* Naidu, R., Kookana, R.S., Oliver, D.P., Rogers, S. and McLaughlin, M.J. (eds), Kluwer Academic Publishers, The Netherlands, pp. 235–236.

Hendrix, P.F., Parmelee, R.V., Crossley, D.A., Jr, Coleman, D.C., Odum, E.P. and Groffman, P. (1986) Detritus food webs in conventional and no-tillage agroecosystems. *BioScience* 36, 374–380.

Henkinet, R., Couteaux, M.M., Billes, G., Bottner, P. and Palka, L. (1990) Acceleration of carbon turnover and stimulation of the priming effect by predation in forest humus. *Soil Biology and Biochemistry* 22, 555–561.

Homma, Y. and Cook, R.J. (1985) Influence of matric and osmotic water potentials and soil pH on the activity of giant vampyrellid amoebae. *Phytopathology* 75, 243–246.

Homma, Y., Yamashita, Y. and Katsube, T. (1983) Distribution of *Arachnula impatiens* Cienk., a mycophagous amoeba in Japan and effect of environmental factors on its activity in soil. *Annals of the Phytopathological Society of Japan* 43, 370–371.

Hunt, H.W., Coleman, D.C., Ingham, E.R., Ingham, R.E., Elliott, E.T., Moore, J.C., Rose, S.L., Ried, C.P.P. and Morley, C.R. (1987) The detrital food web in a shortgrass prairie. *Biology and Fertility of Soils* 3, 57–68.

Hussey, R.S. and Roncadori, R.W. (1981) Influence of *Aphelenchus avenae* on vesicular-arbuscular endomycorrhizal growth response in cotton. *Journal of Nematology* 13, 48–52.

Hyvönen, R. and Huhta, V. (1989) Effects of lime, ash and nitrogen fertilizers on nematode populations in scots pine forest soils. *Pedobiologia* 33, 129–143.

Hyvönen, R. and Persson, T. (1990) Effect of acidification and liming of feeding groups of nematodes in coniferous forest soils. *Biology and Fertility of Soils* 9, 205–210.

Iftode, E., Currgy, J.J., Fleury, A. and Fryd-Versavel, G. (1985) Action of a heavy metal ion, Cd^{++}, on two ciliates, *Tetrahymena pyriformis* and *Euplotes vannus*. *Acta Protozoologica* 24, 273–279.

Ishibashi, N., Maraoko, M., Kondo, E., Yamasaki, H., Kai, H., Iwakiri, T. and Nakahara, M. (1978) Effect of annual applications of herbicides on nematodes, soil mites and springtails in satsuma mandarin orchards. *Agricultural Bulletin Saga University* 44, 43–55

Jairajpuri, M.S. and Ahmad, W. (1992) *Dorylaimida – Free-living, Predaceous and Plant-parasitic Nematodes.* Brill, Leiden, 458 pp.

Jones, F.G.W., Larbey, D.W. and Parrott, D.M. (1969) The influence of soil structure and moisture on nematodes, especially *Xiphinema, Longidorus, Trichodorus* and *Heterodera* spp. *Soil Biology and Biochemistry* 1, 153–165.

Korthals, G.W., Bongers, T., Alexiev, A.D. and Lexmond, T.M. (1993) Influence of copper and pH on a terrestrial nematode community. *Poster SETAC* (1993), Lisbon.

Laminger, H. and Maschler, O. (1986) Auswirkungen von einigen Bioziden auf die Bodenprotozoen im Raum Vill (Tirol/Österreich). *Zoologischer Anzeiger* 216, 109–122.

Lehle, E. and Funke, W. (1989) The microfauna of forest soils. II. Ciliata (Protozoa: Ciliophora). Effects of anthropogenic influences. *Verhandlungen Gesellschaft für Ökologie* 17, 385–390.

Lousier, J.D. and Parkinson, D. (1981) The disappearance of the empty tests of litter- and soil-testate amoebae (Testacea, Rhizopoda, Protozoa). *Archiv für Protistenkunde* . 124, 312–336.

Maggenti, A.R. (1981) *General Nematology*. Springer-Verlag, New York, 372 pp.

Maggenti, A.R., Luc, M., Raski, D.J., Fortuner, R. and Geraert, E. (1988) A reappraisal of Tylenchina (Nemata). 11. List of generic and supra-generic taxa, with their junior synonyms. *Revue de Nématolologie* 11, 177-188.

Magurran, A.E. (1988) *Ecological Diversity and its Measurement.* Croom Helm Ltd, London, UK, 174 pp.

Mahn, E.G. and Kastner, A. (1985) Effects of herbicide stress on weed communities and soil nematodes in agro-ecosystems. *Oikos* 44, 185–190.

Margalef, R. (1958) Information theory in ecology. *General Systematics* 3, 36–71.

McLaren, R.G. and Smith, C.J. (1996) Issues in the disposal of industrial and urban wastes. In: Naidu, R., Kookana, R.S., Oliver, D.P., Rogers, S. and McLaughlin,

M.J. (eds) *Contaminants and the Soil Environment in the Australasia–Pacific Region*. Kluwer Academic Publishers, The Netherlands, pp. 183–212.

Minton, N.A. (1986) Impact of conservation tillage on nematode populations. *Journal of Nematology* 18, 135–140.

Neher, D.A., Peck, S.L., Rawlings, J.O. and Campbell, C.L. (1995) Measures of nematode community structure and sources of variability among and within agricultural fields. *Plant and Soil* 170, 167–181.

Norton, D.C. (1978) *The Biology of Free-living Nematodes*. Wiley, New York, 257 pp.

Norton, D.C. and Niblack, T.L. (1991) Biology and ecology of nematodes. In: Nickle, W.R. (ed.), *Manual of Agricultural Nematology*. Marcel Dekker, New York, USA, pp.47–72.

Norton, D.C. and Oard, M. (1981) Plant-parasitic nematodes in loess toposequences planted with corn. *Journal of Nematology* 13, 314–321.

Odeyemi, O., Salami, A. and Ugoji, E.O. (1988) Effect of common pesticides used in Nigeria on soil microbial population. *Indian Journal of Agricultural Sciences* 58, 624–628.

Old, K.M. (1986) Mycophagous soil amoebae: their biology and significance in the ecology of soil-borne plant pathogens. *Progress in Protistology* 1, 163–194.

Page, F.C. (1976) *An Illustrated Key to Freshwater and Soil Amoebae*. Freshwater Biological Association Scientific Publication No. 34, Freshwater Biological Association, Ambleside, Cumbria, 122 pp.

Page, F.C. (1988) *A New Key to Freshwater and Soil Gymnamoeba*. Freshwater Biological Association, Ambleside, Cumbria.

Parmelee, R.W. and Alston, D. (1986) Nematode trophic structure in conventional and no-tillage agroecosystems. *Journal of Nematology* 18, 403–407.

Patterson, D.J. and Larsen, J. (1991) *The Biology of Free-living Heterotrophic Flagellates*. Oxford University Press, New York.

Petz, W. and Foissner, W. (1989) The effects of mancozeb and lindane on the soil microfauna of a spruce forest: a field study using a completely randomized block design. *Biology and Fertility of Soils* 7, 225–231.

Petz, W., Foissner, W. and Adams, H. (1985) Culture, food selection and growth rate in the mycophagous ciliate *Grossglockneria acuta* Foissner,1980: first evidence of autochthonous soil ciliates. *Soil Biology and Biochemistry* 17, 871–875.

Pielou, E.C. (1977) *Mathematical Ecology*. Wiley, New York, USA, 385 pp.

Pizl, V. (1985) The effect of the herbicide zeazin 50 on the earthworm infection by monocystid gregarines. *Pedobiologia* 28, 399–402.

Pomp, R. and Wilbert, N. (1988) Taxonomic and ecological studies of ciliates from Australian saline soils: colpodids and hymenostomate ciliates. *Australian Journal of Marine and Freshwater Research* 39, 479–495.

Pons, R. and Pussard, M. (1980) Action des herbicides sur les amibes libres (Rhizopoda, Protozoa). Étude préliminaire. *Acta Oecologia* 1, 15–20.

Popham, J.D. and Webster, J.M. (1980) Cadmium toxicity in the free-living nematode, *Caenorhabditis elegans*. *Environmental Research* 20, 183–191.

Popovici, I. (1994) Nematodes as indicators of ecosystem disturbance due to pollution. *Studia Univ. Babes-Bolyai, Biologia* (1992) 37, 15–27.

Popovici, I. and Korthals, G. (1995) Soil nematodes used in the detection of habitat disturbance due to industrial pollution. *Studia Univ. Babes-Bolyai, Biologia* (1993), 38, 37–41.

Popovici, I., Stan, G., Stefan, V., Tomescu, R., Dumae, A., Tarra, A. and Dan, F. (1977) The influence of atrazine on soil fauna. *Pedobiologia* 17, 209–215.

Pradhan, G.B., Senapati, B.K. and Dash, M.C. (1988) Relationship of soil nematode populations to carbon:nitrogen in tropical habitats and their role in the laboratory decomposition of litter amendments. *Revue d'Ecologie et de Biologie du Sol* 25, 59–76.

Pratt, J.R., McCormick, P.V., Pontasch, K.W. and Cairns, J. (1988) Evaluating soluble toxicants in contaminated soils. *Water, Air, and Soil Pollution* 37, 293–307.

Purrini, K. (1983) Studies on pathology and diseases of soil invertebrates of palm and forest soils in Tanzania and Thailand. In: Lebrun, P., Andrén, H.M., De Medts, A., Grégoire-Wio, C. and Wauthy, G. (eds) *New Trends in Soil Biology (Proceedings of the 8th International Colloquium of Soil Zoology).* Dieu-Brichart, Ottingnies-Louvain-la-Neuve, pp. 82–87.

Rauenbusch, K. (1987) Biologie und Feinstruktur (REM-Untersuchungen)-terrestrischer Testaceen in Waldböden (Rhizopoda, Protozoa). *Archiv für Protistenkunde* 134, 191–294.

Robinson, B., Dodson, P., Christy, P. and Hayes, S. (1996) The role of thermal niches in the ecology and biogeography of *Naegleria* spp. *Abstracts of the 7th International Conference on Small Freeliving Amoebae,* January 1996, Adelaide, Australia. p. 53.

Rodriguez-Kábana, R. (1986) Organic and inorganic nitrogen amendments to soil as nematode suppressants. *Journal of Nematology* 18, 129–135.

Roper, M.M. and Gupta, V.V.S.R. (1995) Management practices and soil biota. *Australian Journal of Soil Research* 33, 321–339.

Ruess, L. and Funke, W. (1995) Nematode fauna of a spruce stand associated with forest decline. *Acta Zoologica Fennica* 196, 348–351.

Sály, A. (1989) Evaluation of herbicide influence on the edaphon in vineyards by means of free-living nematodes. *Polish Ecological Studies* 15, 47–54.

Schnürer, J., Clarholm, M. and Rosswall, T. (1986) Fungi, bacteria, and protozoa in soil from four arable cropping systems. *Biology and Fertility of Soils* 2, 119–126.

Schönborn, W., Petz, W., Wanner, M. and Foissner, W. (1987) Observations on the morphology and ecology of the soil-inhabiting testate amoebae *Schoenbornia humicola* (Schönborn, 1964) Decloitre, 1964 (Protozoa, Rhizopoda). *Archiv für Protistenkunde* 134, 315–330.

Schreiber, B. and Brink, N. (1989) Pesticide toxicity using protozoan as test organisms. *Biology and Fertility of Soils* 7, 289–296.

Singh, B.N. (1949) The effect of artificial fertilizers and dung on the numbers of amoebae in Rothamsted soils. *Journal of General Microbiology* 3, 204–210.

Sohlenius, B. (1990) Influence of cropping system and nitrogen input on soil fauna and microorganisms in a Swedish arable soil. *Biology and Fertility of Soils* 9, 168–173.

Sohlenius, B. and Boström, S. (1984) Colonization, population development and metabolic activity of nematodes in buried barley straw. *Pedobiologia* 27, 67–78.

Sohlenius, B., and Boström, S. (1986) Short-term dynamics of nematode communites in arable soil – influence of nitrogen fertilization in barley crops. *Pedobiologia* 29, 183–191.
Sohlenius, B. and Wasilewska, L. (1984) Influence of irrigation and fertilization on the nematode community in a Swedish pine forest soil. *Journal of Applied Ecology* 21, 327–342.
Steinberg, C., Grosjean, M.C., Bossand, B. and Faurie, G. (1990) Influence of PCBs on the predator–prey relation between bacteria and protozoa in soil. *FEMS Microbiology Ecology* 73, 139–148.
Stout, J.D. (1955) Environmental factors affecting the life history of three soil species of *Colpoda* (Ciliata). *Transactions of the Royal Society of New Zealand* 82, 1165–1188.
Stout, J.D. (1978) Effect of irrigation with municipal water or sewage effluent on the biology of soil cores. II. Protozoan fauna. *New Zealand Journal of Agricultural Research* 21, 11–20.
Stout, J.D. and Heal, O.W. (1967). Protozoa. In: Burges, A. and Raw, F. (eds) *Soil Biology.* Academic Press. Inc., New York, pp. 149–195.
Swift, M.J., Heal, O.W. and Anderson, J.M. (1979) *Decomposition in Terrestrial Ecosystems.* Univeristy of California Press, Berkeley.
Thomas, S.H. (1978) Population densities of nematodes under seven tillage regimes. *Journal of Nematology* 10, 24–27.
Tomescu, R. (1987) Influenta poluarii asupra protozoarelor din zona industriala Zlanta. *Studii si Cercetari de Biologie (Seria Biologie Animale) Romania* 39, 167–170.
Twinn, D.C. (1974) Nematodes. In: Dickinson, C.H. and Pugh, G.J.F. (eds) *Biology of Plant Litter Decomposition.* Academic Press, London, pp. 421–465.
van der Knapp, E., Rodriguez, R.J. and Freckman, D.W. (1993) Differentiation of bacterial-feeding nematodes in soil ecological studies by means of arbitrarily-primed PCR. *Soil Biology and Biochemistry* 25, 1141–1151.
Viswanath, G.K. and Pillai, S.C. (1977) Influence of superphosphate on soil protozoa. *Journal of the Indian Institute of Science* 59, 113–120.
Vodkin, M.H., Howe, D.K., Visvesvara, G.S. and McLaughlin, G.L. (1992) Identification of Acanthamoeba at the generic and specific levels using the polymerase chain reaction. *Journal of Protozoology* 39, 378–385.
Wanner, M. (1991) Zur Ökologie von Thekamöben (Protozoa: Rhizopoda) in süddeutschen Wäldern. *Archiv für Protistenkunde* 140, 237–288.
Wardle, D.A., Yeates, G.W., Watson, R.N. and Nicholson, K.S. (1995) The detritus food-web and the diversity of soil fauna as indicators of disturbance regimes in agro-ecosystems. *Plant and Soil* 170, 35–43.
Wasilewska, L. (1995) Differences in development of soil nematode communities in single- and multi-species grass experimental treatments. *Applied Soil Ecology* 2, 53–64.
Weiss, B. and Larink, O. (1991) Influence of sewage sludge and heavy metals on nematodes in an arable soil. *Biology and Fertility of Soils* 12, 5–9.
Westcott, S.W. and Barker, K.R. (1976) Interaction of *Acrobeloides buetschlii* and *Rhizobium leguminosarum* on Wanda pea. *Phytopathology* 66, 468–472.
Willmer, E.N. (1963) *Advances in Soil Science* 20, 119–127. (from Stout, J.D. and Heal. O.W., 1967)

Wodarz, D., Aescht, E. and Foissner, W. (1992) A weighted coenotic index (WCI): description and application to soil animal assemblages. *Biology and Fertility of Soils* 14, 5–13.

Wong, A.T.S., Tylka, G.L. and Hartzler, R.G. (1993) Effects of eight herbicides on *in vitro* hatching of *Heterodera glycines. Journal of Nematology* 25, 578–574.

Yeates, G.W. (1978) Populations of nematode genera in soils under pasture. II. Seasonal dynamics in dryland and effluent irrigated pastures on a central yellow-grey earth. *New Zealand Journal of Agricultural Research* 21, 331–340.

Yeates, G.W. (1981) Nematode populations in relation to soil environmental factors: a review. *Pedobiologia* 22, 312–338.

Yeates, G.W. (1982) Variation of pasture nematode populations over thirty-six months in a summer dry silt loam. *Pedobiologia* 24, 329–346.

Yeates, G.W. (1994) Modification and qualification of the nematode maturity index. *Pedobiologia* 38, 97–101.

Yeates, G.W. and Bird, A.F. (1994) Some observations on the influence of agricultural practices on the nematode faunae of some South Australian soils. *Fundamental and Applied Nematology* 17, 133–145.

Yeates, G.W. and Bongers, T. (1997) Nematode diversity in agro-ecosystems. In: Paoletti, M.G. (ed.) *Biodiversity in Agro-ecosystems – Role for Sustainability and Bioindication.* Lewis Publishing, Boca Raton, USA (in press).

Yeates, G.W. and Hughes, K.A. (1990) Effect of three tillage regimes on plant and soil nematodes in an oats/maize rotation. *Pedobiologia* 34, 379–387.

Yeates, G.W., Barker, G.M. and Pottinger, P.R. (1983) Effect of insecticides on nematode populations below ten grass cultivars. *New Zealand Journal of Experimental Agriculture* 11, 147–151.

Yeates, G.W., Stout, J.D., Ross, D.J., Dutch, M.E. and Thomas, R.F. (1976) Long-term effects of paraquat-diquat and additional weed control measures on some physical, biological and respiratory properties of a soil previously under grass. *New Zealand Journal of Agriculture Research* 19, 51–61.

Yeates, G.W., Bamforth, S.S., Ross, D.J., Tate, K.R. and Sparling, G.P. (1991) Recolonization of methyl bromide sterilized soils under four different field conditions. *Biology and Fertility of Soils* 11, 181–189.

Yeates, G.W., Bongers, T., de Goede, R.G.M., Freckman, D.W. and Georgieva, S.S. (1993) Feeding habits in soil nematode families and genera – an outline for ecologists. *Journal of Nematology* 25, 315–331.

Yeates, G.W., Orchard, V.A., Speir, T.W., Hunt, J.L. and Hermans, M.C.C. (1994) Impact of pasture contamination by copper, chromium, arsenic timber preservative on soil biological activity. *Biology and Fertility of Soils* 18, 200–208.

Zullini, A. and Peretti, E. (1986) Lead pollution and moss-inhabiting nematodes of an industrial area. *Water, Air and Soil Pollution* 27, 403–410.

Zwart, K.B., Kuikman, P.J. and Van Veen, J.A. (1994) Rhizosphere protozoa: their significance in nutrient dynamics. In: Darbyshire, J.D. (ed.) *Soil Protozoa.* CAB International, Wallingford, UK, pp. 93–122.

10

Community Structure of Soil Arthropods as a Bioindicator of Soil Health

N.M. van Straalen
Vrije Universiteit, Department of Ecology and Ecotoxicology, De Boelelaan 1087, 1081 HV Amsterdam, The Netherlands

The Concept of Community Bioindicators

Communities of microarthropods, especially those of forest soils with a well-developed litter and humus layer, are characterized by a great diversity of species. The coexistence of such a great number of species in a relatively small area, combined with the fact that many of them seem to have similar feeding niches, was described by soil ecologists as the 'enigma of soil animal species diversity' (Anderson, 1975). Later research has revealed additional factors that allow for niche differentiation in microarthropod communities, such as seasonal segregation of species by life-history differentiation (Vegter, 1987) and spatial segregation of species caused by differences in the physiological responses to humidity (Vegter, 1983). It has also been shown that the feeding habits of microarthropods are more diverse than they seem on first sight, different species showing a different array of digestive enzymes (Siepel, 1990), and different preferences for specific food items (Hasegawa and Takeda, 1995). The conclusion must be that soil arthropod species richness is supported by an equally great niche diversity, along a variety of niche dimensions.

Considering the intricate relations between microarthropods and their niches in soil, their community structure could be used as a bioindicator of soil 'health'. Many soil arthropods (Collembola, Oribatida, Isopoda, Diplopoda) live a rather sedentary life and therefore reflect the local conditions of a habitat better than organisms with a high dispersal capacity, such as many flying insects. These facts have been recognized for a long time, and relationships between soil types and soil mesofauna have been established in several pedobiological works (Ghilarov, 1977; Rusek, 1978). On the other hand, a true theory of community composition of soil arthropods, comparable to the systems developed in plant ecology (Ellenberg, 1974), still remains to be developed (Usher *et al.*, 1982; Ghabbour, 1991). This seems to be due mainly to a lack of autecological knowledge that would allow the occurrence of soil arthropod species to be understood from their ecophysiological responses to specific soil factors.

The development of a bioindicator system can be seen as the converse of the relationship between an environmental factor and a biological parameter (Fig. 10.1). Various biological endpoints may be defined, such as species composition, symptoms of damage in organisms, body burdens of pollutants, induction or inhibition of enzymes, etc. The terminology used in this field ('biomonitors', 'biomarkers', 'biosensors') is sometimes confusing and calls for a redefinition of terms (Van Gestel and Van Brummelen, 1996; see also Chapter 16, this volume). In the present chapter, emphasis will be placed on bioindicators at the community level, i.e., the composition of the community will be the main biological parameter (Fig. 10.1).

The usefulness of a bioindicator depends on the strength of the relationship between the causative environmental factor and the ecological endpoint. If the structure of the community is not influenced by the abiotic environment, the left-to-right relationships in Fig. 10.1 will be weak and it will be impossible to convert them for the purpose of bioindication. This may be the case when communities are shaped mainly by biotic factors and internal interactions, such as competition and predation.

The most successful bioindicator systems have been developed as quantitative indices that allow specific environmental factors to be estimated under conditions where the latter cannot be measured directly. The use of bioindicators in this way is illustrated by the reconstruction of lake pH histories from diatom communities in dated sediments (Renberg and Hellberg, 1982), and the reconstruction of temperature changes over geological times from dated beetle remains (Atkinson *et al.*, 1987). Such systems have not yet been developed for soil microarthropods.

It may be useful to have a closer look at how the above mentioned successful bioindicators were developed, as some lessons may be learnt for the development of soil microarthropod indicators. The procedures followed in Atkinson *et al.* (1987) and Renberg and Hellberg (1982) may be summarized as follows:

1. A community that contains a sufficient number of species that differ in their reaction to some specific factor is selected.
2. The response of each species to the factor of interest is estimated, either from the present distribution of the species in field gradients, or from laboratory experiments in which the performance of the species under various levels of the factor is established.
3. The responses are summarized in some fashion, e.g. as median tolerances or tolerance ranges for each species. Species showing similar responses may be grouped together.

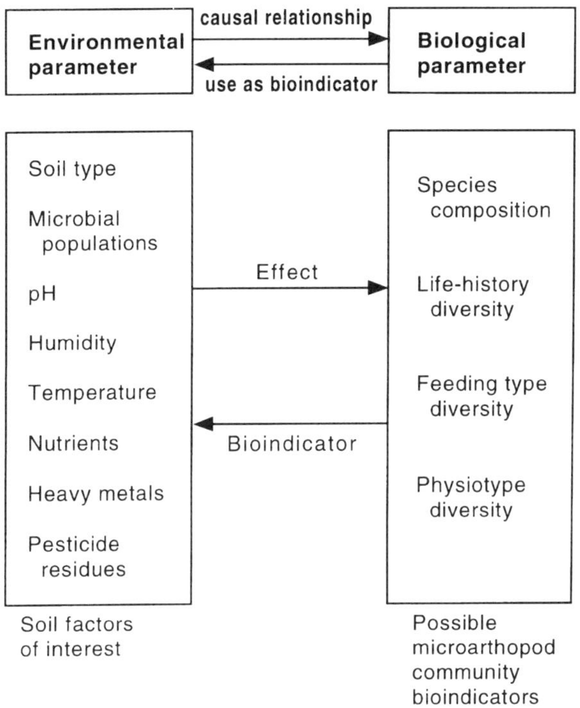

Fig. 10.1. Schematic relationship between soil factors and microarthropod communities. The figure illustrates that soil properties influence the structure of microarthropod communities through various causal relationships (left to right). When community structure is used as a bioindicator, this may be seen as a conversion of the causal relationship (right to left).

4. Based on the summary information, an index is developed that combines the abundance of (groups of) species with their tolerances. The index should be sensitive to the replacement of species with different tolerances; this may be achieved by greater weighting of the abundances of species with extreme preferences compared with those of indifferent species.
5. The index is calibrated using field inventories in which species composition is measured in conjunction with the factor of interest. The calibration will optimize the predictive power of the index, and at the same time will provide information on the achievable level of accuracy.
6. Using the index, quantitative estimates of the factor of interest are made from a survey of the species composition of the community.

The question may be asked under which circumstances may it be useful to estimate an environmental factor from complicated biological information if it can also be measured directly (e.g. temperature). The reconstruction of past temperature changes, as in the case of dated remains of beetles is a clear example of a useful application. Atkinson *et al.* (1987) even maintain that their biological method to estimate temperature variation over 22,000 years is more reliable than some physico-chemical methods that rely on the composition of gases trapped in ice air bubbles. The skeletons of several soil arthropods, especially those of oribatid mites, are often preserved as fossils (Krivolutsky and Druk, 1986), and so a reconstruction of past environmental changes would be possible. It might also be useful to consider situations where species lists have been reported in the older literature, or where preserved collections are still available, and soil factors have not been measured.

Also in present-day applications, bioindicators may be useful if some impact factor is not easily measurable. For example, in agroecosystems, the effects of pesticides, like the synthetic pyrethroid deltamethrin, are easily observed in the epigeic invertebrate fauna, while the chemical determination of the (very transient) residue poses more problems (Everts *et al.*, 1989; Krogh, 1994).

Even if a soil factor can be measured directly, it may still be useful to apply a bioindicator of the kind outlined above and compare its results with the direct measurement of soil factors. For example, the measurement of pH is usually done on samples of more than 10 g of soil and does not provide information on the distribution of microsites with lower or higher acidities; a sensitive bioindicator might reveal information not seen in the bulk pH. This is specifically relevant where pHs show a high degree of spatial variability, e.g. in forest floors (Bringmark, 1989). Another argument supporting the application of bioindicators is that they integrate environmental information over a certain period of time, while physico-chemical factors usually provide only an instantaneous measure.

The present chapter will consider possible microarthropod community indicators based on a classification after species abundance, dominance, life-histories, feeding types, and physiotypes (Fig. 10.1). The soil factors for which these classifications might prove useful are soil type, humus type, microbial populations, pH, humidity, temperature, nutrient status, heavy metals and pesticide residues. At the moment, it is not yet possible to suggest any quantitative system that is sufficiently developed for regulatory use or monitoring purposes. The spirit of this chapter will therefore be to identify promising areas of further development, rather than to discuss the regulatory possibilities.

Stability of Soil Microarthropod Communities

A basic prerequisite for any bioindicator system is that it does not show random fluctuations unrelated to the factor to be indicated. Basically this means that community composition should be site-characteristic and stable in time as long as environmental conditions do not change. Before discussing microarthropod communities as bioindicators, it may be useful to consider their stability.

Fluctuations of single species

Long-term studies on the population dynamics of soil microarthropod communities, comparable to those carried out on carabid beetles over more than 20 years (Den Boer, 1981), are rare. Most studies cover a sampling period of one or two years, enough to analyse life cycles and estimate demographic parameters, but insufficient to answer questions on temporal stability and site-specificity of soil communities. Some of the longer-term studies include the work of Takeda (1987), a 15 year study on Collembola in a Japanese red pine forest soil, Huhta (1979), a 7 year study on successional changes of Finnish forest soil invertebrates following clearcutting, and Hogervorst *et al.* (1993), a 5-year study of the oribatid mite *Platynothrus peltifer* in pine forests in the Netherlands. Most of the work before 1980 has been reviewed by Petersen and Luxton (1982), and the more recent studies by Petersen (1994).

As an index for the analysis of temporal population stability, Connell and Sousa (1983) suggested the standard deviation of the logarithmically transformed population abundances. This was applied by Takeda (1987) to his data on fluctuations of annual densities of eight Collembola species over 15 years; population variability ranged between 0.137 and 0.490. These were considered to be relatively low values compared to data for aquatic invertebrates and terrestrial plants (Takeda, 1987). Somewhat higher values of temporal variability were estimated by Vegter *et al.* (1988) for five species of entomobryid Collembola, sampled over 4 years in eight woodlands in the Netherlands; the standard deviation of log(x+1) transformed abundances varied between 0.3 and 0.9. Temporal variance in this study was clearly correlated with population density; the more abundant a species was, the less it tended to fluctuate in numbers. The conclusion from these and other studies may be that microarthropod populations in soil are rather stable over a number of years, especially when their abundances are high.

In contrast to stability from year to year, fluctuations of microarthropod populations within a year can be considerable. This is largely due to demographic processes, such as recruitment and mortality, which show clear seasonal cycles. Temperature and soil humidity appear to be the main driving variables for seasonal life-history events in microarthropods (Van Straalen, 1985; Stamou *et al.*, 1993; Diekkrüger and Röske, 1995), especially in temperate climates and in climates with dry summer periods.

Muller and Van Straalen (1986) reviewed literature data on population fluctuations of Collembola and estimated seasonal variability for a number of species. Variability was expressed as the 'fluctuation coefficient' (FC), defined as the standard deviation of abundances at different times in a year, as a percentage

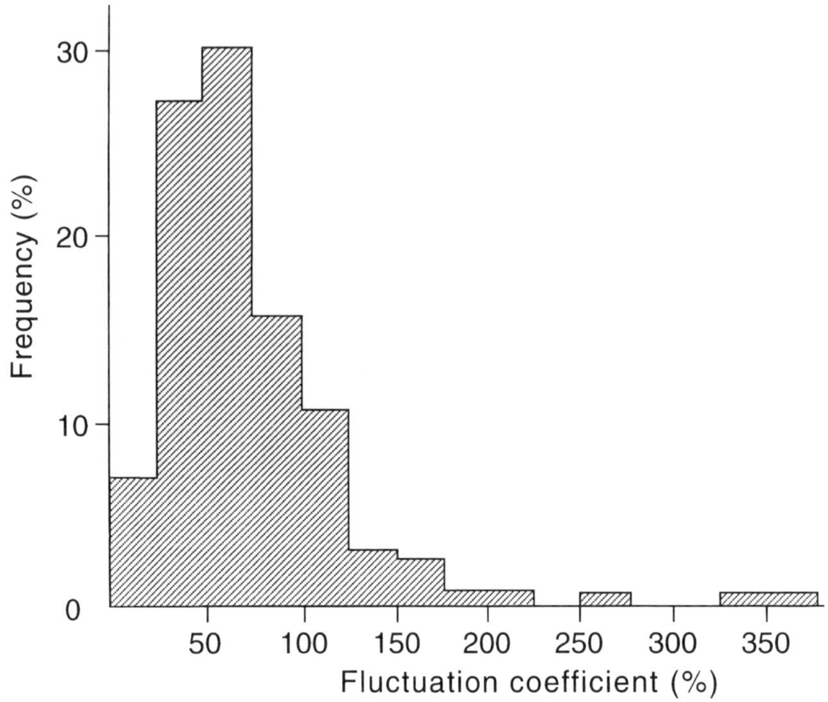

Fig. 10.2. Frequency distribution of fluctuation coefficients (temporal standard deviation relative to annual mean density) for Collembola populations sampled throughout a season. The distribution is based on 207 different values. From Muller and Van Straalen (1986).

of the mean density. This is comparable to the 'coefficient of variation', commonly used in statistics. Kooijman *et al.* (1989) used a similar index to express population consequences of inter-individual variability in physiological parameters of a dynamic energy budget model. From their literature survey Muller and Van Straalen (1986) obtained 207 FC-values. The data, plotted as a frequency distribution in Fig. 10.2, show that the seasonal amplitude is rather variable between species. The median value is 64%, but for some species FC may amount to 350%.

The evidence thus suggests that, although microarthropod populations are rather stable from year to year, densities may exhibit large fluctuations within a year. The within-year fluctuations depend on the life-history of the species; within the Collembola, some species have sharp spring or autumn peak densities, while others fluctuate more gradually throughout the year. Where systematic year-to-year changes in density are apparent, these are driven by successional changes following disturbance or changes in vegetation cover. Annual mean microarthropod densities thus show a high degree of site-specificity, a necessary condition for their use as bioindicators.

Predictability of community composition

The structure of communities is often displayed as a plot of log abundance versus rank. Takeda (1987) found a total of 41 species of Collembola in his long-term study of a Japanese red pine forest; the community structure is reproduced in Fig. 10.3. It appears that there is a linear relationship between rank and log abundance for the first ten species, whose total density amounts to 97% of the community density. The steepness of this line, plus the fact that the first species comprises about 32% of the total number of springtails indicate that

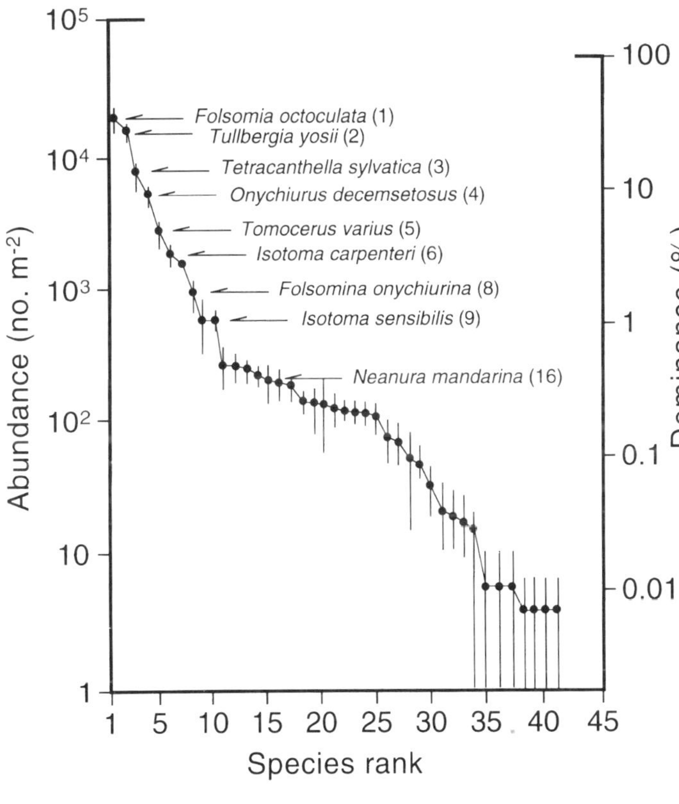

Fig. 10.3. Rank abundance graph of the Collembola community of a Japanese red pine forest, sampled over 15 years. Reproduced from Takeda (1987), with permission of the Society of Population Ecology, Kyoto.

collembolan communities are characterized by a high level of dominance. A similar conclusion was drawn by Kopeszki and Meyer (1994), who, on the basis of an extensive study of the Collembola fauna in six Austrian woodlands, found that the most dominant species comprised 22–50% of the community within a habitat.

Bengtsson (1994) analysed literature data on forest soil communities using Kendall's coefficient of concordance, W, as a measure of temporal stability of communities. The data analysed covered various invertebrate groups (Collembola, mites, beetles, spiders, millipedes, molluscs) and a total of 67 studies or plots. The frequency distribution of W-values (Fig. 10.4) shows that most of the organisms score above $W = 0.75$ (corresponding with a Spearman's rank coefficient of approximately 0.5), which indicates a high degree of correlation between the ranked abundances of species sampled at different times. Consequently, soil invertebrate communities seem to be remarkably stable from year to year.

Figure 10.4 also shows (not surprisingly) that when species are lumped into higher taxa, the correlation through time is even higher; in addition, communit-

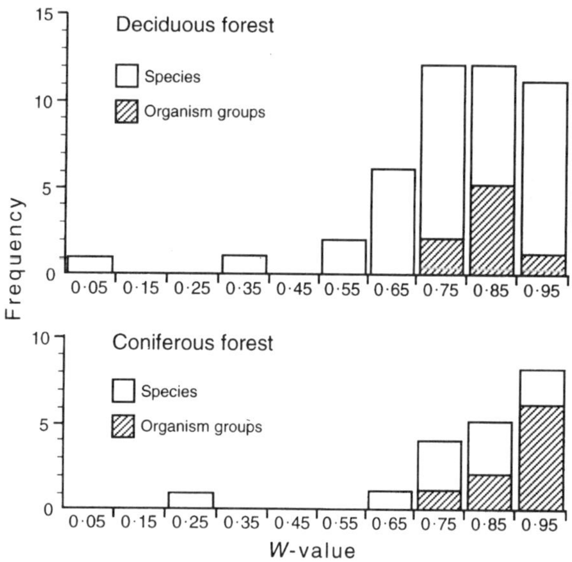

Fig. 10.4. Frequency distributions of Kendall's coefficient of concordance (W-value), measuring temporal similarity of forest soil communities. The data are based on 67 studies, each covering 2 years or more. W-values are given for communities identified to species (open bars) and for communities aggregated into groups (beetles, Collembola, enchytraeids, mites, molluscs, Myriapoda and spiders). Reproduced from Bengtsson (1994), with permission of the British Ecological Society, London.

ies in pine forests seem to be more stable than communities in deciduous forests. The latter effect was attributed to the influence of body-size: organism groups with small-sized species, that dominate in pine forests, tend to have communities that show a high degree of predictability (Bengtsson, 1994). This effect could, however, also be due to density (see the study of Vegter *et al.*, 1988, discussed above). The conclusion drawn from this analysis may be that soil communities in natural ecosystems are quite stable in time, over periods up to 20 years. There is thus a good theoretical basis for considering the use of these communities as bioindicators.

Indicator Value of Species in Communities

The species composition of a community may be summarized using various statistical parameters, often expressed as diversity indices. Sheehan *et al.* (1984) provided a list of 18 parametric and non-parametric diversity indices, and discussed their application in studies of pollution. Despite their obvious appeal, diversity indices of microarthropod communities have not proven to be very useful. An example is the study of Siepel and Van de Bund (1988), who studied the influence of various management practices (fertilization, mowing, grazing) on the community structure of Collembola and Oribatida in grasslands. An effect of management on the microarthropod community could not be detected by means of the diversity indices of Simpson and Shannon-Wiener, while more sophisticated multivariate analysis revealed clear effects. Kuznetzova (1994) also criticized the use of the Shannon-Wiener diversity index, which did not give a good picture of Collembola communities in disturbed urban habitats in Moscow city, compared to natural forests.

The main reason for the failure of diversity indices to indicate disturbance seems to be that they are not sensitive to the nature of the species; if species A is replaced by an equal number of individuals of species B, there is no change of diversity. The following sections will therefore attempt to analyse how information on the ecology of the species can be exploited to optimize the indicator value of the microarthropod community.

Single indicator species

In some cases it has been argued that the abundance of a single species of microarthropod, or a group of related species could be indicative of a change in a soil property. To correct for site-specific effects influencing the microarthropod community as a whole (e.g. amount of litter), Van Straalen *et al.* (1988) suggested that the density of an indicator species should be expressed as a fraction of the total density of the group to which it belongs.

Soil pH may be the only factor for which the single indicator species approach is useful. The evidence for this comes from experimental studies in which acidified water, lime or fertilizers are applied to field plots, and from comparative studies in which a variety of sites, with a range of soil factors, are compared for their species composition.

Hågvar (1984b) and Hågvar and Abrahamsen (1984) studied the densities of Collembola and Acari in seven vegetation types in each of two study areas in Norwegian forests. Correlations between microarthropod abundance and soil chemistry were never very strong, but the oribatid mite *Tectocepheus velatus* and the collembolan *Mesaphorura yosii* showed a consistent response. When abundance of *T. velatus* was plotted against soil pH (Fig. 10.5), it appeared that this species was hardly ever found in soils with a pH above 4.5; high densities were found in soils with pHs below 4. The collembolan *Isotoma notabilis*, on the other hand, showed a positive relationship with soil pH. The indicator value of these species was confirmed in acidification experiments (Hågvar and Amundsen, 1981; Hågvar, 1984a). In these experiments, forest plots were treated with acidified groundwater, sprayed at regular intervals onto the forest floor. Complicated reactions of the soil microarthropod community were observed, some species showing a positive, others a negative response, however, *Mesaphorura yosii* and *Tectocepheus velatus* consistently increased, while *Iso-*

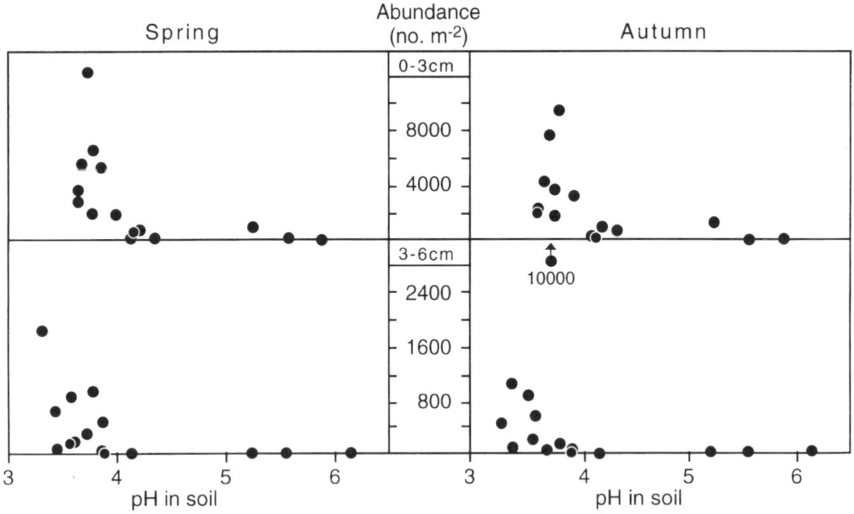

Fig. 10.5. Correlation between the density of the oribatid mite *Tectocepheus velatus* (adults plus juveniles), and soil pH. The data are for 15 different forest locations near Oslo, Norway, sampled in spring (left) and autumn (right), in two different depth ranges. Reproduced from Hågvar (1984b), with permission of Fischer Verlag, Jena.

toma notabilis decreased. Thus it seems that artificial acidification favours those species that already have a high abundance in naturally acid soils.

Microarthropod responses to changing soil pH have also been analysed in acidification/liming experiments in Sweden, Finland, Belgium and Germany (Bååth *et al.*, 1980; Huhta *et al.*, 1983; Heungens and Van Daele, 1984; Schauermann, 1987; Kratz, *et al.*, 1991). Although pH seems to be a main driving variable, it is impossible to attribute observed changes in microarthropod abundance to a single soil chemical factor, because many soil factors are strongly correlated with pH. Different groups of organisms may be affected by different factors; some direct (e.g. pH of pore water, soluble nutrients), others indirect (e.g. changed microflora populations). It is interesting to note that some bark-living microarthropods, in particular the oribatid mite *Humerobates rostrolamellatus*, are extremely sensitive to acidifying air pollutants (SO_2, NO, NO_2), an effect which could be due to these gases lowering the pH of the pore water of their substrate (Lebrun *et al.*, 1976; André *et al.*, 1982; Weigmann and Kratz, 1987).

Through their response to soil factors, microarthropods may be used to indicate not only soil quality, but also ecosystem 'health' in general. This was attempted in a study by Van Straalen *et al.* (1988), who compared Collembola

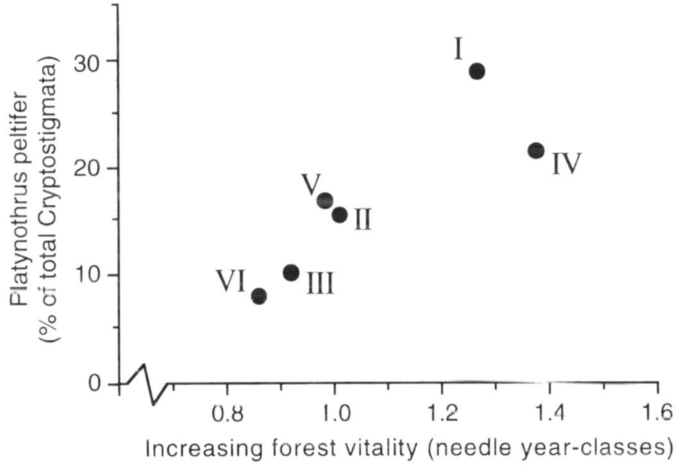

Fig. 10.6. Correlation between the density of the oribatid mite *Platynothrus peltifer* (expressed as a percentage of total Cryptostigmata) and the vitality of pine trees (number of needle year classes), sampled in six pine forests (numbered I to VI) in the Veluwe area, the Netherlands. The data are averages taken over a 5-year period. Correlation coefficient is 0.86. Reproduced from Hogervorst *et al.* (1993), with permission of Springer Verlag, Berlin.

and Oribatida densities in 12 different stands of *Pinus sylvestris* in the Netherlands, selected to form a gradient of forest vitality (judged on the number of needle year-classes). The relative density of the oribatid mite *Platynothrus peltifer* was significantly correlated with forest vitality. Chemical factors differing between sites were K, Mn and N contents of the litter, as well as the C/N ratio of the litter (but not pH). The correlation between tree vitality and microarthropod abundance was subsequently studied in six of the twelve sites over a period of 5 years (Hogervorst *et al.*, 1993). Only small changes in microarthropod abundances occurred over the years, and the correlation between *P. peltifer* and tree vitality also held for the 5-year means (Fig. 10.6). The response of this oribatid mite may be related to its preference for alkaline substrates and its very high content of manganese, one of the mobile elements that easily leaches under acid conditions (Van Straalen *et al.*, 1987).

Properties of community structure

The use of a single indicator species is attractive because of its simplicity, but it does not provide enough resolution for detecting subtle changes in soil properties. It may therefore be useful to take more advantage of the information contained in the community structure as a whole. An interesting approach has been applied by Hågvar (1994), who, inspired by the work of marine benthic ecologists, has looked at the dominance structure of microarthropod communities.

The principle of Hågvar's method is illustrated in Fig. 10.7. Community structure may be summarized by means of a frequency distribution of the number of species falling into a range of abundance classes. The graph will be very skewed, because there are many rare species and few abundant ones; however, it can be made symmetric by expressing the horizontal axis in logarithmic units. Instead of using absolute abundances, Hågvar (1994) used percentage dominance to scale the horizontal axis; for example, the upper histogram of Fig. 10.7 shows that 2% of the species fall in the highest dominance class (dominances between 64 and 32%), 6.5% fall in the class below it (32–16% dominance), etc.

Log-normal dominance distributions were applied to literature data on Collembola and Oribatida community structure in four case-studies: lead pollution of natural origin, copper and zinc pollution by a brass factory, artifical acid rain and liming, and ploughing (Hågvar, 1994). In the stressed communities the log-abundance distribution appeared to be skewed to the right. This effect can be understood as a movement of sensitive species to lower abundance classes (to the left in Fig. 10.7); a stressed community thus has a relatively high frequency of rare species. When these disappear due to extinction, the distribution may become log-normal again. Changes in the dominance structure may thus be considered as an 'early warning' signal for stress.

The approach of Hågvar (1994) is certainly interesting, but its strength is

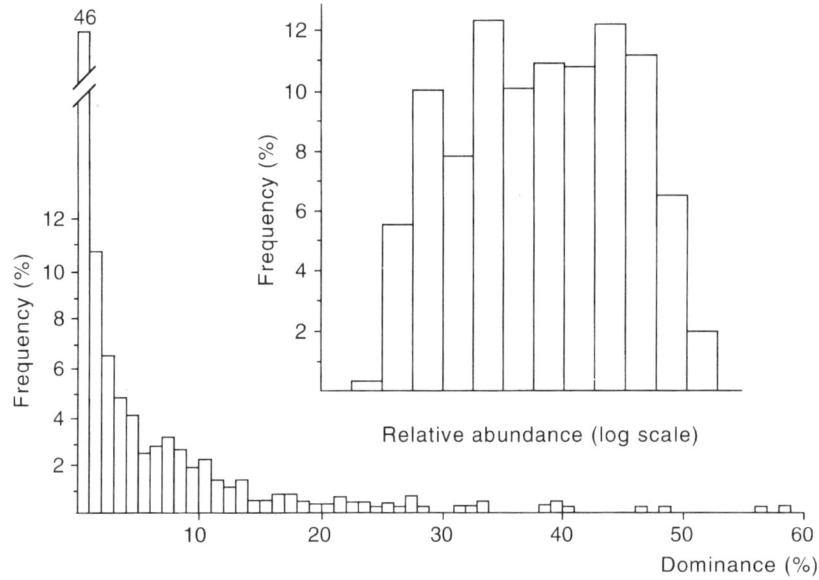

Fig. 10.7. Pooled distribution of Collembola densities sampled from 15 different sites in Norwegian coniferous forests. The distribution specifies the number of species falling into a certain dominance class. Dominance is given on a linear scale in the lower figure and on a logarithmic scale in the upper figure. Reproduced from Hågvar (1994), with permission of the Finnish Zoological and Botanical Publishing Board, Helsinki.

manifested only if large communities are available (preferably more than 100 species); its applicability to soil microarthropod communities where a maximum of 20–30 species of Collembola, and some 50 species of oribatids are usually found at one site, will be limited. A less demanding approach would be to directly analyse the rank-abundance graph of a community, as displayed in Fig. 10.3. The theory of community succession predicts that these graphs will show increased species richness, and a shallower slope with advancing successional stage of the community. Conversely, disturbing factors would bring down species richness and increase the slope of the rank-abundance graph.

The succession paradigm has been confirmed in studies of soil mite communities in grasslands of varying age (Emmanuel *et al.*, 1985), and of ant communities in a successional gradient of coastal dunes (Boomsma and Van Loon, 1982). Collembola communities, however, do not seem to fit this pattern. Vegter *et al.* (1988) argued that young springtail communities are characterized by a linear relationship between rank and logarithmic abundance. In these communities, interactions between species are weak and a log-series distribution or a broken stick distribution is expected, with a low number of species, but a relatively high degree of evenness. When the community develops, resources become partitioned among species in a more hierarchical way, a few species become more and more dominant and others are forced into secondary roles; the total number of species increases, with more and more rare ones. According to this line of reasoning, developed communities would show a split-line of log abundance versus rank. Some support for this is seen in Fig. 10.3, where species 11 to 25 seem to lie on a line with a more shallow slope, compared to species 1 to 10.

Another type of information may be derived from the joint analysis of abundance–rank graphs and biomass–rank graphs. Warwick (1986) has developed a method ('abundance-biomass comparison') for marine benthic invertebrates, in which the cumulative rank–abundance graph is compared with the cumulative rank–biomass graph. In his analysis, stressed communities are characterized by a high abundance of a few small species, so that biomass is more evenly distributed over species, compared to an undisturbed community, which is characterized by the dominant species representing a great deal of the biomass. This type of analysis has not yet been applied to soil microarthropod communities.

Multivariate analysis of communities

Detailed statistical analysis of complex communities can significantly improve our knowledge about relationships between soil factors and soil microarthropod communities (Usher *et al.*, 1982; Cancela da Fonseca *et al.*, 1995). Of particular interest are multivariate methods, in which the data are viewed as a series of cases (sampling sites), each being characterized by several variates (hence multivariate analysis). In soil ecology studies, the variates are of two classes: densities of soil invertebrates, and soil factors. The analysis is often designed to classify the cases based on the correspondence between the two groups of variates, in other words to identify the environmental factors that are the most important determinants of community structure. A number of statistical procedures are available and the use of these has developed into a specialized field, too extensive to summarize here.

The application of multivariate statistical methods to soil communities has been carried forward strongly by the 'French school' of soil ecologists (Cancela da Fonseca, 1980; Ponge, 1983, 1993; Arpin *et al.*, 1984; Poursin and Ponge, 1984); others have, however, also applied these techniques (e.g. Bolger and Curry, 1984; Ghabbour *et al.*, 1985; Siepel and Van de Bund, 1988; Paoletti *et al.*, 1988).

Ponge (1993) analysed a large data set of Collembola communities, obtained over several years from samplings in the forests of Orléans and Sénart, France. The data contained information on 60 different sampling sites, with a total of 104 species (some rare species were deleted from the analysis). A

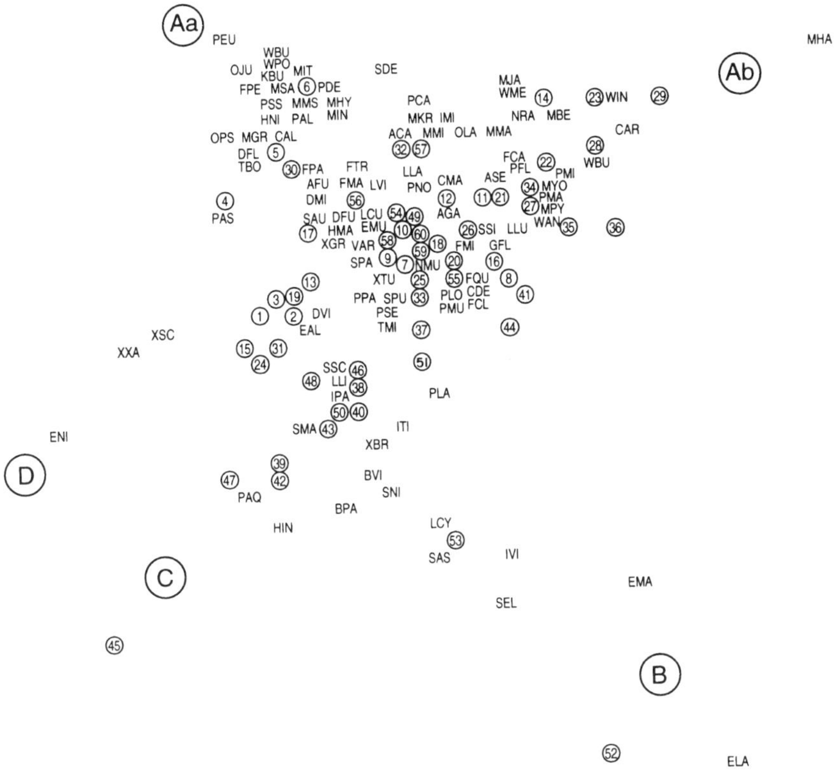

Fig. 10.8. Representation of a correspondence analysis applied to Collembola communities sampled at 60 different biotopes in forests in the neighbourhood of Paris, France. The lay-out is obtained following projection of the data on the plane through ordination axes 1 and 4. Each three-letter code represents a species. Encircled numbers represent sampling locations. The shape of the graph shows that five different groups of species and locations may be distinguished, designated as Aa, Ab, B, C and D. Reproduced from Ponge (1993), with permission of Fischer Verlag, Jena.

correspondence analysis was applied to the data, with the aim of identifying natural assemblages, characteristic for certain habitats. Fig. 10.8 shows a representation of the correspondence, obtained after projecting the data onto the plane through the ordination axes 1 and 4. The closer two species are positioned in such a graph, the higher is their correspondence of occurrence.

It may be seen from Fig. 10.8 that Collembola and sampling locations fall into five groupings; these are denoted by Aa, Ab, B, C and D. Branch Aa groups the species living in soils with a mull humus form and pH above 5, branch Ab consists of species living in soils with a moder humus form and pH lower than

5, branch B comprises some species living in open habitats (grasslands, cultivated fields), branch C consists of hydrophilic species living on the water surface of ponds and in moist habitats, such as bogs, and branch D is a group of forest species occurring on trees, trunks and rocks. Ponge (1993) lists characteristic species, plus accompanying species, for each of these communities.

The conclusion from this analysis is that four main factors determine the nature of Collembola communities: the depth gradient (soil versus litter), the openness of the habitat (grassland versus forest), the humidity conditions (water surface versus dry forest sites), and soil pH (neutral versus acid humus forms). This work is a great step forward compared to the earlier classification system of Gisin (1943), which was chiefly based on subjective arguments.

Statistical data analysis is necessarily a descriptive method. The usefulness of the multivariate approach as a diagnostic and prognostic bioindicator tool still remains to be established. In the analysis reproduced above, a rather broad array of habitats was included; if one were to analyse more subtle effects, e.g. a comparison of similar habitats polluted to different degrees by heavy metals, the analysis could be more difficult. Nevertheless Bolger and Curry (1984) successfully applied detrended correspondence analysis to mite and collembolan communities in grasslands affected by pig slurries. Using the multivariate technique, they were able to show that the influence of pig slurry on the microarthropod communities extended to a distance of 35 m from the source. This and other examples confirm the high degree of resolution that is achievable using multivariate techniques.

Indicator Value of Ecological Groupings

There are several reasons for considering not only the indicator value of species in communities, but also the indicator value of groups of species. First of all, some species may behave similarly with respect to the factor of interest, and for the purpose of bioindication it may not be necessary to distinguish between them. This would save time and make the bioindicator system more easy to apply. Secondly, by assigning ecological properties to species, the bioindicator value might become more specific. Among soil arthropods, one-to-one relationships between taxonomic position and ecological niche are rare; species belonging to the same family or even the same genus may occupy very different niches. In this section I will therefore consider three different ways in which soil arthropods may be classified, in order to further improve the bioindicator value of communities.

Life-history classifications

The potential of using life-history patterns to classify soil invertebrates and to develop a bioindicator index from them is illustrated by the maturity index for

Table 10.1. Summary of life-history patterns of Collembola related to their ecological niche in the soil profile.

Surface-living (epigeon)	Soil-living (euedaphon)
Sexual reproduction	Parthenogenesis
Sexual dimorphism	Sexes alike
Precopula behaviour	Random deposition of spermatophores
High fertility	Low fertility
High mortality	Low mortality
High turnover of population	Low turnover
Development temperature-dependent	Slow development, indifferent to temperature
Life-cycle triggered by season	Reproduction spread throughout the year
Oligopause/diapause	No resting stages

From Petersen (1980) and Van Straalen (1985, 1994a) and references therein.

nematodes (Bongers, 1990; De Goede *et al.*, 1993). In Collembola and oribatid mites, life-history patterns have been subject to a great deal of research, but an index comparable to the maturity index for nematodes has not yet been proposed.

Life-history patterns of Collembola show a strong correlation with the 'life-forms' distinguished by Gisin (1943). This is a system in which Collembola are subdivided into three main groups based on their external morphology. The three groups are designated as 'euedaphon' (species with reduced furca, unpigmented, eyeless, short legs), 'hemiedaphon' (species with pigment equally distributed over the body surface, eyes, furca and legs), and 'epigeon' (species with colour patterns, well-developed furca, eyes, long antennae). The names derive from their location in the soil profile: soil, litter, and surface, respectively. In addition to these main morphological types, Christiansen (1964) distinguished two other groups, the 'troglomorphs' (resembling the euedaphon, but living in caves), and the 'neuston' (resembling the hemiedaphon, but living on water surface). There are numerous variations on this theme, e.g. Rusek (1989) included feeding habits and body-size into Gisin's system, arriving at a classification of 13 groups. The attractiveness of the original life-form system is its simplicity and the possibility to recognize groups on the basis of simple morphological characteristics.

The life-history patterns of Collembola, as related to life-forms, are summarized in Table 10.1. The information contained in this table is based on a number of papers, including Petersen (1980) and Van Straalen (1985, 1994a), and references therein. The main trend is a development towards higher fertility (including sexual behaviour), higher mortality, and a more seasonal life-cycle when going from soil-living to surface-living species.

The relevance of these life-history patterns for the behaviour of populations in a polluted environment was discussed by Van Straalen (1994b) and Crommentuijn *et al.* (1995). Ecotoxicity experiments, in which different arthropods were exposed to food polluted with cadmium, suggest that soil contamination will favour opportunistic species with a high reproductive output, even if these are more sensitive in terms of the median lethal concentration (LC_{50}). With equal exposure, the epigeic life-history pattern would be more suitable to overcome the effects of pollution than the euedaphic strategy. This is in line with an intra-species comparison of populations: both in the springtail *Orchesella cincta* and in the isopod *Porcellio scaber* pollution-induced selection pressure causes the life-history to shift towards a shorter life-cycle with higher reproductive effort earlier in life (Donker *et al.*, 1993; Posthuma *et al.*, 1993). The effects of fertilizers on Collembola communities of grassland can also be interpreted in terms of a shift among life-forms; at higher application levels, the fraction of euedaphic species was lower in favour of hemiedaphic and especially epigeic species (Siepel and Van de Bund, 1986).

The classification system discussed above ignores some peculiarities of microarthropod life-cycles, such as the dispersal through attachment onto other animals (phoresy). Siepel (1994) designed a classification scheme of microarthropod life-histories based on three dimensions: reproduction tactics (parthenogenesis, iteroparity), dispersal tactics (phoresy) and synchronization tactics (diapause). This resulted in 12 different patterns, which are interrelated as shown in Fig. 10.9. These may be seen as adaptations to the persistence of the biotope in space and time. Discontinuity in space calls for phoresy, discontinuity in time calls for diapause.

The system of Fig. 10.9 was applied to various examples from the literature, including studies on the effects of fertilization, cattle grazing, heavy metals and pesticides (Siepel, 1994). Persistent pollutants, irrespective of the kind, always seemed to favour the parthenogenetic (thelytokous) species more than the sexual species. The presence of a dominating, persistent environmental factor can be seen as making the habitat more 'constant', thus favouring thelytoky. The proportion of thelytokously reproducing microarthropods was recommended as a good indicator of persistent pollutants.

Classification of feeding types

Information on the feeding habits of microarthropods may be derived from four types of evidence: morphology of mouthparts (e.g. Kaneko, 1988), gut content analysis (e.g. Hasegawa and Takeda, 1995), activity of digestive enzymes (e.g. Urbásek and Rusek, 1994), and laboratory feeding experiments (e.g. Schultz, 1991). Each of these methods has its merits and demerits; the analysis is further complicated by the fact that most microarthropods do not exhibit a great degree of food specialization. According to rhizotron observations by Gunn and Cherrett (1993) the majority of arthropods interact directly with the detritus resource in the soil.

The classification of feeding types of soil microarthropods is an obvious

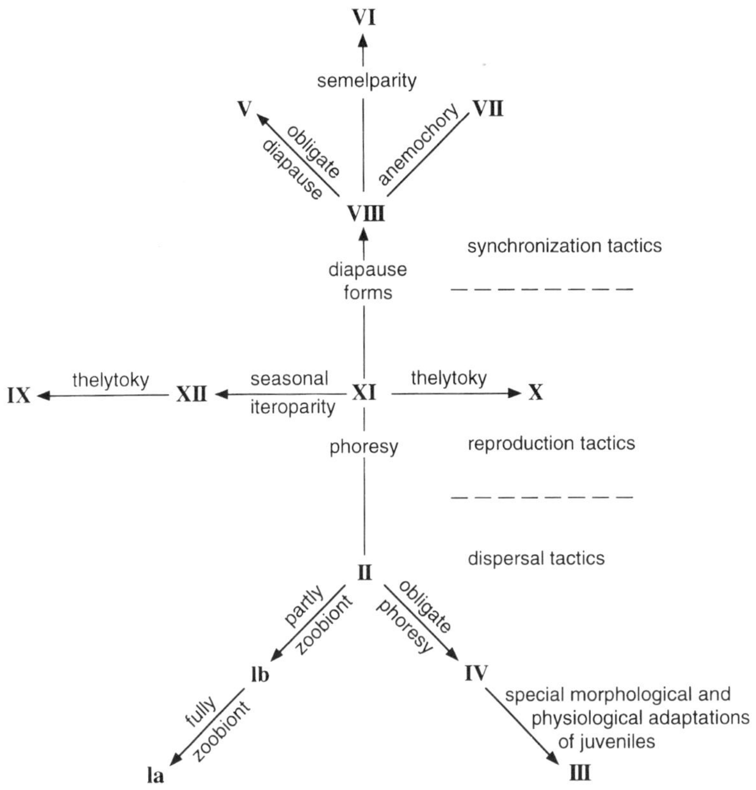

Fig. 10.9. Scheme of interrelationships between 12 different life-history tactics of microarthropods, according to Siepel (1994). Reproduction, phoresy and synchronization are used as discriminating factors. Reproduced from Siepel (1994), with permission.

indicator of the nature of the organic material. For example, a low proportion of fungivores in a community would indicate a low dominance of fungi in the microflora. This type of application was used by Kaneko (1995), in a study on the organization of oribatid mite communities in six different forest soils in Japan. When the forests were placed in a mull-moder-mor series (Fig. 10.10), the proportion of macrophytophagous mites (feeding on dead organic matter) remained more or less constant, but the mor forests had more microphytophagous mites (feeding on fungal hyphae and bacteria) and less fragment feeders (feeding on fine particulate matter including macrofauna faeces). This reflects the dominance of fungus-dominated decomposition processes in mor humus types, and the relatively greater importance of microarthropod fragmentation processes in mull type humus forms.

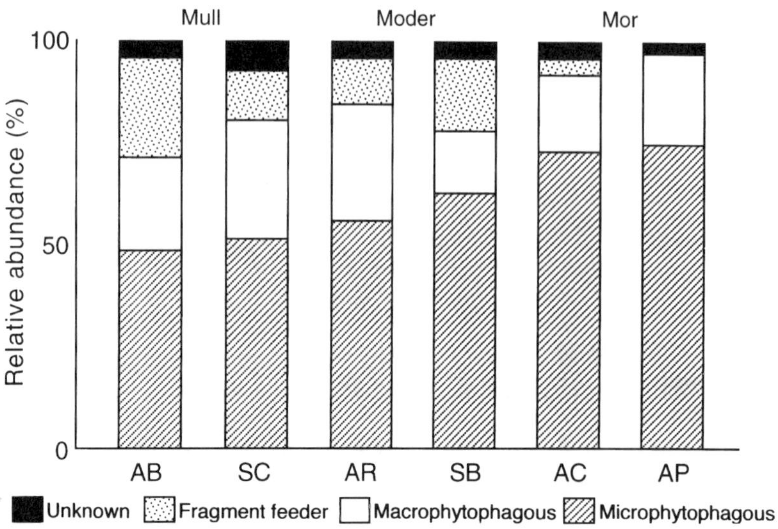

Fig. 10.10. Relative abundance of four different feeding types of oribatid mites in six Japanese forests (designated AB to AP), forming a gradient from mull to mor. Feeding types were identified on the basis of gut contents and mouth parts. Reproduced from Kaneko (1995), with permission of Kyoto University Press.

Classification of microarthropod feeding types is also very important for the establishment of trophic relations underlying soil food-webs. Food-webs may be used to quantify the contribution of microarthropods to energy flow and nutrient mineralization in soils (O'Neill, 1974; Moore *et al.*, 1988; Schaefer, 1990; Verhoef and Brussaard, 1990; De Ruiter *et al.*, 1994; Petersen, 1995). Basal building blocks for food-web models are functional groups, loosely defined as groups of organisms that exploit or process a food resource in a similar manner. Faber (1991) criticized the conventional concept of functional groups on the ground that it ignores the vertical stratification aspect of the community. Microarthropods feeding on fungi will have a totally different role in the decomposition process depending on whether they live on the soil surface or deep into the soil. 'Fungivores', as a functional group, thus comprises a heterogeneous group of different functions. The proposal of Faber (1991) is to split fungivores into euedaphic, hemiedaphic and epigeic representatives (Table 10.2). This proposal does not necessarily increase the number of compartments in a food-web model, because the criterion can also be used to lump taxonomic groups often kept separate (e.g. hemiedaphic fungivore Collembola and hemiedaphic fungivore Oribatida).

When theory further develops, the food-web approach might produce new indices that summarize the trophic structure of a community and that could be

Table 10.2. Functional classification of soil fauna according to Faber (1991), applied to fungivores in stratified forest soils.

Group	Microhabitat	Feeding habits	Functional significance
Epigeic fungivores	Fresh leaf litter	Ingest saprophytic fungi	Interact with fungal growth by selective grazing
Hemiedaphic fungivores	Raw humus, fragmented litter	Ingest saprophytic fungi	Affect net mineralization and nutrient mobility
Euedaphic fungivores	Humus, mineral soil	Ingest mycorrhizal fungi	Affect plant growth by interfering with root nutrient uptake

used for bioindication. Candidates are parameters such as the 'connectedness' and the 'interaction strength' of the food-web. De Ruiter *et al.* (1995) recently showed that in seven different below-ground food-web models interaction strength varied from a dominance of top-down effects at low trophic levels to a dominance of bottom-up effects at high trophic levels. This pattern contributes a great deal to the stability of the food-web, and changes in this pattern might be indicative for a loss of stability.

Ecophysiological classifications

When a bioindicator system is seen as an inverted relationship between an environmental factor and a biological endpoint (cf. Fig. 10.1), it is clear that classifications based on the physiological response of a microarthropod towards a soil factor hold the greatest promise for the development of bioindicators of soil health. When bioindicator systems are derived solely from field data, there is an argument of circularity in their application, e.g. 'species A is acidophilous because it occurs in acid soil', and 'because species A is present, the soil is acid'. This objection may be raised against the way the 'maturity index' for nematode communities was originally introduced by Bongers (1990). It would therefore be worthwhile to try and set up an indicator system based on autecological (ecophysiological) observations, which then would allow an independent test on field data to be made. Unfortunately, of all data for soil microarthropods, ecophysiological data are lacking most.

Soil factors which have been studied in ecophysiological and ecotoxicological experiments include: temperature, humidity, pH, pesticides and heavy metals. Temperature responses have been studied in sufficient detail to allow the establishment of a good bioindicator system, comparable to the one developed for carabid beetles (Atkinson *et al.*, 1987). Temperature bioindicators

Table 10.3. Overview of median pH preferences and indicator values of 20 species of soil arthropods.

Species	Median preference (pH)	Indicator value
Collembola		
Orchesella cincta	5.7	1
Orchesella flavescens	5.6	1
Entomobrya corticalis	7.3	1
Tomocerus minor	4.7	1
Tomocerus flavescens	2.9	2
Lepidocyrtus cyaneus	6.6	1
Folsomia candida	5.6	0
Isotoma notabilis	7.0	1
Cryptostigmata		
Nothrus silvestris	–	0
Platynothrus peltifer	7.6	2
Rhysotritia duplicata	–	0
Pelops occultus	7.1	2
Odontocepheus elongatus	–	0
Hypochthonius rufulus	4.1	1
Adoristes ovatus	4.4	1
Isopoda		
Porcellio scaber	6.0	1
Oniscus asellus	5.1	1
Trichoniscus pusillus	5.9	1
Philoscia muscorum	6.1	1
Armadillidium vulgare	7.0	1

The data are derived from standardized behavioural tests in which arthropods were introduced in a circular walkway of which the bottom was divided into 16 compartments filled with sand and adjusted to a pH in the range 2 to 9, using iso-osmolar phosphate–citrate buffers. Indicator value expresses the specificity of the response: 2 for species with at least 50% of the distribution within two adjacent pH-classes, 1 for species with less than 50% in two classes, 0 for species showing no response at all. From van Straalen and Verhoef (1997).

might be useful to monitor the ecological effects of climate change. As to humidity, a similar argument holds, although the necessary data are more scarce.

For soil pH, a bioindicator index was recently developed by van Straalen and Verhoef (1997). This was based on the preferences of microarthropods exhibited in experimental systems consisting of a circular walkway, of which the substrate had an increasing and decreasing gradient of pHs between 2 and 9. A total of 20 species were tested and each was given a median preference (the median of the observed distribution of animals over substrate pHs), and an indicator value (to denote the specificity of the response). This information is summarized in Table 10.3.

To position a community of soil arthropods, sampled in the field, on a scale of pHs van Straalen and Verhoef (1996) proposed an 'arthropod acidity index', φ, as follows:

$$\varphi = \frac{\sum_{i=1}^{s} p_i f_i v_i}{\sum_{i=1}^{s} f_i v_i} \tag{10.1}$$

where: φ is the arthropod acidity index for a community of $i = 1, 2, \ldots s$ species of arthropods, p_i is the median preferred pH of species i (Table 10.3), v_i is the indicator value of species i (Table 10.3) and f_i is the abundance score for species i (a number indicating how abundant species i is).

The index is constructed so that it can be interpreted as the median preferred pH of the community, taking into account that species with a strong preference (high indicator value), and species with a high abundance, are given a heavier weight. The index would, after a calibration exercise, in principle allow the reconstruction of soil pH from the community of arthropods present at a site; it would thus be able to serve the same purpose as the diatom index and the beetle index mentioned in the introduction of this chapter.

The ecophysiological approach in the construction of bioindicators is also applicable to soil pollutants. Bioindicators for pollutants are of various types, ranging from the measurement of residues in field organisms to bioassays applied to a field-collected community. A recent review is given in van Straalen and Krivolutsky (1996). The reader is also referred to other chapters in this book.

Evaluation and Conclusions

There are various ways in which the microarthropod bioindicators discussed above may be evaluated. One aspect would be the prospect for application in monitoring programmes conducted by regulatory authorities, e.g. as part of a soil quality inventory, an assessment of contaminated sites, or a crop management system. None of the approaches outlined has, however, reached a stage where such an application would be feasible. Despite the number of proposals made, there is still insufficient insight into which bioindicators would serve which purpose. The aim of this section is to shortly evaluate the material discussed above, so as to shed more light on the 'what for' question. This will be done by evaluating two aspects of each bioindicator, specificity and resolution.

An indicator with a high specificity responds to one soil factor much more

Table 10.4. Evaluation of soil microarthropod community bioindicators with regard to the extent to which they respond to a single soil factor (specificity: +), or to many different factors (specificity: –), and with regard to their ability to detect small changes (resolution: +) or only large changes (resolution: –).

Microarthropod community	Specificity	Resolution
Species diversity indices	– –	–
Single indicator species	+	– –
Ratios between species	+	–
Dominance structure	–	+
Multivariate statistics	–	++
Life-history patterns	–	+
Feeding types	++	+
Functional groupings	++	+
Food-web parameters	?	?
Ecophysiological classifications	++	+

strongly than to other factors. Hence, when such an indicator gives a signal, it is immediately clear which factor is changing and appropriate measures can be taken if considered necessary. It must be noted however, that there is often an indirect relationship between management practices and the soil factor that elicits a response in a bioindicator (see the examples discussed in Curry, 1994). Application of fertilizer will affect the nutrient status of the soil, but it might also influence pH and increase the heavy metal content; herbicides leave residues in soil, but may also increase the amount of litter available for saprotrophic arthropods. Bioindicators responding with a high degree of specificity to a certain soil factor may not show the same degree of specificity when defined in terms of management.

Specificity of bioindicators is not necessarily always desirable. If one has no *a priori* idea about the soil factors that are subject to change, it may be more profitable to apply a non-specific, general bioindicator that will react to a broad range of factors, and that gives an indication of 'soil health' in general. Often one would prefer to combine both specific and general indicators.

Table 10.4 summarizes the evaluation with regard to the question of specificity. The highest specificity is expected from those indicators that are built on the responses of species to a single factor (ecophysiological classifications and feeding types). General indicators, such as species diversity and dominance structure do not give a cue to a certain soil factor, because they react to many different factors in a similar way.

The second aspect of evaluation concerns the degree of resolution (Table 10.4). A high degree of resolution means that the bioindicator readily reacts to small changes in the soil factor concerned. This is not necessarily correlated with specificity; for example, multivariate techniques probably allow the highest resolution, because they use all information in the community, but being descriptive, they do not reveal information on the soil factors underlying community change, unless community measurements are combined with measurements of the soil factors themselves.

Based on Table 10.4, it may be recommended that a combination of ecophysiological approaches and multivariate statistics would hold the greatest promise. Classification of a community according to physiotypes will provide the necessary degree of specificity, while a multivariate community analysis will provide the greatest resolution. The status of food-web parameters is hard to evaluate at the moment, because the theory is still developing. Potentially this is an area where the greatest degree of ecological relevance to ecosystem processes and sustainable productivity may be achieved.

The material discussed above has clearly indicated that communities of soil arthropods contain a wealth of information on soil properties. Techniques to sift and sort this information are in full development. Relationships between soil factors and community structure must be known, before one can convert them into a bioindicator system (cf. Fig. 10.1). Bioindicator research is a field where fundamental and applied ecology should go hand in hand.

Acknowledgements

Comments on an earlier version of the text received from Ger Ernsting, Steve Hopkin, Sigmund Hågvar, and the editors of this book, were very helpful and are greatly acknowledged.

References

Anderson, J.M. (1975) The enigma of soil animal species diversity. In: Vanek, J. (ed.) *Progress in Soil Zoology*. Junk, B.V., The Hague, pp. 51–58.

André, H.M., Bolly, C. and Lebrun, P. (1982) Monitoring and mapping air pollution through an animal indicator: a new and quick method. *Journal of Applied Ecology* 19, 107–111.

Arpin, P., Ponge, J.-F., Dabin, B. and Mori, A. (1984) Utilisation des nématodes Mononchida et de Collemboles pour caractériser des phènomènes pédobiologiques. *Revue d'Écologie et de Biologie du Sol* 21, 243–268.

Atkinson, T.C., Briffa, K.R. and Coope, G.R. (1987) Seasonal temperatures in Britain during the past 22,000 years, reconstructed using beetle remains. *Nature* 325, 587–592.

Bååth, E., Berg, B., Lohm, U., Lundgren, B., Lundkvist, M., Rosswall, T., Söderström, B. and Wiren, A. (1980) Effects of experimental acidification and liming on soil organisms and decomposition in a Scots pine forest. *Pedobiologia* 20, 85–100.

Bengtsson, J. (1994) Temporal predictability in forest soil communities. *Journal of Animal Ecology* 63, 653–665.

Bolger, T. and Curry, J.P. (1984) Influences of pig slurry on soil microarthropods in grassland. *Revue d'Écologie et de Biologie du Sol* 21, 269–281.

Bongers, T. (1990) The maturity index: an ecological measure of environmental disturbance based on nematode species composition. *Oecologia* 83, 14–19.

Boomsma, J.J. and van Loon, A.J. (1982) Structure and diversity of ant communities in successive coastal dune valleys. *Journal of Animal Ecology* 51, 957–974.

Bringmark, E. (1989) Spatial variation in soil pH of beech forests in relation to buffering properties and soil depths. *Oikos* 54, 165–177.

Cancela da Fonseca, J.P. (1980) Quelques approaches analytiques en dynamique de populations édaphiques (Acariens Oribates). *Revue d'Écologie et de Biologie du Sol* 17, 445–477.

Cancela da Fonseca, J.P., Boudjema, G., Sarkar, S. and Julien, J.M. (1995) Can statistical analyses improve our knowledge about relationships between environmental factors and soil microarthropod communities. In: Edwards, C.A., Abe, T. and Striganova, B.R. (eds) *Structure and Function of Soil Communities.* Kyoto University Press, Kyoto, pp. 42–51.

Christiansen, K. (1964) Bionomics of Collembola. *Annual Review of Entomology* 9, 147–178.

Connell, J.H. and Sousa, W.P. (1983) On the evidence needed to judge ecological stability or persistence. *American Naturalist* 121, 789–824.

Crommentuijn, T., Doodeman, C.J.A.M., Van der Pol, J.J.C., Doornekamp, A., Rademaker, M.C.J. and Van Gestel, C.A.M. (1995) Sublethal sensitivity index as an ecotoxicity parameter measuring energy allocation under toxicant stress: application to cadmium in soil arthropods. *Ecotoxicology and Environmental Safety* 31, 192–200.

Curry, J.P. (1994) *Grassland Invertebrates.* Chapman and Hall, Inc., London.

De Goede, R.G.M., Bongers, T. and Ettema, C.H. (1993) Graphical presentation and interpretation of nematode community structure: c-p triangles. *Mededelingen van de Faculteit Landbouwwetenschappen van de Rijksuniversiteit Gent* 58/2b, 743–750.

De Ruiter, P.C., Neutel, A.-M. and Moore, J.C. (1994) Modelling food webs and nutrient cycling in agro-ecosystems. *Trends in Ecology and Evolution* 9, 378–383.

De Ruiter, P.C., Neutel, A.M. and Moore, J.C. (1995) Energetics, patterns of interaction strength, and stability in real ecosystems. *Science* 269, 1257–1260.

Den Boer, P.J. (1981) On the survival of populations in a heterogeneous and variable environment. *Oecologia* 50, 39–53.

Diekkrüger, B. and Röske, H. (1995) Modelling the dynamics of *Isotoma notabilis* (Collembola) on sites of different agricultural usage. *Pedobiologia* 39, 58–73.

Donker, M.H., Zonneveld, C. and van Straalen, N.M. (1993) Early reproduction and increased reproductive allocation in metal-adapted populations of the terrestrial isopod *Porcellio scaber. Oecologia* 96, 316–323.

Ellenberg, H. (1974) Zeigerwerte der Gefäßpflanzen Mitteleuropas. *Scripta Geobotanica* 9, 1–97.

Emmanuel, N., Curry, J.P. and Evans, G.O. (1985) The soil Acari of barley plots with different cultural treatments. *Experimental and Applied Acarology* 1, 101–113.

Everts, J.W., Aukema, B., Hengeveld, R. and Koeman, J.H. (1989) Side-effects of pesticides on ground-dwelling predatory arthropods in arable ecosystems. *Environmental Pollution* 59, 203–225.

Faber, J.H. (1991) Functional classification of soil fauna: a new approach. *Oikos* 62, 110–117.

Ghabbour, S.I. (1991) Towards a zoosociology of soil fauna. *Revue d'Écologie et de Biologie du Sol* 28, 77–90.

Ghabbour, S.I., Cancela da Fonseca, J.P., Mikhail, W.Z.A. and Shakir, S.H. (1985) Differentiation of soil fauna in desert agriculture of the Mariut region. *Biology and Fertility of Soils* 1, 9–14.

Ghilarov, M.S. (1977) Why so many species and so many individuals can coexist in the soil. In: Lohm, U. and Persson, T. (eds) *Soil Organisms as Components of Ecosystems*. Ecol. Bull., Stockholm. pp. 593–597.

Gisin, H. (1943) Ökologie und lebensgemeinschaften der Collembolen im Schweizerischen Exkursionsgebiet Basels. *Revue Suisse de Zoologie* 50, 131–224.

Gunn, A. and Cherrett, J.M. (1993) The exploitation of food resources by soil meso- and macro-invertebrates. *Pedobiologia* 37, 303–320.

Hågvar, S. (1984a) Effects of liming and artificial acid rain on Collembola and Protura in a coniferous forest. *Pedobiologia* 27, 341–354.

Hågvar, S. (1984b) Six common mite species (Acari) in Norwegian forest soils: Relations to vegetation types and soil characteristics. *Pedobiologia* 27, 355–364.

Hågvar, S. (1994) Log-normal distribution of dominance as an indicator of stressed soil microarthropod communities? *Acta Zoologica Fennica* 195, 71–80.

Hågvar, S. and Abrahamsen, G. (1984) Collembola in Norwegian forest soils. III. Relations to soil chemistry. *Pedobiologia* 27, 331–339.

Hågvar, S. and Amundsen, T. (1981) Effects of liming and artificial acid rain on the mite (Acari) fauna in coniferous forest. *Oikos* 37, 7–20.

Hasegawa, M. and Takeda, H. (1995) Changes in feeding attributes of four collembolan populations during the decomposition process of pine needles. *Pedobiologia* 39, 155–169.

Heungens, A. and Van Daele, E. (1984) The influence of some acids, bases and salts on the mite and Collembola population of a pine litter substrate. *Pedobiologia* 27, 299–311.

Hogervorst, R.F., Verhoef, H.A. and van Straalen, N.M. (1993) Five-year trends in soil arthropod densities in pine forests with various levels of vitality. *Biology and Fertility of Soils* 15, 189–195.

Huhta, V. (1979) Evaluation of different similarity indices as measures of succession in arthropod communities of the forest floor after clear-cutting. *Oecologia* 41, 11–23.

Huhta, V., Hyvönen, R., Koskenniemi, A. and Vilkamaa, P. (1983) Role of pH in the effect of fertilization on Nematoda, Oligochaeta and microarthropods. In: Lebrun, Ph., André, H.M., De Medts, A., Grégoire-Wibo, C. and Wauthy, G. (eds) *New Trends in Soil Biology*. Dieu-Brichart, Ottignies-Louvain-la-Neuve, pp. 61–73.

Kaneko, N. (1988) Feeding habits and cheliceral size of oribatid mites in cool temperate forest soils in Japan. *Revue d'Écologie et de Biologie du Sol.* 25, 353–363.

Kaneko, N. (1995) Community organization of oribatid mites in various forest soils.

In: Edwards, C.A., Abe, T. and Striganova, B.R. (eds) *Structure and Function of Soil Communities.* Kyoto University Press, Kyoto, pp. 21–33.

Kooijman, S.A.L.M., Van der Hoeven, N. and Van de Werf, D.C. (1989) Population consequences of a physiological model for individuals. *Functional Ecology* 3, 325–336.

Kopeszki, H. and Meyer, E. (1994) Artenzusammenstellung und Abundanz von Collembolen in Waldböden Voralbergs (Österreich). *Berichte des Naturwissenschaftlich-Medizinischen Vereins in Innsbruck* 81, 151–166.

Kratz, W., Brose, A. and Weigmann, G. (1991) The influence of lime application in damaged pine forest ecosystems in Berlin (FRG): soil chemical and biological aspects. In: Ravera, O. (ed.) *Terrestrial and Aquatic Ecosystems: Perturbation and Recovery.* Ellis Horwood Ltd, Chichester, pp. 464–471.

Krivolutsky, D.A. and Druk, A.Y. (1986) Fossil oribatid mites. *Annual Review of Entomology* 31, 533–545.

Krogh, P.H. (1994) Microarthropods as bioindicators. PhD thesis, University of Arhus, Arhus, Denmark.

Kuznetzova, N.A. (1994) Collembolan guild structure as an indicator of tree plantation conditions in urban areas. *Memorabilia Zoologica* 49, 197–205.

Lebrun, Ph., Wauthy, G., Leblanc, C. and Goossens, M. (1976) Tests écologiques de toxitolérance au SO_2 sur l'oribate corticole *Humerobates rostrolamellatus* (Grandjean 1936) (Acari: Oribatei). *Annales de Société Royale Zoologique Belgique* 106, 193–200.

Moore, J.C., Walter, D.E. and Hunt, H.W. (1988) Arthropod regulation of micro- and mesobiota in below-ground detrital food webs. *Annual Review of Entomology* 33, 419–439.

Muller, S. and van Straalen, N.M. (1986) Seasonal fluctuations of Collembola populations. Unpublished report, Department of Ecology and Ecotoxicology, Vrije Universiteit, Amsterdam.

O'Neill, R.V. (1974) Systems approaches to the study of forest floor arthropods. *Systems Analysis and Simulation in Ecology* 1, 441–477.

Paoletti, M.G., Iovane, E. and Cortese, M. (1988) Pedofauna indicators and heavy metals in five agroecosystems in north-east Italy. *Revue d'Écologie et de Biologie du Sol* 25, 33–58.

Petersen, H. (1980) Population dynamics and metabolic characterization of Collembola species in a beech forest ecosystem. In: Dindal, D.L. (ed.) *Soil Biology as Related to Land Use Practices.* EPA, Washington, pp. 806–833.

Petersen, H. (1994) A review of collembolan ecology in ecosystem context. *Acta Zoologica Fennica* 195, 111–118.

Petersen, H. (1995) Energy flow and trophic relations in soil communities: state of knowledge two decades after the International Biological Programme. In: Edwards, C.A., Abe, T. and Striganova, B.R. (eds) *Structure and Function of Soil Communities.* Kyoto University Press, Kyoto, pp. 111–130.

Petersen, H. and Luxton, M. (1982) A comparative analysis of soil fauna populations and their role in decomposition processes. *Oikos* 39, 287–388.

Ponge, J.F. (1983) Les collemboles, indicateurs du type d'humus en milieu forestier. Résultats obtenus au sud de Paris. *Acta Oecologia. Oecologia Generalis* 4, 359–374.

Ponge, J.-F. (1993) Biocenoses of Collembola in atlantic temperate grass–woodland ecosystems. *Pedobiologia* 37, 223–244.

Posthuma, L., Verweij, R.A., Widianarko, B. and Zonneveld, C. (1993) Life-history patterns in metal-adapted Collembola. *Oikos* 67, 235–249.

Poursin, J.M. and Ponge, J.F. (1984) Étude des peuplements de microarthropodes (Insectes Collemboles et Acariens Oribates) dans trois humus forestiers acides de la Forêt d'Orléans (Loiret, France). *Pedobiologia* 26, 403–414.

Renberg, I. and Hellberg, T. (1982) The pH history of lakes in Southwestern Sweden, as calculated from the subfossil diatom flora of the sediments. *Ambio* 11, 30–33.

Rusek, J. (1978) Pedozootische Sukzessionen während der Entwicklung von Ökosystemen. *Pedobiologia* 18, 426–433.

Rusek, J. (1989) Ecology of Collembola. In: Dallai, R. (ed.) *Proceedings of the 3rd International Seminar on Apterygota.* University of Siena, pp. 271–281.

Schaefer, M. (1990) The soil fauna of a beech forest on limestone: trophic structure and energy budget. *Oecologia* 72, 230–234.

Schauermann, J. (1987) Tiergesellschaften der Wälder im Solling unter dem Einfluβ von Luftschadstoffen und künstlichem Säure- und Düngereintrag. *Verhandlungen der Gesellschaft für Ökologie* 16, 53–62.

Schultz, P.A. (1991) Grazing preferences of two collembolan species *Folsomia candida* and *Proisotoma minuta,* for ectomycorrhizal fungi. *Pedobiologia* 35, 313–325.

Sheehan, P.J., Miller, D.R., Butler, G.C. and Bourdeau, P. (1984). *Effects of Pollutants at the Ecosystem Level.* John Wiley and Sons, Chichester.

Siepel, H. (1990) Niche relationships between two panphytophagous soil mites, *Nothrus silvestris* Nicolet (Acari, Oribatida, Nothridae) and *Platynothrus peltifer* (Koch) (Acari, Oribatida, Camisiidae). *Biology and Fertility of Soils* 9, 139–144.

Siepel, H. (1994) Structure and function of soil microarthropod communities. PhD Thesis, Agricultural University, Wageningen, The Netherlands.

Siepel, H. and van de Bund, C.F. (1986) The use of life history strategies of microarthropods in grassland management. In: Velthuis, H.H.W. (ed.) *Proceedings of the 3rd European Congress of Entomology,* NEV, Amsterdam, pp. 481–484.

Siepel, H. and van de Bund, C.F. (1988) The influence of management practises on the microarthropod community of grassland. *Pedobiologia* 31, 339–354.

Stamou, G.P., Asikidis, M.D., Argyropoulou, M.D. and Sgardelis, S.P. (1993) Ecological time versus standard clock time: the asymmetry of phenologies and the life history strategies of some soil arthropods from Mediterranean ecosystems. *Oikos* 66, 27–35.

Takeda, H. (1987) Dynamics and maintenance of collembolan community structure in a forest soil system. *Researches on Population Ecology* 29, 291–346.

Urbásek, F. and Rusek, J. (1994) Activity of digestive enzymes in seven species of Collembola (Insecta: Entognatha). *Pedobiologia* 38, 400–406.

Usher, M.B., Booth, R.G. and Sparkes, K.E. (1982) A review of progress in understanding the organization of communities of soil arthropods. *Pedobiologia* 23, 126–144.

Van Gestel, C.A.M. and van Brummelen, T.C. (1996) Incorporation of the biomarker concept in ecotoxicology calls for a redefinition of terms. *Ecotoxicology* 5, 217–225.

Van Straalen, N.M. (1985) Comparative demography of forest floor Collembola populations. *Oikos* 45, 253–265.

Van Straalen, N.M. (1994a) Adaptive significance of temperature responses in Collembola. *Acta Zoologica Fennica* 195, 135–142.

Van Straalen, N.M. (1994b) Ecotoxicological responses at the population level. In: Widianarko, B., Vink, K. and van Straalen, N.M. (eds) *Environmental Toxicology in South East Asia.* VU University Press, Amsterdam, pp. 33–47.

Van Straalen, N.M., and Krivolutsky, D.M. (eds) (1997) *Bioindicator Systems for Soil Pollution.* NATO ASI Series, Vol. 16, Kluwer Academic Publishers, Dordrecht.

Van Straalen, N.M. and Verhoef, H.A. (1997) The development of a bioindicator system for soil acidity based on arthropod pH preferences. *Journal of Applied Ecology* 34 (in press).

Van Straalen, N.M., Geurs, M. and Van der Linden, J.M. (1987) Abundance, pH preference and mineral content of Oribatida and Collembola in relation to vitality of pine forests in the Netherlands. In: Perry, R., Harrison, R.M., Bell, J.N.B. and Lester, J.N. (eds) *Acid Rain: Scientific and Technical Advances.* Selper Ltd, London, pp. 674–679.

Van Straalen, N.M., Kraak, M.H.S. and Denneman, C.A.J. (1988) Soil microarthropods as indicators of soil acidification and forest decline in the Veluwe area, the Netherlands. *Pedobiologia* 32, 47–55.

Vegter, J.J. (1983) Food and habitat specialization in coexisting springtails (Collembola, Entomobryidae). *Pedobiologia* 25, 253–262.

Vegter, J.J. (1987) Phenology and seasonal resource partitioning in forest floor Collembola. *Oikos* 48, 175–185.

Vegter, J.J., Joosse, E.N.G. and Ernsting, G. (1988) Community structure, distribution and population dynamics of Entomobryidae (Collembola). *Journal of Animal Ecology* 57, 971–981.

Verhoef, H.A. and Brussaard, L. (1990) Decomposition and nitrogen mineralization in natural and agro-ecosystems: the contribution of soil animals. *Biogeochemistry* 11, 175–211.

Warwick, R.M. (1986) A new method for detecting pollution effects on marine macrobenthic communities. *Marine Biology* 92, 557–562.

Weigmann, G. and Kratz, W. (1987) Oribatid mites in urban zones of West Berlin. *Biology and Fertility of Soils* 3, 81–84.

11

Can the Abundance or Activity of Soil Macrofauna be used to Indicate the Biological Health of Soils?

B.M. Doube[1] and O. Schmidt[2]

[1]*CSIRO Land and Water, Private Bag No. 2, Glen Osmond, South Australia, 5064, Australia;* [2]*Department of Environmental Resource Management, Faculty of Agriculture, University College Dublin, Belfield, Dublin 4, Ireland*

Introduction

The principal challenge facing the development of effective indicator(s) of the biological health of soils is to identify strong positive links between an index of soil health (probably an aspect of plant performance) and some measure of soil biology. Soil macrofauna (animals >2 mm) are the most conspicuous soil animals and include ants, termites, amphipods, isopods, centipedes, millipedes, adult and larval stages of root-feeding insects, earthworms, enchytraeid worms, slugs and snails. They fragment and redistribute organic residues, thereby increasing microbial activity, which enhances organic matter decomposition and nutrient availability throughout the root zone, and improve soil structure (Linden *et al.*, 1994). In this chapter we concentrate on agricultural soils and consider to what extent different macrofaunal groups fulfil the requirements for useful bioindicators. We venture the opinion that there are few situations in which the abundance of individual species (or the composition of communities) may indicate the health of a soil. More optimistically, soil macrofauna are at the top of the soil food web and so are ideally positioned to function as bioaccumulators of toxic chemicals in soil.

Soil Macrofauna, Food Webs and Indicators of Soil Health

Soil macrofauna occur near the top of the food chain in soil. Commonly their diet consists of primary decomposers (e.g. bacteria, fungi and actinomycetes) and secondary consumers (such as protozoa) and so their abundance reflects an integration of a range of biological processes occurring in soil (Pankhurst *et al.*, 1995; Doube and Brown, 1996). Although soil organisms occupy only a small fraction of the soil volume (<0.5%), they are primarily associated with organic residues, or locations previously occupied by such residues. Soil food webs comprise autotrophic algae (a few hundred to over 1000 kg ha^{-1}; Atlavinyte and Pociene, 1973); the microbial heterotrophs, which commonly occur at biomasses of 0.5 to 20 t ha^{-1} (Jenkinson and Ladd, 1981; Lee and Pankhurst, 1992) and the soil micro and macrofauna, which can occur at biomasses of 0.5 to 20 t ha^{-1} (Killham, 1994; Edwards and Bohlen, 1996).

The case for using soil macrofauna (and earthworms in particular) as bioindicators of soil health rests on their prominent place in the community of soil organisms and their function in promoting processes which are considered to be linked to soil health (e.g. organic decomposition, beneficial soil structures, nutrient turnover, disease control) (Lee and Foster, 1991; Doube *et al.*, 1994a; Pankhurst *et al.*, 1995, 1996; Edwards and Bohlen, 1996). Those promoting this view naturally cite examples in which there is an overwhelming impact of soil fauna. For example, in some situations in the temperate grassland biome, earthworms account for over 50% of the soil faunal biomass and about 40% of the soil faunal respiration (Persson and Lohm, 1977). Similarly up to 60% of the soil faunal biomass in some temperate forest biomes is earthworms (Petersen and Luxton, 1982). In some tropical soils the top 20 cm is primarily earthworm casts (Lavelle *et al.*, 1992).

Earthworms (notably lumbricids) are the most widely studied group of beneficial soil invertebrates, both in the classical disciplines (anatomy, physiology, biogeography) and in more modern fields (ecology, ecotoxicology), and so there is a substantial body of information available on which to base an evaluation of their potential as indicators. Although there is little doubt that there are many situations in which soil macrofauna play a key role in soil ecosystems (they have been called the 'ecosystem engineers' (Lavelle *et al.*, 1992; Lawton, 1994)), the putative links between them and soil health are subjective and intuitive and need to be quantified. Situations with healthy, productive soils but little or no soil macrofauna occur and need also to be considered.

The effect of soil fauna on soil processes has been extensively reviewed (Lavelle, 1988; Curry, 1994; Edwards and Bohlen, 1996). The capacity of soil fauna to act as indicators of soil quality has been examined by Linden *et al.* (1994) – they considered the influences of soil fauna on soil functional properties which influence soil quality (health). They explored the utility of soil fauna as indicators of how well these functions were performed and how changes in soil fauna may reflect changes in soil quality (e.g. water infiltration). They indicated a number of problems (discussed further in this chapter) with the use of soil fauna as indicators of soil health, but concluded with qualified support for the development of a suite of biologically related factors (including burrowing soil fauna, some measure of faunal faecal deposition and a measure of fauna-related organic matter decomposition and nutrient release) as bioindicators of soil health. Earthworms have been proposed as indicators of soil health by some authors (Oades and Walters, 1994) and rejected by others (e.g. Wylie, 1994).

If the factors which determine the abundance or activity of soil-dwelling macrofauna are those (or are correlated with those) that determine the health of soils, there is a good *a priori* case for considering that the abundance/biomass of the macrofauna may serve as indicators of soil health. For such biological measures to act as an index of soil health: (i) measures of soil health must be available; (ii) potential indicators (e.g. earthworms) must be widely distributed; (iii) indicator species must be readily recognized and sampled; (iv) measures of potential indicators must be strongly linked with soil health; (v) numerical responses need to occur at an appropriate time scale; and (vi) reference (base line) values are required.

These considerations provide the structure for the following review. First we ask whether measures of soil health are available. We consider the various ways in which one might define and measure soil health, and conclude that the most relevant and measurable parameter is long-term plant productivity (without environmental damage), and that this is a species specific concept, although it might also apply to communities. Plant productivity could be constrained by foliar diseases, insect attack or by herbivores, but these constraints are not considered further in this chapter.

We then consider the habitat associations and regional distributions of candidate indicator groups and conclude that the loss of trees and surface litter in most agricultural environments renders many groups (those dependent on trees and surface litter) of little value as potential indicators of the health of agricultural soils. It may be that separate sets of bioindicators of soil health could be developed for agricultural and wooded environments. Ants, termites and litter-dwelling macrofauna are not sufficiently abundant in most agricultural soils to warrant further attention. Further, their abundances are not clearly linked to soil productivity. Slugs and snails are agricultural pests. Earthworms are therefore the sole remaining potential soil health indicator among the soil macrofauna.

Next, we consider the ease with which earthworms may be sampled and quantified and conclude that suitable, but laborious, methods are available. These indicate that earthworms are very patchily distributed and that changes in abundance occur at a time scale appropriate for an indicator (1–3 years). The species composition of earthworm communities depends upon the nature of the habitat and the opportunity for new species to colonize it. The primary factors which limit the abundance of earthworms are climate, food, soil moisture and soil composition (pH, texture, etc.), while secondary factors (e.g. land management) induce correlated changes in a suite of these primary factors.

Broad agreement between measures of soil condition and apparent soil health suggests that soil health and earthworm abundance might be promoted by common factors.

The associations between earthworm abundance and soil biological factors (especially plant productivity) indicate that strong linkages are by no means universal. These associations are investigated by considering data from a series of long-term agricultural trials. While there is clear evidence that earthworms have the capacity to increase plant productivity in some circumstances, it seems that the limits to plant productivity are so multi-factorial that dominant unitary associations are rare. We conclude that there are limited situations in which earthworm abundance might indicate soil health, but even in these situations, the absence of reference (or base line) values precludes effective comparisons between localities. However, their capacity to act as bioaccumulators of toxic soil chemicals may make them useful indicators of poor soil health.

Definition and Measurement of Soil Health

One essential characteristic of soil health is that it must be measurable. This is frequently overlooked in the quest for a conceptually appealing definition of the meaning, rather than the measure, of soil health.

Practical assessment of soil health requires consideration of the multiple functions of soil and their variation in time, space and intensity (Doran *et al.*, 1994), and requires a holistic, not a reductionist, approach (the meaning). Nevertheless, specific measures of soil health are required and these are necessarily reductionist (the measures).

Definitions of soil health and soil quality (which, according to some authors, e.g. Doran and Parkin, (1994), are synonyms) are legion (Parr *et al.*, 1992; Larson and Pierce, 1994; Doran and Parkin 1994; Yakovchenko *et al.*, 1996), but both usually embody the sentiment: 'the capacity of a soil to function within ecosystem boundaries to sustain biological productivity, maintain environmental quality and promote plant and animal health' (Doran and Parkin, 1994). We, like Doran and Parkin, prefer the term soil health because it more clearly portrays the soil as a living, dynamic entity that functions in a holistic way. Furthermore, we suggest that the term quality should refer to an instantaneous measure whereas soil health, with its sustainable component, has a time dimension. One could say that soil health, unlike soil quality, has a future!

As with most previous definitions of soil health, ours has two components: productivity and sustainability. In other words, a productive soil is only healthy if that level of productivity is sustainable in the medium to long term. The time frame over which productivity is sustained is a matter for debate but we have chosen a relatively short time frame of one or two decades (that contained within the scope of long-term field trials). Obviously ecosystem health should have a time frame of centuries, and be bounded by climatic changes. The absence of environmental degradation is an essential element of sustainability. Healthy soils are those in which the capacity of plants to convert sunlight and CO_2 to structural carbohydrate (i.e. the capacity to grow) remains near the upper limits of potential productivity (set by environmental constraints) over an extended period (years). This definition inexorably links soil health with plant species and implies that the characteristics of healthy soils will vary with plant species. This is intrinsically sensible, even though complications ensue.

Critics of this species-based view hold that such definitions become an inextricable tangle of contrasting requirements for different species (and varieties within species) which spawns an unmanageable multiplicity of estimates of soil health. This is a real problem, but what is left if one ignores plant species in a definition of soil health? Very little, we think. Estimates of the optimum long-term net primary productivity of a soil cannot be independent of plant species. A species-independent definition of soil health is a concept of little practical use because it infers that one should consider only those species (or communities) which fix most carbon in a given environment. This is not useful.

Extreme soil conditions provide another conceptual difficulty. The natural plant communities associated with extreme environments are well adapted to those extreme conditions (e.g. acidic or saline soils), and are sustainable in the longer term and so are clearly 'healthy' soils for those plant species. Nevertheless the gross productivity (tonnes of carbon fixed per ha) of these soils can be substantially increased by altering the environment with physical and chemical ameliorants (e.g. ploughing and liming) and changing plant species (e.g. from heath to cereal). Following such amelioration, the health of the soil decreases for the native species and increases for cereal crops. At first sight this may appear contradictory, but at least our definition allows soil health to be measured, an essential prerequisite for developing indicators.

Once one accepts that soil health is a plant-driven concept, the prospect for identifying biological indicators of soil health is substantially improved because many of the constraints, which make difficulties for other definitions, cease to exist. The search for a universal definition of a healthy soil which is independent of plant species creates such an array of questions of overwhelming complexity that authors are forced to retreat into platitudes, analogies and generalities. With a species-based definition, soil health can be quantified, and so the measure is useful, even if only in a limited way.

Using our species-based definition, soil health can be measured in terms of sustained plant productivity. The soil conditions which favour growth will vary with species (e.g. wheat, peas, ryegrass, saltbush) and, because of season–season variability in rainfall etc., numbers of seasons will be needed to assess whether production is sustainable (i.e. sustained average yields without environmental degradation). The requirements of some species (e.g. temperate cereals or pasture grasses) may be so similar that a common set of factors determines soil health for those species. In these cases we can describe and measure the factors that give rise to 'healthy cereal soils', 'healthy pasture soils' and so on, suggesting the possibility of a definition of soil health for certain plant communities.

Further conceptual challenges arise when we consider the relationships between root disease, plant productivity and soil health. Root pathogens can be important constraints upon plant productivity but clearly it is not sensible to designate a soil containing severely diseased plants as healthy soil for that plant species. Nevertheless it is extremely unlikely that measures of soil macrofauna will respond directly to level of abundance/activity of root pathogens and so act as a surrogate indicator of the disease status of soil. However, it is possible that root disease levels and some measures of soil macrofauna are autocorrelated (i.e. both respond to a similar set of factors, e.g. organic matter), and so soil macrofauna could act as an indicator of the disease status of soil, even though the two factors are not functionally linked to each other.

The Spatial Distribution of Candidate Indicator Groups

Indicator groups which function in wooded as well as agricultural systems are needed. Most agricultural practices (with some exceptions, e.g. alley cropping, some forestry) remove both the native vegetation and the majority of the surface litter, thereby destroying the habitats of many of the macrofaunal elements common in undisturbed environments. For example, the availability of food for many species of termites (dead structural plant tissue), ants (prey, honeydew, food for fungus farms, etc.) and amphipods, isopods, centipedes and millipedes (prey and organic residues) is obviously reduced substantially by clearing and by most cropping practices. Abbott *et al* (1979) found that cultivation almost completely eliminated the larger soil animals such as ants, termites and beetles, compared with nearby native vegetation. Clearly, these groups have limited potential to act as indicators in agricultural systems, except perhaps where minimal tillage and trash retention are practised. Dung beetles are also important soil-dwelling insects amongst which functional groups are readily identified (Doube, 1990) but their dependence on dung-producers makes them unsuitable as indicators of soil health.

In contrast, adult and larval root-feeding insects, earthworms, enchytraeid worms, slugs and snails can be abundant in both agricultural and undisturbed habitats and at times make up a large proportion of the biomass of the soil macrofauna in agricultural habitats. Further, their abundance commonly increases with conservation soil management practices (no till, stubble retention, etc.) and this suggests that they might act as general indicators of soil health. However, root-feeding insects, slugs and snails are common agricultural pests and control measures commonly modify their abundance, regardless of soil health. Enchytraeid worms make up only a tiny proportion of the biomass of soil fauna (Brussaard *et al.*, 1990), but their abundance in acidic soils (where earthworms are scarce) give them some value as potential indicators in restricted soil conditions. These considerations make these groups unsuitable as general indicator species and leaves earthworms as the sole remaining potential indicator group.

Patterns of Regional and Local Distribution of Earthworms

Apart from extreme environments (polar and desert regions, highly sandy, saline or extremely acidic or alkaline soils etc.) earthworms have a natural worldwide distribution. The regional distribution of earthworms (approximately 3000 species from 16 families) has been reviewed (Gates, 1972; Sims, 1980; Lee, 1985; Hendrix, 1995; Edwards and Bohlen, 1996).

Between 10 (Sims, 1980) and 16 (Gates, 1972) species of European lumbricids are believed to have greatly extended their natural (i.e. Holartic) family range during the past few hundred years by following European settlers into previously uncolonized temperate and other suitable climatic regions of the world (Michaelsen, 1928; Gates, 1972). Some species from other regions (e.g. *Pontoscolex* spp. and *Microscolex* spp.) have also dispersed widely through the agricultural regions of the world (Lee, 1985; Lavelle, 1988; Edwards and Bohlen, 1996). The anthropogenic dependency of these species has often been discussed (Gates, 1972; Lee, 1985).

Largely the same set of species appear to have colonized agricultural soils worldwide. They have been termed peregrine (Michaelsen, 1928), anthropochorous and haemerophilic (Gates, 1972) and form a group of great agricultural importance. They are the new fauna of many agricultural regions throughout the world, even in some parts of Europe (Terhivuo, 1988). Their ubiquity in cultivated land and other disturbed environments predestines them as potential bioindicators in agricultural lands. Hence agricultural soils, worldwide, are likely to contain a similar suite of peregrine species although the faunas will vary with climate (e.g. tropical and temperate fauna will have different species compositions). This suggests that peregrine species, but not native species, may have the potential to act as indicators of the health of agricultural soils.

The common peregrine species have not yet reached the natural limits to their distribution and numerous reports deal with the natural spread of peregrine species (Hendrix, 1995) and their deliberate introduction to areas lacking them (Curry and Boyle, 1995; Baker *et al.*, 1994; Lee, 1985; Scheu and Parkinson, 1994; Temple-Smith and Pinkard, 1995). However, it is clear that most peregrine species have a wide distribution and must reach the natural limits to their distributions in the foreseeable future. That time interval will depend upon their natural rate of dispersal, which ranges from metres per night (Schwert, 1980; Mather and Christensen, 1992) to metres per year (Lee, 1985) in association with deliberate introductions.

Sampling Considerations for Earthworms

Earthworms are an ecologically diverse group of organisms which is differentiated in terms of body size, vertical stratification in the soil profile, feeding behaviour and food preferences, reproductive potential, and adaptive and survival strategies (Bouche, 1977; Satchell, 1980; Lee, 1985). All these factors affect the choice of sampling procedures and timing of sampling.

Earthworms are essentially sessile. The capability of earthworms to migrate is generally quite limited, which simplifies the approach to sampling populations. Nevertheless, field edges have been identified as potential sources of earthworms immigrating into fields with low earthworm numbers (Hemmann, 1994; Curry and Boyle, 1995), active migration of lumbricids has been observed from grass verges into trial plots and between small trial plots in field trials (Westernacher-Dotzler, 1988) and mass migrations have been observed. Despite this, large species and density differences have persisted between plots separated by 0.5 m (Doube *et al.*, 1995) and the persistent presence of treatment effects in numerous field trials involving plots (see later) testifies to the sedentary nature of earthworm populations.

A variety of measures are used to assess 'earthworms' in relation to soil health. The most obvious measure is that of abundance (measured numerically or as biomass of individual species or pooled over species). Surface casting could also be used but is unlikely to produce a reliable index of earthworm activity because most casting by the peregrine species takes place underground, and the rates of surface casting vary with earthworm species and the bulk density and water content of the soil (Kretzschmar, 1991; Doube *et al.*, 1994b; Kang *et al.*, 1994). The age structure of populations or species composition of a community could be used to describe the effect of the environment on earthworms but these measures are far too precise to be of general use.

There is no ideal method for sampling earthworms to provide accurate population estimates for all sampling conditions and communities (Baker and Lee, 1993). The most reliable method for extracting earthworms is hand sorting of soil: this is laborious but non-selective. Dynamic methods (e.g. formalin or electrical expulsion) are more selective and their effectiveness varies with soil conditions. Most peregrine earthworms live in moist surface soil and retreat to depth to avoid hostile (e.g. dry) environmental conditions: response rates to these stimuli vary between species and sampling needs to occur when all species are present in the top soil.

Compared to many other soil invertebrates, earthworms have relatively long life-spans and generation times and so populations are relatively stable and their abundance can be estimated with one or two samplings per year.

Owing to their size, earthworms can be easily observed, handled and identified (at least as adult specimens). Their taxonomy is relatively simple and does not require specialized technical equipment and so are accessible to the non-taxonomist: farmers can quickly learn to identify their local species

(J. Buckerfield, personal communication). Further, the number of species in any one location is commonly low (less than ten), and communities are usually dominated by one to three species. However, some workers consider that differentiation at the species level is too precise to be a useful tool for monitoring soil health: one should use total numbers or, preferably, total biomass, as used by some workers (e.g. Buckerfield and Auhl, 1994).

Critics of the use of pooled data suggest that it is not appropriate to combine

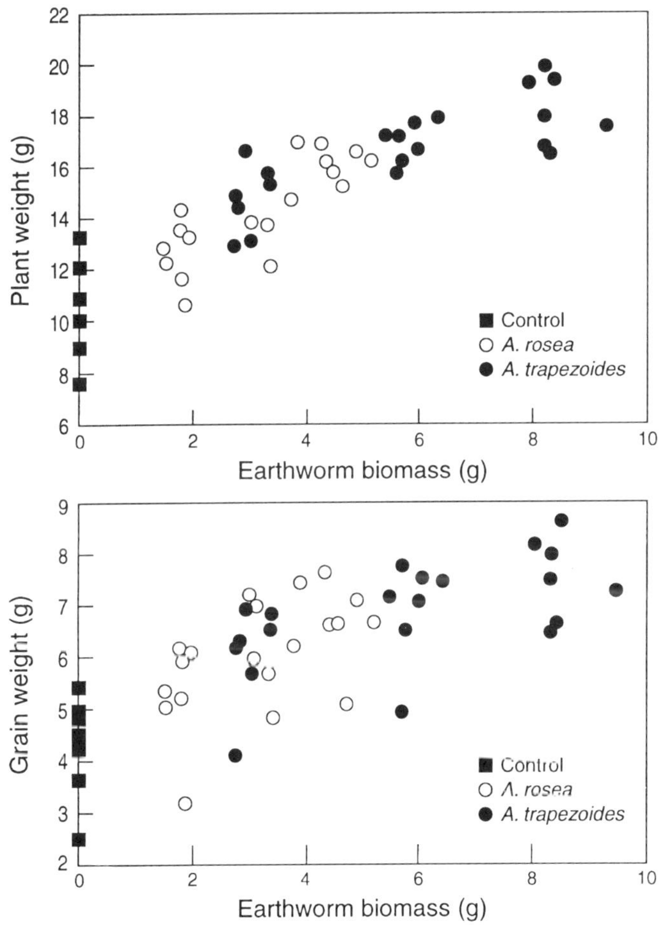

Fig. 11.1. The relationship between plant performance (grain yield and total plant biomass) and the biomass of the earthworms *Aporrectodea trapezoides* and *A. rosea* in a red-brown earth soil with a clay subsoil (P. Williams and B.M. Doube, unpublished).

data from species with very different life styles, especially since there are species-specific responses to environmental parameters (see below). However, samples from field sites may contain individuals whose mean weight may vary 10-fold between adults of different species and up to 100-fold between juveniles and adults within species. Thus the argument comes down to whether the impact of earthworms is best represented by a given change (e.g. a 10-fold change) in earthworm numbers or biomass.

The use of total biomass avoids the difficulties associated with earthworm size and taxonomy, and is more stable than numerical abundance as a measure of earthworm impact. For example, Williams and Doube (unpublished) examined the relationship between numbers and biomass of two species of earthworm (*Aporrectodea trapezoides* and *A. rosea*) and barley yields in a red-brown earth. They found that biomass was far superior to abundance as a predictor of the impact of earthworms on plant weight ($r = 0.88$, $P<0.001$) and grain yield ($r = 0.75$, $P<0.001$) (Fig. 11.1), supporting the view that species with similar ecological requirements (as above) may be pooled and that biomass, rather than numerical abundance, should be used as a measure of abundance. Whether it is useful to pool the biomass of species from different ecological categories is not known.

Earthworm abundance in intensively cultivated land can be very low, rendering their use as bioindicators impractical for methodical and statistical reasons (Knüsting, 1992).

Once agricultural soils have their full complement of peregrine species, can the abundances of species or groups of species (or even the composition of communities) act as an indicator of soil health? To answer this question we need to examine the factors determining the abundance of earthworms and establish whether the same factors limit sustainable crop productivity.

Determinants of Earthworm Abundance

Factors that influence the abundance of earthworms in agroecosystems include the organic matter status of the soil, soil type and depth, pH, moisture holding capacity, rainfall, temperature, cultivation, crop type and crop residues, and predation and parasitism (Curry, in press). Amongst these, the availability of organic matter is commonly the most important (Edwards and Bohlen, 1996), but other factors (e.g. pH, soil type, habitat type or predation) can also be of overriding importance in particular circumstances. Furthermore, agricultural practices (tillage, fertilizer, etc.) can also have a major influence on earthworm abundance.

Added organic matter commonly increases earthworm populations. For example, the abundance of *A. trapezoides, A. rosea* and *M. dubius* increased 2- to 10-fold when sheep dung was added to a pasture soil (Hughes *et al.*, 1994a). The abundance of the same three species increased 2- to 6-fold in response to added cereal straw in a cropping soil (Doube *et al.*, 1995) (Fig. 11.2). The type of organic matter is also important. For example, Barley (1959a) showed that *Aporrectodea caliginosa* gained weight when fed dung, maintained weight on a diet of plant foliage and lost weight when fed roots. On a regional basis, Hendrix *et al.* (1992) reported a significant positive but weak correlation between earthworm numbers and soil organic matter content across a number of ecosystem types on the Georgia piedmont, USA.

Soil pH can have an overriding influence on earthworm abundance. Jefferson (1956) measured the abundance of ten earthworm species in a field trial in which soil pH in adjacent plots differed radically. Five species made up 94.5% of those sampled and *Aporrectodea tuberculata* made up 60% of all earthworms. Earthworms were virtually absent from the highly acidic soils (pH<4) and alkaline soils (pH>8) and were most abundant in soils in the pH range 6.0–7.0 (Fig. 11.3). Earthworms were absent from plots treated with ammonium salts but were present at ca. 700 m^{-2} in adjacent untreated plots (Jefferson, 1955). It is also important to note that there were major differences between species. For example, *A. tuberculata* and, to a lesser extent, *Lumbricus festivus* and *Lumbricus rubellus* were tolerant of acidic soil conditions whereas *Aporrectodea chlorotica*, *Eisenia* (*Aporrectodea*) *rosea* and *Octalasion cyaneum* were greatly reduced in acidic soils and abundant in neutral soil (Fig. 11.3). Similarly, Edwards (1974) examined earthworm populations in the Park Grass Plots at Rothamsted, UK (trial established 1856) where liming acidic soils had created soils whose pH ranged from 3 to 8. Again earthworms were scarce or absent from soils of extreme pH. The earthworm *L. terrestris* was most abundant in soils in the range pH 7–8 whereas the other species were most abundant in the range pH 5–6.

Soil type also appears to influence earthworm numbers, dramatically in some cases (Lee, 1985; Edwards and Bohlen, 1996). Earthworms are commonly scarce in sandy soils, relatively abundant in loamy soils and scarce in some clays (e.g. krasnozems, Wylie, 1994). Plant productivity does not follow the same pattern in relation to soil type, but it is possible that, within the caveat of soil type, earthworm numbers might act as indicators of soil health.

The species composition of earthworm communities depends upon the nature of the habitat and the opportunity for new species to colonize it. For example, in New Zealand, replacing natural vegetation with pasture caused the loss of virtually all native species: these pastures remained worm-free until colonized, decades later, by introduced European pasture species (Lee, 1985). Similarly in tropical, subtropical and temperate regions, the conversion of native lands to agricultural land has seen the local demise of populations of native earthworm species and colonization by an introduced exotic species (Lavelle, 1988; Lavelle *et al.*, 1992).

Predation on earthworms can have dramatic effects on local populations (Edwards and Bohlen, 1996). For example, plovers, in a hayfield in Iceland, reduced earthworm populations (238 m^{-2}) by nearly 50% in 22 days (Bengston

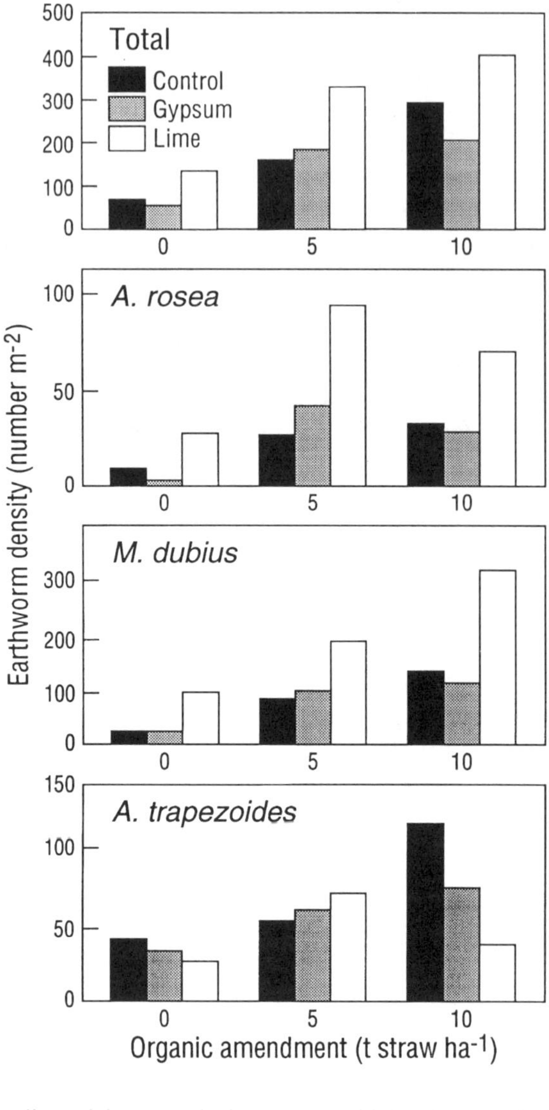

Fig. 11.2. The effect of the rate of wheat straw addition (0, 5 and 10 t ha^{-1}) and calcium (0, 2.0 t ha^{-1} $year^{-1}$ lime, 0, 2.4 t ha^{-1} $year^{-1}$ gypsum) on the abundance of three species of earthworm in a trial conducted in a red-brown earth at the Waite Agricultural Research Institute (from Doube *et al.*, 1995).

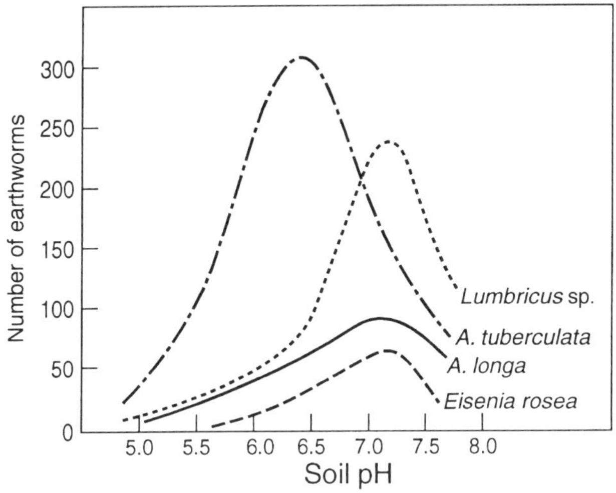

Fig. 11.3. A schematic illustration of the effect of soil pH on the relative abundance of four species of earthworm in experimental plots at Bingley, UK (redrawn from Jefferson, 1956).

et al., 1976) and Blackshaw (1990) contends that the New Zealand planarian, *Artioposthia triangulata* was responsible for the total elimination of earthworm populations from a grass field near Belfast, Northern Ireland. This species threatens earthworm populations across northern Europe (Boag *et al.*, 1995). Clearly earthworm abundance will be a poor indicator of soil productivity in the presence of the flatworm or other effective predators.

The effect of agricultural practices on earthworm abundances has recently been reviewed by Fraser (1994). She concludes that the influential factors are pastoral management techniques, the sequence of crop rotation, the timing, frequency and type of cultivation used, irrigation, whether or not mulched or crop residues are returned to the soil and the use of agricultural fertilizers and/or pesticides. Commonly the extreme treatments induce a 2- to 5-fold change in earthworms abundance, but there is no evidence for corresponding changes in crop productivity.

The speed with which earthworm populations respond to favourable conditions is a vital aspect of their capacity to act as indicators. Some earthworms have low fecundity and slow growth rates (e.g. the giant Gippsland earthworm). By contrast, large changes in the abundance of the peregrine agricultural species can occur within one season. Given favourable conditions, single earthworms can produce up to several hundred cocoons per year and juveniles can reach maturity within a few months (Edwards and Bohlen, 1996). Further cocoons may remain viable in the soil, even under adverse conditions, for several years

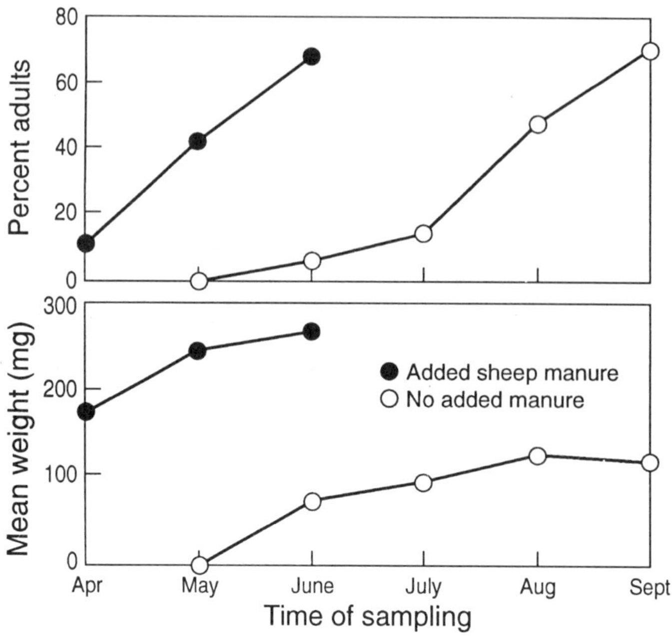

Fig. 11.4. The effect of added sheep manure on the growth and maturity of the earthworm *Microscolex dubius* in the field in the Adelaide Hills, South Australia (from Hughes *et al.*, 1994b).

(Edwards and Bohlen, 1996). Clearly peregrine earthworm species have the potential to respond rapidly when conditions for growth and reproduction change (Fig. 11.4). For example, within half a season (8 weeks) the numbers of three peregrine species had increased 2- to 10-fold following organic enrichment of the soil (Hughes *et al.*, 1994a). Similarly, Buckerfield (1996) found that earthworm populations in potato growing soils were reduced in the year immediately following potatoes but recovered to densities greater than those in adjacent pastures in the following year. Boström and Lofs-Holmin (1988) found that earthworm populations were severely reduced by rotary hoeing but their numbers had recovered one year later. In Lismore, New Zealand (Francis and Knight, 1993) and Temora, Australia (Doube *et al.*, 1994c) initial earthworm numbers following ploughing of pasture were high (850 and 410 m^{-2}, respectively) but populations decreased by 50–70% in the following 1–2 years, presumably due to a reduced food supply (Fig. 11.5) (Gupta and Kirkegaard, personal communication).

From the above it is clear that earthworm populations respond positively to many of the conservation practices in agriculture and have a capacity to respond in a time scale appropriate for an indicator (1–2 years). Numerical responses in earthworms are more intense than crop responses to environmental changes and

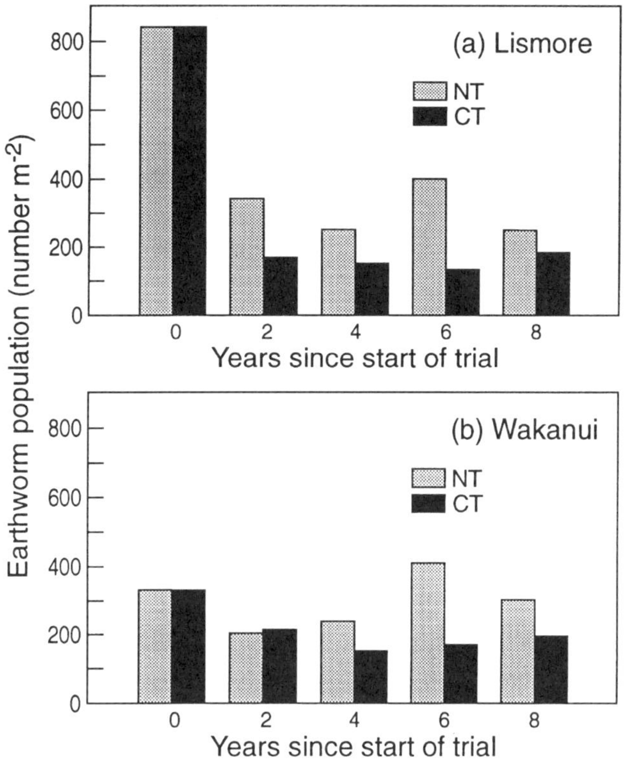

Fig. 11.5. The effects of cultivation and time on populations of earthworms at two locations on the Canterbury Plains, New Zealand (from Francis and Knight, 1993). NT, no tillage; CT, conventional tillage.

so measures of earthworm abundance could provide a sensitive indicator of soil health.

Linkages between Earthworm Abundance, Soil Properties and Plant Growth

Evaluation of the links between earthworm abundance and plant productivity (an index of soil health) is possible using data from long term field trials found in agricultural regions throughout the world. These provide ideal situations in which to investigate the linkages between soil biological parameters and soil health because long term measures of productivity and the biological condition of the soil are often available and comparisons are not confounded by differences in soil type or climatic conditions. In this next section we consider data from long term trials in North America, Europe and Australasia, with special emphasis on agronomic situations.

The majority of these trials include treatments which can be termed conservation management. These include treatments such as no tillage (or minimum tillage), high levels of organic residue inputs (e.g. retention of stubble or addition of animal manures), liming of acidic soils, the use of slow-release organic fertilizers, etc. All of these practices improve the sustainability of agriculture (though they may not necessarily increase short-term yields) through a wide variety of mechanisms including reduced surface soil erosion, increased water storage and infiltration rates, improved soil structure and decreased incidence of root diseases (Kladivko and Temmenga, 1990; Roget, 1995). Many of these practices also cause a corresponding increase in earthworm populations due to increased availability of organic residues (food). Intuitively, this suggests that earthworms might have the capacity to function as indicators of sustainable land-use practices.

Agronomic systems

In stable native vegetation ecosystems, the links between normal soil processes and soil fertility are likely to remain intact (biological fertility) whereas in highly disturbed environments the normal linkages between soil biological processes (nutrient mineralization, nitrification, etc.) and soil fertility and plant growth/productivity may become uncoupled (e.g. by added fertilizer, i.e. chemical fertility) so that plant growth is stimulated without the corresponding increases in the biological functions of soil (microbial biomass, etc.). However, in persistently disturbed environments it seems reasonable that the linkages between soil biological processes and soil fertility might become re-established over time thereby restoring soil health.

In the following ten case studies, each largely within one soil type, we consider the capacity of earthworms to act as indicators of crop productivity and sustainable land-use practices. The emphasis on productivity (rather than sustainability) is a consequence of the type of data available and the difficulty in measuring sustainability in regions with strong year-to-year variability in climate.

Case study 1. A clay loam at Woodslee, Ontario, Canada

The trial examined the effects of crop rotation (continuous corn, soybeans and wheat, and rotations of corn–soybean, wheat–soybean and wheat–corn–soybean) and weed control treatments on grain yields (established in 1982). Earthworms were sampled (formalin extraction) on four occasions during 1990–1991

(Tomlin *et al.*, 1995). In the spring of both years very few earthworms were collected (due to dry conditions) but autumn samples gave moderate numbers of *Lumbricus terrestris* and *Aporrectodea turgida* (range 2–38 m^{-2} and 21–59 m^{-2} respectively, and a total liveweight of 11–42 g m^{-2}; Tomlin *et al.*, 1995). Earthworm populations were not affected by weed control measures (including the herbicide, atrazine). The numbers and biomass of *L. terrestris* were significantly lower in one treatment (continuous soybean) on only one sampling occasion and the abundance of *A. turgida* was unaffected by rotation. Crop yields are commonly reduced by competition with weeds. Although earthworms were considered to benefit soil structure (hence sustainability), there was no consistent association between their numbers (or biomass) and plant productivity.

Case study 2. A silt-loam at Wayne County, Ohio, USA

The effects of earthworms on corn yield in a highly productive soil was examined over a 3-year period by artificially elevating and decreasing earthworm abundances associated with three fertilizer treatments (160 kg inorganic N ha^{-1}, a legume/rye cover crop, added cow manure) (Blair *et al.*, 1995). Substantial amounts of corn stubble were returned to the soil annually. In 1991/92, earthworm populations (mainly *L. terrestris* and *Aporrectodea tuberculata* with some *Lumbricus rubellus*) were decreased by 67% (279 m^{-2}) using electroshocking, increased by 12% by adding large numbers of field-collected earthworms (245 m^{-2}) or left unmodified (587 m^{-2}, about 35 g m^{-2} live biomass) (Blair *et al.*, 1995). Similar effects were recorded in 1993 (Bohlen *et al.*, 1997). Field sampling and preliminary experiments indicate that no 'non-target' soil invertebrates were affected by electroshocking (Blair *et al.*, 1995). Earthworm numbers and biomass were reduced in the inorganic-N treatments compared with those receiving organic nutrient inputs. Despite the persistent differential in earthworm populations, there was no significant effect of earthworms on corn production (1992–1995) (C.A. Edwards and R.W. Parmelee, personal communication). Obviously earthworm numbers and biomass were poor indicators of crop production but they had a positive influence on soil structure.

Case study 3. A sandy clay loam, Horshoe Bend, Georgia, USA

The effects of conventional tillage (CT) and no tillage (NT) on detrital food webs were examined (Coleman *et al.*, 1994). Earthworms, principally *L. rubellus* and *Aporrectodea caliginosa*, were about five times more abundant in the NT soils (*c.* 1000 m^{-2}), accounted for up to 30% of the heterotrophic respiration at these sites (Coleman *et al.*, 1994) and had a significant effect on the comminution of the coarse particulate organic matter in these soils (Parmelee *et al.*, 1990). Earthworms comprised 2.2% (NT) and 13.5% (CT) of the total carbon in the soil biomass (Brussaard *et al.*, 1990). In contrast, crop yields were substantially lower in the NT than in the CT treatments, due largely to strong competition from weeds in the NT treatments (P.F. Hendrix, personal communication). Earthworm abundance or biomass was not a useful indicator of crop yield, despite contributing to the superior structure of the NT soil (and its sustainability).

Case study 4. Silt-loams in Wisconsin, USA

The effects of tillage (mouldboard ploughing, chisel-discing and no-tillage) and corn residue management (removed, unaltered and doubled stubble) on crop production and earthworm abundance were examined in a trial (established 1981) in productive silt-loam soils in Wisconsin (Karlen *et al.*, 1994a,b). After 10–12 years, in an attempt to identify indicators of soil quality, earthworm abundance (by formalin extraction), microbial biomass and soil respiration were assessed. Earthworm (unnamed) abundance was increased by *c.* 50% in the added stubble treatments (which did not differ from each other at 78 m^{-2}) and were higher in the no tillage treatments (78 m^{-2}) than in the moldboard (53 m^{-2}) or chisel-disc (52 m^{-2}) treatments. The earthworm response to added stubble (+40%) and no-till (+40%) was much more muted than that of microbial biomass C (*c.* 3-fold increases) and soil respiration (*c.* 5-fold increases) (Karlen *et al.*, 1994a,b). Grain yields varied with season from 4 to 11 t ha^{-1}, and the long-term average ranged from 8.0 to 8.6 t ha^{-1} depending on treatment. In the no till treatments, the presence of stubble decreased crop yield in two years early in the trial, increased crop yield in one year later in the trial, and had no effect at other times, including the time of earthworm sampling. No-till practices improved the biological, chemical and physical condition of the silt-loam soil (Karlen *et al.*, 1994b) but did not increase grain yields. Clearly, earthworm numbers did not predict yield even though the presence of earthworms was correlated with improvements in soil parameters.

Karlen *et al.* (1994a) developed a soil quality index that gave ratings of 0.45, 0.68 and 0.86 in removal, normal and double residue treatments and 0.49, 0.48, 0.68 for plough, chisel and no-till treatments, respectively (Karlen *et al.*, 1994a). Earthworm abundance was correlated with the indices from the tillage but not from the stubble treatments. Neither earthworm abundance nor the soil quality index was correlated with grain yield.

Case study 5. Silty clay loams in Iowa, USA

The effects of tillage on grain yield and earthworms were examined (Berry and Karlen, 1993). Five species of earthworms were recovered and total numbers varied from 34 to 192 m^{-2}, depending upon occasion and treatment (*Octalasion tytaeum* was the most abundant species in all treatments). Earthworm densities generally decreased as the amount of tillage increased but tillage had little effect on plant productivity. The responses to treatments and climate differed between species, with *A. tuberculata* being least variable, and *A. trapezoides* being most variable. The ratio of maximum:minimum mean abundance levels was 25 for *A. trapezoides,* 20 for *Aporrectodea turgida,* 10 for *L. terrestris,* 14 for O. *tytaeum,* and 3 for *A. tuberculata.* This differs from Tomlin (personal communication) who found in Canada that *A. turgida* appeared to be resistant to cultivation and disturbance.

Earthworm abundances were also compared in nearby conventional and alternative farms on silty clay loams (Berry and Karlin, 1993). *Octalasion tytaeum* and *A. turgida* were absent, *A. rosea* was present (along with *L. terrestris*) and *A. tuberculata* predominated (*c.* 90% of total abundance). Both farms produced reasonable crops but earthworms were virtually absent (0–9 m^{-2}) from the conventional farming systems while high densities (up to 837 m^{-2}) were observed in the alternative system, which were attributed to the addition of 45 t ha^{-1} of manure and municipal sludge during the previous three years. Further, the total numbers in the alternative system decreased nearly 5-fold over the study period (1989–1990).

Clearly there was no way to relate earthworm abundance or species composition to plant productivity in any of these examples, nor was one system clearly more sustainable than another, despite very clear differences in earthworm numbers.

Case study 6. Acidic red-brown earth soils in South Australia

Three rotations (wheat–fallow (WF), wheat–pasture (WP), permanent pasture (PP)) in the Waite Long-term Trial (established 1925) (Grace *et al.*, 1995) were sampled for earthworms 26 years apart (1956/57, Barley, 1959b; Buckerfield, 1993): treatment effects were similar on both occasions. Three species (*A. trapezoides, A. rosea, Microscolex dubius)* were present, with *A. rosea* making up 50–60% of the total biomass of earthworms (Barley, 1959b). *Microscolex dubius* was virtually absent from the pasture plots, but made up to 20% of the biomass on some cultivated plots (Barley, 1959b). In 1956/57 total earthworms in WF (20 m^{-2}, 2 g m^{-2}) were substantially lower than in the WP and PP (500–600 m^{-2}, 75–80 g m^{-2}). In 1957, wheat yields were 50% greater in the WP than the WF rotation. Over the duration of the trial, inclusion of a short term pasture phase in the rotation maximized wheat yields (Grace *et al.,* 1995) and also increased earthworm numbers (Barley, 1959b; Buckerfield, 1993). A similar result was obtained by Rovira *et al.* (1987) in a red-brown earth at Kapunda, South Australia. In a trial at Tarlee, South Australia (Schultz, 1995), the wheat phase of three rotations (wheat–wheat, wheat–beans and wheat–pasture) was sampled for earthworms (J.C. Buckerfield and K. Webster, personal communication). Neither total numbers nor biomass were affected by rotation.

However, there were statistically significant species-specific patterns. *A. trapezoides* was most numerous under wheat–wheat, *A. rosea* was most numerous under wheat–pasture and *M. dubius* was most abundant under wheat–beans, but the reasons for this are not clear. The best wheat yields were always in rotations that included a grain legume or a legume pasture (Schultz, 1995).

Earthworm populations under direct drill were nearly double those under conventional cultivation (Rovira *et al.,* 1987), but there were no corresponding trends in grain yields (D. Roget, personal communication). In a trial at nearby Tarlee (Schultz, 1995), tillage had no effect on earthworm numbers (Pankhurst *et al.*, 1995) or grain yield (Schultz, 1995). In contrast, Buckerfield (1996) surveyed nearby paddocks and found that cultivation was significantly associated with reduced earthworm density ($r = -0.70$) and reduced grain yields ($r = -0.61$) and earthworm density ($r = 0.48$), grain yield (0.79) and grain protein ($r = 0.77$) increased with added N-fertilizer. However, the mean earthworm density for trial plots (Buckerfield, 1993, 1996) were 5–20 m^{-2} (Tarlee, crop), 40–130 m^{-2} (Halbury, crop), 210–800 m^{-2} (Kapunda, crop), 50–180 m^{-2} (Back Valley, pasture) and 140–280 m^{-2} (Birdwood, pasture): clearly there is no common base line density.

The effects of stubble management and fertilizer in the Tarlee trial were reported by Pankhurst *et al.* (1995) and the numbers and biomass of earthworms (5–20 m^{-2}) were reduced (by *c.* 60%) by removal of stubble although stubble management had little effect on grain yields (Schultz, 1995). Additional yield increases in all rotations came from the use of N-fertilizer, but over the course of the experiment this caused a substantial decrease in soil pH (up to 1 pH unit) with a corresponding decrease in the suitability of the soil for the growth of earthworms (Fig. 11.6).

Varying the levels of lime (but not gypsum) and organic matter (cereal straw) added to plots in a red-brown earth (adjacent to the Waite trial, Doube *et al.*, 1995) changed both earthworm abundance and the balance of species (Fig. 11.2). Four species (*A. trapezoides*, *A. rosea*, *M. dubius* and *M. phosphoreus*) were recovered. In the absence of ameliorants, the earthworm population (*c.* 80 m^{-2}) was dominated by *A. trapezoides* (73% of total biomass) but the addition of lime (which increased soil pH from 4.1 to 6.7) and straw resulted in 4-fold increases in the abundances of both *A. rosea* and *M. dubius* which together made up 80% of the total biomass in the limed plots with 10 t straw ha^{-1}.

From these studies it is clear that earthworm populations increase in acidic red-brown earths when organic matter or lime (but not gypsum) were added, when tillage was minimized and when a pasture phase was included in the rotation sequence. These procedures are considered to promote sustainability but were not necessarily linked to increased productivity. Direct linkages between earthworm abundance and grain yield were not evident but higher numbers in one locality may indicate the more sustainable farming practices.

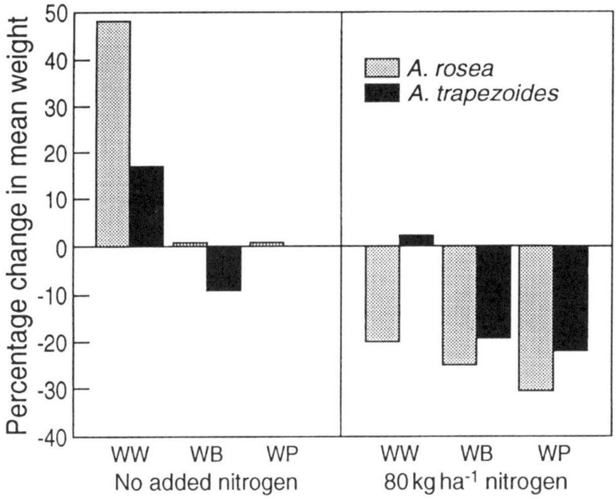

Fig. 11.6. The effect of fertilizer and rotation on the growth of two species of peregrine lumbricid in a red-brown earth at Tarlee. Growth of juvenile earthworms (after 9 weeks at 20°C) was assessed in the laboratory in soil taken from long-term field trials at Tarlee South Australia. (WW, wheat–wheat; WB, wheat–beans; WP, wheat–pasture) (B.M. Doube, unpublished).

Cast study 7. Red-brown earth at Temora and Harden, New South Wales, Australia (established 1990)

The effects of tillage and stubble management practices on soil structure and crop yield were examined at Temora and Harden, New South Wales, Australia (Doube *et al.*, 1994c; Kirkegaard *et al.*, 1994). The introduced lumbricid *A. trapezoides* was the dominant species at both Harden (99.4%) and Temora (97.0%): four other species were also recovered (*A. rosea, M. dubius, M. phosphoreus* and a native species) (Doube *et al.*, 1994c). In the first year of the trial *A. trapezoides* was substantially more abundant at Temora (425 m^{-2}, 115 g m^{-2}) than at Harden (84–146 m^{-2}, 34–62 g m^{-2}) but over the subsequent three years the earthworm densities at Temora decreased to about 150 m^{-2} (V.V.S.R. Gupta and J.A. Kirkegaard, personal communication) while those at Harden remained largely unaltered.

At Harden, reduced seedling growth and yield of wheat were associated with conservation tillage in the first year of the trial (1990) (Kirkegaard *et al.*, 1994). In contrast, the direct drilled plots had a greater population density and biomass of earthworms and cocoons than did tilled plots, and the adult earthworms were larger as was the number of cocoons per adult (Doube *et al.*, 1994c). Clearly, conservation tillage practices improved soil conditions for *A. trapezoides* and yet wheat yields were reduced.

Waterlogging at Temora had a major effect on canola yields but little effect on earthworm numbers (Doube *et al.,* 1994c). A subjective index of the effect of waterlogging on the canola plants (0 = no effect, high yield to 5 = severe stunting, no yield, plant death) was used in relation to earthworm and cocoon abundance. Severe waterlogging reduced earthworm numbers by 50% (to about 280 m^{-2}) only in the most extreme conditions but these conditions killed most canola plants (Doube *et al.*, 1994c).

Case study 8. Silt-loam soils on the Canterbury Plains, New Zealand

The effects of mixed cropping practices (alternating grazing and cropping in 3-year phases) on grain yields were examined (Francis and Knight, 1993; Fraser *et al.*, 1993; Haynes *et al.*, 1993; Fraser, 1994) (trials established 1978, 1980). At one site (Wakanui) cultivation of a previously cropped soil did not significantly alter earthworm populations. At a second site (Lismore), cultivation of an area previously under grazed pasture resulted in a pronounced reduction (over two years) in earthworm populations from *c.* 830 m^{-2} under grazed pasture to about 350 m^{-2} under a direct drilled treatment and 180 m^{-2} under conventional cultivation. Similarly, when an arable field with a population of <100 m^{-2} was converted to a grazed grass–clover pasture, the population had increased to between 400 m^{-2} and 600 m^{-2} within two years (Fraser, 1994). Haynes *et al.* (1993) observed that populations averaged about 800–900 m^{-2} under long term pasture, less than 200 m^{-2} under long term arable and about 500 m^{-2} under rotations of three years arable followed by three years pasture (Fraser, 1994). These practices clearly produce rapid and large systematic fluctuations in the numbers of earthworms, probably related to changes in their food supply (Fraser, 1994). Although earthworm numbers and grain yields tended to be higher under no till than under conventional cultivation (Francis and Knight, 1993), there was no correspondence between earthworm abundance and plant productivity.

Case study 9. A calcareous marine silt-loam, Noordoospolder, Netherlands

Soil organisms in integrated and conventional cropping systems were studied in reclaimed polders (Brussaard *et al.*, 1990; Brussaard, 1994). Earthworms were not recovered from the conventional farming plots, and formed a minor part of the soil fauna in the integrated plots (1.65% of the total organism biomass carbon) and they are known to improve the structure and fertility of polder soil (Lee, 1985). Nevertheless, both farming systems were productive. Clearly, earthworms were of little value as indicators of soil health.

Case study 10. Several soil types in England and Ireland

Earthworm populations in a conventional winter wheat production system were compared with those in a novel low-input bicrop system (perennial white clover under wheat) in field-scale plots at two sites in England and two in Ireland (Clements, 1996). Earthworms were sampled in spring and autumn using the electrical octet method, and monthly at one Irish site by handsorting (O. Schmidt, unpublished data). Grain yields in the first season ranged from 5.6 to 8.6 t ha^{-1} in the conventional system and from 2.0 to 4.4 t ha^{-1} in the bicrop system. Similarly, the conventional system yielded more whole plant silage than the bicropped fields (3.86–16.40 t ha^{-1} and 9.50–12.00 t ha^{-1}, respectively). In contrast, earthworm densities under no-tillage bicropping ranged from 278 to 484 individuals m^{-2} and were 2- to 11-fold greater than under conventional cropping. Similarly, earthworm biomass was 3- to 7-fold greater (48.8 and 148.7 g m^{-2}) under the bicrop system than under conventional management. Clearly, in the short term, a dramatic increase in earthworm populations coincided with reduced productivity in the low-input production system. Nevertheless, earthworms have an important function in this system (e.g. incorporating large amounts of plant residues into the soil preventing accumulation of litter mat, alleviating soil compaction, promoting nitrogen mineralization) and so are important biological indicators of soil health in this system.

Pasture systems

Much has been written about earthworms in pastures and there are outstanding examples of earthworm-induced permanent increases in plant productivity (hence soil health) which have followed the introduction of peregrine species to polders in the Netherlands, to reclaimed peat bogs in Ireland and to pastures in New Zealand and Tasmania (Lee, 1985; Edwards and Bohlen, 1996; Temple-Smith and Pinkard, 1995). Furthermore, there are numerous examples where liming of acidic pasture soil, adding fertilizer (both N and P, Fraser, 1994), or adding organic residues (e.g. dung) can result in increased earthworm densities (Lee, 1985; Fraser, 1994; Edwards and Bohlen, 1996). However, there are also numerous instances in which earthworm populations are unaffected by such ameliorants, or where species are affected differentially. Furthermore, earthworm populations in apparently healthy pasture soils vary enormously (> 100 fold) between localities and so there is clearly no base line data which could be viewed as a threshold level for a healthy soil. Thus, while earthworms may promote soil health, their abundance cannot be used as an indicator of soil health.

Bioaccumulation and ecotoxicology

The use of earthworms as biomonitors of soil toxicity and bioaccumulators of pollutants is considered by Ebing and Haque (1979), Greig-Smith *et al.* (1992), and others. Importantly, these organisms, like protozoa (Gupta and Yeates, Chapter 9, this volume), will measure levels of bioavailable pollutant, rather than absolute levels in the soil. Functional differentiation within groups of earthworms (Lee, 1985) provides the capacity to use different types of the same animal group to monitor pollutants in different soil niches. For example, epigeic (surface living) earthworms are directly exposed to biocides and contaminants applied to the soil surface, whereas endogeic (soil-dwelling) species are exposed to accumulated substances in the soil matrix. Their capacity for bioaccumulation varies widely with earthworm and pollutant species (Greig-Smith *et al.*, 1992). The potential to damage members of higher trophic levels which feed upon earthworms (e.g. mammals, birds, reptiles and arthropods) is discussed by Lee (1985) and Edwards and Bohlen (1996) but is not considered further here.

The sensitivity of earthworms to agrochemicals in laboratory tests is used as a basis for determining whether the product can be registered for use in agriculture (Greig-Smith *et al.*, 1992). Earthworms are a particularly useful group because their relative sensitivity to toxic substances is higher than most for arthropods but lower than microorganisms, gastropods and molluscs (Edwards and Bohlen, 1996). Mortality of mature *Eisenia fetida,* determined as an LC_{50} (median lethal concentration), in the Artificial Soil Test (OECD, 1984, in Greig-Smith *et al.*, 1992) is common practice but gives little recognition to the higher sensitivity of juveniles or the benefits of measuring sublethal doses using ecologically relevant earthworm species. There is a need to develop new methods of increased sensitivity which are relevant to earthworms of agricultural soils.

Conclusions

Earthworms are well recognized as having the capacity to increase plant growth in temperate (Lee, 1985; Hendrix 1995; Edwards and Bohlen, 1996) and tropical environments (Tian *et al.*, 1993; Kang *et al.*, 1994) by favourably influencing a vast array of soil biological, physical and chemical factors. Such changes to soil characteristics are commonly considered to be associated with improved plant productivity and sustainability (soil health) but recent reviews (e.g. Kirkegaard, 1995) have challenged this belief. Similarly, we challenge the belief that earthworms can function as indicators of soil health.

Analysis of data from 33 agronomic experiments which measured trends in wheat yield responses to conservation cropping in Australia, (Kirkegaard, 1995) have shown that the effects of tillage (direct drilled vs cultivated) were small in all regions (–0.18 to +0.06 t ha^{-1}) while stubble retention reduced yields in all regions (–0.31 to –0.02 t ha^{-1}). There was also little evidence that the yield of direct drilled or stubble-retained treatments increased relative to cultivated or stubble-burnt treatments, despite the improvement in soil conditions reported at many sites (Kirkegaard, 1995). In contrast, earthworm numbers commonly respond to these treatments, but not always in a consistent manner (see above case studies).

There are numerous instances where the same species of earthworm can have positive, negative or no effect on plant performance, depending upon circumstances (Edwards and Bohlen, 1995; Temple-Smith and Pinkard, 1995; Doube *et al.*, 1996). Furthermore, there are highly productive soils (e.g. kraznozems, Wylie, 1994) which do not contain earthworms. Thus it is clear that earthworms are not a necessary component of productive and sustainable systems, even though their activity results in soil characteristics which promote soil health.

Positive effects of earthworms on plant growth occur only in circumstances in which constraints to plant growth are alleviated by earthworm activity. This is not always so. The agencies responsible for soil health are multi-factorial and individual factors are rarely likely to have an overriding effect and thereby have the capacity to predict the health of soils. We hold that earthworms are likely to be important components of the soil fauna in many locations and that their activity generally (but not always) contributes to the health of soils, but they are of little use as indicators of soil health.

One test of the usefulness of a parameter is to consider how one might interpret a hypothetical set of data. Large numbers of peregrine earthworms in a soil sample suggests a recent history of substantial inputs of crude organic residue in a moist soil of near-neutral pH. These are not prerequisite conditions for healthy soils. Similarly, low earthworm numbers might indicate a number of conditions (e.g. high or low soil pH, shallow sandy soil, low organic inputs) and again these are not prerequisites for unhealthy soils.

It is clear that earthworm abundance cannot be used as a universal indicator of soil health because the key agronomic factors which determine plant yield and soil conservation are not necessarily those which influence earthworm abundance. There are some factors (e.g. chemical fertilizer, waterlogging, root disease, etc.) which can be of overriding importance in determining yield and which do not have a corresponding effect on earthworm abundance. Conversely, there are other factors (organic residues, soil pH, rotations) which can be of overriding importance in determining earthworm abundance and which do not have a corresponding effect on yield. Clearly, there are also instances in which management practices promoted both earthworm abundance and soil health, but there are many other examples when the two were disassociated. Further, the development of useful reference values (base line levels of earthworm abundance or biomass) is extremely unlikely because of the enormous variability in abundance between locations and with management.

Perhaps the requirement for strong positive linkages between a biological parameter (e.g. earthworm abundance or biomass) and crop productivity is too strict a demand for a biological indicator of soil health (or quality). After all, there is abundant evidence of the beneficial effects of earthworms in agriculture (see above) and many of these benefits are positively associated with factors which promote soil health. Others have suggested that earthworms should be one of a suite of soil parameters used to assess soil health, and this may be a useful step which recognizes the complexity of the biological processes leading to healthy soils.

References

Abbott, I., Parker, C.A. and Sills, I.D. (1979) Changes in abundance of large soil animals and physical properties of soil following cultivation. *Australian Journal of Agricultural Research* 30, 343–353.

Atlavynite, O. and Pociene, C. (1973) The effect of earthworms and their activity on the amount of algae in the soil. *Pedobiologia* 13, 445–455.

Baker, G.H. and Lee, K.E. (1993) Earthworms. In: Carter, M.R. (ed.), *Soil Sampling and Methods of Analysis.* Canadian Society of Soil Science, Lewis Publishers, Boca Raton, Florida, pp. 359–371.

Baker, G.H., Carter, P.J., Barrett, V.J., Kilpin, G.C., Buckerfield, J.C. and Dalby, P.R. (1994) The introduction and management of earthworms to improve soil structure and fertility in south-eastern Australia. In: Pankhurst, C.E., Doube, B.M., Gupta, V.V.S.R. and Grace, P.R. (eds) *Soil Biota: Management in Sustainable Farming Systems.* CSIRO Press, Melbourne, Australia. pp. 42–49.

Barley, K.P. (1959a) The influence of earthworms on soil fertility. II. Consumption of soil and organic matter by the earthworm *Allolobophora caloginosa* (Savigny). *Australian Journal of Agricultural Research* 10, 179–185.

Barley, K.P. (1959b) The influence of earthworms on soil fertility. I. Earthworm populations found in agricultural land near Adelaide. *Australian Journal of Agricultural Research* 10, 171–178.

Bengston, S.A., Nilsson, A., Nordstrom, S. and Rundgren, S. (1976) Effects of bird predation on lumbricid populations. *Oikos* 27, 9–12.

Berry, E.C. and Karlen, D.L. (1993) Comparison of alternative farming systems: II Earthworm population density and species diversity. *American Journal for Alternative Agriculture* 8, 21–26.

Blackshaw, R.P. (1990) Studies on *Artioposthia triangulata* (Dendy) (Trichladida: Terricola) *Annals of Applied Biology* 118, 689–694.

Blair, J.M., Bohlen, P.J., Edwards, C.A., Stinner, B.R., McCartney, D.A. and Allen, M.F. (1995) Manipulation of earthworm populations in field experiments in agroecosystems. *Acta Zoologica Fennica* 196, 48–51.

Boag, B., Evans, K.A., Neilson, R., Yeates, G.W., Johns, P.M., Mather, J.G., Christensen, O.M. and Jones, H.D. (1995) The potential spread of terrestrial planarians *Artioposthia triangulata* and *Australoplana sanguinea* var *alba* to continental Europe. *Annals of Applied Biology* 127, 385–390.

Bohlen, P.J., Parmelee, R.J., Blair, J.M., Edwards, C.A. and Stinner, B.R. (1996) Responses of earthworm populations to experimental manipulation in agroecosystems receiving organic or inorganic nutrient inputs. *Soil Biology and Biochemistry* (in press).

Boström, U. and Lofs-Holmin, A. (1988) Earthworm population dynamics and flows of carbon and nitrogen through *Aporrectodea caliginosa* (Lumbricidae) in four cropping systems. In: *Ecology of Earthworms in Arable Land: Population Dynamics and Activity in Four Cropping Systems.* Institutionen for ekologi och miljovard, Report no. 34. Swedish University of Agricultural Sciences, Uppsala.

Bouche, M.B. (1977) Strategies lombriciennes. In: Lohm, U. and Persson, T. (eds) *Soil Organisms as Components of Ecosystems Ecological Bulletin* (Stockholm) 25, 122–132.

Brussaard, L., Bouwman, L.A., Geurs, M., Hassink, J. and Zwart, K.B. (1990) Biomass, composition and temporal dynamics of soil organisms of a silt loam soil under conventional and integrated management. *Netherlands Journal of Agricultural Science* 38, 283–302.

Brussaard, L. (1994) The ecology of soil organisms in reduced input farming. In: Pankhurst, C.E., Doube, B.M., Gupta, V.V. S.R. and Grace, P.R. (eds) *Soil Biota: Management in Sustainable Farming Systems.* CSIRO Press, East Melbourne, Australia. pp. 197–203.

Buckerfield, J.C. (1993) Pastures in crop rotations encourage earthworms. In: *Tillage Systems, Rotations, Nutrition and Associated Root Diseases.* Report of a Workshop, CRC for Soil and Land Management, Adelaide. pp. 66–67.

Buckerfield, J.C. (1996) Earthworms as indicators of sustainable production. *Ecological Economics* 7, 333–339.

Buckerfield, J.C. and Auhl, L.H. (1994) Earthworms as indicators of sustainable production in intensive cereal cropping. In: Pankhurst, C.E. (ed.) *Soil Biota: Management in Sustainable Farming Systems, Poster Papers.* CSIRO Press, East Melbourne, Australia. pp. 169–172.

Clements, R.O. (1996) (ed.) *Exploitation of a Sustainable Low-input and Reduced-output System for Arable Crops.* Second 12 months report to EU, Contract No. AIR 3 CT93-0893, 350 pp.

Coleman, D.C., Hendrix, P.F., Beare, M.H., Crossley, D.A., Hu, S. and van Vleit, P.C.J. (1994) The impact of management and biota on nutrient dynamics and soil structure in subtropical agro-ecosystems: Impacts on detrital food webs. In: Pankhurst, C.E., Doube, B.M., Gupta, V.V.S.R. and Grace, P.R. (eds) *Soil Biota: Management in Sustainable Farming Systems.* CSIRO Press, East Melbourne, Australia. pp. 133–143.

Curry, J.P. (1994) *Grassland Invertebrates. Ecology, Influence on Soil Fertility and Effects on Plant Growth.* Chapman and Hall, London, 437 pp.

Curry, J.P. (1996) Factors affecting earthworm abundance. In: *The Ecology of Earthworms* (in press).

Curry, J.P. and Boyle, K.E. (1995) Restoring soil organisms in reclaimed land, with particular reference to earthworms in cutover peat. *Acta Zoologica Fennica* 196, 371–375.

Doran, J.W. and Parkin, T.B. (1994) Defining soil quality. In: Doran, J.W., Coleman, D.C., Bezdicek, D.F. and Stewart, B.A. (eds) *Defining Soil Quality for a Sustainable Environment.* Soil Science Society of America Special Publication no. 35, Madison, Wisconsin, pp. 3–21.

Doran, J.W., Sarrantonio, M. and Janke, R. (1994) Strategies to promote soil quality and soil health. In: Pankhurst, C.E., Doube, B.M., Gupta, V.V.S.R. and Grace, P.R. (eds) *Soil Biota: Management in Sustainable Farming Systems.* CSIRO Press, East Melbourne, Australia, pp. 230–237.

Doube, B.M. (1990) A functional classification for the analysis of the structure of dung beetle assemblages. *Ecological Entomology* 15, 371–383.

Doube, B.M. and Brown, G.G. (1996) Life in a complex community: functional interactions between earthworms, organic matter, microorganisms and plants. In: *The Ecology of Earthworms* (in press).

Doube, B.M., Stephens, P.M., Davoren, C.W. and Ryder, M.H. (1994a) Interactions between earthworms, beneficial soil microorganisms and root pathogens. *Applied Soil Ecology* 1, 3–10.

Doube, B.M., Davoren, C.W., Hindell, R., Long, N. and Cass, A. (1994b) Surface cast production by the earthworm *Aporrectodea trapezoides* and the influence of organic matter on cast structural stability. In: Pankhurst, C.E. (ed.) *Soil Biota: Management in Sustainable Farming Systems, Poster Papers.* CSIRO Press, East Melbourne, Australia, pp. 123–125.

Doube, B.M., Buckerfield, J.C. and Kirkegaard, J.A. (1994c) Short-term effects of tillage and stubble management on earthworm populations in cropping systems in southern New South Wales. *Australian Journal of Agricultural Research* 45, 1587–1600.

Doube, B.M., Buckerfield, J.C., Barrett, V.J., Carter, P.J., Baldock, J.A. and Oades, J.M. (1995) The effects of lime, gypsum, and wheat straw on four species of earthworms in a red-brown earth in South Australia. *Acta Zoologica Fennica* 196, 150–155.

Doube, B.M., Williams, P.M.L. and Willmott, P.J. (1997) The influence of two species of earthworm (*Aporrectodea trapezoides* and *Aporrectodea rosea*) on the growth of wheat, barley and faba beans in three soils in the greenhouse. *Soil Biology and Biochemistry* (in press).

Ebing, K.W. and Haque, A. (1979) Summarising report of previous studies concerning earthworms to test the ecological effects of organic chemicals on soil organisms. *3rd Meeting of the OECD Expert Group C 'Degradation/Accumulation'*, Tokyo.

Edwards, C.A. (1974) The invertebrate fauna of the Park Grass Plots. *Rothamsted Report* for 1974, Part 2, pp. 133–145.

Edwards, C.A. and Bohlen, P.J. (1996) *Biology and Ecology of Earthworms.* Chapman and Hall, London.

Francis, G.S. and Knight, T.L. (1993) Long-term effects of conventional and no-tillage on selected soil properties and crop yield in Canterbury, New Zealand. *Soil and Tillage Research* 26, 193–210.

Fraser, P.M. (1994) The impact of soil and crop management practices on soil macrofauna. In: Pankhurst, C.E., Doube, B.M., Gupta, V.V.S.R. and Grace, P.R. (eds) *Soil Biota: Management in Sustainable Farming Systems.* CSIRO Press, East Melbourne, Australia. pp. 125–132.

Fraser, P.M., Haynes, R.J. and Williams, P.H. (1993) Effect of pasture improvement and intensive cultivation on size of microbial biomass, enzyme activities and composition and size of earthworm populations. *Biology and Fertility of Soils* 17, 185–190.

Gates, G.E. (1972) Burmese earthworms: an introduction to the systematics and biology of megadrile oligochaetes with special reference to Southeast Asia. *Transactions of the American Philosophical Society, New Series* 62, 1–326.

Grace, P.R., Oades, J.M., Keith, H. and Hancock, T.W. (1995) Trends in wheat yield and soil organic carbon in the permanent rotation trial at the Waite Agricultural Research Institute, South Australia. *Australian Journal of Experimental Agriculture* 35, 857–864.

Greig-Smith, P.W., Becker, H., Edwards, P.J. and Heimbach, R. (1992) *Ecotoxicology of Earthworms.* Intercept Ltd, Andover, UK.

Haynes, R.J., Francis, P.M. and Williams, P.H. (1993) Earthworm population size and composition and microbial biomass: effect of pastoral and arable management in Canterbury, New Zealand. In: *Functional Significance and Regulation of Soil Biodiversity. Proceedings of Symposium of the Soil Ecological Society,* Michigan State University, Kluwer Academic Publishers, Dordrecht.

Hemman, C. (1994) Untersuchungen zum Bestand an Lumbriciden in ackerbaulich genutzten Böden und deren Randbereichen. PhD Thesis, University of Halle-Wittenberg, Germany.

Hendrix, P.F. (1995) (ed.) *Earthworm Ecology and Biogeography in North America.* CRC Press, Florida.

Hendrix, P.F., Mueller, B.R. and Bruce, R.R. (1992) Abundance and distribution of earthworms in relation to landscape factors on the Georgia Piedmont, USA. *Soil Biology and Biochemistry* 24, 1357–1361.

Hughes, M.S., Bull, C.M. and Doube, B.M. (1994a) The use of resource patches by earthworms. *Biology and Fertility of Soils* 18, 241–344.

Hughes, M.S., Bull, C.M. and Doube, B.M. (1994b) The effect of sheep manure on the survival and growth of the earthworm *Microscolex dubius* (Annelida: Acanthodrilidae). *Applied Soil Ecology* 1, 291–298.

Jefferson, P. (1955) Studies on the earthworms of turf. A. The earthworms of experimental turf plots. *Journal of the Sports Turf Research Institute* 9, 6–27.

Jefferson, P. (1956) Studies on the earthworms of turf. B. Earthworms and soil. *Journal of the Sports Turf Research Institute* 10, 276–289.

Jenkinson, D.S. and Ladd, J.N. (1981) Microbial biomass in soil, measurement and turnover. In: Paul, E.A. and Ladd, J.N. (eds) *Soil Biochemistry,* Vol. 5. Marcel Dekker, Inc., New York. pp. 415–472.

Kang, B.T., Akinnifesi, F.K. and Pleysier, J.L. (1994) Effect of agroforestry woody species on earthworm activity and physicochemical properties of worm casts. *Biology and Fertility of Soils* 18, 192–199.

Karlen, D.L., Wollenhaup, N.C., Edbach, D.C., Berry, E.C., Swan, J.B., Eash, N.S. and Jordahl, J.L. (1994a) Crop residue effects on soil quality following 10 years of no-till. *Soil and Tillage Research* 31, 149–167.

Karlen, D.L., Wollenhaup, N.C., Edbach, D.C., Berry, E.C., Swan, J.B., Eash, N.S. and Jordahl, J.L. (1994b) Long-term tillage effects on soil quality. *Soil and Tillage Research* 32, 313–327.

Killham, K. (1994) *Soil Ecology.* Cambridge University Press, Cambridge, UK.

Kirkegaard, J.A. (1995) A review of trends in wheat yield responses to conservation cropping in Australia. *Australian Journal of Experimental Agriculture* 35, 835–848.

Kirkegaard, J.A., Angas, J.F., Gardner, P.A. and Muller, W. (1994) Reduced growth and yield of wheat with conservation cropping. I. Field studies in the first year of the cropping phase. *Australian Journal of Agricultural Research* 45, 511–528.

Kladivko, E.J. and Temmenga, H.J. (1990) Earthworms and agricultural management. In: Box, J.E. and Hammond (eds) *Rhizosphere Dynamics.* Westview Press Inc., Boulder, Colorado, pp. 192–216.

Knüsting, E. (1992) Regenwürmer auf Ackerflächen mit abgestufter Bewirtschaftungsintensität. PhD Thesis, University of Braunschweig, Germany.

Kretzschmar, A. (1991) Burrowing ability of the earthworm *Aporrectodea longa* limited by soil compaction and water potential. *Biology and Fertility of Soils* 11, 48–51.

Larson, W.E. and Pierce, F.J. (1994) The dynamics of soil quality as a measure of sustainable management. In: Doran, J.W., Coleman, D.C., Bezdicek, D.F. and Stewart, B.A. (eds) *Defining Soil Quality for a Sustainable Environment.* Soil Science Society of America, Special Publication no. 35, Madison, Wisconsin, pp. 37–51.

Lavelle, P. (1988) Earthworms activities and the soil system. *Biology and Fertility of Soils* 6, 237–251.

Lavelle, P., Spain, A.V., Blanchart, E., Martin, A. and Martin, p S. (1992) Impact of soil fauna on the properties of soils in the humid tropics. In: Sanchez, P.A. and Lal, R. (eds) *Myths and Science of Soils in the Tropics.* Soil Science Society of America, Special Publication no. 29, Madison, Wisconsin. pp. 157–185.

Lawton, J.H. (1994) What do species do in ecosystems? *Oikos* 71, 367–374.

Lee, K.E. (1985) *Earthworms: Their Ecology and Relationships with Soils and Land Use.* Academic Press, Sydney, 411 pp.

Lee, K.E., and Foster, R.C. (1991) Soil fauna and soil structure. *Australian Journal of Soil Research* 29, 745–775.

Lee, K.E. and Pankhurst, C.E. (1992) Soil organisms and sustainable productivity. *Australian Journal of Soil Research* 30, 855–892.

Linden, D.R., Hendrix, P.F., Coleman, D.C. and van Vleet, P. (1994) Faunal indicators of soil quality. In: *Defining Soil Quality for a Sustainable Environment.* Soil Science Society of America, Special Publication no. 39, pp. 91–106.

Mather, J.G. and Christensen, O. (1992) Surface migration of earthworms in grassland. *Pedobiologia* 36, 51–57.

Michaelsen, W. (1928) Oligochaeta. In: Kükenthal, W. and Krumbach, T. (eds) *Handbuch der Zoologie vol. II (2) pt 8.* Walter de Gruyter, Berlin and Leipzig, pp. 1–118.

Oades, J.M. and Walters, L.J. (1994) Indicators for sustainable agriculture: Policies to paddock. In: Pankhurst, C.E., Doube, B.M., Gupta, V.V.S.R. and Grace, P.R. (eds) *Soil Biota: Management in Sustainable Farming Systems.* CSIRO Press, East Melbourne, Australia. pp. 219–223.

Pankhurst, C.E., Hawke, B.G., McDonald, H.J., Kirkby, C.A., Buckerfield, J.C., Michelson, P., O'Brien, K.A., Gupta, V.V.S.R. and Doube, B.M. (1995) Evaluation of soil biological properties as potential bioindicators of soil health *Australian Journal of Agricultural Research* 35, 1015–1028.

Pankhurst, C.E., Ophel-Keller, K., Doube, B.M. and Gupta, V.V.S.R. (1996) Biodiversity of soil microbial communities in agricultural systems. *Biodiversity and Conservation* 5, 197–209.

Parmelee, R.W., Beare, M.H. and Cheng, W. (1990) Earthworms and enchytraeids in conventional and no-tillage agroecosystems: a biocide approach to assessing their role in organic matter breakdown. *Biology and Fertility of Soils* 10, 1–10.

Parr, J.F., Papendick, R.I., Hornick, S.B. and Meyer, M.E. (1992) Soil quality: attributes and relationship to alternative and sustainable agriculture. *American Journal of Alternative Agriculture* 7, 5–11.

Persson, T. and Lohm, U. (1977) Energetical significance of the annelids and arthropods in Swedish grassland soil. *Ecological Bulletins* 23, 1–211.

Petersen, H. and Luxton, M. (1982) A comparative analysis of soil fauna populations and their role in decomposition processes. *Oikos* 39, 287–388.

Roget, D.K. (1995) Decline in root rot (*Rhizoctonia solani* AG-8) in wheat in a tillage and rotation experiment at Avon, South Australia. *Australian Journal of Agricultural Research* 35, 1009–1014.

Rovira, A.D., Smettem, K.R.J. and Lee, K.E. (1987) Effects of rotation and conservation tillage on earthworms in a red-brown earth under wheat. *Australian Journal of Experimental Agriculture* 31, 509–513.

Satchell, J.E. (1980) *r* worms and *K* worms: A basis for classifying lumbricid earthworm strategies. In: Dindal, D.L. (ed.) *Soil Biology as Related to Land Use Practices*. EPA, Washington DC, pp. 848–864.

Scheu, S. and Parkinson, D. (1994) Effects of invasion of an Aspen forest (Canada) by *Dendrobaena octaedra* (Lumbricidae) on plant growth. *Ecology* 75, 2348–2361.

Schultz, J.E. (1995) Crop production in a rotation trial at Tarlee, South Australia. *Australian Journal of Experimental Agriculture* 35, 865–876.

Schwert, D.P. (1980) Active and passive dispersal of lumbricid earthworms. In: Dindal D.L. (ed.) *Soil Biology as Related to Land Use Practices*, EPA, Washington DC, pp. 182–188.

Sims, R.W. (1980) A classification and the distribution of earthworms, suborder Lumbricina (Haplotaxida: Oligochaeta). *Bulletin of the British Museum (Natural History), Zoology Series* 39, 103–124.

Temple-Smith, M.G. and Pinkard, T. (1995) (eds) The role of earthworms in agriculture and land management. Report of a National Workshop, June 1993, Dept. Primary Industry and Fisheries, Tasmania.

Terhivuo, J. (1988) The Finnish Lumbricidae (Oligochaeta) fauna and its formation. *Annales Zoologici Fennici* 25, 229–247.

Tian, G., Brussaard, L. and Kang, B.T. (1993) Biological effects of plant residues with contrasting chemical compositions under humid tropical conditions: effects of soil fauna. *Soil Biology and Biochemistry* 25, 731–737.

Tomlin, A.D., Tu, C.M. and Miller, J.J. (1995) Response of earthworms and soil biota to agricultural practices in corn, soybean and cereal rotations. *Acta Zoologica Fennica* 196,195–199.

Westernacher-Dotzler, E. (1988) Abundanz von Regenwörmern (Lumbricidae, Oligochaeta) unter verschiedenen Kulturpflanzen. PhD Thesis, University of Gieıen, *Giessener Bodenkundliche Abhandlungen* 4, 149 pp.

Wylie, P. (1994) Indicators of sustainable cropping systems. In: Pankhurst, C.E., Doube, B.M., Gupta, V.V.S.R. and Grace, P.R. (eds) *Soil Biota: Management in Sustainable Farming Systems*. CSIRO Press, East Melbourne, Australia, pp. 224–229.

Yakovchenko, V., Sikora, L.J. and Kaufman, D.D. (1996) A biologically based indicator of soil quality. *Biology and Fertility of Soils* 21, 245–251.

12

Biodiversity of Soil Organisms as an Indicator of Soil Health

C.E. Pankhurst
CSIRO Land and Water, Private Bag No. 2, Glen Osmond, South Australia 5064, Australia

Introduction

The importance of the biodiversity of soil biota to the integrity, function and long term sustainability of natural and managed terrestrial ecosystems is becoming increasingly recognized (Altieri, 1991a; Groombridge, 1992; Hawksworth and Ritchie, 1993; Elliott and Lynch, 1994; Freckman, 1994). However, whilst our knowledge of the functional processes carried out by the soil biota (e.g. the biogeochemical cycling of nutrients, the maintenance of soil structure, the degradation of pollutants) has steadily improved over the last half century, our knowledge of the biodiversity of the soil biota which contribute to these functional processes remains obscure. This is mainly due to technical difficulties associated with sampling and quantifying the biodiversity of soil organisms, and to a lack of techniques to evaluate the contribution of different components of the soil biota to ecosystems (Pankhurst *et al.*, 1996).

Biodiversity is an expression of the variety of living things, at genetic, species and ecosystem levels (Harper and Hawksworth, 1994). Enumeration of the number of species or genotypes in a biological community is a simple measure of biodiversity. However, the relative abundance of the different components within a community and the interactions that occur between the components are also important attributes of biodiversity. Thus studies of community structure underpin much of the current work on biodiversity. If biodiversity of soil organisms is to be used as a bioindicator of soil health, defined as 'the continued capacity of soil to function as a vital living system, within ecosystem boundaries, to sustain biological productivity, promote the quality of air and water environments, and maintain plant, animal and human health' (see Doran and Safley, Chapter 1, this volume), then we must be able to define relationships between biodiversity, plant productivity and soil health. Several recent reviews have

(eds C.E. Pankhurst, B.M. Doube and V.V.S.R. Gupta)

addressed the issue of biodiversity as it relates to ecosystem functioning (Potter and Meyer, 1990; Ehrlich and Wilson, 1991; Walker, 1992; Lawton and Brown, 1993; Pimm, 1993; Swift and Anderson, 1993; Vitousek and Hooper, 1993; Anderson, 1994; Elliott and Lynch, 1994; Bianchi and Bianchi, 1995) and the close relationship between ecosystem function and soil health is reviewed by Rapport *et al.* in Chapter 2 of this volume. The complexity of the issues and viewpoints brought forward in these reviews does little to resolve the question as to whether high biodiversity confers resilience on ecosystem function, or whether there is a minimal biodiversity required for proper ecosystem functioning, beyond which most species are redundant in their roles (Walker, 1992; Lawton and Brown, 1993; Pimm, 1993). In view of the current unprecedented and rapid decline in biodiversity in many parts of the world (Ehrlich and Wilson, 1991; Lawton and Brown, 1993), there is clearly some urgency in being able to define what kinds of biodiversity and how much is needed in order for ecosystems to function.

In this review, the relationship between the biodiversity of soil organisms and soil functioning is explored and the potential for using the biodiversity of soil organisms as an indicator of soil health is examined. To be useful as a bioindicator, biodiversity will need to meet a number of criteria that have been proposed as minimum requirements for bioindicators (Doran *et al.*, 1994; Pankhurst, 1994). The bioindicator should: (i) reflect the structure and/or function of ecological processes in soils and be generally applicable to all soil types and geographical locations; (ii) respond to changes in soil health; (iii) have available methodologies; and (iv) be interpretable. For practical purposes, it is also desirable that samples can be taken by a non-scientist at a reasonable cost.

The issues that are specifically addressed in the following discussion include: (i) the biodiversity of the soil biota; (ii) new methods for measuring biodiversity, particularly of soil microbial communities; (iii) biodiversity indices and their use; (iv) scale effects; (v) the relationship between biodiversity and ecosystem function; and (vi) the practical use of soil organism biodiversity as a bioindicator of soil health.

Biodiversity of the Soil Biota

The soil biota includes representatives of all groups of microorganisms, algae and nearly all animal phyla. Current estimates of the number of species of some groups include bacteria (30,000), fungi (1,500,000), algae (60,000), protozoa (10,000), nematodes (500,000) (Hawksworth and Mound, 1991), and earthworms (3000) (Lee, 1985). Individual organisms range in size from <1 μm diameter, with a weight of $<10^{-12}$ g for the smallest bacteria, to >1 m in length, >20 mm in diameter, with a weight of >500 g for the largest earthworms. Soil organism biomass in a fertile soil may exceed 20 t ha^{-1} (Lee and Pankhurst, 1992). Linkages and interactions between the various components of the soil

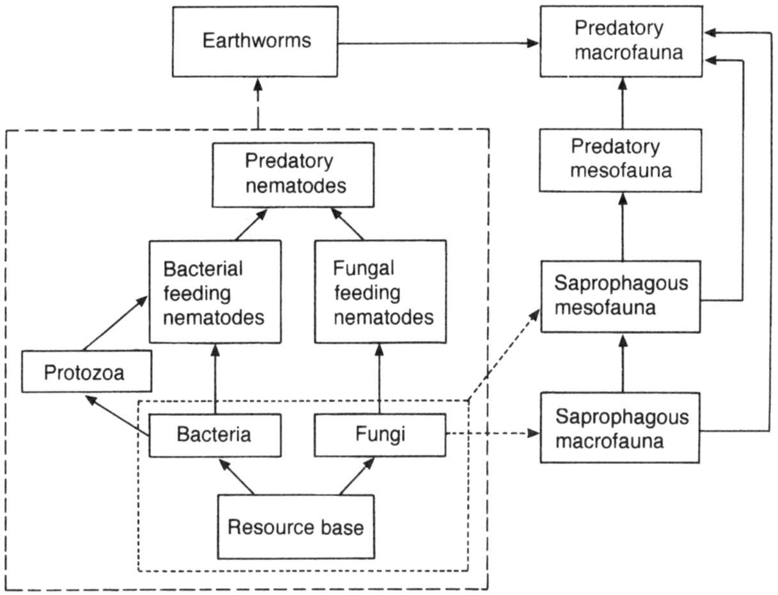

Fig. 12.1. Simplified functional food web for detritus-based systems in soil. Organisms in the dotted and dashed boxes may be fed upon by organisms which consume substrate containing material of more than one trophic level. From Wardle (1995).

biota are represented as a decomposer-based (detritus) food web (Fig. 12.1). These linkages and interactions between the soil biota components are critical in regulating the nutrient cycling and energy flow in the soil (Pankhurst and Lynch, 1995). In functional terms, the detritus food web in the soil is made up of primary decomposers (bacteria and fungi), herbivores (e.g. plant parasitic nematodes), consumers of bacteria and fungi (e.g. protozoa, bacterivorous and fungivorous nematodes, collembola), saprophagous mesofauna and macrofauna and predators (e.g. predacious nematodes, predacious mites) (Moore and de Ruiter, 1991; Wardle, 1995).

Currently, very little is known about the vast majority of soil organisms and the ecological role they play in the soil. Some of the reasons for this paucity of knowledge have been discussed by Freckman (1994) and include the following:

1. The opaqueness of the soil which makes the *in situ* identification of most soil organisms impractical.
2. The large range in size of soil organisms (microbes to earthworms) makes their interactions and ecological roles difficult to assess.
3. The morphology of many organisms changes during their life cycle making them difficult to identify.

4. Methods for extracting and quantifying many microorganisms, fungi and mesofauna from soils have not been determined and techniques for culturing them have not been developed.
5. The temporal and spatial scale of the niches occupied by different organisms in the soil varies greatly.
6. The activity and abundance of different organisms changes in response to changes in the physico-chemical properties of the soil, the quality of organic matter, climate and geography, resulting in few comparisons of the ecological roles of soil taxa in different ecosystems.

Because of these problems, many of the ecological roles of the soil biota are attributed to trophic groups, or groups of species with similar morphology (Siepel and de Ruiter-Dijkman, 1993; Yeates *et al.*, 1993).

Approaches to sampling and measuring the biodiversity of soil organism communities have developed steadily in recent years. Soil macrofauna can usually be collected quantitatively, many may be identified to species and their ecological roles in the soil are generally known. These roles include: (i) direct processing of organic matter (e.g. snails, earthworms, enchytraeids, millipedes, ants, termites); (ii) predation (e.g. spiders, ants); and (iii) creation of soil structure (e.g. earthworms, termites). Micro- and mesofauna (protozoa, nematodes, collembola, mites) can be extracted from soil with varying degrees of success, but knowledge of their ecological function in the soil is less well understood. Many are predators of bacteria and fungi and involved in the decomposition of organic matter. Some groups of collembola and mites are well characterized (see van Straalen, Chapter 10, this volume) but less is known taxonomically about many groups of the protozoa, rotifers, tardigrades and nematodes (Gupta and Yeates, Chapter 9, this volume). Their ecological roles and, to a certain extent, their biodiversity are based on trophic group analysis. In contrast, significant advancements have been made during the last decade with methods for the assessment of bacterial and fungal biodiversity (Pankhurst *et al.*, 1996). Some of these methods are based on community function analysis and have provided insight into the ecological roles of bacteria and fungi in soils and how these are affected by various disturbances.

Approaches to measuring the biodiversity of soil microbial communities

The large number of microorganisms present in some soils (estimates of 10^9 bacteria, 10^7 actinomycetes, 10^6 fungi, 10^4 algae and 10^5 protozoa g^{-1} soil (Miller, 1990)), their high diversity (estimates of 4000 different genomes g^{-1} soil (Torsvik *et al.*, 1990)), their difficulty to culture and the problems of adequately defining species of different microorganisms, has led to approaches to measuring biodiversity based on communities rather than species. Some recently developed approaches include analysis of community functioning based on utilization of substrates, analysis of community structure based on extraction or amplification of soil DNA and analysis of community structure based on extraction and chromatography of fatty acids.

Functional diversity (metabolic profiling)

A simple approach to examining the functional diversity of microbial communities is based on the number of different substrates that are utilized by a microbial community in a defined habitat. Using the BIOLOG microplate identification system (BIOLOG, Inc., Haywood, California, USA) which tests the capacity of bacteria to utilize 95 different substrates, Garland and Mills (1991), Grayston *et al.* (1994) and Zak *et al.* (1994) were able to obtain an assessment of functional differences between soil bacterial communities from a variety of habitats. Employing alternative approaches to data analysis, including comparison of the number and diversity of substrates used, cluster analysis based on the presence or absence of utilized substrates and principal component analysis based on rate of substrate use with time, Zak *et al.* (1994) were able to descibe differences in functional diversity among communities of soil bacteria from six different sites (each under a different plant community) located along an elevational and moisture gradient in the northern Chihuahuan desert in the USA. Results presented in Fig. 12.2 showed that the bacterial communities at three of the sites were indistinguishable, whilst at the other three sites, the communities were dissimilar from each other, as well as from the other three communities. They also analysed the data with respect to the communities' capacity to utilize substrates within different biochemical classes or guilds, providing information at another level of functional resolution. In other studies, BIOLOG plates have been used to examine the functional diversity of bacterial communities associated with soil aggregates (Winding, 1994), plant rhizospheres (Garland, 1995), rice straw decomposition (Bossio and Scow, 1995) and dairy wastes decomposition (Insam *et al.*, 1995). It is acknowledged that application of the BIOLOG system to community analysis has some limitations, including the use of substrates biased towards simple carbohydrates, the dependence of patterns on inoculum density and its inability to determine fungal activity (Haack *et al.*, 1995). Nonctheless, the technique has the capacity to produce a rich data set that is ideal for detecting site-specific differences in the functional diversity of soil bacteria and for evaluating the relationship between biodiversity and the expression of function in a natural ecosystem. There is clearly an opportunity for the application of this kind of approach for other soil microorganisms (fungi, actinomycetes, protozoa).

DNA approaches

Several approaches based on DNA technology have been successfully used for measuring different aspects of the biodiversity of soil organism communities.

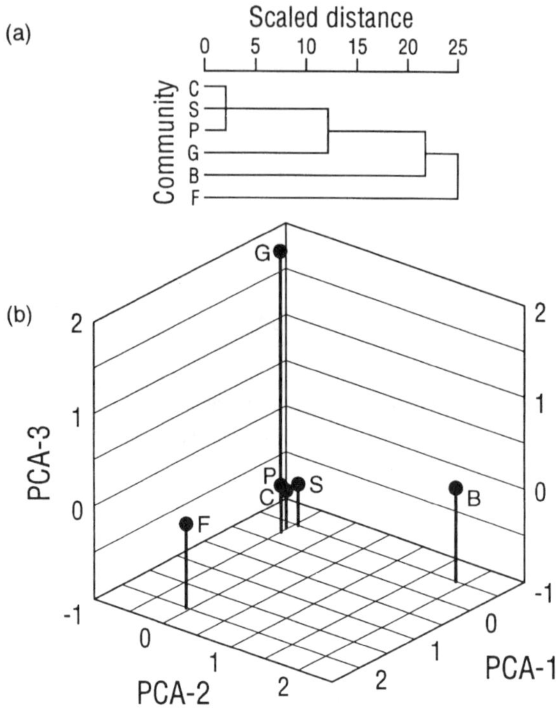

Fig. 12.2. Compositional relationship among soils collected from six different plant communities from the Jornada Long-Term Ecological Research transect, northern Chihuahuan Desert, USA, based on overall substrate utilization (60 h incubation) using (a) cluster analysis (presence or absence of activity) and (b) principal component analysis (levels of activity). The abbreviations correspond to plant communities: F = mesquite–playa fringe, B = herbaceous bajada, G = black grama grassland, C = creosote-bush bajada, P = playa grassland, S = *Sporobolus* grassland. From Zak *et al.* (1994).

These include: (i) use of DNA probes to identify community members; (ii) estimations of total genetic diversity via DNA reassociation kinetics; (iii) estimations of genetic diversity within groups of organisms via the construction and analysis of ribosomal DNA libraries; and (iv) a wide range of DNA hydridization techniques. All techniques have been made possible by advances in the quantitative extraction of DNA from soil (Holben, 1994). A novel use of DNA probes to look at community structure was proposed by Voordouw *et al.* (1991) in a study of sulphate-reducing bacteria from oil-fields. Whole soil DNA was labelled and used to probe filters to which DNA from bacterial standards had been applied. This allowed rapid detection of a range of bacteria which were present in the community. A rapid method for estimating total genetic diversity of soil DNA was developed by Torsvik *et al.* (1990). The method involves bulk DNA extraction from soil, its denaturation, and spectrophotometric measurement of its reassociation kinetics. These values are measured relative to a homogeneous DNA standard and an estimate can then be made of the number of different genomes present in the environmental sample and thus its genetic diversity. A comparison of this technique with estimates of phenotypic diversity using bacterial isolation and identification using the API 20B system (API System S.A., Montalieu Vercieu, France) gave a good correlation (Torsvik *et al.*, 1994). A similar technique based on the degree of cross-hybridization between DNA extracted from contrasting soil microbial communities has been developed by Ritz and Griffiths (1994).

In contrast to the above approaches, analysis of community structure based on ribosomal libraries, relies on the cloning either directly or after amplification via polymerase chain reaction (PCR) of DNA coding for rRNA from whole soil DNA. These cloned rRNA fragments create a 'library' of the organisms present in soil and theoretically constitute an unbiased representation of the community. The cloned rRNA fragments can be sequenced and compared to an extensive data base of rRNA sequence information (Olsen *et al.*, 1991) or they can be more rapidly and cheaply screened via dot blot hybridization to genus or species-specific rRNA probes. Several studies have used this method to look at whole soil communities (Liesack and Stackebrandt, 1992; Stackebrandt *et al.*, 1993). The power of the technique to reveal unculturable members of a community was shown by the fact that most of the soil libraries isolated rRNA sequences which had not been found in cultured bacteria. In a study which compared an rRNA library with conventional isolation from soil (Stackebrandt *et al.*, 1993), genera which were commonly isolated and dominant in the soil were not represented in the rRNA sequences. This points out not only the severe limitations of our ability to culture a representative part of the community but also raises the possibility that amplification and cloning contain their own biases. A study of different methods for DNA isolation and cloning revealed method-dependent differences in the composition of cyanobacterial libraries (Ward *et al.*, 1990). Because the relative abundance of rRNA in uncultivated species is not known, the method has limited quantitative use. Also, PCR may preferentially amplify some sequences making quantitative comparison of sequence abundance unreliable (Embley and Stackebrandt, 1994).

Fatty acid analysis

Community structures in different soils have been examined by measurement of ester-linked fatty acids extracted from soil. The fatty acid methyl esters (FAMEs) are measured using gas chromatography and analysed using the MIDI system (Microbial ID, Inc., Newark, Delaware, USA). The soil FAME profiles tend to be complex because many fatty acids are common to different

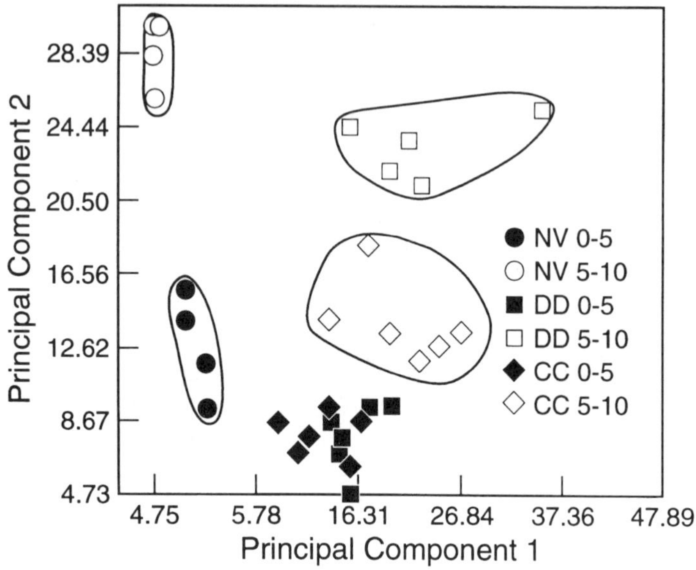

Fig. 12.3. Principal component analysis of GC-FAME profiles of soil from two depths (0–5 and 5–10 cm) under native vegetation (NV), and adjacent cropped soil subjected to conventional cultivation (CC) and direct-drilling (DD). C.E. Pankhurst (unpublished results).

microorganisms and interpretation may be difficult. In a study of FAME profiles of soil samples taken across a conventionally tilled corn field, Cavigelli *et al.* (1995) found a high degree of reproducibility between samples. A comparison of the FAME profiles of the microorganism populations cultured from several of the soil samples and the soil FAME profiles showed many differences, indicating that profiles of FAMEs extracted from soil reveal portions of the microbial community not culturable on laboratory media. However, a small number of plated communities were found to be distinct and to share profile characteristics that identified soil community FAME profiles as distinct. This suggests that plated communities are not solely the result of selection by the growth medium, but in fact do reflect the distribution, *in situ*, of dominant, culturable soil microbial populations (Cavigelli *et al.*, 1995).

In other studies Pankhurst *et al.* (unpublished results) have used the MIDI system to analyse the FAME profiles of soils subjected to different agricultural management practices. Principal component analysis showed the soils to have quite different FAME profiles (Fig. 12.3) indicative of management-induced changes in the composition of the microbial communities present. Fatty acids derived from soil organic matter may be a complicating factor in such comparisons but this problem can be partially overcome by removal of the soil fraction during the extraction process (Pankhurst *et al.*, unpublished results). Restricting the analysis of soil fatty acids to only those derived from phospholipids (PLFA) and/or lipopolysaccharides in the soil markedly increases the capacity to detect differences in soil microbial community structure in soils subjected to different agricultural practices (Zelles *et al.*, 1992) and soil exposed to various pollutants (Frostegard *et al.*, 1993). In their study, Frostegard *et al.* (1993) followed the effects of five heavy metals on microbial biomass, activity and community structure in two soil types over a 6 month period. Gradual changes in community structure revealed by PLFA analysis were found in response to increasing amounts of metal contamination, suggesting the development of a metal tolerant community. These changes were, in general, detected at similar or even lower metal concentrations than those at which effects on ATP and respiration occurred, indicating that community analyses could be a sensitive way of detecting environmental perturbations. Further fractionation and analysis of individual fatty acids ('signature fatty acids') permits identification of specific functional groups of microorganisms (bacteria, fungi, actinomycetes) from the fatty acid profiles and provides information about the nutritional status (degree of stress) of the microbial community (Frostegard *et al.*, 1996; Pennanen *et al.*, 1996; also see White and Macnaughton, Chapter 15, this volume).

Diversity Indices

Irrespective of the methods used to obtain information about the biodiversity of selected taxa, functional groups or whole communities of soil organisms, there is a need to be able to describe this biodiversity in some way. A large number of indices have been developed that attempt to describe different facets of biodiversity (Magurran, 1988).

In classical terms, biodiversity is described as a function of two components: (i) the total number of species present, i.e. species richness or species abundance; and (ii) the distribution of individuals among those species i.e. species evenness or species equitability (Margalef, 1958). However, the incorporation of these two components into diversity indices has led to much controversy among ecologists (Atlas, 1984; Peet, 1984; Magurran, 1988). The Shannon diversity index, known with slight mathematical variations as the Shannon–Weaver and Shannon–Wiener indices, is probably the most widely used index for measuring species diversity (Shannon and Weaver, 1949; Atlas, 1984; Magurran, 1988). But this index, as the many others that have been proposed, has its limitations. A major difficulty is that both richness and evenness play a role in determining the value of the index. Quite different communities, as a consequence, can have the same index (Atlas, 1984). It must be remembered that any diversity index is a single value and thus cannot indicate the total make up of a community. It is recommended therefore that evenness, richness and diversity values should be calculated in order to obtain an objective assessment of community structure (Atlas, 1984).

Despite their limitations and the large number of diversity measures that have been proposed (Magurran, 1988; Kvalseth, 1991), different measures of biodiversity have been used successfully to describe and quantify the effects of pollution and environmental disturbance on communities of organisms. Most of the studies that have employed diversity indices in this context have involved aquatic ecosystems and a range of different communities including diatoms, blue-green algae and phytoplankton (Atlas, 1984). Some important generalizations to emerge from these studies (Atlas, 1984; Magurran, 1988) include the following:

1. Enriched or polluted systems display a reduction in biodiversity.
2. Stable equilibrium communities usually have a log normal pattern of species abundance whilst the species abundance distribution of a disturbed or polluted community shifts backwards through succession to a less equitable log or geometric series distribution.
3. Species dominance is directly related to productivity and inversely related to diversity and stability.
4. A species rich community requires a lower amount of energy to maintain its diversity but has a lower rate of production per unit of biomass.
5. Biodiversity within a community can be tightly regulated by trophic interactions between different components of the community and by changes in the physicochemical environment.

These generalizations may be directly applicable to soil-based ecosystems, but because of the difficulties in assessing soil biodiversity (considered above), few studies have attempted to use biodiversity indices (particularly at the community level) to assess the impact of pollution or disturbance on soil health.

Examples of the use of diversity indices to describe the impact of soil management practices and soil pollution on communities of soil organisms

Soil nematode communities have provided a focus for several studies where biodiversity indices have been used as indicators to document the impact of some perturbation on soil biological activity (Freckman and Ettema, 1993; Yeates *et al.*, 1993; Neher *et al.*, 1995). Soil nematodes (free-living and plant parasitic) occupy an important position in the soil detritus food web, (e.g. they graze on bacteria and fungi, Fig. 12.1) and are thus significant regulators of decomposition and nutrient mineralization. Any change to the soil habitat of nematodes that influences their food source or environment, such as soil and crop management practices, should be reflected in the biodiversity of the nematode community. Freckman and Ettema (1993) used a number of indices including species richness, absolute abundance, number of trophic groups, fungal feeder/bacterial feeder ratio, trophic diversity, the Shannon diversity index, the Simpson diversity index (Simpson, 1949), a maturity index and a plant parasitic index (Bongers, 1990) to describe the effects of various agricultural treatments on nematode community structure. The agricultural treatments were described as: high chemical input (conventional and no-tillage treatments), organic (low and zero input treatments), perennial (poplar and alfalfa treatments) and successional (abandoned after tillage and never tilled treatments). Species diversity was greatest in successional treatments (least human intervention), whereas nematode abundance was highest in the high input and organic systems. The Shannon diversity index, which gives more weight to rare species, was effective in showing differences in diversity at the system and crop level, whereas Simpson's diversity index, which gives more weight to common species, was not discriminatory because only a few taxa were dominant in the community (Freckman and Ettema, 1993). The maturity index was found to be particularly useful and showed consistent differences at treatment, system and cropping levels. A canonical discriminate analysis based on absolute abundance of nematode taxa (Fig. 12.4) clustered treatments into four systems of similar agricultural management and proved to be an effective approach for demonstrating the effect of the different management practices on soil nematode communities. Freckman and Ettema (1993) concluded that the nematode maturity index combined with multivariate data analysis of nematode community data were good predictors for assessing human intervention in soil systems. Similar conclusions were reached by Neher *et al.* (1995) who also found that the nematode maturity index was the most successful index to use for assessing the health of soils based on nematode community structure.

In a study of the impact of pasture contamination by the heavy metals Cu, Cr and As, on nematode communities, Yeates *et al.* (1994) found that diversity (described using the Shannon–Weaver index) declined with increasing levels of metal contamination. They also described a shift in dominance from plant-feeding nematodes in control uncontaminated soil to bacterial-feeding nematodes in the highly contaminated soil and a general increase in the proportion of predatory nematodes as the soils became increasingly contaminated. Weiss and Larink (1991) found a similar increase in the abundance of predatory nematodes in soil following the addition of sewage sludge and heavy metals to soil.

In another example, Kennedy and Smith (1995) used a variety of richness, evenness and diversity indices to compare heterotrophic bacterial communities in soil from a prairie grassland and from an adjacent cultivated wheat land. Bacterial isolates were ranked with respect to their ability to utilize a range of substrates and their ability to withstand stress (e.g. grow in the presence of antibiotics or heavy metals). The rankings were used to calculate the Hill (Hill, 1973), Shannon and Simpson diversity indices. Microbial biomass C, phosphatase, dehydrogenase, denitrification and nitrification potential were all significantly higher in the prairie system. Inorganic N values were higher in the

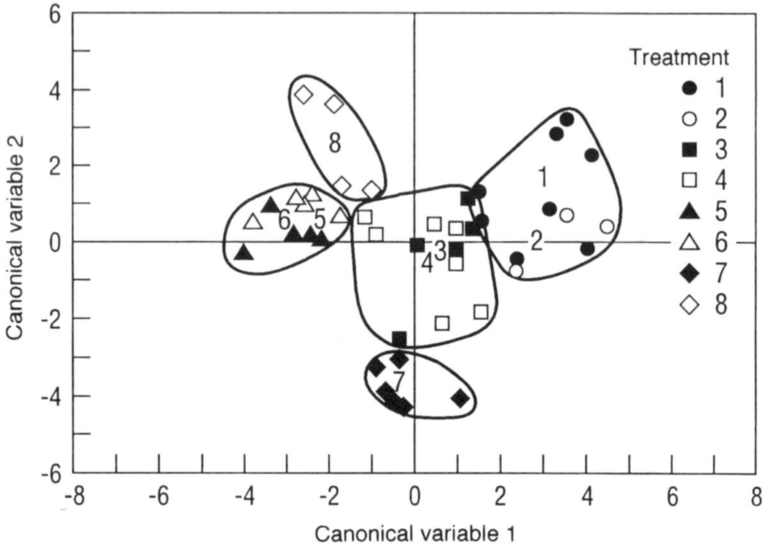

Fig. 12.4. Scatter plot analysis of nematode abundance data for eight treatments at the Kellogg Biological Station Long Term Ecological Research site, Michigan, USA. Treatments are: (1) conventional tillage, (2) no tillage, (3) low input, (4) zero input, (5) poplar, (6) alfalfa, (7) succession-A, abandoned after years of historical tillage, and (8) succession-N, never tilled. Treatments 1 and 2 were corn/soybean rotation, 3 and 4 were corn/soybean/wheat rotations. From Freckman and Ettema (1993).

cultivated soils and pH was not different between the two systems. Values derived for the various diversity indices were different for the two systems, but overall diversity was greatest in the cultivated soil. The greater diversity among the bacterial isolates from the cultivated soil was due to their broader range of substrate utilization and stress resistance.

Wardle (1995) has used the Shannon diversity index to assess the influence of tillage (conventional tillage compared with no-tillage) on the biodiversity of several taxonomic and functional groups of soil organisms including fungi, nematodes, collembola, mites, earthworms, spiders and carabid beetles. The data used was contained in 20 individual studies published between 1974 and 1993. From this limited analysis Wardle concluded that there was less effect of tillage on species diversity with the smaller soil organisms (i.e. the fungi and the nematodes) than there was on the meso- and macro-fauna. However, generally poor taxonomic resolution within the studies of the fungi and nematodes used in this analysis may have been a factor in the lack of response shown. The responses of the meso- and macrofauna were unpredictable with some studies showing higher biodiversity in response to tillage and others showing the reverse. This unpredictability probably reflects the multiple effects tillage can have on the soil. Similar analyses applied to data from studies evaluating the effect of herbicides on the soil biota (Wardle, 1995), showed little or no evidence for effects on soil fungal biodiversity species but large effects on larger soil organisms, notably collembola. These effects of herbicides on collembola were thought to be indirect and due largely to modification of the plant species composition of the habitat.

Scale effects

A major difficulty in describing the biodiversity of soil communities and linking this to soil health is the large difference in spatial and temporal scale within which different soil organisms operate (Anderson, 1994). For example, many soil organisms are opportunistic in their feeding and reproductive behaviour in that they exist for part of the time in resting stages but are capable of changing quickly to an active feeding and reproducing mode when suitable food and favourable environmental conditions become available.

Bacteria are the most successful and functionally diverse of opportunistic soil biota; they commonly have the ability to survive as spores or in similar resting stages for long periods, but respond quickly to the availability of a suitable substrate and have a high intrinsic rate of population increase. Jenkinson and Ladd (1981) estimated that in an English soil bacterial cells divide, on average, once every 2.5 years, spending the remainder of the time in resting stages. Their activities in soil must be seen as so disjunct in time and space that the concepts of individuals making up populations of a particular species, or of aggregations of species to form coherent communities that apply to most organisms cannot easily be applied to soil bacteria. Similar problems exist with soil fungi, many of which produce an extensive network of hyphae capable of potentially unlimited growth and persistence (Swift, 1987). As the fungal mycelium grows the mycelial network will usually remain connected, so that there is communication between various segments of the mycelium in response to differences in the local environments encountered. Thus, fungal hyphae can convert organic material and nutrients in the soil solution into fungal biomass at scales that range from several mm to distances that may encompass entire tracts of forest (Smith *et al.*, 1992). Appreciation of the spatio-temporal scales at which a fungal species can occur and function, and the growth form of the organism is thus critical for understanding how fungal biodiversity is organized in the soil and the importance of this organization to ecosystem processes (Rayner *et al.*, 1994).

The scale at which organisms influence soil processes can therefore vary widely from microsites such as a particle of organic matter to large patches which are influenced by the plant species present. Beare *et al.* (1995) have proposed that the soil is composed of a number of biologically relevant spheres of influence that define much of their spatial and temporal heterogeneity. These 'spheres of influence' include the detritusphere (litter layer on the soil surface), the drilosphere (the portion of soil which is influenced by mucilages or nitrogenous wastes excreted by earthworms), the porosphere (the region of water films channels between soil aggregates), the aggregatusphere (the interstices between micro-aggregates) and the rhizosphere (the zone of soil directly influenced by plant roots and/or their exudates and exfoliates. Although not mutually exclusive, each sphere has fairly distinct properties that regulate the interactions among organisms and the soil processes they mediate.

Biodiversity and ecosystem function

Given that our prime objective is to minimize the loss of biodiversity in soils and the above ground ecosystems they support, the important question is: what kinds of soil biodiversity are most significant to the maintenance of ecosystem function? Current ecological knowledge supports a wide range of views on this topic (Ehrlich and Ehrlich, 1981; Walker, 1992; Lawton and Brown, 1993; Swift and Anderson, 1993; Vitousek and Hooper, 1993; Anderson, 1994; Lawton, 1994), but empirical tests are few (Vitousek and Hooper, 1993). If for example the species of an entire functional group, e.g. all the plants, or all the nematodes were removed, the functioning of the ecosystem would be drastically impaired. Thus there is a certain minimal diversity required for proper ecosystem functioning. Beyond this minimum, most species may be redundant in their roles (Walker, 1992; Lawton and Brown, 1993). Above this minimum, adding or deleting species may have no detectable effect on the process or processes in question. A second, contrasting view suggests that all species make a contribution to ecosystem processes, which therefore decline progressively as species are lost (Ehrlich and Ehrlich, 1981). A third view, suggests that ecosystem function changes when diversity changes, but that the magnitude and direction of change is unpredictable, because the roles of the individual species are complex and varied (Swift and Anderson, 1993; Vitousek and Hooper, 1993; Lawton, 1994).

Experimental evidence that biodiversity is intimately linked with soil and ecosystem function is provided by Naeem *et al.* (1995) who monitored selected ecosystem processes (community respiration, decomposition rates, nutrient retention, plant productivity) in a series of artificially constructed terrestrial microcosms in a controlled environment facility, the Ecotron. Plant and animal diversity was manipulated to create low, intermediate, and high diversity microcosms, with 9, 15 and 31 species, structured so that the lower diversity microcosm contained a subset of the species present in the higher-diversity microcosm. This manipulation produced a set of microcosms in which the lower-diversity communities functionally resembled depauperate descendants of higher-diversity communities that had lost species uniformly across all trophic categories. Most of the processes monitored varied significantly but unpredictably with species richness. Community respiration and plant productivity, however, both declined as species richness declined, indicating that reduced biodiversity may indeed alter the performance of an ecosystem.

Even if high species richness does not always play a significant role in maintaining ecosystem processes under normal environmental conditions, it may be important when conditions change. If dominant species are sensitive to a particular perturbation (e.g. a pollutant), rare species with different environmental tolerances may show compensatory changes in abundance, and maintain ecosystem processes (Walker, 1992; Lawton and Brown, 1993). There is growing evidence for such effects in the field; for example Tilman and Downing (1994) showed that the primary productivity of a grassland with high plant species richness was more resistant to and recovered more fully from a major perturbation (drought) than a grassland with lower plant species diversity. Thus, species redundancy under normal environmental conditions, manifest by suites of species with apparently similar roles, many of them rare, may be critical for the maintenance of ecosystem processes in the face of extreme perturbations such as severe drought, fire or contamination by a pollutant.

Biodiversity, Soil Functional Processes and Soil Health

The principles discussed above at the ecosystem level will apply equally well to the maintenance of functional processes in the soil mediated by the soil biota. For any single function in the soil, e.g. cellulose decomposition, nitrogen mineralization or sulphur oxidation, there is a diverse range of soil organisms and pathways available to carry out the function. Any one group of organisms or any one pathway may be sufficient for the adequate expression of the function. What is not known, however, is what level of diversity of soil organisms may be required to ensure that the function will proceed when disturbances, such as tillage or heavy metal pollution, are applied. In some cases, as for example in the decomposition of organic matter, the majority of heterotrophic soil microorganisms have some involvement in the process of decomposition so that the loss of species that may be sensitive to a particular disturbance (e.g. to the application of a particular herbicide, or a change in plant species) will be of little consequence as less sensitive organisms will take their place. In the case of processes that have a more restricted involvement of microorganisms, such as symbiotic nitrogen fixation by *Rhizobium/Bradyrhizobium* spp., or phosphorus uptake by arbuscular mycorrhizal fungi, loss of diversity within the population may have more serious implications. Some examples of the linkages between biodiversity of individual functional groups or communities of soil organisms and maintenance of soil functional processes are given below.

Rhizobium and N_2-fixation

Rhizobium and *Bradyrhizobium* species are an important functional group of soil bacteria through their capacity to fix atmospheric nitrogen in symbiosis with legumes. This process can provide significant N inputs into the soil and provide significant advantages to legumes as early colonizers of new habitats. The diversity of *Rhizobium* species has been shown to decline with decreasing soil pH (Harrison *et al.*, 1988) and in the presence of heavy metals (Giller *et al.*, 1989). In the latter study, strains of *Rhizobium leguminosarum* bv. *trifolii* isolated from a metal-contaminated soil were found to be wholly ineffective in N_2-fixation on white and red clovers. The plasmid profiles and restriction fragment length polymorphism patterns of these isolates were all very similar indicating a lack of genetic diversity in the population surviving in high concentrations of heavy metals (Giller *et al.*, 1989; Hirsch *et al.*, 1993). Isolates from comparable field plots of uncontaminated soil were effective in N_2-fixation, and showed high levels of genetic diversity. Thus the population of *R. leguminosarum* bv. *trifolii* had been radically altered by long-term exposure to heavy metal contamination, apparently losing those agronomically-important strains capable of fixing nitrogen with white and red clover.

Mycorrhizal fungi and phosphorus uptake

The majority of vascular plants associate with arbuscular mycorrhizal (AM) fungi or ectomycorrhizal fungi (EM) and benefit from an increased capacity to extract phosphorus and other nutrients from the soil. Mycorrhizal fungi thus have an important role in plant community development, nutrient cycling and the maintenance of soil structure (Miller and Jastrow, 1994). Factors affecting the diversity of both AM and EM fungi are poorly understood, but are likely to be tightly linked with the host plant. It is known that certain crop plants preferentially form mycorrhizae with certain species of AM fungi; thus cropping sequences can have a major influence on the density and species composition of AM fungal communities (Johnson and Pfleger, 1994). This selectivity may be mediated by the capacity of the plant host to affect the sporulation rate of AM fungi (Hendrix *et al.*, 1995; Bever *et al.*, 1996). An important outcome of these observations is to consider how the AM fungal species that proliferate within a particular cropping system might affect crop production, as the dominant species within a system may not be the most beneficial mutualists. Schenck and Sequeira (1987), for example have suggested that decline-type diseases occurring in a number of tree crops may be related to a shift from beneficial to predominantly non-beneficial species of AM fungi in the soil. Similarly, Johnson *et al.* (1992), in a study of AM fungi in soil under corn and soybean monocultures obtained evidence that suggested declining crop yields were positively associated with spore densities of non-beneficial species of AM fungi. Species composition of AM fungi may also be affected by addition of fertilizers and pesticides to soil (Johnson and Pfleger, 1994).

Biological control of plant pests and diseases

One of the major factors governing the health of terrestrial ecosystems is the on-going interaction between plants and pests. Plant breeding programmes throughout the world continue to produce new pest and disease resistant varieties in order to keep pace with the appearance of new pests and new pathogenic genotypes of known pathogens. It is, however, the management practices that are used in many agroecosystems, e.g. the growth of crops as monocultures, the extensive use of tillage and chemical inputs (fertilizers and pesticides) that break down the fragile web of community interactions between pests and their natural enemies that lead to increased pest and disease problems (Altieri, 1991b). This situation can be reversed if attempts are made to conserve biodiversity in agriculture by maintaining natural vegetation adjacent to crops (van Emden, 1990), reducing chemical inputs and developing more diverse cropping strategies such as traditional polycultures and agroforestry (Letourneau, 1987). In other systems, such as dryland farming agroecosystems in Australia, conservation farming practices based on reduced tillage, crop rotation and the retention of crop residues can augment a shift in the biomass and biodiversity of the soil microbiota and lead to the development of natural biological suppression (control) of plant root diseases (Roget, 1995).

Protozoa and nutrient cycling

Protozoa are important bacterial grazers because they can enter pore spaces within soil aggregates unavailable to nematodes. They also have the capacity to encyst which enables them to withdraw from environmental stresses with minimum mortality, and furnish a large active population immediately upon return of favourable conditions. When dormant bacterial populations respond to favourable soil conditions, their encysted protozoan predators excyst to quickly graze them, increasing the amount of soluble nutrients and decreasing the competitive abilities of bacteria, thus making nutrients more available to plants (Bamforth, 1985; Foissner, 1987; Gupta, 1994). Protozoa enhance nutrient recycling out of proportion to their biomass (Bamforth, 1985). Most protozoan predation of bacteria is carried out by the numerous and moderately species diverse flagellates and naked amoebae, and the less abundant but diverse ciliates (Bamforth, 1995). In a study of ciliate biodiversity, Bamforth (1995) found that a few species (members of the Polyhymenophora and Colpodida) had broad overlapping niches and were dominant in most soils. The ratio of Colpodida (C) to Polyhymenophora (P) was characteristic of different ecosystems and useful as a bioindicator. Gupta (1996) also found the C/P ratio useful for describing differences in the diversity of ciliates in agricultural soils in Australia; high input agricultural systems usually had a C/P ratio >1, whereas native (low input) systems had a C/P ratio <1. In an investigation of the effect of protozoan diversity on organic residue (wheat stubble) decomposition, Gupta (1996) showed that the rate of decomposition was significantly higher in the presence of a more diverse protozoan community (containing flagellates, amoebae and ciliates) than a less diverse community (containing flagellates only).

Earthworms and soil structure

Earthworms are the most common species of the soil macrofauna and are widespread, being found in all but the driest and coldest land areas of the world (Lee, 1985). The beneficial effects that earthworms can have on soil structure, fertility and plant production through their feeding and burrowing activities are well documented (Lee, 1985). For example, earthworms can improve the incorporation of organic matter below the soil surface, increase the numbers of water stable soil aggregates, improve water infiltration, aeration and root penetration and increase microbial activity (Lee and Foster, 1991). They can also have a major influence on the composition and diversity of soil organism communities (Brown, 1995). Earthworms occupy selective niches within the soil volume, e.g. epigeic and epiendogeic species occupy the litter and surface soil layers, whilst meso- and oligohumic endogeic species inhabit mineral soil within the rhizosphere and beyond, and their distribution and abundance is influenced by climatic and edaphic factors. Commonly less than a half-dozen earthworm species are found in a given soil (Lee, 1985) with individual species usually occupying one of the niches described above. Reduced earthworm diversity may not strongly affect soil processes, but more diverse assemblages may more effectively exploit soil resources and influence a wider array of processes (Brown, 1995). In a study of the distribution and biodiversity of earthworm fauna in agricultural soils of southern Australia Baker *et al.* (1994) found that introduced Lumbricidae (mostly from northern Europe) predominate with different species dominating different soil types and land uses. Native earthworm species (Megascolecidae) are generally rarer than introduced species and found in only some locations. Introduction of two exotic lumbricids, *Aporrectodea caliginosa* and *A. longa* to pastures in northern Tasmania increased pasture production by up to 75% within 3 years (Temple-Smith, 1991) clearly demonstrating the beneficial effects of earthworms to soil processes and plant growth. Further introductions of earthworm species into selected niches and where further economic benefits may be realized are the targets of an on-going programme to capture the benefits from earthworms in southern Australia (Baker *et al.*, 1994).

Community interactions

The examples above represent only a sampling of the rich and varied ways soil organisms contribute to the functioning of soils. Although the functional diversity of different groups of organisms is very high, it is exceeded to a large degree by the biodiversity of the organisms themselves. It is for this reason that many authors have argued that simple accounting of species richness contributes little to our understanding of soil functioning because of the high degree of functional redundancy assumed for many functional groups of soil organisms (Walker, 1992; Lawton and Brown, 1993). It is generally inferred from this that a loss of species will not necessarily be reflected by a change in rates of biologically mediated processes and biogeochemical transformations (Franklin, 1993). Though redundancy of a single function may be common among many soil organisms, the suite of functions attributable to any one species is unlikely to be redundant. Furthermore, functionally similar organisms often have different environmental tolerances, physiological requirements and microhabitat preferences. As such, they are likely to play quite different roles in the soil system.

Whilst there is little doubt that some organisms such as earthworms have an overriding influence on soil functional processes, the importance of interactions between different soil organism groups in regulating these processes should not be overlooked (Brown, 1995). Biotic interactions can be either positive (e.g. mutualistic, associative) or negative (e.g. competitive, predatory) in function. The importance of negative interactions in regulating soil processes has been the subject of much research (Lee and Pankhurst, 1992). More recently there has been interest shown in understanding the positive interactions in soils and whether they may confer greater stability and resilience to soils. Overall, the greater the complexity of the biotic interactions the greater the probability that indirect effects will be important to regulating ecosystem function (Price, 1988). Thus, the diversity that supports these complex interactions will be important to regulating the processes such as nutrient cycling that contribute to the maintenance of soil health. Greater focus on understanding the complexity and specificity of biotic interactions will therefore be more important to determining the significance of biodiversity to soil and ecosystem functioning than will pursuing the concepts of redundancy and the importance of selected 'keystone' species (Mills *et al.*, 1993).

Practical Use of Biodiversity as an Indicator of Soil Health

It is clear, from the above discussion, that there is some good evidence that links biodiversity with ecosystem function and stability, and some limited evidence that links the functional biodiversity of soil organisms with the maintenance of soil functional processes within ecosystems. The question of how much biodiversity within a given group of organisms capable of carrying out a specific function is needed to ensure continuance of that function in the face of perturbations will continue to be a topic of research for many years to come. It is also clear that interactions between the various functional groups of soil organisms, as depicted by detritus food webs, has an important influence in regulating soil functional processes.

The question therefore is, can biodiversity of soil organisms be used as a bioindicator of soil health, and if so does the measurement meet the requirements for a good bioindicator (see introduction)? The answer is yes, but with qualifications. Based on the examples and discussion given above, there would appear to be a number of different levels at which soil biodiversity can be used as an indicator of soil health. These can be summarized as follows.

Individual taxa or species

Here biodiversity within an individual soil organism taxon or species would be used as the bioindicator. The best example here would be the mutualist/dependent associations between plant species and soil organisms, such as the mycorrhizal fungi or *Rhizobium*. Loss of biodiversity of these organisms (at both species and genotype levels) within a soil, as a consequenece of soil and crop management or some perturbation such as heavy metal pollution, may reduce the capacity of these organisms to carry out their respective functions of phosphorus uptake or nitrogen fixation (Giller *et al.*, 1989; Johnson *et al.*, 1992). This in turn will affect the establishment and growth of certain plant species and have flow effects for system stability.

Functional groups of soil organisms

This is the most important and complex area for the use of biodiversity as a bioindicator, but it remains confounded with the issue of what level of biodiversity within a functional group is necessary/important for continuance of the function and whether or not assays for measurement of the function *per se* would not suffice and be far simpler to perform. The problem with simple measurement of the function is that it will not reveal any change in the biodiversity within the community of organisms carrying out the function and hence possible vulnerability of the soil's capacity to continue the function in the face of some perturbation. Use of newly developed assays such as metabolic profiling using BIOLOG (Bossio and Scow, 1995), DNA reassociation kinetics (Torsvik *et al.*, 1994), or phospholipid fatty acid analysis (Zelles *et al.*, 1992), for quantifying diversity within functional groups or communities, may be useful as bioindicators in this regard. For other functional groups of soil organisms such as the protozoa, nematodes and microarthropods, analysis of biodiversity based on functional attributes such feeding behaviour or physiotype (Freckman and Ettema, 1993; Bamforth, 1995; van Straalen, Chapter 10, this volume) coupled with multivariate analysis would appear to have bioindicator potential.

Community analysis

This is an extension of diversity analysis within functional groups to whole soil communities. As it is presently not possible to characterize or quantify the biodiversity of all functional groups of soil organisms within a soil community, techniques such as the phospholipid analysis or DNA hybridization of soils, which provide a 'snap shot' picture of the broad composition of soil organism communities, may have potential as bioindicators if a relationship between this type of analysis and soil health could be established. As discussed above, these techniques have been used successfully to characterize shifts in microbial community diversity associated with soil perturbations such as heavy metal pollution (Frostegard *et al.*, 1996) or agricultural practices (Zelles *et al.*, 1992). For larger soil organisms such as the meso- and macrofauna, there is significant evidence linking changes in abundance and diversity of species within communities of these organisms to changes in soil health (Wardle, 1995; van Straalen, Chapter 10 and Doube and Schmidt Chapter 11, this volume). The major limitations to using such changes in species abundance as a bioindicator are the unpredictability of the response of the community to the perturbation (Wardle, 1995), taxonomic difficulties and the time required for sampling, extraction and sorting the organisms. Also the choice of diversity index to quantify or describe the change in community diversity may pose difficulties (see discussion above). However, if the changes in the community structure are based simply on the loss of sensitive species or increase in abundance of tolerant/resistant species in response to a soil perturbation, then this form of analysis has potential for bioindication.

Conclusions

Assessment of biodiversity of soil organisms has the potential to provide useful insight into the health and functioning of soil. As a measure it has the advantages of being seen as integrative and responsive to a wide range of factors in the soil. It has public appeal and is compatible with the ideals of biodiversity conservation and ecologically sustainable development. It is also widely used as an indicator in environmental monitoring, especially in aquatic systems (Atlas, 1984). However, there are major conceptual and practical issues that need to be addressed, and significant knowledge gaps that need to be further researched before we can usefully use biodiversity as a purposeful bioindicator of soil health. Some of these issues and knowledge gaps are listed below.

1. Techniques for sampling and measuring biodiversity of some soil organisms (e.g. bacteria, fungi, and protozoa) require further development.
2. We require standardized approaches to the analysis and interpretation of soil biodiversity data.
3. We require more information about the linkages between soil biodiversity and the biodiversity of above ground plant communities and plant production.
4. We require more information concerning the response of soil biodiversity to perturbations (e.g. soil pollution).
5. We need more information concerning the relationship between soil biomass, biodiversity and soil biotic activity.
6. We need to obtain a better understanding of the complex interactions between soil organisms and how this affects soil biodiversity and soil functional processes.
7. We need to understand how biodiversity at one spatio-temporal scale may affect biodiversity at other levels of organization.
8. We require more information about the minimum requirements for biodiversity and ecosystem functioning and how to maintain biodiversity.
9. We need to be able to predict what degree of human-induced disturbance on the soil is needed to reduce biodiversity to the point where soil functional processes proceed at reduced efficiency.

References

Altieri, M.A. (1991a) How best can we use biodiversity in agroecosystems? *Outlook on Agriculture* 20, 15–23.

Altieri, M.A. (1991b) Increasing biodiversity to improve insect pest management in agro-ecosystems. In: Hawksworth, D.L. (ed.) *The Biodiversity of Microorganisms and Invertebrates: its Role in Sustainable Agriculture.* CAB International, Wallingford, UK, pp. 165–182.

Anderson, J.M. (1994) Functional attributes of biodiversity in land use systems. In: Greenland, D.J. and Szabolcs, I. (eds) *Soil Resilience and Sustainable Land Use.* CAB International, Wallingford, UK, pp. 267–290.

Atlas, R.M. (1984) Diversity of microbial communities. *Advances in Microbial Ecology* 7, 1–47.

Baker, G.H., Carter, P.J., Barnett, V.J., Kilpin, G.P., Buckerfield, J.C. and Dalby, P.R. (1994) The introduction and management of earthworms to improve soil structure and fertility in South-Eastern Australia. In: Pankhurst, C.E., Doube, B.M., Gupta, V.V.S.R. and Grace, P.R. (eds) *Soil Biota: Management in Sustainable Farming Systems.* CSIRO Press, Melbourne, pp. 42–49.

Bamforth, S.S. (1985) The role of protozoa in litters and soils. *Journal of Protozoology* 32, 404–409.

Bamforth, S.S. (1995) Interpreting soil ciliate biodiversity. *Plant and Soil* 170, 159–164.

Beare, M.H., Coleman, D.C., Crossley, D.A., Jr, Hendrix, P.F. and Odum, E.P.

(1995) A hierarchial approach to evaluating the significance of soil biodiversity to biogeochemical cycling. *Plant and Soil* 170, 5–22.

Bever, J.D., Morton, J.B., Antonovics, J. and Schultz, P.A. (1996) Host dependent sporulation and species diversity of arbuscular mycorrhizal fungi in a mown grassland. *Journal of Ecology* 84, 71–82.

Bianchi, A. and Bianchi, M. (1995) Bacterial diversity and ecosystem maintenance: an overview. In: Allsopp, D., Colwell, R.R. and Hawksworth, D.L. (eds) *Microbial Diversity and Ecosystem Function.* CAB International, Wallingford, UK, pp. 185–198.

Bongers, T. (1990) The maturity index: An ecological measure of environmental disturbance based on nematode species composition. *Oecologia* 83, 14–19.

Bossio, D.A. and Scow, K.M. (1995) Impact of carbon and flooding on the metabolic diversity of microbial communities in soils. *Applied and Environmental Microbiology* 61, 4043–4050.

Brown, G.G. (1995) How do earthworms affect microfloral and faunal community diversity? *Plant and Soil* 170, 209–231.

Cavigelli, M.A., Robertson, G.P. and Klug, M.J. (1995) Fatty acid methyl ester (FAME) profiles as measures of soil microbial community structure. *Plant and Soil* 170, 99–113.

Doran, J.W., Sarrantonio, M. and Janke, R. (1994) Strategies to promote soil quality and health. In: Pankhurst, C.E., Doube, B.M., Gupta, V.V.S.R. and Grace, P.R. (eds), *Soil Biota: Management in Sustainable Farming Systems.* CSIRO Press, Melbourne, pp. 230–237.

Ehrlich, P.R. and Ehrlich, A.H. (1992) *Extinction. The Causes and Consequences of the Disappearance of Species.* Random House, New York.

Ehrlich, P.R. and Wilson, E.O. (1991) Biodiversity studies: Science and policy. *Science* 253, 758–762.

Elliott, L.F. and Lynch, J.M. (1994) Biodiversity and soil resilience. In: Greenland, D.J. and Szabolcs, I. (eds) *Soil Resilience and Sustainable Land Use.* CAB International, Wallingford, UK, pp. 353–364.

Embly, T.M. and Stackebrandt, E. (1996) The use of 16S ribosomal RNA sequences in microbial ecology. In: Pickup, R.W. and Saunders, J.R. (eds) *Ecological Approaches to Environmental Microbiology.* Prentice Hall, London, pp. 39–62.

Foissner, W. (1987) Soil protozoa: Fundamental problems, ecological significance, adaptions in ciliates and testaceans, bioindicators, and guide to the literature. *Progress in Protistology* 2, 69–212.

Franklin, J.F. (1993) Preserving biodiversity: species, ecosystems, or landscapes? *Ecological Applications* 3, 202–205.

Freckman, D.W. (ed.) (1994) *Life in the Soil, Soil Biodiversity: Its Importance to Ecosystem Processes.* Report of a workshop held at the Natural History Museum, London, UK.

Freckman, D.W. and Ettema, C.H. (1993) Assessing nematode communities in agroecosystems of varying human intervention. *Agriculture, Ecosystems and Environment* 45, 239–261.

Frostegard, A., Tunlid, A. and Bååth, E. (1993) Phospholipid fatty acid composition, biomass and activity of microbial communities from two soil types experimentally exposed to different heavy metals. *Applied and Environmental Microbiology* 59, 3605–3617.

Frostegard, A., Tunlid, A. and Bååth, E. (1996) Changes in microbial community structure during long-term incubation in two soils experimentally contaminated with metals. *Soil Biology and Biochemistry* 28, 55–63.

Garland, J.L. (1995) Patterns of potential C source utilization by rhizosphere communities. *Soil Biology and Biochemistry* 28, 223–230.

Garland, J.L. and Mills, A.L. (1991) Classification and characterization of heterotrophic microbial communities on the basis of patterns of community-level sole-carbon-source utilization. *Applied and Environmental Microbiology* 57, 2351–2359.

Giller, K.E., McGrath, S.P. and Hirsch, P.R. (1989) Absence of nitrogen fixation in clover grown on soil subject to long-term contamination with heavy metals is due to survival of only ineffective *Rhizobium. Soil Biology and Biochemistry* 21, 841–848.

Grayston, S.J., Campbell, C.D. and Vaughan, D. (1994) Microbial diversity in the rhizospheres of different tree species. In: Pankhurst, C.E. (ed.) *Soil Biota: Management in Sustainable Farming Systems, Poster Papers.* CSIRO Press, Melbourne, pp. 155–157.

Groombridge, B. (ed.) (1992) *Global Biodiversity: Status of the Earth's Living Resources.* Chapman and Hall, London.

Gupta, V.V.S.R. (1994) The impact of soil and crop management practices on the dynamics of soil microfauna and mesofauna. In: Pankhurst, C.E., Doube, B.M., Gupta, V.V.S.R. and Grace, P.R. (eds) *Soil Biota: Management in Sustainable Farming Systems.* CSIRO Press, Melbourne, pp. 107–124.

Gupta, V.V.S.R. (1996) Diversity of protozoa – ecological significance in terrestrial ecosystems. In: *Proceeding of the 7th International Symposium on Microbial Ecology.* Santos, Brazil (in press).

Haack, S.K., Garchow, H., Klug, M.J. and Forney, L.J. (1995) Analysis of factors affecting the accuracy, reproducibility, and interpretation of microbial community carbon source utilization patterns. *Applied and Environmental Microbiology* 61, 1458–1468.

Harper, J.L. and Hawksworth, D.L. (1994) Biodiversity: measurement and estimation. Preface. *Philophosical Transactions of the Royal Society of London B.* 345, 5–12.

Harrison, S.P., Jones, D.G., Shunmann, P.H.D., Forster, J.W. and Young, J.P. (1988) Variation in *Rhizobium leguminosarum* bv. *trifolii* Sym plasmids and the association with effectiveness of nitrogen fixation. *Journal of General Microbiology* 134, 2721–2730.

Hawksworth, D.L. and Mound, L.A. (1991) Biodiversity databases: the crucial significance of collections. In: Hawksworth, D.L. (ed.) *The Biodiversity of Microorganisms and Invertebrates: its Role in Sustainable Agriculture.* CAB International, Wallingford, UK, pp. 17–29.

Hawksworth, D.L. and Ritchie, J.M. (1993) (eds) *Biodiversity and Biosystematic Priorities: Microorganisms and Invertebrates.* CAB International, Wallingford, UK.

Hendrix, J.W., Guo, B.Z. and An, S.Q. (1995) Divergence of mycorrhizal communities in crop production systems. *Plant and Soil* 170, 131–140.

Hill, M.O. (1973) Diversity and evenness: a unifying notation and its consequences. *Ecology* 54, 427–431.

Hirsch, P.R., Jones, M.J., McGrath, S.P. and Giller, K.E. (1993) Heavy metals from past applications of sewage sludge decrease the genetic diversity of *Rhizobium leguminosarum* bv. *trifolii* populations. *Soil Biology and Biochemistry* 25, 1485–1490.

Holben, W.E. (1994) Isolation and purification of bacterial DNA from soil. In: Weaver, R.W., Angle, S., Bottomley, P., Bezdicek, D., Smith, S., Tabatabai, A. and Wollum, A. (eds) *Methods in Soil Analysis. Part 2. Microbiological and Biochemical Properties.* Soil Science Society of America, Madison, Wisconsin, pp. 727–751.

Insam, H., Amor, K., Renner, M. and Crepaz, C. (1995) Changes in functional abilities of the microbial community during composting of manure. *Microbial Ecology* 31, 77–87.

Jenkinson, D.S. and Ladd, J.N. (1981) Microbial biomass in soil: measurement and turnover. In: Paul, E.A. and Ladd. J.N. (eds) *Soil Biochemistry.* Marcel Dekker, New York, pp. 415–471.

Johnson, N.C. and Pfleger, F.L. (1994) Vesicular-arbuscular mycorrhizae and cultural stresses. In: Pfleger, F.L. and Linderman, R.G. (eds) *Mycorrhizae in Plant Health.* American Phytopathological Society, St Paul, pp. 71–99.

Johnson, N.C., Copeland, R.K., Crookston, R.K. and Pfleger, F.L. (1992) Mycorrhizae: possible explanation for yield decline with continuous corn and soybean. *Agronomy Journal* 84, 387–390.

Kennedy, A.C. and Smith, K.L. (1995) Soil microbial diversity and the sustainability of agricultural soils. *Plant and Soil* 170, 75–86.

Kvalseth, T.O. (1991) Note on biological diversity, evenness, and homogeneity measures. *Oikos* 62, 123–127.

Lawton, J.H. (1994) What do species do in ecosystems? *Oikos* 71, 1–8.

Lawton, J.H. and Brown, V.K. (1993) Redundancy in ecosystems. In: Schulze, E.-D. and Mooney, H.A. (eds) *Biodiversity and Ecosystem Function.* Springer-Verlag, Berlin, pp. 255–270.

Lee, K.E. (1985) *Earthworms: their Ecology and Relationships with Soils and Land Use.* Academic Press, Sydney.

Lee, K.E. and Foster, R.C. (1991) Soil fauna and soil structure. *Australian Journal of Soil Research* 29, 745–775.

Lee, K.E. and Pankhurst, C.E. (1992) Soil organisms and sustainable productivity. *Australian Journal of Soil Research* 30, 855–892.

Letourneau, D.K. (1987) The enemies hypothesis: tritrophic interactions and vegetational diversity in tropical agroecosystems. *Ecology* 68, 1616–1622.

Liesack, W. and Stakebrandt, E. (1992) Occurrence of novel groups of the domain *Bacteria* as revealed by analysis of genetic material isolated from an Australian terrestrial environment. *Journal of Bacteriology* 174, 5074–5078.

Magurran, A.E. (1988) *Ecological Diversity and its Measurement.* Croom Helm Ltd., London, 179 pp.

Margalef, R. (1958) Information theory in ecology. *Genetic Systematics* 3, 36–71.

Miller, R.H. (1990) Soil microbiological inputs for sustainable agricultural systems. In: Edwards, C.A., Lal, R., Madden, P., Miller, R.H. and House, G. (eds) *Sustainable Agricultural Systems.* Soil Water Conservation Society, Ankey, Iowa, USA, pp. 614–623.

Miller, M.A. and Jastrow, J.D. (1994) Vesicular-arbuscular mycorrhizae and biogeochemical cycling. In: Pfleger, F.L. and Linderman, R.G. (eds) *Mycorrhizae in*

Plant Health. American Phytopathological Society, St Paul, Minnesota, pp. 189–212.

Mills, L.S., Soule, M.E. and Doak, D.F. (1993) The keystone-species concept in ecology and conservation. *Bioscience* 43, 219–224.

Moore, J.C. and de Ruiter, P.C. (1991) Temporal and spatial heterogeneity of trophic interactions within below-ground food webs. *Agriculture, Ecosystems and Environment* 34, 371–397.

Naeem, S., Thompson, L.J., Lawler, S.P., Lawton, J.H. and Woodfin, R.M. (1995) Declining biodiversity can alter the performance of ecosystems. *Nature* 368, 734–737.

Neher, D.A., Peck, S.L., Rawlings, J.O. and Campbell, C.L. (1995) Measures of nematode community structure and sources of variability among and within agricultural fields. *Plant and Soil* 170, 167–181.

Olsen, G.J., Overbech, R., Larsen, N. and Woese, C.R. (1991) The ribosomal RNA database project. *Nucleic Acids Research* 19 Supp, 2017–2021.

Pankhurst, C.E. (1994) Biological indicators of soil health and sustainable productivity. In: Greenland, D.J. and Szabolcs, I. (eds) *Soil Resilence and Sustainable Land Use.* CAB International, Wallingford, UK, pp. 331–351.

Pankhurst, C.E. and Lynch, J.M. (1995) The role of soil microbiology in sustainable intensive agriculture. *Advances in Plant Pathology* 11, 229–247.

Pankhurst, C.E., Ophel-Keller, K., Doube, B.M. and Gupta, V.V.S.R. (1996) Biodiversity of soil microbial communities in agricultural systems. *Biodiversity and Conservation* 5, 197–209.

Peet, R.K. (1984) The measurement of species diversity. *Annual Review of Ecology and Systematics.* 5, 285–307.

Pennanen, T., Frostegard, A., Fritze, H. and Bååth, E. (1996) Phospholipid fatty acid composition and heavy metal tolerance of soil microbial communities along two heavy metal-polluted gradients in coniferous forests. *Applied and Environmental Microbiology* 62, 420–428.

Pimm, S.L. (1993) Biodiversity and the balance of nature. In: Schulze, E.-D. and Mooney, H.A. (eds) *Biodiversity and Ecosystem Function.* Springer-Verlag, Berlin, pp. 347–359.

Potter, C.S. and Meyer, R.E. (1990) The role of soil biodiversity in sustainable dryland farming systems. *Advances in Soil Science* 13, 241–251.

Price, W.P. (1988) An overview of organismal interactions in ecosystems in evolutionary and ecological time. *Agriculture Ecosystems and Environment* 2, 369–377.

Rayner, A.D.M., Griffith, G.S. and Ainsworth, A.M. (1994) Mycelial interconnectedness. In: Gow, N.A.R. and Gadd, G.M. (eds) *The Growing Fungus.* Chapman and Hall, London, pp. 22–40.

Ritz, K. and Griffiths, B.S. (1994) Potential application of a community hybridization technique for assessing changes in the population structure of soil microbial communities. *Soil Biology and Biochemistry* 26, 963–971.

Roget, D.K. (1995) Decline in root rot (*Rhizoctonia solani* AG-8) in wheat in a tillage and rotation experiment at Avon, South Australia. *Australian Journal of Experimental Agriculture* 35, 1009–1013.

Schenck, N.C. and Siqueira, J.O. (1987) Ecology of VA mycorrhizal fungi in temperate agroecosystems. In: *Mycorrhizae in the Next Decade – Practical Applica-*

tions and Research Priorities. University of Florida, Gainsville, Florida, pp. 2–4.

Shannon, C.E. and Weaver, W. (1949) *The Mathematical Theory of Communication.* University Illinois Press, Urbana, Illinois, 117 pp.

Siepel, H. and de Ruiter-Dijkman, E. (1993) Feeding guilds of oribatid mites based on their carbohydrase activities. *Soil Biology and Biochemistry* 25, 1491–1497.

Simpson, H. (1949) Measurement of diversity. *Nature* 163, 688.

Smith, M.L., Bruhn, J.N. and Anderson, J.B. (1992) The fungus *Armillaria bulbosa* is among the largest and oldest living organisms. *Nature* 356, 428–431.

Stackebrandt, E., Liesack, W. and Goebel, B.M. (1993) Bacterial diversity in a soil sample from a subtropical Australian environment as determined by 16S rRNA analysis. *FASEB Journal* 7, 232–236.

Swift, M.J. (1987) Organisation of assemblages of decomposer fungi in space and time. In: Giller, P. and Gee, J. (eds) *Organisation of Ecological Communities: Past and Present.* Blackwell, Oxford, pp. 229–253.

Swift, M.J. and Anderson, J.M. (1993) Biodiversity and ecosystem function in agricultural systems. In: Schulze, E.-D. and Mooney, H.A. (eds) *Biodiversity and Ecosystem Function.* Springer-Verlag, Berlin, pp. 15–41.

Temple-Smith, M.G. (1991) Earthworm populations and their effects on pasture production. In: Haines, P.J. (ed.) *Earthworms. Improving Soil for Agriculture.* Australian Institute of Agricultural Science, Occasional Publication No. 62, pp. 11–15.

Tilman, D. and Downing, J.A. (1994) Biodiversity and stability in grasslands. *Nature* 367, 363–365.

Torsvik, V., Goksoyr, J. and Daae, F.L. (1990) High diversity in DNA of soil bacteria. *Applied and Environmental Microbiology* 56, 782–787.

Torsvik, V., Goksoyr, J., Daae, F.L., Sorheim, R., Michalsen, J. and Salte, K. (1994) Use of DNA analysis to determine the diversity of microbial communities. In: Ritz, K., Dighton, J. and Giller, K.E. (eds) *Beyond the Biomass.* John Wiley and Sons, Chichester, UK, pp. 39–48.

Van Emden, H.F. (1990) Plant diversity and natural enemy efficiency in agroecosystems. In: Mackauer, M., Ehler, L.E. and Roland, J. (eds) *Critical Issues in Biological Control.* Intercept, Andover, Hants, pp. 63–80.

Vitousek, P.M. and Hooper, D.U. (1993) Biological diversity and terrestrial ecosystem biogeochemistry. In: Schulze, E.-D. and Mooney, H.A. (eds) *Biodiversity and Ecosystem Function.* Springer-Verlag, Berlin, pp. 3–14.

Voordouw, G., Voordouw, J.K., Karkhoff-Schweizer, R.R., Fedorak, P.M. and Westlake, D.W.S. (1991). Reverse sample genome probing, a new technique for identification of bacteria in environmental samples by DNA hybridization, and its application to the identification of sulfate-reducing bacteria in oil field samples. *Applied and Environmental Microbiology* 57, 3070–3078.

Walker, B.H. (1992) Biodiversity and ecological redundancy. *Conservation Biology* 6, 18–23.

Ward, D.M., Weller, R. and Bareson, M.M. (1990) 16S rRNA sequences reveal numerous uncultured microorganisms in a natural community. *Nature* 345, 63–65.

Wardle, D.A. (1995) Impacts of disturbance on detritus food webs in agroecosystems of contrasting tillage and weed management practices. *Advances in Ecological Research* 26, 105–185.

Weiss, B. and Larink, O. (1991) Influence of sewage sludge and heavy metals on nematodes in an arable soil. *Biology and Fertility of Soils* 12, 5–9.

Winding, A. (1994) Fingerprinting bacterial soil communities using Biolog microtitre plates. In: Ritz, K., Dighton, J. and Giller, K.E. (eds) *Beyond the Biomass.* John Wiley and Sons, Chichester, pp. 85–94.

Yeates, G.W., Bongers, T., de Goede, R.G.M., Freckman, D.W. and Georgieva, S.S. (1993) Feeding habits in nematode families and genera – an outline for soil ecologists. *Journal of Nematology* 25, 315–331.

Yeates, G.W., Orchard, V.A., Speir, T.W., Hunt, J.L. and Hermans, M.C.C. (1994) Impact of pasture contamination by copper, chromium, arsenic timber preservative on soil biological activity. *Biology and Fertility of Soils* 18, 200–208.

Zak, J.C., Willig, M.R., Moorhead, D.L. and Widman, H.G. (1994) Functional diversity of microbial communities: a quantitative approach. *Soil Biology and Biochemistry* 26, 1101–1108.

Zelles, L., Bai, Q.Y., Beck, T. and Beese, F. (1992) Signature fatty acids in phospholipids and lipopolysaccharides as indicators of microbial biomass and community structure in agricultural soils. *Soil Biology and Biochemistry* 24, 317–323.

Biomonitoring of Soil Health by Plants

13

T. Pandolfini, P. Gremigni and R. Gabbrielli
Dipartimento di Biologia Vegetale, Università di Firenze, Via Micheli 1, I-50121 Firenze, Italy

Introduction

Changes in the composition of undisturbed and arable soils can be the consequence of climatic variation or of human activity. The input of industrial contaminants to soils by atmospheric deposition produces long term changes at both global and regional scales, whereas the use of phytochemicals (fertilizers and pesticides) and the application of sewage sludge to agricultural lands are responsible for changes in soil composition at smaller spatial scales (Jones, 1991). Such disturbances can adversely affect soil health resulting in the loss of plant productivity and pose a risk to human health.

Soil quality/health can be assessed by chemical, physical and biological analysis (see Doran and Safley, Chapter 1, this volume). Biological monitoring is the measurement of the response of living organisms to changes in their environment. Wittig (1993) has analysed the differences between instrumental and biological monitoring. In comparison to instrumental monitoring (which provides information on discrete attributes of the soil), the use of living organisms as bioindicators provides information that integrates many environmental factors, although the natural variability in biological systems can be greater than that of many physical and chemical attributes of soil. Moreover, biological monitoring has special relevance to human health because it evaluates the effects of environmental changes on key elements of the food chain. A further advantage of monitoring by means of living organisms is the potential for tracing chronological changes in soil condition, for example by analysis of growth rings of trees. However, distinguishing between airborne and soilborne pollutants is more difficult using bioindicators than when using instrumental monitoring (Wittig, 1993).

Plants as Biomonitors

Both plants and animals can be used as bioindicators of soil health. However, plants are commonly more suitable than animals for revealing the characteristics of soil in a specific area because they are sessile and so their roots cannot avoid absorbing soluble inorganic compounds which are present in the soil.

The use of plants as bioindicators of soil health in agriculture and in geobotanical prospecting has its origin in ancient times (Ernst, 1993; Wittig, 1993). The presence of plant species can be used to diagnose a particular soil condition, e.g. the acidic or basic nature of the substrate (qualitative bioindication) or can allow us to quantify a particular soil component or contaminant (quantitative bioindication). The floral composition of whole plant communities can act as a bioindicator of soil condition. For instance, soils associated with calcareous rocks have a characteristic flora (a calciphilous flora), whilst soils rich in sodium chloride, sodium carbonate and sodium sulphate have a halophyte flora (Brooks, 1972). In serpentine soils which contain high levels of Mg, Ni, Co and Cr, plant communities are markedly different from surrounding vegetation (Brooks, 1972). Individual species, as well as varieties, are also used to assess soil health. Wittig (1993) reported that some agricultural weeds can be used to indicate soil conditions. For example, the presence of *Teesdalia nudicaulis* indicates a very acid sandy soil, *Mercurialis annua* indicates alkaline soils and the presence of *Ranunculus repens* on arable fields indicates compacted soil. Many other species could be described as bioindicators of soil condition, for instance, *Phragmites australis* thrives in humid habitats, species of the genera *Dianthus* prosper in calciphilous conditions, and halophytes such as *Atriplex* spp. and *Salicornia herbacea* indicate salt-affected soils.

A distinction is usually made between passive and active biomonitoring. Passive biomonitoring is defined as the use of native and cultivated vegetation present in the study area, whereas active biomonitoring is carried out by introducing test plants which are commonly selected on the basis of their sensitivity to the soil contaminant to be monitored. Selected test plants are also used for laboratory tests in which case they are grown on defined substrates under controlled environmental conditions (temperature, light, humidity).

The choice of the biomonitoring method depends on the aim of the investigation. Passive biomonitoring is the most suitable method for retrospective analysis of soil, as well as being a relatively cheap process. Active biomonitoring is a much more instructive procedure because selected plants of genetic uniformity which possess specific sensitivities can be grown under standardized and known conditions. Plants used as active bioindicators should initially be healthy, without visible parasitic infestation and the plant parts sampled should be homogeneous as regard to age and physiology. These constraints are rarely feasible with passive biomonitoring where it is particularly important to have information about the soil properties and other plants in the study area, as well as cultivation methods if agricultural soils are present.

Vegetation, as a means of biomonitoring soil health can be studied at a variety of spatial scales, namely at the ecosystem, population, organism, tissue, organ or cell scale. Changes in plant distribution patterns are caused by alteration in climatic factors or soil conditions, sometimes leading to variations in vegetation structure within the landscape. These changes in plant community composition are usually monitored by means of mapping the relative abundance of single plant species (Weinert, 1991) or mapping vegetation changes by remote sensing. Two processes modify the floral composition of plant communities, namely: (i) the extinction or recession of sensitive species (sensitive bioindicators); or (ii) colonization or increased dominance by tolerant species, sometimes not native to the particular area under study (neophyte). Changes in soil quality/health are in general accompanied by changes in the diversity of plant communities (Weinert, 1991).

Sensitive bioindicators

The reaction of plants which are sensitive to changes in the soil can be evaluated not only in terms of the presence or absence of the plant, but also through the quantification of visible injuries to the plant. These injuries may include anatomical and morphological abnormalities and/or an alteration in growth rate. Visible symptoms are the result of complex metabolic changes and some biochemical alterations may be detectable before the symptoms become visible. Hauschild (1993a) has used the term 'biomarkers' to describe specific compounds that are produced, accumulated or activated by a plant in response to stress. These biomarkers can take the form of enzymes or metabolites, and represent a promising research field for the early detection of changes in soil character. The main drawback of biomarkers, however, is that they often lack specificity; the same metabolic alterations can be induced by different stress factors. The ideal sensitive bioindicator should possess the following characteristics: (i) it should be highly sensitive to specific key soil factor(s) and unaffected by other stress factors (for instance to atmospheric pollutants); (ii) the response should be specific, reproducible and easy to detect; and (iii) the intensity of the response must be easily quantified and strictly related to the extent of soil alteration.

Accumulative bioindicators

If the aim of biomonitoring is to determine the level of a contaminant or trace element in the soil, the bioindicator should be a plant species which has the ability to accumulate the contaminant without showing toxic symptoms. These accumulative bioindicators should give a quantitative measurement of the level of soil contamination.

It is important that the efficiency of biological accumulation of the contaminant is documented over a broad range of contaminant concentrations in the soil. In higher plants, the relationship between plant and soil concentration of an element is seldom linear.

Depending on the purpose of the monitoring activity, different parts of the plant can be used as bioaccumulators of the contaminant or trace element. Roots and rhizomes are suitable organs for determining soilborne substances, although contamination by soil particles may be a problem. Biomonitoring by means of leaf analysis is widespread but, in some circumstances, contamination by atmospheric deposition adds complexity to interpreting such analyses. Washing of the leaves before analysis partially overcomes this problem but direct absorption into the leaves may occur. Using unwashed plant material is preferable when the contribution of aerial pollution to soil pollution is being evaluated or when the effect of a contaminant on the trophic chain is being considered. Further, plant bioaccumulation is a useful tool for evaluating contaminant bioavailability and mobility in soil.

A variety of different groups of plants (lichens, mosses, fungi, higher plants) have been used as biomonitors of environmental changes. Lichens are generally recognized as useful biomonitors of environmental pollution (Garty, 1993) and can be sensitive bioindicators of atmospheric pollution as well as accumulative bioindicators of trace element pollution in soil. Mosses have been widely exploited as accumulative indicators in passive and active biomonitoring of metal pollution (Steinnes, 1993; Tuba and Csintalan, 1993). Both lichen and mosses have several advantages: they possess a marked capacity for bioaccumulation of a variety of compounds, they are widely distributed in nature and they show minimal seasonal variation in biomass and morphology. Moreover, there is strong evidence that the response of cryptogam plants (non-flowering plants) to environmental contaminants is more sensitive and rapid than that of most flowering plant species (Rasmussen *et al.*, 1980). On the other hand, the physiology and ecology of the majority of higher plants are better characterized than that of cryptogram plants (Wittig, 1993). Higher plants are routinely cultivated in the laboratory and genetically uniform material is readily obtained using *in vitro* techniques (Wittig, 1993). Further, species or cultivars of plants which are tolerant to different stress factors (e.g. atmospheric pollutants or pathogens) are available.

Assessment of Trace Element Toxicity

The biomonitoring of trace element concentration in plants, with special reference to heavy metals, will be discussed in detail as an example of soil health assessment. Soil pollution (increase in heavy metal concentrations, acidification, contamination from trace organic compounds) is a major problem in natural and agricultural soils in all industrialized countries. These changes in soil health are significant both at a regional and at a local scale (Jones, 1991). The presence of an excessive concentration of trace elements in the soil and the use of accumulative indicators for the quantification of soil contamination have been the subject of numerous publications (Albaway *et al.*, 1988; Kabata-Pendias *et al.*, 1993; Wagner, 1993; Farago, 1994).

Trace element burden in the soil and bioavailability

The term 'trace element' commonly refers to those mineral elements which are present in soil in low concentration, relative to the more abundant elements in both the soil solution and the plant. Amongst the trace elements are many heavy metals (metallic elements with a specific mass higher than 4.5) and some other elements, e.g. Al. From a biological point of view and based upon their ligand affinity, metal ions can be divided into three classes: class A metals forming complexes preferentially with ligands containing oxygen; class B metals forming complexes preferentially with ligands containing nitrogen or sulphur; and borderline metals showing intermediate behaviour between class A and class B metals (Nieboier and Richardson, 1980). Here the terms 'metal' and 'heavy metal' will be used to indicate metallic trace elements (i.e. chemically defined heavy metals plus Al). Some of these metals are essential plant micronutrients since they are required for plant growth and development (Fe, Mn, Cu, Zn , Ni, Mo, Co), while others (for instance Cd, Cr, Pb, Al), although not essential for growth, may be present in plant tissues. Both essential and non-essential heavy metals are considered as soil pollutants if they are present in plant tissues in concentrations which produce undesirable effects.

An excessive concentration of heavy metals in the soil may be due to natural rock mineralization processes which produce metal-enriched soils. Alternatively, the heavy metal contamination of soil may be derived from various anthropogenic activities such as mining, metal smelting, other industrial processes, agricultural practices (e.g. the use of fertilizers, and pesticides; sewage sludge disposal), vehicle exhausts and the waste products of electric power plants (see Mhatre and Pankhurst, Chapter 14, this volume).

It is not easy to use chemical analysis to assess whether the measured concentration of a trace element in soil is out of line with acceptable baseline values. Davies (1994) has discussed the criteria used to define the upper acceptable limits for the concentrations of the various elements in soil. It is important to recognize that the total concentration of a trace element in the soil is not always the relevant parameter for biomonitoring, because only a proportion of that amount may be available for plant absorption. The estimation of availablity of a trace element for plant uptake is commonly obtained using soil extractants, such as dilute solutions of mineral or organic acids and/or metal complexing agents (Davies, 1994). Mild extractants such as water or $CaCl_2$ are used to evaluate soil solution composition, while readily exchangeable metals are

Table 13.1. Soil factors affecting metal bioavailability.

Soil phases	Factors
Solid phase	Superficial negative charges of: clays metal oxides organic material
Liquid phase (soil solution)	pH Redox potential Presence of adsorbing or complexing agents Water flux
Living phase	Microorganism action Plant exudate

extracted with NH_4^+. The chelating agent EDTA (ethylenediaminetetraacetic acid) acts as an organic ligand giving information on the organically bound fraction (Streit and Stumm, 1993).

All metals which are present in a soluble form in the soil solution can be taken up by plants. The concentration of a metal in the soil solution depends on a number of factors (Table 13.1). Firstly, interactions with organic and inorganic soil components can remove metals from the soil solution by their adsorption to organic matter and clays and the formation of insoluble complexes which precipitate. These processes (metal adsorption and complexation) are always more intense in neutral or alkaline soils (compared with acidic soils) and are dependent on the soil's cation exchange capacity (CEC). Metals can exist in the soil solution in different chemical forms depending partially on the concentration of the other mineral elements in the soil solution. Metal speciation affects bioavailability, as the absorption capacity of roots is dependent on the form of the inorganic complexation of the metal in the soil solution. The metal availability is also influenced by the biological activity of the rhizosphere, since both microorganisms and plant roots can modify the form of metals.

Biomonitoring metal toxicity using plants

The most common methods of assessing metal toxicity in soil are those listed by Streit and Stumm (1993): (i) monitoring the presence or absence of specific plant ecotypes, plant species or plant associations (indicator plant or vegetation); (ii) measuring the tissue metal concentration of selected species (accumulative bioindicators); and (iii) recording physiological and biochemical responses (biomarkers) in sensitive bioindicators. These topics are discussed briefly for lower plants (lichens, bryophytes and fungi) and more extensively for higher plants.

Lichens

Lichens are plant-like organisms composed of a green alga or cyanobacterium (photobiont) and a fungus (mycobiont). Their thalli lack cuticles, stomata and functional roots. They do, however, have an efficient nutrient uptake system and a great ability to absorb and lose water rapidly.

Lichens are generally accepted as being valuable bioindicators of a wide range of aerial pollutants, such as metals, non-metals, radionucleotides, and organic substances (Richardson, 1988; Henderson, 1990). Toxic pollutants can poison the more sensitive species causing a change in the population growing on a particular substrate (Richardson, 1991). However, those which have evolved a survival strategy can rapidly absorb, and simultaneously accumulate, various pollutants in their thalli. Samples of lichens collected near urban or industrial emission sources show elevated concentrations of pollutants. The highest elemental levels can be found in lichens collected close to the emission source, with values falling off logarithmically as proximity decreases (Henderson, 1990). Most aerial pollutants finish up entering the soil and living organisms. The suitability of lichens as bioindicators in the assessment of aerial pollution has led to the search for terricolous species that could be useful for biomonitoring soil contaminants. Many reports concerning metal accumulation capacity and the metal resistance of lichen species of mineralized and metal-contaminated soils have been published in recent decades and are reviewed by Nash (1990).

Seaward *et al.* (1978) investigated seven lichen species (*Peltigera canina, P. rufescens, Cladonia furcata, C. implexa, C. uncialis, Cornicularia aculeata* and *C. muricata*) occurring on mineralized soils in northern England and concluded that they were good bioindicators for Pb and Zn. Three different species collected from this site were able to absorb from 520 to 880 $\mu g\ g^{-1}$ Zn. Several other lichen species growing on Zn rich soils were found to contain Zn levels from 3000 to 10,000 $\mu g\ g^{-1}$ Zn (ash weight) (Lounamaa, 1965; Shimwell and Laurie, 1972). In experiments of Zn and Cd uptake from water solutions, the thalli of *Cladonia uncialis* and *Lasalia populosa* were found to reach internal concentrations of around 3000 $\mu g\ g^{-1}$ Zn and 7500 $\mu g\ g^{-1}$ Cd (Nash, 1975). A great capacity to accumulate Cu, Ni and Zn was observed in the rhizinae of several terricolous lichens (*Peltigera* spp.) growing on soils polluted by heavy metals.

Lecanora cascadensis, Umbilicaria phaea, Acarospora chlorophana and *L. melanophthalma* are well known bioindicators of Cu-rich substrates, accumulating the metal to concentrations between 9000 and 23,300 $\mu g\ g^{-1}$ Cu (Czehura, 1977). Purvis (1984) studied lichens collected from different sites with Cu-sulphide mineralization and observed *Lecidea inops* to have a particular ability to concentrate Cu (as Cu-oxalate), thus constituting a specific indicator of Cu mineralization in soil. On gneissic rocks in the Disko Island, West Greenland, Alstrup and Hansen (1977) found three lichen species (*Umbilicaria lynge*, *Alectoria pubescens* and *Lecanora polytropa*) which were tolerant of the high Cu concentrations in the rock substrate.

Iron can provide a good chemical substrate for lichens (Brightman and Seaward, 1977). Puckett *et al.* (1973) examined the competitive uptake of metals by lichens and found that they have a high affinity for Fe, with their affinity for other metals decreasing in the order Fe $\gg$ Pb $\gg$ Cu $\gg$ Ni > Zn > Co. Some species of lichens growing on waste heaps of iron ore can accumulate high concentrations of Fe (Noeske *et al.*, 1970).

Several lichen species colonized serpentine soils (well known for their naturally high content of heavy metals), and their thalli contained high Ni levels which, nevertheless, did not affect plant health. Wilson *et al.* (1981) investigated the weathering of a serpentine soil by *Lecanora atra* and found high concentrations of Mg and Fe in thalli, together with traces of Ni and Mn. On ultramafic rocks rich in Cr, several lichen species were better able than other plants to concentrate Cr in their tissues, but tissue levels did not exceed the concentration of this element in the substrate. Studies on the Cr content of lichens in relation to that of the soil have produced highly variable ratios of Cr(soil) : Cr(lichen) (Garty, 1985). The Cd concentration in lichens growing on different substrates has also not shown a clear correlation with the metal concentration in that substrate (Garty, 1993). Similarly no clear relationship has been found between the Hg content of soil and the total Hg content of lichens (Siegel and Siegel, 1976).

Terricolous lichens could function as bioindicators of soil pollution by providing a means of detecting heavy metals in contaminated soils, but such studies need to be further developed with special emphasis on understanding of the physiology of metal absorption in lichens. The development of improved sampling methods for lichens should consider the fundamental physiological differences between broad groups of lichens, especially the differences between green algae and cyanobacteria-containing lichens. Some factors to consider include: the non-homogeneous distribution of metals in lichen thalli, the ideal size of the thalli for sampling (Gailey and Lloyd, 1986a), the exposure times in active biomonitoring (Gailey and Lloyd, 1986b), the variability between populations collected at the same site and the contamination of lichen by terrigenous material (Bargagli, 1990).

Bryophytes

Bryophytes are a group of non-vascular, mainly terrestrial plants, lacking cuticle and roots. They can absorb large amounts of water but they also have the capacity to survive periods of drought. Despite their infrequent use, bryophytes have great potential as bioindicators. Bryophytes are widespread in metal-contaminated environments, where they can easily survive due to their high reproductive capability (Gilbert, 1968). Several species of bryophyte occur on mine tailings (Ernst, 1974) and on natural soils containing high concentrations of mineralized metals (Shacklette, 1965; Empain, 1985). The so called copper mosses (including at least seven species of moss and liverwort endemic to metalliferous soils) are a specialized group of bryophytes which are commonly associated with heavy metal deposits (Shaw, 1990). While higher plants tend to accumulate high concentrations of macronutrients, bryophytes tend to accumulate elevated concentrations of trace elements such as Cd, Hg, Cu, Pb, Zn, Cr and Ni, even when these elements are not detectable in the soil (Shaw, 1990). Only one bryophyte species, *Pohlia nutans* (a ubiquitous species occurring also on mine tailings), was found in a highly polluted soil close to a foundry emitting copper and zinc in southern Sweden (Folkeson and Andersson-Bringmark, 1988). The same species was the most abundant bryophyte collected on naturally copper-enriched soil in Ontario, Canada (Dykeman and DeSousa, 1966). While the Cu concentration in the gametophytic plants of this species reached 36000 $\mu g\ g^{-1}$ (3.6%) Cu, no other species was found in this area with a concentration of Cu exceeding that found in the soil (Dykeman and DeSousa, 1966).

In most experiments using lichens and mosses as bioindicators of metal pollution, the passive biomonitoring technique has been applied, providing reliable information about contaminant distribution. However, more recently, active biomonitoring using transplantations has been extensively applied. Here lichens and mosses of an optimal size have been collected and transplanted into exposure boxes, then attached for suitable exposure times to the soil surface in both control and contaminated areas (Tuba and Csintalan, 1993). Good results can been obtained from analysis of heavy metal concentrations in the thalli.

Fungi

Higher fungi can readily accumulate heavy metals from soil, because their vegetative cells have unprotected surfaces and their mycelia have a large surface area to volume ratio (Wondratschek and Röder, 1993). Fruiting bodies (with high rates of growth and short lives above ground) are also suitable for biomonitoring soil pollution. Hundreds of different fungal species have been examined for their capability to accumulate heavy metals from soil. In a recent review on this topic, Mejstrík and Lepsová (1993) concluded that many species of fungi can function as indicator species, despite their different sensitivities to heavy metals and their different abilities in accumulating them. On the other hand, Wondratschek and Röder (1993) propose extreme caution in the use of higher fungi for biomonitoring soil pollution because different species show highly variable responses to heavy metals. Even if the level of substrate contamination by heavy metals is related to their uptake by some fungus species, the use of fungi as bioindicators is not always advisable. They are most useful in distinguishing between polluted and non-polluted areas and in determining sources of heavy metal emission. Reliable results require samples from a number of different species collected from a range of sites in the study area. Further research is required on the metal tolerance of higher fungi, as well as improved standardized methods for sampling and data analysis.

Higher plants

Indicator plants and vegetation associations

In soils characterized by toxic concentrations of naturally mineralized metals, plant cover is sparse and biological diversity is low when compared with non-mineralized soils in the surrounding areas. The vegetation of metal-enriched areas commonly comprises herbs and shrubs, but few trees, and it is characterized by the presence of endemic plant species (i.e. metal resistant species restricted to metal-enriched soils) and metal-resistant ecotypes of plant species which also occur on the surrounding non-toxic substrates. Some endemic species occupy the same niche as species in non-mineralized environments, e.g. the Cu-indicator *Becium homblei* occupies the same niche as *Becium ovatum* from non-mineralized soils in Central Africa (Ernst, 1993). Endemic species (see above) have been extensively used as metal indicators in geobotanical prospecting (Brooks, 1972, 1983; Ernst, 1993). On the other hand, resistant ecotypes of common plant species are not useful as a bioindicator of soils with high metal levels because they are usually morphologically identical to the corresponding non-tolerant genotypes found in the surrounding non-mineralized soils. Some examples of specific indicator plants are reported in Table 13.2.

Table 13.2. Some examples of metal indicator species.

Species	Element	Geographical distribution
Silene cobalticola	Co	Central Africa
Armeria maritima	Cu	United Kindom
Becium homblei	Cu	Africa
Gypsophila patrinii	Cu	Russia
Minuartia verna	Cu	Germany
Viscaria alpina	Cu	Scandinavia
Alyssum bertolonii	Ni	Italy
Alyssum murale	Ni	Balkans
Homalium austrocaledonicum	Ni	New Caledonia
Eriachne mucronata	Pb	Australia
Astragalus spp.	Se	Western USA
Thlaspi calaminare	Zn	Germany
Viola calaminaria	Zn	Germany

After Brooks (1972).

These species can function as metal indicators at a regional scale (e.g *Minuartia verna* in the colline and montane zone of Europe) or at a local scale (e.g. *Viola calaminaria*).

When plants were exposed to high concentrations of heavy metals in areas contaminated by mining activity, smelting and other industrial activities, changes to plant communities took several years to become evident (Verkleij, 1993). Accompanying the disappearance of sensitive plant species was the colonization of the area with metal-tolerant populations of a few grass species (Verkleij, 1993). In the case of less serious contamination, such as that produced by agricultural activities or by long distance aerial deposition, a detectable change in vegetation structure and floristic composition of the plant communities may not occur, although some plants may show symptoms of metal toxicity. Metal contamination can therefore be monitored by assessment of symptoms such as chlorosis, stunted growth, dwarfism, malformation (e.g. epinasty), and abnormal biochemical responses. Additionally, contamination of the soil by heavy metals may be revealed by analysing the tissue of suitable accumulative bioindicator plants.

Metal accumulative bioindicators

Plants take up essential mineral elements from the soil and accumulate them in concentrations above that present in the soil solution. Non-essential mineral elements in the soil solution are also taken up by roots and therefore plant analysis can be used to predict the soil concentration of trace elements. This is the rationale for the widespread use of plant analysis in biogeochemical prospecting.

Plants of metalliferous soils have been classified by Baker (1981) into three groups on the basis of their capacity to accumulate metals in their shoots.

1. Accumulators, which are plants with the capacity to concentrate metal in their shoots from soils containing both low and high levels of trace elements.
2. Indicators, which are plants whose shoot metal concentrations reflect those in soil.
3. Excluders, which are plants whose shoot metal concentrations remain unaffected by soil concentrations up to a critical soil value. Above this limit the translocation of metal to the shoots is no longer restricted.

Based on the Baker classification (above), indicator plants of metalliferous soils should fulfil the demand for accumulative bioindicators. The most suitable organ for tissue sampling from accumulative indicator species needs to be selected in relation to the type of pollution being assessed. For example, leaves may not always be the most suitable tissue. Root analysis may work well when assessing soilborne contamination or when metals with a high affinity for cell-wall components (e.g. Cr and Pb) are studied (Ernst, 1993). On the other hand, aerial parts may be suitable for sampling airborne contamination and are also useful where metals are readily translocated from the roots to the shoots (e.g.

Cu and Zn). If the aim of the monitoring activity is to obtain a chronological profile of metal content in the soil, tree bark represents the best organ (Dion *et al.*, 1993).

Among the flora which can tolerate high metal concentrations in soils, accumulator species deserve special attention. Minguzzi and Vergnano (1948) first reported an exceptional Ni accumulating ability in *Alyssum bertolonii*, endemic to the serpentine soils in Tuscany, Italy. In this species, the foliar Ni concentration can be as high as 10000 $\mu g\ g^{-1}$ dry weight (DW). Brooks *et al.* (1977) introduced the term 'hyperaccumulator' to refer to plants containing more than 1000 $\mu g\ g^{-1}$ DW of Ni. Following this discovery, many other species which hyperaccumulate Ni have been identified (Baker and Brooks, 1989). Hyperaccumulation of Zn, Co and Cu in plant tissue is now also well documented (Baker and Brooks, 1989).

Hyperaccumulator species (and other species of naturally metal-enriched soils) have potential as metal indicators, as well as an important role in the revegetation and decontamination of metal polluted sites. For instance, some plants of the Italian and California serpentine soils are good candidates for biomonitoring of Ni contamination (Pandolfini, 1987; Arianoutsou *et al.*, 1993). They also have potential as active biomonitors, even though these species are poor competitors in the absence of high levels of soil contamination.

The assessment of anthropogenic metal contamination in natural and agricultural soils has been extensively carried out through plant analysis. Wild and cultivated plant species have been used as passive accumulative bioindicators for large-scale and local soil evaluation (Otte *et al.*, 1991; Kabata-Pendias *et al.*, 1993; Wagner, 1993; Ward and Savage, 1994). The aim of these investigations is to establish a relationship between metal concentrations in the soil and in plants growing in that soil. This relationship is often expressed by calculating concentration factors, that is the ratio between the concentration of the metal in the plant and that in the soil. To compare the extent of metal contamination in different regions, it is useful to calculate the enrichment ratio (ratio of metal concentration of plants in polluted areas to those in similar plants from non-polluted regions) (Kabata-Pendias *et al.*, 1993). Metal bioaccumulation is influenced by both soil (Table 13.1) and plant factors (e.g. species, variety or cultivar, age, organ), as well as by microclimatic conditions. As a consequence, standardized methods for sampling and analysis of both soils and plants are necessary. Pot experiments under standard environmental conditions permit evaluation of effects of metal concentration on plant growth. Such studies have been used to determine permissible levels of metals in the soil (Piotrowska and Dudka, 1994). For instance, cereal plants were used to determine the maximum permissible levels of Cd in light soils (3 $\mu g\ g^{-1}$ DW) (Piotrowska and Dudka, 1994). In his recent review, Wagner (1993) evaluated standardized methods for use of metal bioindicators in large-scale screening projects. He recommended *Populus nigra* 'Italica' as an effective accumulative bioindicator because of its genetic homogeneity (due to vegetative propagation), its worldwide distribution (in agricultural and urbanized areas) and its large capacity to absorb trace metals. Other trees, such as *Picea abies and Pinus sylvestris* have also been proposed as metal indicators (Wagner, 1993). A more complete list of potential accumulative bioindicator species for detection of metal contamination is reported by Wittig (1993).

Crop plants, because of their importance in the food chain, have been used frequently to assess the level of metal contamination in soils (Panda *et al.,* 1992; Kabata-Pendias *et al.*, 1993). Barley has been shown to be a useful indicator of Hg pollution (Panda *et al.,* 1992). In various cereal plants, the concentration of Cd was related to the concentrations of the exchangeable form of Cd in the soil (Piotrowska and Dudka, 1994). There was a marked accumulation of Cd in the foliage of pasture in situations in which soil Cd levels were elevated (Albway *et al.*, 1988). Kabata-Pendias *et al.* (1993) compared the levels of metal accumulation in different crop plants and observed that legumes were more sensitive indicators of soil metal levels than were cereals. They reported a high correlation between foliar metal concentrations in *Trifolium pratense* and those in soil, particularly for the metals Cd and Zn. Studies of Pb levels in crop plants revealed that Pb accumulation in aerial plant parts is often the result of atmospheric deposition onto leaf surfaces and Ward and Savage (1994) observed that 40–80% of Pb associated with untreated foliar samples can be removed by washing. Kabata-Pendias *et al.* (1993) studied Pb accumulation in some legumes and monocotyledons concluding that foliar Pb levels were not reliable indicators of the level of Pb contamination in soil.

Physiological and biochemical markers

High concentrations of metals cause visible damage to plants, which principally takes the form of stunted growth. Growth parameters, e.g. root elongation and biomass of plants, have been successfully used to compare the tolerance of different populations and species to increasing metal dose rates (Fig. 13.1). The same parameters can be employed to assess metal toxicity.

However, before visible symptoms appear and also in the presence of low degrees of contamination, metabolic dysfunctions can occur. Metals may alter the activity and isoform pattern of some enzymes and the production of metabolites (Van Assche and Clijsters, 1990a). Moreover, proteins and polypeptides with metal binding capacity are induced as a result of increased metal concentration in tissues (Grill *et al.*, 1987; Rauser, 1990; Steffens, 1990). Physiological responses of sensitive plants might be used in biomonitoring, once a relationship between the response and the extent of soil contamination has been established. The main disadvantage of using such parameters as bioindicators is their low specificity, due to the fact that similar responses occur to other stress factors, e.g. atmospheric pollution or drought. In laboratory experiments where test plants are

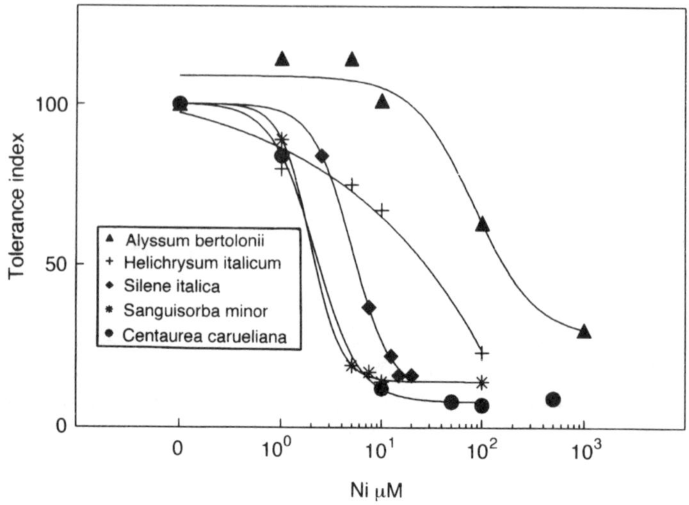

Fig. 13.1. Tolerance index [(root length in metal solution : root length in control solution) ×100] of various plant species from Italian serpentine soils. The seedlings were treated for two weeks with different concentrations of Ni (from Pandolfini, 1987).

grown under controlled environmental conditions, metabolic changes can be more clearly related to soil conditions.

Phytochelatins

Most metals, such as Cd, Ag, Bi, Pb, Zn, Cu and Hg induce the synthesis of heavy metal-binding polypeptides in plants (Grill *et al.*, 1987). These induced compounds are low molecular mass (3000–10,000), cysteine-rich (40–47%) polypeptides which have been designated 'phytochelatins' (PCs) by Grill *et al.* (1985). PCs are also referred to by their structural name, e.g. poly(γ-glutamyl-cysteinyl)glycine or (γ-EC)nG, or class III metallothioneins (MTs) (Rauser, 1990; Steffens, 1990; Tomsett *et al.*, 1992). Class I and II MTs are gene-encoded polypeptides found in all animals examined (Hamer, 1986) and in some fungi (Winge *et al.*, 1985; Munger *et al.*, 1987), while class III MTs or PCs are not gene-encoded but are synthesized from glutathione (GSH) by means of phytochelatin synthetase, a specific enzyme activated by metal cations. Nevertheless, the absence of gene-encoded MTs from plants is not certain. In *Mimulus guttatus*, both PCs and Cu-binding peptides have been found; the latter were not related to PCs and have an amino acid composition similar to gene-encoded MTs (Tomsett *et al.*, 1992).

Heavy metals entering plant cells can be bound by PCs through the cysteine thiol binding of metals by PCs, as has been demonstrated for Cd, Cu, Pb and Zn (Grill *et al.*, 1987). PCs are probably also involved in metal transport from the cytoplasm to the vacuole (Vogeli-Lange and Wagner, 1990). PCs play an important role in detoxification by the removal of excess metal from plant tissues and possibly also in the homeostatic control of metal ion concentrations in plant tissue in response to changes in external concentration. Activation of PC synthase cannot occur in the absence of metal ions (Loeffler *et al.*, 1989). Recent reports indicate that PCs do not influence the metal tolerance of plants (de Knecht *et al.*, 1994, 1995). Metal-tolerant plants or cells do not produce more PCs than do those sensitive to metals (De Vos *et al.*, 1992; Schat and Kalff, 1992; de Knecht *et al.*, 1994). In tolerant and non-tolerant populations of *Silene vulgaris,* exposed to toxic Cu concentrations which cause an equal degree of root length reduction, the same amount of PCs were produced (Schat and Kalff, 1992). The use of PC levels in plants has been proposed as a biomarker of metal stress, rather than as a measure of metal-tolerance (Ernst *et al.*, 1992).

Enzymatic changes

Metals inhibit the activity of many enzymes, mainly because of their high affinity for the sulphydryl groups present in the active site of the enzymes and also by competitive replacement of essential elements of metalloproteins by the toxic metals. On the contrary, the activity of other enzymes (e.g. peroxidase, catalase, superoxide dismutase, malate dehydrogenase and malic enzyme) are stimulated as an indirect consequence of metal stress (Van Assche and Clijsters, 1990a). Most of these reports concern the enzyme peroxidase which is highly responsive to metal stress. Increased peroxidase activity (Fig. 13.2) (together with a variation in the isoperoxidase pattern) has been reported as a phytotoxic effect of several metals, including Cd, Cu, Pb, Ni, Zn (Van Assche *et al.*, 1988; Van Assche and Clijsters, 1990a; Pandolfini *et al.*, 1992; Pandolfini and Gabbrielli, 1993). However, peroxidase activity is also stimulated by other stress factors such as SO_2, O_3, salinity, etc. and it is considered to be a non-specific indicator of environmental stress (Castillo, 1986).

Van Assche and Clijsters (1990b) have used enzymatic parameters to establish a biological test to evaluate soil phytotoxicity. Van Assche *et al.* (1988) observed a positive relationship between both Zn and Cd concentrations and the corresponding level of enzyme activity in the primary leaves of *Phaseolus vulgaris*. They grew *P. vulgaris* on contaminated and clean soils under controlled environmental conditions and measured various stress parameters (growth, enzyme activity, isoperoxidase pattern). Soil samples were grouped into phytotoxicity classes on the basis of the response of the plants (Van Assche and Clijsters, 1990b). The same procedure was also used to evaluate the

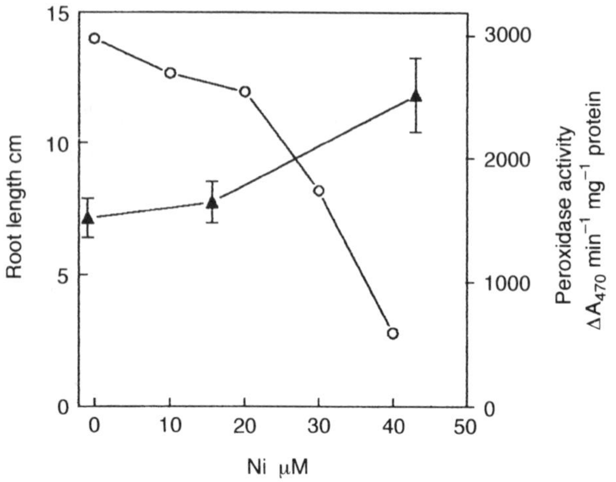

Fig. 13.2. Effect of Ni on root length (○) and peroxidase activity (▲) in *Triticum aestivum* (from Pandolfini *et al.*, 1992).

success of reclamation of heavily polluted soil in Belgium (Vangronsveld *et al.*, 1991).

Other biochemical markers

The polyamines (PAs), putrescine, spermidine and spermine are products of secondary plant metabolism (compounds not involved in the normal metabolic processes essential for plant growth and development) and they are ubiquitous in plants (Smith, 1985). Several reports suggest that the remarkable changes in the production and level of PAs observed under a variety of types of stress may be a fundamental response of plants to adverse conditions (Smith, 1985; Flores, 1990). The metabolism of PAs is particularly elevated by ionic stress, notably in K^+ and Mg^{2+} deficiency, by acidic conditions, by osmotic stress and by elevated salinity. The use of putrescine concentration in leaves as a bioindicator of K^+ deficiency has been proposed. Legumes are particularly sensitive, and accumulation of putrescine occurs before symptoms of K^+ deficiency can be observed (Basso and Smith, 1974).

Unbound putrescine has been found to accumulate in the leaves of several plants following heavy metal stress (Wettlaufer *et al.*, 1991; Hauschild, 1993a). Oat plants exposed to stressful levels of Cd in hydroponic cultures showed increased levels of diamine putrescine in their foliage (Weinstein *et al.*, 1986; Huschild, 1993a). The induction of putrescine synthesis also occurred in barley and rape plants at the early stages of Cr(VI) and Cr(III) stress. In barley plants exposed to Cr(III), the foliar concentration of Cr was more sensitive than free putrescine as an indicator of metal toxicity, and thus foliar Cr may be used to reveal the potential phytotoxic effects of Cr-contaminated soils (Hauschild, 1993b).

The amino acid proline is another metabolite that accumulates in plants subject to a variety of stresses including water deficit, salinity and low temperature (Flowers *et al.*, 1977; Morgan, 1984; Naidu *et al.*, 1991). Increased levels of proline have been detected in plants of *Lemna minor* grown in a nutrient solution containing Zn (10 ppm) or Cu (5 ppm) (Bassi and Sharma, 1993). In the shoots and roots of the plants *Cajanus cajan, Vigna mungo* and *Triticum aestivum*, levels of proline increased proportionally to the concentrations of Cd, Co, Pb and Zn in the culture medium (Alia and Pardha Saradhi, 1991). Equimolar concentration of K^+ did not affect proline content of shoots and roots. Thus the authors inferred that proline accumulation can be used as a marker of heavy metal pollution. Further research is necessary to test this.

Conclusions

Modifications of soil health by industrial and agriculture activities can result in reduced plant productivity and potential risks for human health. Simple and reliable methods for large- and local-scale assessment of soil health are therefore required.

Biomonitoring soil health through plant analysis represents an effective alternative to chemical and physical analysis. This approach appears competitive in terms of costs and has the advantage in that it can integrate the effects of changes in soil conditions as they affect plant growth and biological activity.

Lower plants provide a promising opportunity for both passive and active large-scale biomonitoring primarily because the concentration of contaminants in these plants strictly reflect that of soil. To date, a limitation to this approach is represented by the incomplete knowledge of the physiology and ecology of these organisms. For example standardized methods of growth and reproduction of lichens are necessary in order to obtain genetically uniform material suitable for active biomonitoring.

Genetically uniform material and standardized methods of exposure and sampling are available for several species of higher plants employed as accumulative bioindicators of soil contamination. However, it has been suggested that higher plants should only be used in local screening when substrate conditions are fairly constant (Kabata-Pendias *et al.*, 1993) since root uptake is a very complex process affected by microclimate and soil characteristics.

Biochemical markers of soil contamination have been giving interesting results, especially when stress marker enzymes, such as peroxidases, are used.

The identification of metabolic responses specific for single soil factors would further improve the effectiveness of biochemical markers in the assessment of soil health.

Acknowledgements

The authors are grateful to Professor R.R. Brooks for his helpful suggestions.

References

Albway, B.J., Thorton, L., Smart, G.A., Sherlock, J.C. and Quinn, M.J. (1988) Metal availability. *Science and Total Environment* 75, 41–69.

Alia and Parda Saradhi, P. (1991) Proline accumulation under heavy metal stress. *Journal of Plant Physiology* 138, 554–558.

Alstrup, V. and Hansen, E.S. (1977) Three species of lichens tolerant of high concentrations of copper. *Oikos* 29, 290–293.

Arianoutsou, M., Rundel, P.W. and Berry, W.L. (1993) Serpentine endemics as biological indicators of soil elemental concentrations. In: Markert, B. (ed.) *Plants as Biomonitors. Indicators for Heavy Metals in the Terrestrial Environment.* VCH, Weinheim, pp. 179–189.

Baker, A.J.M. (1981) Accumulators and excluders strategies in the response of plants to heavy metals. *Journal of Plant Nutrition* 3, 643–654.

Baker, A.J.M. and Brooks, R.R. (1989) Terrestrial higher plants which hyperaccumulate metallic elements – A review of their distribution, ecology and phytochemistry. *Biorecovery* 1, 81–126.

Bargagli, R. (1990) Assessment of metal air pollution by epiphytic lichens: the incidence of crustal materials and of possible uptake from substrate barks. *Studia Geobotanica* 10, 97–103.

Bassi, R. and Sharma, S.S. (1993) Changes in proline content accompanying the uptake of zinc and copper by *Lemna minor. Annals of Botany* 72, 151–154.

Basso, L.C. and Smith, T.A. (1974) Effect of mineral deficiency on amine formation in higher plants. *Phytochemistry* 13, 833–875.

Brightman, F.H. and Seaward, M.R.D. (1977) Lichens on man-made substrates. In: Seaward, M.R.D. (ed.) *Lichen Ecology.* Academic Press, London, pp. 253–293.

Brooks, R.R. (1972) *Geobotany and Biogeochemistry in Mineral Exploration.* Harper and Row, London.

Brooks, R.R. (1983) *Biological Methods of Prospecting for Minerals.* John Wiley, New York.

Brooks, R.R., Lee, J., Reeves, R.D. and Jaffré, T. (1977) Detection of nickeliferous rocks by analysis of herbarium specimens of indicator plants. *Journal of Geochemical Exploration* 7, 49–77.

Castillo, F.J. (1986) Extracellular peroxidases as markers of stress? In: Greppin, H., Penel, C. and Gaspar, T. (eds) *Molecular and Physiological Aspects of Peroxidases.* University of Geneva, Geneva, pp. 419–426.

Czehura, S.J. (1977) A lichen indicator of copper mineralization, Lights Creek District, Plumas Country, California. *Economical Geology* 72, 796–803.

Davies, B.E. (1994) Soil chemistry and bioavailability with special reference to trace elements. In: Farago, M.E. (ed.) *Plants and the Chemical Elements.* VCH, Weinheim. pp. 1–30.

de Knecht, J.A., van Dillen, M., Koevoets, P.L.M., Schat, H., Verkleij, J.A.C. and Ernst, W.H.O. (1994) Phytochelatins in cadmium-sensitive and cadmium-tolerant *Silene vulgaris. Plant Physiology* 104, 255–261.

de Knecht, J.A., van Baren, N., Ten Bookum, W.M., Wong Fong Sang, H.W., Koevoets, P.L.M., Schat, H. and Verkleij, J.A.C. (1995) Synthesis and degradation of phytochelatins in cadmium sensitive and cadmium tolerant *Silene vulgaris. Plant Science* 106, 9–18.

De Vos, C.H.R., Vonk, M.J., Vooijs, R. and Schat, H. (1992) Glutathione depletion due to copper-induced phytochelatin synthesis causes oxidative stress in *Silene cucubalus. Plant Physiology* 98, 853–858.

Dion, M., Loranger, S., Kennedy, G., Courchesne, F. and Zayed, J. (1993) Evaluation of black spruce (*Picea mariana*) as a bioindicator of aluminium contamination. *Water, Air, and Soil Pollution* 71, 29–41.

Dykeman, W.R. and De Sousa, A.S. (1966) Natural mechanisms of copper tolerance in a copper-swamp forest. *Canadian Journal of Botany* 44, 871–878.

Empain, A. (1985) Heavy metals in bryophytes from Shaba Province. In: Brooks, R.R., Malaisse, F. and Balkema, A.A. (eds) *The Heavy Metal-tolerant Flora of South Central Africa.* Boston, pp. 103–117.

Ernst, W.H.O. (1974) *Scwermetallvegetation der Erde.* Gustav Fischer Verlag, Stuttgart.

Ernst, W.H.O. (1993) Geobotanical and biogeochemical prospecting for heavy metals deposits in Europe and Africa 1993. In: Markert, B. (ed.) *Plants as Biomonitors. Indicators for Heavy Metals in the Terrestrial Environment.* VCH, Weinheim, pp. 107–126.

Ernst, W.H.O., Verkleij, J.A.C. and Schat, H. (1992) Metal tolerance in plants. *Acta Botanica Neerlandica.* 41, 229–248.

Farago, M.E. (1994) Plants as indicators of mineralisation and pollution. In: Fargo, M.E. (ed.) *Plants and the Chemical Elements.* VCH, Weinheim, pp. 221–240.

Flores, H.E. (1990) Polyamines and plant stress. In: Alscher, R.G. and Cumming, J.R. (eds) *Stress Responses in Plants: Adaptation and Acclimation Mechanisms.* Wiley-Liss Inc., New York, pp. 217–239.

Flowers, T.J., Troke, P.F. and Yeo, A.R. (1977) The mechanism of salt tolerance in halophytes. *Annual Review of Plant Physiology* 28, 89–121.

Folkeson, L. and Andersson-Bringmark, E. (1988) Impoverishment of vegetation in a coniferous forest polluted by copper and zinc. *Canadian Journal of Botany* 66, 417–428.

Gailey, F.A.Y. and Lloyd, O.L. (1986a) Methodological investigations into low technology monitoring of atmospheric metal pollution: Part I – The effects of sampler size on metal concentration. *Environmental Pollution (Ser. B)* 12, 41–59.

Gailey, F.A.Y. and Lloyd, O.L. (1986b) Methodological investigations into low technology monitoring of atmospheric metal pollution: Part 2 – The effects of length of exposure on metal concentrations. *Environmental Pollution (Ser. B)* 12, 61–74.

Garty, J. (1985) The amounts of heavy metals in some lichens of the Negev desert. *Environmental Pollution (Ser. B)* 10, 287–300.

Garty, J. (1993) Lichens as biomonitors for heavy metal pollution. In: Markert, B. (ed.) *Plants as Biomonitors. Indicators for Heavy Metals in the Terrestrial Environment.* VCH, Weinheim, pp. 193–263.

Gilbert, O.L. (1968) Bryophytes as indicators of air pollution in the Tyne Valley. *New Phytologist* 67, 15–30.

Grill, E., Winnacher, E.L. and Zenk, M.H. (1985) Phytochelatins: The principle heavy metal complexing peptides in higher plants. *Science* 230, 674–676.

Grill, E., Winnacker, E.L. and Zenk, M.H. (1987) Phytochelatins, a class of heavy metal binding peptides from plants are functionally analogues to metallothioneins. *Proceedings of the National Academy of Sciences (USA)* 84, 439–443.

Hamer, D.H. (1986) Metallothionein. *Annual Review of Biochemistry* 55, 913–951.

Hauschild, M.Z. (1993a) Chromium content of leaves reveals chromium(III)-stress of higher plants before sensitive biomarkers do. *Science in the Total Environment suppl.* 1345–1352.

Hauschild, M.Z. (1993b) Putrescine (1,4-diaminobutane) as an indicator of pollution-induced stress in higher plants: barley and rape stressed with Cr(III) or Cr(VI). *Ecotoxicology and Environmental Safety* 26, 228–247.

Henderson, A. (1990) Literature on air pollution and lichens XXXI. *Lichenologist* 22, 173–182.

Jones, K.C. (1991) Contaminants trends in soils and crops. *Environmental Pollution* 69, 311–325.

Kabata-Pendias, A., Piotrowska, M. and Dudka, S. (1993) Trace metals in legumes and monocotyledons and their suitability for the assessment of soil contamination. In: Markert, B. (ed.) *Plants as Biomonitors. Indicators for Heavy Metals in the Terrestrial Environment.* VCH, Weinheim, pp. 485–494.

Loeffler, S., Hochberger, A., Grill, E., Winnacker, E.L. and Zenk, M.H. (1989) Termination of the phytochelatin synthase reaction through sequestration of heavy metals by the reaction product. *FEBS Letters* 258, 42–46.

Lounamaa, J. (1965) Studies on the content of iron, manganese and zinc in macrolichens. *Annales Botanici Fennici*, 2, 127–137.

Mejstrík, V. and Lepsová, A. (1993) Applicability of fungi to the monitoring of environmental pollution by heavy metals. In: Markert, B. (ed.) *Plants as Biomonitors. Indicators for Heavy Metals in the Terrestrial Environment.* VCH, Weinheim, pp. 365–378.

Minguzzi, C. and Vergnano, O. (1948) Il contenuto di nichel nelle ceneri di *Alyssum bertolonii* Desv. *Atti della Società di Toscana Scienze Naturali Mem. B* 55, 49–71.

Morgan, J.M. (1984) Osmoregulation and water stress in higher plants. *Annual Review of Plant Physiology* 35, 299–348.

Munger, K., Germann, U.A. and Lerch, K. (1987) The *Neurospora crassa* metallothionein gene. *Journal of Biological Chemistry* 262, 7363–7367.

Naidu, B.P., Paleg, L.G., Aspinall, D., Jennings, A.C. and Jones, G.P. (1991) Amino acid and glycine betaine accumulation in cold-stressed wheat seedlings. *Phytochemistry* 30, 407–409.

Nash, III T.H. (1975) Influence of effluents from a zinc factory on lichens. *Ecological Monographs* 45, 183–198.

Nash, III T.H. (1990) Metal tolerance in lichens. In: Shaw, A.J. (ed.) *Heavy Metal Tolerance in Plants: Evolutionary Aspects.* CRC, Boca Raton, Florida, pp. 119–131.

Nieboier, E. and Richardson, D.H.S. (1980) The replacement of the nondescript term 'heavy metals' by a biologically and chemically significant classification of metal ions. *Environmental Pollution (Ser.B)* 1, 3–35.

Noeske, O., Lauchli, A., Lange, O.L., Vieweg, G.H. and Ziegler, H. (1970) Konzentration und Localisierung von Schwermetallen in Flechten der Erzschlacken halden des Hartzes. *Deutsche Botanische Gesellschaft* 4, 67–79.

Otte, M.L., Bestebroer, S.J., van der Linden, J.M., Rozema, J. and Broekman, R.A. (1991) A survey of zinc, copper and cadmium concentrations in salt marsh plants along the Dutch coast. *Environmental Pollution* 72, 175–189.

Panda, K.K., Lenka, M. and Panda, B.B. (1992) Monitoring and assessment of mercury pollution in the vicinity of a chloralkali plant. II Plant availability, tissue concentration and genotoxicity of mercury from agricultural soil contaminated with solid waste assessed in barley (*Hordeum vulgare* L.). *Environmental Pollution* 76, 33–42.

Pandolfini, T. (1987) Rapporto tra presenza di metalli pesanti nelle piante e nell'ambiente e possibilità di individuazione di indicatori biologici. *Archivio Botanico e Biogeografico Italiano* 63, 22–31.

Pandolfini, T. and Gabbrielli, R. (1993) Changes in the activity and in the isozyme pattern of peroxidases from different cellular fractions in Ni-treated plants. In: Welinder, K.G., Rasmussen, S.K., Penel, C. and Greppin, H. (eds) *Plant Peroxidases: Biochemistry and Physiology.* University of Geneva, Geneva, pp. 417–421.

Pandolfini, T., Gabbrielli, R. and Comparini, C. (1992) Nickel toxicity and peroxidase activity in seedlings of *Triticum aestivum* L. *Plant Cell Environment* 15, 719–725.

Piotrowska, M. and Dudka, S. (1994) Estimation of maximum permissible levels of cadmium in a light soil by using cereal plants. *Water, Air and Soil Pollution* 73, 179–188.

Puckett, K.J., Nieboer, E., Gorzynski, M.J. and Richardson, D.H.S. (1973) The uptake of metal ions by lichens: a modified ion-exchange process. *New Phytologist* 72, 329–342.

Purvis, O.W. (1984) The occurrence of copper oxalate in lichens growing on copper sulphide-bearing rocks in Scandinavia. *Lichenologist* 16, 197–204.

Rasmussen, L., Pilegaard, K. and Gydesen, H. (1980) The application of cryptogams as monitoring organisms of metal air pollution in Denmark. *Botanisk Tidsskrift* 75, 93–99.

Rauser, W.E. (1990) Phytochelatins. *Annual Review of Biochemistry* 59, 61–86.

Richardson, D.H.S (1988) Understanding the pollution sensitivity of lichens. *Botanical Journal of the Linnean Society* 96, 31–43.

Richardson, D.H.S. (1991) Lichens as biological indicators. Recent developments. In: Jeffrey, D.W. and Madden, B. (eds) *Bioindicators and Environmental Management.* Academic Press, London, pp. 263–272.

Schat, H. and Kalff, M. (1992) Are phytochelatins involved in differential metal tolerance or do they merely reflect metal-imposed strain? *Plant Physiology* 99, 1475–1480.

Seaward, M.R.D., Goyal, R. and Bylinska, E.A. (1978) Heavy metal content of some terricolous lichens from mineral enriched sites in Northern England. *Naturalist* 103, 135–141.

Shacklette, H.T. (1965) Bryophytes associated with mineral deposits and solutions in Alaska. *United States Geological Survey Bulletin* 1198-C, 1–18.

Shaw, A.J. (1990) Metal tolerance in bryophytes. In: Shaw, A.J. (ed.) *Heavy Metal Tolerance in Plants: Evolutionary Aspects.* CRC, Boca Raton, Florida, pp. 133–152.

Shimwell, D.H. and Laurie, A.E. (1972) Lead and zinc contamination of vegetation in the Southern Pennines. *Environmental Pollution* 3, 291–301.

Siegel, B.Z. and Siegel, S.M. (1976) Unusual mercury accumulation in lichen flora of Montenegro. *Water, Air and Soil Pollution* 5, 335–337.

Smith, T.A. (1985) Polyamines. *Annual Review of Plant Physiology* 36, 117–143.

Steffens, J.C. (1990) The heavy metal-binding peptides in plants. *Annual Review of Plant Physiology* 41, 553–575.

Steinnes, E. (1993) Some aspects of biomonitoring of air pollutants using mosses, as illustrated by the 1976 Norwegian survey. In: Markert, B. (ed.) *Plants as Biomonitors. Indicators for Heavy Metals in the Terrestrial Environment.* VCH, Weinheim, pp. 381–394.

Streit, B. and Stumm, W. (1993) Chemical properties of metals and the process of bioaccumulation in terrestrial plants. In: Markert, B. (ed.) *Plants as Biomonitors. Indicators for Heavy Metals in the Terrestrial Environment,* VCH, Weinheim, pp. 31–62.

Tomsett, A.B., Sewell, A.K., Jones, S.J., de Miranda, J.R. and Thurman, D.A. (1992) Metal-binding proteins and gene-regulated gene expression in higher plants. In: Wray, J.L. (ed.) *Inducible Plant Proteins: their Biochemistry and Molecular Biology.* Cambridge University Press, Cambridge, pp. 1–27.

Tuba, Z. and Csintalan, Z. (1993) The use of moss transplantation techniques for bioindication of heavy metal pollution. In: Markert, B. (ed.) *Plants as biomonitors. Indicators for Heavy Metals in the Terrestrial Environment.* VCH, Weinheim, pp. 403–411.

Van Assche, F., Cardinaels, C. and Clijsters, H. (1988) Induction of enzyme capacity in plants as a result of heavy metal toxicity: dose-response relations in *Phaseolus vulgaris* L., treated with zinc and cadmium. *Environmental Pollution* 52, 103–115.

Van Assche, F. and Clijsters, H. (1990a) Effects of metals on enzyme activity in plants. *Plant Cell Environment* 13, 195–206.

Van Assche, F. and Clijsters, H. (1990b) A biological test system for the evaluation of the phytotoxicity of metal contaminated soils. *Environmental Pollution* 66, 157–172.

Van Gronsveld, J., Cammaer, C., Van Assche, F. and Clijsters, H. (1991) Reclamation of a 'desert like' site in the north east of Belgium: evolution of the metal pollution and experiments *in situ. Proceedings of an International Conference on Heavy Metals in the Environment,* CEP Consultant, Edinburgh, pp. 58–61.

Verkleij, J.A.C. (1993) The effects of heavy metal stress on higher plants and their use as biomonitors. In: Markert, B. (ed.) *Plants as Biomonitors. Indicators for Heavy Metals in the Terrestrial Environment,* VCH, Weinheim, pp. 417–424.

Vogeli-Lange, R. and Wagner, G.J. (1990) Subcellular localization of cadmium and cadmium-binding peptides in tobacco leaves. Implications of a transport function for cadmium-binding peptides. *Plant Physiology* 92, 1086–1093.

Wagner, G. (1993) Large-scale screening of heavy metal burdens in higher plants. In: Markert, B. (ed.) *Plants as Biomonitors. Indicators for Heavy Metals in the Terrestrial Environment.* VCH, Weinheim, pp. 425–434.

Ward, N.I. and Savage, J.M. (1994) Metal dispersion and transportational activities using food crops as biomonitors. *Science in the Total Environment* 146/147, 309–319.

Weinert, H. (1991) Biomonitoring of environmental change using plant distribution patterns. In: Jeffrey, D.W. and Madden, B. (eds) *Bioindicators and Environmental Management.* Academic Press, London, pp. 179–190.

Weinstein, L.H., Kaur-Sawhney, R., Rajam, M.V. and Wettlaufer, S.H. (1986) Cadmium-induced accumulation of putrescine in oat and bean leaves. *Plant Physiology,* 82, 641–645.

Wettlaufer, S.H., Osmeloski, J. and Weinstein, L.H. (1991) Response of polymines to heavy metal stress in oat seedlings. *Environmental Toxicology and Chemistry* 10, 1083–1088.

Wilson, M.J., Jones, D. and McHardy, W.J. (1981) The weathering of serpentinite by *Lecanora atra. Lichenologist* 13, 167–176.

Winge, D.R., Nielson, K.B., Gray, W.R. and Hamer, D.H. (1985) Yeast metallothionein: sequence and metal binding properties. *Journal of Biological Chemistry* 260, 14464–14470.

Wittig, R. (1993) General aspects of biomonitoring heavy metals by plants. In: Markert, B. (ed.) *Plants as Biomonitors. Indicators for Heavy Metals in the Terrestrial Environment.* VCH, Weinheim, pp. 3–27.

Wondratschek, I. and Röder, U. (1993) Monitoring of heavy metals in soils by higher fungi. In: Markert, B. (ed.) *Plants as Biomonitors. Indicators for Heavy Metals in the Terrestrial Environment.* VCH, Weinheim, pp. 345–363.

14

Bioindicators to Detect Contamination of Soils with Special Reference to Heavy Metals

G.N. Mhatre[1] and C.E. Pankhurst[2]

[1]304 A, Kohinoor Tower, Bal Govindas Marg, Dadar, Bombay 400 028, India; [2]CSIRO Land and Water, Private Bag No. 2, Glen Osmond, South Australia 5064, Australia

Introduction

Bioindication has been defined as the use of an organism, a part of an organism or a community of organisms to obtain information about the quality of the environment (Wittig, 1993). Traditionally, it is a community of organisms, rather than an individual organism that is most often used as the bioindicator. The concept also embraces the responses of organisms or communities of organisms to specific stresses (Chaphekar, 1978). Thus changes in the composition of a grassland community which provides information about the degree of overgrazing, or the age of lodgepoles in a pine plantation which provides information about the past occurrence of forest fires, are examples of bioindicators.

With the increase in anthropogenic influences on the environment, a whole new set of bioindicators has emerged. Industrial pollutants, such as heavy metals, generate many types of stresses on the environment resulting in many organisms and communities of organisms displaying specific injury symptoms or shifts in community composition (Chaphekar, 1978). The magnitude of this phenomenon has resulted in the need to broaden the scope of bioindication to include 'indicator species' (organisms which are highly sensitive to pollutants) and 'indicator processes' (processes in living organisms which are highly sensitive to pollutants) (Salanki, 1986). Additionally, accumulation of hazardous substances by microbes, plants and animals has become an important component of bioindication, as this allows the presence of low levels of such chemicals in the environment to be identified and quantified.

Scales of Bioindication

Weinert (1986) concluded that bioindication is possible at various scales including macromolecules, cell, organ, organism, population and biocoenosis. The form the bioindication could take would include: (i) biochemical and physiological reactions; (ii) anatomic, morphologic, biorhythmic and behavioural deviation from the norm; (iii) floristic, faunistic and chronological changes of populations; (iv) changes of the biocoenosis and the biocoenosis combinations; (v) changes of the structure and function of ecosystems; and (vi) changes in the character of the landscape.

Bioindicators are likely to be most useful when the environmental stress affecting organisms or communities of organisms is sublethal, and when organisms differ in their levels of sensitivity and tolerance to the same stress.

Stress and Evolution

Exposure of organisms to sublethal doses of stress (e.g. a pollutant chemical) over a long time period results in many interactions between the organisms and the stress factor. Initially these interactions may occur at the level of biochemical and cellular processes. These initial interactions or 'shock-absorbing' events lead to physiological effects such as disruption of respiratory, excretory, locomotory, feeding, circulatory, reproductive and neural phenomena in animals and photosynthetic, transpiratory, respiratory, growth and reproductive processes in plants and microorganisms. These physiological effects in turn may affect the structure of the DNA and chromosomes in the organism leading to modifications and eventual evolution of organisms which are capable of withstanding the stresses. The well-documented evolution of melanism in moths (Ford, 1964) is a good example of the response of an individual organism to industrial pollution of the environment. This pattern of evolution of resistance or tolerance to the stress factor/s also occurs in entire communities of organisms, e.g. shifts in the composition of plant communities in the vicinity of industries (Bell *et al.,* 1982), mining areas (Bradshaw, 1952, 1976), polluted rivers (Mhatre *et al.*, 1980; Chaphekar and Mhatre, 1986). In many instances these shifts in plant community composition could result in the evolution of plant species that accumulate heavy metals (Antonovics *et al.*, 1971; Ernst, 1976; Mhatre, 1991, 1995).

The present article gives an insight into the importance of bioindicators of contamination of soil with special reference to heavy metals.

History of Environmental Contamination by Heavy Metals

Environmental contamination with heavy metals is as ancient as the discovery of fire. Small quantities of metals are released into the air by burning firewood and the metal enriched wood ash is generally discarded into the environment. With the discovery of mining and metal-working techniques in ancient times, close links between metals, metal pollution and human culture were established. During the Roman Empire, large quantities of metals, especially lead (aptly known as Roman Metal) were required to sustain the high standard of living (Nriagu, 1983). The mines were operated on a small scale, but uncontrolled smelting in open fires often resulted in severe local contamination. The banning of mining operations near ancient cities and the Roman edict forbidding any mining activities in Italy have been attributed to concern for environmental quality (Nriagu, 1990).

With the development of large furnaces equipped with tall stacks during the sixteenth century, the sphere of influence of smelters and industrial installations was extended drastically. Heavy metal profiles in peats, lake sediments, and ice sheets show that, by the end of the seventeenth century, the metal pollutants released by industries in Britain and central Europe were reaching most regions of Scandinavia (Livett, 1988). Accelerated accumulation of lead pollution in Greenland ice fields began around the turn of the eighteenth century (Murozumii *et al.*, 1969). Historical records in soils, peats, ice sheets, sediments, and tree rings in most parts of the Northern Hemisphere document pervasive heavy metal pollution of increasing severity since the beginning of the Industrial Revolution. Between 1850 and 1900, worldwide industrial emissions of cadmium, copper, lead, nickel and zinc to the atmosphere averaged about 380, 1800, 22,000, 240 and 17,000 tonnes per year respectively (Nriagu, 1979).

By the turn of this century, ever-expanding technological development has sharply increased the industrial consumption and discharge of toxic metals. Between 1900 and 1980, mine production of aluminium, nickel, chromium, copper and zinc increased 114-, 35-, 18-, 5- and 4-fold respectively, while mine outputs of copper, lead, nickel and zinc amounted to 250, 160, 17 and 185 million tonnes, respectively (Nriagu, 1988). Industrial emissions of cadmium, copper, lead, nickel and zinc to the air increased 8-, 6-, 9-, 51- and 8-fold respectively (Nriagu, 1979). These figures are certainly indicative of the large quantity of toxic metals being discharged into the environment. Sooner or later, traces of all the 'new' metals that are mined will be discharged into the biosphere. Once dispersed in the biosphere, the metals cannot be degraded or recovered using current technology. Any environmental effects of heavy metal pollution, therefore, tend to be permanent.

Heavy Metals as Pollutants

Heavy metals and their salts occur naturally in the environment. They are metals with atomic numbers greater than 23 and densities more than 5. Until recently, heavy metals were defined only on the basis of their densities and Passow *et al.* (1961) lists 38 elements with densities greater than five. However, Nieboer and

Richardson (1980) proposed that this classification be abandoned in favour of a system which classifies metal ions as oxygen seeking (Class A), nitrogen/sulphur seeking (Class B) and intermediate (Border Line). This classification is related to the atomic properties and solution chemistry of the metal ions.

Heavy metals are non-degradable, persist in nature for long periods and are toxic to living organisms at fairly low concentrations (Mhatre, 1991).

Industrial Emissions of Metals

Metals are discharged into the air, water and soils from a wide variety of natural and industrial sources (Mhatre, 1995). Emissions of heavy metals from primary natural sources to the atmosphere are shown in Table 14.1. Windblown soil particles account for more than 50% of the manganese, chromium and vanadium and for 30–50% of the antimony, nickel, molybdenum, and zinc. Volcanoes contribute more than 60% of the cadmium, 40–50% of the mercury and nickel, and 30–40% of the arsenic, chromium and copper fluxes from natural sources. Biogenic sources are the leading contributors of arsenic, mercury and selenium, while fires and sea salt sprays generally account for less than 15% of the total natural emissions of each element. Of course, some of the metals emitted by

Table 14.1. Worldwide emissions of trace metals from natural sources into the atmosphere (thousands of tonnes year^{-1}).

Element	Wind-borne soil particles	Sea salt spray	Volcanoes	Forest fires	Biogenic sources	Total[a]
Antimony	0.78	0.56	0.71	0.22	0.29	2.6
Arsenic	2.6	1.7	3.8	0.19	3.9	12
Cadmium	0.21	0.06	0.82	0.11	0.24	1.4
Chromium	27	0.07	15	0.09	1.1	43
Cobalt	4.1	0.07	0.96	0.31	0.66	6.1
Copper	8.0	3.6	9.4	3.8	3.3	28
Lead	3.9	1.4	3.3	1.9	1.7	12
Manganese	221	0.86	42	23	30	317
Mercury	0.05	0.02	1.0	0.02	1.4	2.5
Molybdenum	1.3	0.22	0.40	0.57	0.54	3.0
Nickel	11	1.3	14	2.3	0.73	29
Selenium	0.18	0.55	0.95	0.26	8.4	10
Vanadium	16	3.1	5.6	1.8	1.2	28
Zinc	19	0.44	9.6	7.6	8.1	45

[a] Totals are rounded.
From Nriagu (1989b).

Table 14.2. Worldwide atmospheric emissions of heavy metals from anthropogenic sources (thousand tonnes $year^{-1}$).

Element	Energy production	Mining	Smelting and refining	Manufacturing processes	Commercial uses[a]	Waste incineration	Transportation	Total[b]
Antimony	1.30	0.10	1.42			0.67		3.5
Arsenic	2.22	0.06	12.3	1.95	2.02	0.31		19
Cadmium	0.79		5.43	0.60		0.75		7.6
Chromium	12.7			17.0		0.84		31
Copper	8.04	0.42	23.2	2.01	1.58	35		
Lead	12.7	2.55	46.5	15.7	4.50	2.37	248	332
Manganese	12.1	0.62	2.55	14.7	8.26	38		
Mercury	2.26		0.13			1.16		3.6
Nickel	42.0	0.80	3.99	4.47	0.35	52		
Selenium	3.85	0.16	2.18	0.11	6.3			
Thallium	1.13			4.01				5.1
Tin	3.27		1.06			0.81		5.1
Vanadium		84.0	0.06	0.74		1.15		86
Zinc	16.8	0.46	72.0	33.4	3.25	5.90	132	

[a]Including agricultural uses.
[b]Totals are rounded.
From Nriagu and Pacyna (1988).

Table 14.3. Worldwide inputs of heavy metals into soils (thousand tonnes $year^{-1}$).

Source	Antimony	Arsenic	Cadmium	Chromium	Copper	Lead	Manganese	Mercury	Molybdenum	Nickel	Selenium	Vanadium	Zinc
Agricultural and animal wastes	4.9	5.8	2.2	82	67	26	158	0.85	34	45	4.6	19	316
Logging and wood wastes	2.8	1.7	1.1	10	28	7.4	61	1.1	1.6	13	1.6	5.5	39
Urban refuse	0.76	0.40	4.2	20	26	40	24	0.13	2.3	6.1	0.33	0.2	60
Municipal sewage and organic waste	0.18	0.25	0.18	6.5	13	7.1	8.1	0.44	0.43	15	0.11	1.3	39
Solid wastes from metal fabrication	0.08	0.11	0.04	1.5	4.3	7.6	2.6	0.04	0.08	1.7	0.10	0.12	11
Coal ashes	12	22	7.2	289	214	144	1076	2.6	441	68	32	39	298
Fertilizers and peat	0.25	0.28	0.20	0.32	1.4	2.9	12	0.01	0.46	2.2	0.27	0.97	2.5
Discarded manufactured products[a]	2.4	38	1.2	458	592	292	300	0.68	1.9	19	0.15	1.7	465
Atmospheric fallout	2.5	13	5.3	22	25	232	27	2.5	2.3	24	2.0	60	92
Total input[b]	26	82	22	898	971	759	1669	8.3	87	294	41	128	1322

[a]Metals used for industrial installations and 'durable' goods are assumed to have a definite life span and to be released into the environment at a constant rate.

[b]Totals are rounded.

Note: These inputs exclude mine tailings and slags at the smelter sites.

From Nriagu and Pacyna (1988).

natural sources are derived anthropogenically; for instance, metals in wind blown dusts often are of industrial origin. Thus, the data in Table 14.1 probably overstate the actual natural fluxes of metals into the pre-industrial era atmosphere.

The three main industrial sources of atmospheric metal pollution are mining, smelting and refining of metals, the burning of fossil fuels and the production and use of metallic commercial products (Table 14.2). The burning of fossil fuels accounts for more than 95% of vanadium and 80% of nickel. Hence, scientists use these two elements to identify air samples coming from power plants. Fossil fuel combustion also accounts for about 60% of anthropogenic emissions of mercury, selenium and tin and significant amounts of antimony, arsenic and copper. Smelters represent the leading source of arsenic, copper and zinc, while the manufacture of steel is responsible for the largest fractions of manganese and chromium released to the atmosphere by human activities. In spite of the reduced use of leaded gasoline in many countries, the automobile tail pipe still accounts for roughly two-thirds of lead emissions.

A comparison of Table 14.1 and 14.2 leads to the conclusion that industrial emissions have become dominantly responsible for most of the heavy metals in the air. Anthropogenic emissions of lead, cadmium, vanadium and zinc exceed the fluxes from natural sources by 28-, 6-, 3- and 3-fold, respectively. Industrial contributions of arsenic, copper, mercury, nickel and antimony amount to 100–200% of emissions from natural sources. In many urban areas and around some point sources, the natural emissions are insignificant in comparison with the anthropogenic metal burden.

Modern societies generate large quantities of various metal-containing wastes that are discarded on land (Table 14.3). The disposal of coal and wood ashes, industrial installations and commercial products (which corrode and decompose) on land, together account for about 55–80% of the metal pollution in soils. The large volumes of wastes associated with animal husbandry, logging, agriculture and food production may also affect quantities of heavy metal in many soils (Nriagu, 1990). Agricultural soils generally receive most of their metal pollution from the atmosphere and from fertilizers, pesticides and manure. Because of their very high metal contents, both municipal and industrial sewage sludges are often considered unsafe for disposal on land and clearly represent one of the most important sources of metal contamination in soils.

Worldwide Distribution of Metals in Soil

Soils represent the ultimate sink for heavy metals in continental areas. Because metals are fairly immobile in soils (except under the influence of acid rain), metal pollution tends to accumulate primarily in the surface layers. The build up of toxic metals in the most biologically active part of the soil, the organic topsoil, makes the metals readily accessible to some crops and vegetables. The chemistry of heavy metals in soils thus plays an important role in the transfer of contaminant metals to the food supply.

Human influence on heavy metals in soils is demonstrated dramatically by the highly elevated levels of metals that now characterize the soils in urban areas and around major industries (Adriano, 1986; Purves, 1985). Most of these metals are delivered via the atmosphere. The median values (and typical ranges) reported for atmospheric fallout of heavy metals in urban areas of North America are 160 (20–980) g ha^{-1} $year^{-1}$ for copper; 910 (140–3500) for lead; 18 (7–36) for cadmium and 3200 (80–4,800) g ha^{-1} $year^{-1}$ for zinc (Jeffries and Schneider, 1981). The values for urban areas of Europe are 320 (78–500) for copper; 400 (180–600) for lead; 310 (33–530) for nickel; 15 (3–28) for cadmium and 1000 (220–5850) g ha^{-1} $year^{-1}$ for zinc (Bergkvist *et al.*, 1989). At these deposition rates, the levels of most heavy metals in surface soils will double in 2–10 years. Atmospheric fallout of metals in rural areas also has risen sharply; the rates in rural Europe have been estimated to be 150 (14–320) g ha^{-1} $year^{-1}$ for copper; 550 (38–3900) for zinc; 220 (63–550) for lead; 32 (7–100) for nickel; and 4 (2–13) for cadmium (Jeffries and Schneider, 1981).

Since the atmosphere is just one of the media for industrial emissions, the human perturbation of the distribution of heavy metals in urban soils is obviously substantial. Indeed, numerous studies have demonstrated that soils in urban areas – whether in parks or domestic gardens, cities or small towns – have become so highly contaminated with a wide variety of heavy metals that any base-line differences originating from the local geology is difficult to trace (Purves, 1985; Adriano, 1986). Typical concentrations of many toxic metals in both urban and rural soils are now 2 to more than 10 times higher than the levels in uncontaminated soils (Purves, 1985). Moreover, the 'available' levels defined as the fraction extractable using mild reagents also have increased markedly in urban soils (Purves, 1985). In terms of the elevated levels and chemical forms of heavy metals, soils in urban areas have become quite unique and are unlike anything produced by natural weathering processes.

Soils downwind of urban and industrialized centres also receive large doses of heavy metal pollution. For example, the concentrations of lead, cadmium, arsenic, and antimony in surface soils of southern Norway are about 10 times higher than the levels in the northern part of the country (Steinnes, 1987). The regional difference has been attributed to the exposure of pollutants from Britain and central Europe. The lead contamination of forests in remote parts of the north-eastern United States is primarily caused by long-range transport of industrial emissions from New England and the Mid-west (Friedland *et al.*, 1984). For arable soils, primary sources of metal pollution also include fertilizers, agricultural chemicals, and liquid and solid wastes (see Table 14.3). It has been estimated that the average cadmium input into agricultural lands in Europe is about 8 g ha^{-1} $year^{-1}$ from the atmosphere and 5 g ha^{-1} $year^{-1}$ from the application of phosphate fertilizer (Hutton, 1982). Metal contamination of agricultural soils in Belgium from fertilizers and the atmosphere has been estimated to average 16, 20, 260 and 3800 g ha^{-1} $year^{-1}$ for arsenic, cadmium, lead and zinc, respectively (Navarre *et al.*, 1980). At such loading rates, many arable soils in Europe may be close to exceeding their carrying capacity for heavy metal pollution. In Japan, on the other hand, the problem is no longer hypothetical. About 9.5% of rice paddy soils have been rendered unsuitable for growing rice for human consumption because of excessive metal contamination (Asami, 1983).

Global Effects

Environmental metal pollution may either result in direct assaults on human health or have indirect effects on human welfare by interfering with the integrity and vitality of the life support systems. Expectedly, concern about the wanton fouling of the biosphere with toxic metals has centred around the effects on human health and, specifically, on acute rather than chronic effects of toxic metals. Reported case histories of acute poisoning in the general population by metals in the environment are few; familiar examples are the Itai-itai (cadmium poisoning) and Minamata (mercury poisoning) diseases of Japan. On the other hand, the chronic effects of toxic metals on human and animal populations and on ecosystem health have yet to receive adequate attention.

Ecotoxic Effects of Heavy Metals on Plants

When sensitive plant species are exposed to pollutants, numerous symptoms or types of damage to the plant may result. The type and extent of damage to the susceptible plants can provide us with a means of monitoring pollution and also help in determining the geographical distribution of a pollutant over a large area (Brennan *et al.*, 1967).

The most common adverse effect of heavy metals on plants is inhibition of root growth followed by stunting or dwarfing of shoots (Foy *et al.*, 1978; Wong and Bradshaw, 1982). Germination and early growth of seedlings were also found to be sensitive to heavy metals such as mercury, lead and arsenic (Mhatre and Chaphekar, 1982). The form of leaf necrosis may be specific to a particular metal, but there is also a general chlorosis of the younger leaves common to all metals (Gemmell, 1977). A distinct relationship between the amount of injury to plant leaves (Leaf Injury Index (LII)) with mercury and the photosynthetic capacity and the biomass produced by plants was established (Mhatre and Chaphekar, 1984, 1985). The LII was thus suggested as a simple index that could be used for biomonitoring. Some specific symptoms elicited by heavy metals on plants are listed in Table 14.4.

In contrast, some plants do not show any signs of damage in spite of high levels of heavy metals in their tissues (Ernst, 1976, see also Pandolfini *et al.* Chapter 13 this volume). A classical example of heavy metal accumulation with-

Table 14.4. Effects of heavy metals on plants*.

Aluminium	Shortening of roots; leaf scorch
Boron	Stunting; deformation, blotching and browning of leaves
Chromium	Chlorosis of leaves
Cobalt	Chlorosis of leaves and/or decrease of chlorophyll
Copper	Chlorosis of leaves and dwarfism; reduction of size of seeds and corollas
Iron	Darkening of leaves
Manganese	Chlorosis of leaves with white blotching
Molybdenum	Abnormally coloured shoots; chlorosis and vulnerability to insect attack
Nickel	Chlorosis and necrosis of leaves
Uranium	Variation in flower colour; abnormal fruits; extra chromosomes; stimulated growth
Zinc	Chlorosis of leaves; symptoms of manganese deficiency

* From Tiagi and Aery (1985).

Table 14.5. Heavy metals in sediments and in the tissues of *Pyreus macrostachyos* (Lamack) J. Raynal at polluted (P) and non-polluted (NP) areas along the Kalu River Estuary.

	μg g^{-1} dry weight					
	Sediment		Leaf		Rhizome	
Metals	P	NP	P	NP	P	NP
Hg	1.5–140	39–52	3.3–110	1.4	6.9–53.3	2.9
Pb	5.4–10.6	3.8–4.4	100	3.0	100	2.8
Cd	0.62–12.6	0.42–6.8	2.5–10	2.5	0.98–10.0	2.1
Cu	91–864	78–89	9.0–26.4	23.0	2.03–307	89.0

out injury was found for plants growing along the banks of the Kalu River Estuary in India (Table 14.5). This river is polluted with heavy metals that are discharged into it from the industries located along its banks. The plant in question, *Pyreus macrostachyos* (Lamack) J. Raynal, was the only plant able to thrive along the the banks of this river (Mhatre *et al.*, 1980) and so acted as a clear bioindicator of pollution.

Plants as Bioindicators of Heavy Metals

Plant communities as indicators

It is generally accepted that plant communities serve as better, more effective and more reliable bioindicators than single plant species. Chaphekar *et al.* (1973)

reported a reduction in the diversity of a plant community along the Ulhas River in an industrial suburb near Bombay as a consequence of pollution of the water. A similar phenomenon of highly reduced plant community diversity was shown at a heavily metal polluted spot along the Kalu River Estuary, also near Bombay (Mhatre *et al.*, 1980). Such reduced plant diversity is a direct reflection of the quality of water and sediment present in the polluted areas.

Non-angiosperm plants

The use of epiphytes, e.g. lichens and mosses for monitoring metal pollution of the air has been reported by many workers (see Pandolfini *et al.*, Chapter 13, this volume). Lichens have been shown to be useful as bioindicators of heavy metal deposition (Nieboer *et al.*, 1972). Little and Martin (1974) used *Sphagnum* moss for detection of zinc, lead and cadmium at 47 sampling points over a large area around a zinc and lead smelting complex at Avonmouth, near Bristol, UK. Parkarinen and Tolonen (1976) also used the peat moss *Sphagnum fuscum* for monitoring different heavy metals (lead, zinc, mercury, cadmium, nickel) at 15 sites in Finland and two sites in Western Continental Canada. Peat mosses, especially *S. fuscum*, were also used for extensive mapping of heavy metal deposition within boreal forest vegetation around mining areas and specific metal indicator species were identified (Parkarinen and Tolonen, 1976). Chaphekar and Kulkarni (1979) have also reported differential sensitivity of some commonly found bryophytes in Western India to mercury.

Angiosperm plants

Plants that are largely restricted to or are particularly abundant on metal contaminated soils have been cited as bioindicators of metal containing soils (Antonovics *et al.*, 1971). Many such plants accumulate heavy metals in their tissues (see Pandolfini *et al.*, Chapter 13, this volume). A universally available plant from the tropics, *Alternanthera sessilis* was identified as an accumulator of aluminium and its roots were found to contain up to 7100 ppm of the metal whereas neighbouring plants (different species) contained less than 700 ppm (Nadkarni and Chaphekar, 1977). Khalap (1986) has investigated *Eichhornia crassipes*, *Typha angustata* as well as *Lemna minor* and found them to accumulate large quantities of chromium in their tissues. Similarly, accumulation of heavy metals like Zn, Cd, Cu, Pb and Cr in the tissues of *Ipomoea carnea* was reported by Puri (1988). All of these investigations suggest the possibility of using these plants as bioindicators or biomonitors for heavy metals.

Soil Microbiological Properties as Bioindicators of Heavy Metals in Soils

So far the discussion of bioindicators for heavy metal pollution of the environment has focused on plants, but soil microorganisms and soil functional processes mediated by soil microorganisms have also been used extensively as indicators of heavy metal contamination of soils (Bååth, 1989; Hicks *et al.*, 1990; Pankhurst, 1994; Brookes, 1995). As with plants, soil microorganisms are particularly well suited for monitoring soil pollution. They are in intimate contact with the soil microenvironment, are sensitive to low concentrations of heavy metals and respond rapidly (Bååth, 1989). They also have the potential to reveal the bioavailability of heavy metals in the soil (see Paton *et al.*, Chapter 16, this volume).

Microbiological properties that have potential as bioindicators of heavy metal pollution of soil fall into two main groups. The first is those that measure the activity of the whole microbial population, e.g. microbial respiration, soil nitrogen mineralization and soil enzyme activity. The second type measures the size of the microbial population, at the single organism level, at the functional group level and in some more recent examples, at the community level.

Microbial activity

Measurement of microbial respiration (CO_2-C) is the most commonly measured parameter in all studies of the effect of heavy metal pollution on soil microbiological properties. It is usually measured in conjunction with measurement of microbial biomass C (Jenkinson and Powlson, 1976) and the total organic C content of the soil. Microbial biomass C is usually lower in metal contaminated soil compared to uncontaminated soil, whereas microbial respiration is generally not affected. Consequently a decrease in the ratio of biomass C/total organic C and an increase in the specific respiratory activity (ratio of CO_2-C/biomass C), an indicator of metabolic stress (Killham, 1985), is commonly observed in metal contaminated soil (Brookes, 1995). These observations indicate that the activity of soil microorganisms are adversely affected by heavy metals (especially Cu and Zn); effects which may persist for many years (Chandler and Brookes, 1993). Because of these effects on microbial activity, the amount of soil organic matter in metal contaminated soils may accumulate (Chandler and Brookes, 1991), although this is difficult to demonstrate in natural environments because appropriate uncontaminated soils are not readily avaliable for comparison.

Soil enzyme activity

Enzymes are present in two general locations in soils; those associated with living cells (intracellular) and those that are extracellular or abiontic. Because it

Table 14.6. Soil enzyme activity in pasture soils (0–5 cm) contaminated with heavy metals.

Property	Degree of contamination			LSD (95%)
	Control	Medium	High	
Cu (mg kg^{-1})	38	140	900	
Cr (mg kg^{-1})	50	132	842	
As (mg kg^{-1})	8	148	1124	
pH (H_2O)	5.6	5.5	5.4	
Organic C (%)	6.4	6.0	6.7	
Soil respiration (l CO_2 g^{-1} soil h^{-1})	5.05	4.72	3.38	0.50
Dehydrogenase (pmol formazan g^{-1} s^{-1})	12.4	12.2	2.2	1.2
Urease (nmol NH_4^+ –N g^{-1} s^{-1})	1.32	0.89	0.41	0.08
Phosphatase (nmol p-nitrophenol g^{-1} s^{-1})	4.19	3.12	2.95	0.15
Sulphatase (nmol p-nitrophenol g^{-1} s^{-1})	0.56	0.57	0.14	0.02

From Yeates *et al.* (1994).

is difficult to extract enzymes from soil, their activity is generally measured in laboratory assays under defined and optimal conditions. A major problem with assaying soil enzyme activity is that abiontic activity cannot readily be separated from that of the intracellular enzymes of living cells. However, despite these difficulties soil enzyme activity has been used as a sensitive indicator of the effect of heavy metals on soils (see Dick, Chapter 6, this volume). In general, most soil enzymes are inhibited by high concentrations of heavy metals in the soil (Bååth, 1989; Hicks *et al.*, 1990). An example of this is shown in Table 14.6, where heavy metal contamination of a pasture soil in New Zealand was shown to significantly inhibit the activity of four enzymes (Yeates *et al.*, 1994). Measurement of soil enzyme activity also has potential for assessing the impact of accidental spills of hydrocarbons onto soils and following the effectiveness of remediation (Song and Bartha, 1990).

Nitrogen fixation

Biological N_2 fixation carried out by by free-living autotrophic and heterotrophic bacteria and by symbiotic microorganisms (e.g. *Rhizobium/Bradyrhizobium* and *Frankia*) has frequently been suggested as a suitable indicator of heavy metal pollution in soil. Using a laboratory assay, Brookes *et al.* (1986) demonstrated a decline in N_2 fixation by cyanobacteria in response to a gradient of heavy metal pollution but this could not be repeated in other soils, possibly due to the low amounts of cyanobacteria present (Lorenz *et al.*, 1992). Similarly, N_2 fixation by *Rhizobium leguminosarum* biovar *trifolii* in symbiotic association with *Trifolium repens* (white clover) was found to be reduced in to soils containing high levels of Zn, Cu, Ni and Cd (McGrath *et al.*, 1988). This was shown to be due to the survival of metal-tolerant rhizobia that were ineffective in N_2 fixation in the metal contaminated soils (Giller *et al.*, 1989). These observations have led to a lot of interest in using the *Rhizobium*–legume symbiosis as an indicator of soil pollution by heavy metals. However, results from other work have been conflicting (Kinkle *et al.*, 1987) suggesting the need for more research.

Microbial community structure

Measuring the effects of heavy metals on the growth of selected functional groups of soil microorganisms has been attempted in many studies. However, problems with the large variability associated with such methods have made them unreliable. In general, algae and photosynthetic bacteria appear to be more susceptible to heavy metals than other groups. Actinomycetes, saprophytic fungi and heterotrophic bacteria appear to be less affected, possibly because the population density of predatory microorganisms such as protozoa was suppressed (Hicks *et al.*, 1990).

In contrast, a range of methodologies, several of which do not require the culturing of microorganisms, have been used to assess the effect of heavy metal contamination on the structure or composition of soil microbial communities (Pankhurst, Chapter 12, this volume). These methodologies include evaluation of the fatty acid composition of phospholipids extracted directly from soil (Frostegård *et al.*, 1993) and the range of aromatic substances used by microbial communities as C sources (Reber, 1992). In each case significant shifts in the composition or functional capabilities of the microbial community in the contaminated soil compared to uncontaminated soil were detected. More recently, Pennanen *et al.* (1996) using analysis of phospholipid fatty acids of soils were able to detect progressive shifts in the composition of microbial communities in response to a gradient of heavy metal pollution of soils at increasing distances from a smelter in south-west Finland. These methods, along with other methods of detecting and quantifying the composition or biodiversity of soil microbial communities (e.g. DNA hydridization, Voordouw *et al.* (1991)), may provide an array of sensitive new tools for analysis of the effects of heavy metals on soils and their functioning.

Soil Fauna as Bioindicators of Heavy Metals in Soils

The abundance and activity of most groups of soil fauna (e.g. protozoa, nematodes, microarthropods (e.g. collembola) and macrofauna (e.g. earthworms)) are affected by heavy metal pollution of soils. However, despite their sensitivity to heavy metals, the expertise required for enumeration and identification of the smaller soil fauna has made them less attractive as general bioindicators than the larger fauna such as earthworms.

Protozoa and nematodes

Because of their rapid growth and delicate external membranes, protozoa, in particular the ciliate protozoa and testate amoebae, have been found to be sensitive bioindicators of heavy metals in soil (Foissner, 1994). In laboratory assays, the growth of the ciliate *Colpoda steinii* was found to be strongly related to the concentration of heavy metal in its available form rather than the total concentration, suggesting that this protozoan could be used as a bioassay of metal availability (Forge *et al*., 1993). Nematodes generally appear to be little affected by heavy metals, and in some instances (e.g. after the addition of metal contaminated sewage sludge to soils) their numbers may increase substantially (Weiss and Larink, 1991). Closer examination of nematode trophic group structure in metal contaminated soils, however, does reveal that some groups, e.g. the predatory nematodes and the bacterial feeders may increase in abundance whilst other groups, e.g. the plant parasitic nematodes and the omnivores may show little change or decrease in abundance (Weiss and Larink, 1991; Yeates *et al*., 1994). The changes in relative abundance of different nematode trophic groups in metal contaminated soils appears to be largely related to changes in the amount of available nutrients in the soil.

Collembola

Collembola as a group are less sensitive to heavy metals than other soil animals because of their efficient excretion or avoidance mechanisms (Bengtsson and Tranvik, 1989). There are, however, distinct differences in the sensitivity of single species which enable their use as bioindicators of heavy metal pollution. For example, the abundance of *Mesaphorura krausbaueri* and *Onychiurus armatus* increased in soil along a Cu/Zn gradient, whilst *Folsomia quadrioculata* disappeared and *Isotomiella minor* was reduced (Bengtsson and Rundgren, 1988). Filser *et al*. (1995) also found *M. krausbaueri* and *O. armatus* to be adundant in soil containing high Cu levels whilst *F. quadrioculata*, *I. minor* and several other species were either lacking or greatly reduced. It was concluded that the abundance of *F. quadrioculata* and *I. minor* could be used as a bioindicator of Cu contamination of soils if suitable uncontaminated control sites were available for comparison (Filser *et al*., 1995).

Earthworms

Earthworms are sensitive to several heavy metals and may also accumulate metals in their tissues (Lee, 1985). They are therefore potentially useful as

Table 14.7. Effect of heavy metal contamination of pasture soils on earthworm abundance, biomass and heavy metal content.

Contamination	*Lumbricus rubellus*					*Aporrectodea rosea*					Casts		
	no. m^{-2}	g m^{-2}	Cu	Cr	As	no. m^{-2}	g m^{-2}	Cu	Cr	As	Cu	Cr	As
Control	67.0	52.3	7.0	2.9	4.4	413.0	764.0	7.0	2.9	8.0	21.5	39.5	12.5
Low	120.0	85.4	17.6	9.8	41.4	383.0	64.4	13.3	2.9	40.0	90.7	119	122
Medium	0	0	No sample			7.0	7.0	No sample			160	202	225
High	0	0	No sample			0	0	No sample			No sample		

Earthworm abundance (number m^{-2}), fresh weight (g m^{-2}), heavy metal contents (mg kg^{-1}), and heavy metal contents of surface casts (mg kg^{-1}).
No sample (treatments in which there was either no material or insufficient for heavy metal analysis).
From Yeates *et al.* (1994).

bioindicators of high levels of metal contamination (lower abundance) and of low levels of contamination (concentration of metal in their tissues or casts). For example, Yeates *et al.* (1994) found that the abundance and biomass of two species of earthworm was reduced by contamination of pasture soils with Cu, Cr and As (Table 14.7). At high levels of contamination, no earthworms were present. The concentration of these metals in the tissues and casts of both earthworm species was significantly correlated ($P<0.05$) with the heavy metal contents of the topsoil providing evidence of accumulation of metals by these earthworms (Yeates *et al.*, 1994). Bengtsson and Tranvik (1989) showed a decline in earthworm abundance along a Cu/Zn gradient with surface dewelling species (e.g. *Lumbricus rubellus*) at highest risk of extinction. Filser *et al.* (1995) also reported a marked decline in abundance of lumbricids in soil under previous intensive management (with a high Cu content) compared with soil under previous conventional management (low Cu content).

Conclusions

The continued heavy metal pollution of soils is of major concern in most countries of the world. Plants whose growth is sensitive to these metals and those that have the capacity to accumulate metals in their tissues have potential as bioindicators for the detection of these pollutants in the environment. Similarly, soil microbiological properties, in particular soil microbial biomass C, soil respiration, soil enzymes and the community structure of microbial communities, have potential as bioindicators of heavy metals, and also provide an insight into the long-term impact these pollutants may have on soil functioning. Abundance of soil fauna, notably earthworms and selected species of Collembola may also be useful as bioindicators, whilst soil protozoa have potential to be used as a sensitive bioassay of the bioavailability of some metals. The mere fact that there is considerable selection of possible bioindicators for heavy metal pollution of soils is indicative of the significant consequences these pollutants potentially have on the sustainability of the soil resource.

References

Adriano, D.C. (1986) *Trace Elements in the Terrestrial Environment.* Springer-Verlag, New York.

Antonovics, J., Bradshaw, A.D. and Turner, R.G. (1971) Heavy metal tolerance in plants. *Advances in Ecological Research* 7, 1–85.

Asami, I. (1983) Pollution of soils by cadmium. In: Nriagu, J.O. (ed.) *Changing Metal Cycles and Human Health.* Springer-Verlag, Berlin, pp. 95–111.

Bååth, E. (1989) Effects of heavy metals in soil on microbial processes and populations (a review). *Water, Air and Soil Pollution* 47, 335–379.

Bell, J.N.B., Ayazloo, M. and Wilson, G.B. (1982) Selection for sulphur dioxide tolerance in grass populations in polluted areas. In: *Urban Ecology*. Second European Ecology Symposium, Blackwell Scientific Publications, London.

Bengtsson, G. and Rundgren, S. (1988) The Gusum case: a brass mill and the distribution of soil Collembola. *Canadian Journal of Zoology* 66, 1518–1526.

Bengtsson, G. and Tranvik, L. (1989) Critical metal concentrations for forest soil invertebrates. *Water, Air and Soil Pollution* 30, 87–108.

Bergkvist, B., Folkeson, L. and Berggren, D. (1989) Fluxes of Cu, Zn, Pb, Cd, Cr, and Ni in temperate forest ecosystems. *Water, Air and Soil Pollution* 47, 217–86.

Bradshaw, A.D. (1952) Populations of *Agrostis tenuis* resistant to lead and zinc poisoning. *Nature* 169, 1098–1100.

Bradshaw, A.D. (1976) Pollution and evolution. In: Mansfield, T.A. (ed.) *Effects of Air Pollutants on Plants*. Cambridge University Press, London, pp. 134–159.

Brennan, E.C., Leone, I.A. and Daines, R.H. (1967) Characterisation of plant damage by air pollution in New Jersey. *Plant Disease Reporter* 51, 10.

Brookes, P.C. (1995) The use of microbial parameters in monitoring soil pollution by heavy metals. *Biology and Fertility of Soils* 19, 269–279.

Brookes, P.C., McGrath, S.P. and Heijnen, C. (1986) Metal residues in soils previously treated with sewage-sludge and their effects on growth and nitrogen fixation by blue-green algae. *Soil Biology and Biochemistry* 18, 345–353.

Chandler, K. and Brookes, P.C. (1991) Effects of heavy metals from past applications of sewage sludge on microbial biomass and organic matter accumulation in a sandy loam and a silty loam UK soil. *Soil Biology and Biochemistry* 25, 927–932.

Chandler, K. and Brooks, P.C. (1993) Effects of Zn, Cu, and Ni in sewage sludge on microbial biomass in a sandy loam soil. *Soil Biology and Biochemistry* 25, 1231–1239.

Chaphekar, S.B. (1978) Biological indicators. The concept of new addition. *International Journal of Ecology and Environmental Science* 4, 45–52.

Chaphekar, S.B. and Kulkarni, M.S. (1979) Effects of mercury on some cryptogams. *Geobios* 6, 218–220.

Chaphekar, S.B. and Mhatre, G.N. (1986) *Human Impact on the Ganga River Ecosystem*. Concept Publishing Company, New Delhi, pp. 1–186.

Chaphekar, S.B., Divekar, D.P., Entee, D.F., Jayaraman, K., Jere, M.G., Khan, S.M. and Tipnis, P.P. (1973) Effect of industrial pollution on the angiosperm flora along the Ulhas river at Kalyan, near Bombay. *The Botanique* 4, 85–92.

Ernst, W. (1976) Physiological and biochemical aspects of metal tolerance. In: Mansfield, T.A. (ed.) *Effects of Air Pollution on Plants*. Cambridge University Press, London, pp. 115–133.

Filser, J., Fromm, H., Nagel, R.F. and Winter, K. (1995) Effects of previous intensive agricultural management on microorganisms and the biodiversity of soil fauna. *Plant and Soil* 170, 123–129.

Foissner, W. (1994) Soil protozoa as bioindicators in ecosystems under human influence. In: Darbyshire, J.F. (ed.) *Soil Protozoa*. CAB International, Wallingford, UK, pp. 147–194.

Ford, F.B. (1964) *Ecological Genetics*. Methuen, London.

Forge, T.A., Berrow, M.L., Darbyshire, J.F. and Warren, A. (1993) Protozoan bioassays of soil amended with sewage sludge and heavy metals using the common soil ciliate *Colpoda steinii*. *Biology and Fertility of Soils* 16, 282–286.

Foy, C.D., Chaney, R.L. and White, M.C. (1978) The physiology of metal toxicity in plants. *Annual Review of Plant Physiology* 29, 511–566.

Friedland, A.J., Johnson, A.H., Siccama, T.C. and Mader, D.L. (1984) Trace metal profiles in the forest floor of New England. *Soil Science Society of America Journal* 48, 422–425.

Frostegård, A., Tunlid, A. and Bååth, E. (1993) Phospholipid fatty acid composition, biomass, and activity of microbial communities from two soil types experimentally exposed to different heavy metals. *Applied and Environmental Microbiology* 59, 3605–3617.

Gemmell, R.P. (1977) Colonisation of industrial wasteland. *Studies in Biology*, Vol. 80. Edward Arnold, London.

Giller, K.E., McGrath, S.P. and Hirsch, P.R. (1989) Absence of nitrogen fixation in clover grown in soil subject to long-term contamination with heavy metals is due to survival of only ineffective *Rhizobium*. *Soil Biology and Biochemistry* 21, 841–848.

Hicks, R.J., Stotzky, G. and Van Voris, P. (1990) Review and evaluation of the effects of xenobiotic chemicals on microorganisms in soil. *Advances in Applied Microbiology* 35, 195–253.

Hutton, M. (1982) *Cadmium in the European Communities*. Report no. 26., Monitoring and Assessment Research Centre, University of London.

Jeffries, D.S. and Schneider, W.R. (1981) Atmospheric deposition of heavy metals in central Ontario. *Water, Air and Soil Pollution* 158, 127–152.

Jenkinson, D.S. and Powlson, D.S. (1976) The effects of biocidal treatments on metabolism in soil. V. A method for measuring soil biomass. *Soil Biology and Biochemistry* 8, 209–213.

Khalap, K.P. (1986) Plants and waterborne pollutants – studies of their interactions. PhD Thesis, University of Bombay.

Killham, K. (1985) A physiological determination of the impact of environmental stress on the activity of the microbial biomass. *Environmental Pollution (Ser A)* 38, 204–283.

Kinkle, B.K., Angle, J.S. and Keyser, H.H. (1987) Long-term effects of metal-rich sewage sludge application on soil populations of *Bradyrhizobium japonicum*. *Applied and Environmental Microbiology* 53, 315–319.

Lee, K.E. (1985) *Earthworms – their Ecology and Relationships with Soils*. Academic Press, Sydney.

Little, P. and Martin, M.H. (1974) Biological monitoring of heavy metal pollution. *Environmental Pollution* 6, 1–19.

Livett, E.A (1988) Geochemical monitoring of atmospheric heavy metal pollution: theory and application. *Advances in Ecological Research* 18, 65–176.

Lorenz, S.E., McGrath, S.P. and Giller, K.E. (1992) Assessment of free-living nitrogen fixation activity as a biological indicator of heavy metal toxicity in soil. *Soil Biology and Biochemistry* 24, 601–606.

McGrath, S.P., Brooks, P.C. and Giller, K.E. (1988) Effects of potentially toxic metals in soil derived from past applications of sewage sludge on nitrogen fixation by *Trifolium repens* L. *Soil Biology and Biochemistry* 20, 415–424.

Mhatre, G.N. (1991) Bioindicators and biomonitoring of heavy metals. *Journal of Environmental Biology* 12, 201–209.

Mhatre, G.N. (1995) Heavy metal pollution in biosphere and bioindicators and biomoniting of heavy metals. In: Cheremisinoff, P.N. (ed.) *Encyclopaedia of Environmental Control Technology* Vol. 8, 427–454.

Mhatre, G.N. and Chaphekar, S.B. (1982) Effect of heavy metals on seed germination and early growth. *Journal of Environmental Biology* 3, 53–64.

Mhatre, G.N. and Chaphekar, S.B. (1984) Response of young plants to mercury. *Water, Air and Soil Pollution* 21, 1–8.

Mhatre, G.N. and Chaphekar, S.B. (1985) Effect of mercury on some aquatic plants. *Environmental Pollution* 39, 207–216.

Mhatre, G.N., Chaphekar, S.B., Ramani Rao, I.V., Patil, M.R. and Haldar, B.C. (1980) Effect of industrial pollution on the Kalu River ecosystem. *Environmental Pollution* 23, 67–78.

Murozumi, M., Chow, T.J. and Patterson, C.C. (1969) Chemical concentrations of pollutant lead aerosols, terrestrial dusts and sea salts in Greenland and Antarctic snow strata. *Geochimica et Cosmochimica Acta* 33, 1247–1294.

Nadkarni, R.A and Chaphekar, S.B. (1977) A plant species of suspected accumulator behaviour. *Experientia* 33, 34.

Navarre, J.L., Ronneanu, C. and Priest, P. (1980) Deposition of heavy elements on Belgian agricultural soils. *Water, Air and Soil Pollution* 14, 207–213.

Nieboer, E., Ahmed, H.M., Puckett, K.J. and Richardson, D.H.S. (1972) Heavy metal content of lichens in relation to distance from a nickel smelter in Sudbury, Ontario. *Lichenologist* 5, 292–304.

Nieboer, E. and Richardson, D.H.S. (1980) The replacement of the nondescript term 'heavy metals' by a biologically and chemically significant classification of metal ions. *Environmental Pollution Ser. B* 1, 326–330.

Nriagu, J.O. (1979) Global inventory of natural and anthropogenic emissions of trace metals to the atmosphere. *Nature* 279, 409–411.

Nriagu, J.O. (1983) *Lead and Lead Poisoning in Antiquity.* Wily Interscience, New York.

Nriagu, J.O. (1988) A silent epidemic of environmental metal poisoning? *Environmental Pollution* 50, 139–161.

Nriagu, J.O. (1989a) Effects of atmospheric trace metal deposition on aquatic ecosystems (Background paper prepared for a workshop on The Effects of Atmospheric Contaminants on Aquatic and Terrestrial Ecosystems), Centre for Clean Air Policy, Washington DC.

Nriagu, J.O. (1989b) A global assessment of natural sources of atmospheric trace metals. *Nature* 338, 47–49.

Nriagu, J.O. (1990) Global metal pollution – poisoning the biosphere. *Environment* (32) No. 7.

Nriagu, J.O. and Pacyna, J.M. (1988) Quantitative assessment of worldwide contamination of air, water and soils with trace metals. *Nature* 333, 134–139.

Parkarinen, P. and Tolonen, K. (1976) Regional survey of heavy metals in peat mosses (*Sphagnum*). *Ambio* 5, 38–40.

Passow, H., Rohtstein, A. and Clarkson, T.W. (1961) General pharmacology of heavy metals, *Pharmacology Reviews* 13, 185.

Pennanen, T., Frostegård, A., Fritze, H. and Bååth, E. (1996) Phospholipid fatty acid composition and heavy metal tolerance of soil microbial communities along two heavy metal-polluted gradients in coniferous forests. *Applied and Environmental Microbiology* 62, 420–428.

Puri, A.R. (1988) Studies to explore bioindicator potentials of angiospermic plants. PhD Thesis, University of Bombay.

Purves, D. (1985) *Trace Element Contamination of the Environment.* Elsevier, Amsterdam.

Reber, H.H. (1992) Simultaneous estimates of the diversity and the degradative capability of heavy-metal-affected soil bacterial communties. *Biology and Fertility of Soils* 13, 181–186.

Salanki, J. (1986) *Biological Monitoring of the State of the Environment: Bioindicators.* IUBS Monograph series No. 1 for ICSU IRL Press, Oxford.

Scheider, W.A., Jeffries, D.S. and Dillon, P.J. (1979) Effects of acid precipitation of Pre-cambrian Fresh-Water in southern Ontario. *Journal of Great Lakes Research* 5, 45–51.

Song, H.G. and Bartha, R. (1990) Effects of jet fuel spills on the microbial community of soil. *Applied and Environmental Microbiology* 56, 646–651.

Steinnes, E. (1987) Impact of long range atmospheric transport of heavy metals to the terrestrial environment in Norway. In: Hutchinson, T.C. and Meema, K.M. (eds) *Lead, Mercury, Cadmium and Arsenic in the Environment.* John Wiley & Sons, New York, pp. 107–117.

Tiagi, Y.D. and Aery, N.C. (1986) Plant indicators of heavy metals. In: Salanki, J. (ed.) *Biological Monitoring of the State of the Environment: Bioindicators.* IRL Press, Oxford, pp. 207–222.

Voordouw, G., Voordouw, J.K., Karkhoff-Schweizer, R.R., Fedorak, P.M. and Westlake, D.W.S. (1991) Reverse sample genome probing, a new technique for identification of bacteria in environmental samples by DNA hybridisation, and its application to the identification of sulfate-reducing bacteria in oil field samples. *Applied and Environmental Microbiology* 57, 3070–3078.

Weinert, E. (1986) Selected methods of bioindication in terrestrial ecosystems. In: Salanki, J. (ed.) *Biological Monitoring of the State of The Environment : Bioindicators.* IRL Press, Oxford, pp. 67–70.

Weiss, B. and Larink, O. (1991) Influence of sewage sludge and heavy metals on nematodes in an arable soil. *Biology and Fertility of Soils* 12, 5–9.

Wittig, R. (1993) General aspects of biomonitoring heavy metals by plants. In: Market, B. (ed.) *Plants as Biomonitors.* VCH, Weinheim, pp. 345–363.

Wong, M.H. and Bradshaw, A.D. (1982) A comparison of the toxicity of heavy metals using elongation of ryegrass, *Lolium perenne. New Phytologist* 91, 255–261.

Yeates, G.W., Orchard, V.A., Speir, T.W., Hunt, J.L. and Hermans, M.C.C. (1994) Reduction in soil biological activity following pasture contamination by copper, chromium, arsenic timber preservative. *Biology and Fertility of Soils* 18, 200–208.

15

Chemical and Molecular Approaches for Rapid Assessment of the Biological Status of Soils

D.C. White[1] and S.J. Macnaughton[2]

[1]*Center for Environmental Biotechnology, University of Tennessee, 10515 Research Drive, Knoxville, Tennessee 37932-2575, USA;* [2]*Microbial Insights, Inc., 201 Center Park Drive, Suite 1140, Knoxville, Tennessee 37922-2105, USA*

Introduction

Soil, although largely ignored by the general public, is a most precious resource. The sustainability of our civilization depends on its 'quality'/'health'. Soil health has been defined as the 'continued capacity of a soil to function as a vital living system within ecosystem and land-use boundaries to sustain biological productivity, promote the quality of air and water environments, and maintain plant, animal and human health' (see Doran and Safley, Chapter 1, this volume). Maintenance of soil health is the key to sustainable agriculture (Doran *et al.*, 1994) and the multitude of interacting soil physical, chemical, and biological properties that affect soil quality/health have been clearly related to the potential fitness or capacity to produce healthy and nutritious crops (Doran *et al.*, 1994). Is it possible to develop a rapid, automatable analytical system with a potential for field utilization to correlate the ecology of the soil microbiota to the soil quality/health?

Many soil biologists know in their hearts that the unseen microbes in their ecosystem must be very important. Microbes tend to be ignored because they are difficult to study. The classical methods of isolation and culture of microbes that are taught in most microbiology courses have been enormously successful in clinical medicine where isolation of specific pathogens establishes the diagnosis of disease and the *in vitro* sensitivities to antimicrobials can often predict

(eds C.E. Pankhurst, B.M. Doube and V.V.S.R. Gupta)

the success of treatments. The success of cultural isolation and subsequent characterization in public health microbiology has carried over into examinations of the soil microbiota. Well developed rapid methods of identification are now in widespread usage with isolated and culturable microbes.

Microbiologists have traditionally related the biomass of microbiota in a soil to the number of cells detected by viable count. While this is sufficient for monocultures of microorganisms which are readily grown in the laboratory, it is not satisfactory for soils and sediments where the viable counts often represent 0.1 to 10% of the cells detected using acridine orange direct counts (AODC) or biochemical measures (Olsen and Bakken, 1987; Albrechtsen and Winding, 1992; Bååth *et al.,* 1992). In soil and subsurface sediments, the problems are intensified as the heterogeneous distribution of the microbial community can significantly increase the variability of the accuracy of biomass determination by AODC. Direct microscopic counts present problems when autofluorescence of sediment and clay granules obscures detection of bacteria *in situ.* This problem is often ameliorated by inducing detachment of the microbes from soil granules with solutions containing multicharged ions such as polyphosphate followed by recovery on membrane filters for microscopic counting. However, based on signature lipid biomarker analysis, there is evidence that in subsurface sediments, the de-adherence of the microbes is selective and often not quantitative (White and Ringelberg, unpublished results). Also, with *in situ* direct counting of bacteria at densities of less than 10^4 cells g^{-1} dry weight of sediment, the accuracy of the counts is very low and the error large, even when counting 20 fields with adequate numbers of cells and replicate subsampling (Montagna, 1982). The morphology of the microbes does not often reflect the function or activity so very little insight into the community structure or nutritional status is possible from morphometrics.

Measurements of metabolic processes are complicated by the fact that most microbes in the soil are inactive, but poised for activity when nutrients appear. Adding labelled substrates to determine rates of metabolic activity induces major disturbance artefacts yielding much higher rates than those that actually exist in the environment. This is best exemplified in studies of the deep subsurface microbiota where oxygen and inorganic carbon are found in groundwater with a groundwater age of greater than 1.1×10^5 years (Phelps *et al.,* 1994). Measurements of metabolic activity based on isotope incorporation experiments in subsurface sediments were between 10^3 and 10^6 times greater than those that geochemical evidence would predict. The metabolic activities of the subsurface microbiota indicate growth rates on the order of centuries (Phelps *et al.,* 1994). Furthermore, microbes may be metabolically active even though they are not culturable by traditional methods. Classical microbial tests provide little indication of the nutritional status of microbes or the toxicity which can affect metabolic activities and can be crucial to studies of the ecology of microbial communities.

This review will focus on four non-traditional methods for rapid assessment of the biological status of soils. These methods are: (i) the Biolog system, which measures metabolic activities of isolates or the microbial community on a suite of substrates; (ii) the MIDI fatty acid pattern analysis of isolated culturable microbes; and (iii) two methods that focus on *in situ* analysis of components directly extracted from the soils. The two *in situ* methods are signature lipid biomarker analysis (SLB), and DNA based analyses.

Biolog

The Biolog system (Biolog, Inc., Hayward, California) correlates patterns of substrate utilization with identification of bacterial species. The system is comprised of a 95 well plate of which each well contains a different sole carbon source and nutrients. The 96th well contains neither carbon source nor nutrient and, as such, is a control. Oxidation of the carbon source is indicated by reduction of 2,3,5-triphenyl tetrazolium chloride into a coloured insoluble formazan. After a period of incubation, the plate is scanned colorimetrically with a reader and the resulting patterns compared to a comprehensive data base (Boucher and Savageau, 1977). Microbial communities have also been submitted to this form of sole-carbon source testing producing a community level physiological profile (CLPP) (Garland and Mills, 1991). Unfortunately, this form of analysis requires a transparent carbon-free inoculum. Although many groundwater samples can be assayed directly, soils and subsurface sediments need to be blended, extracted with sodium pyrophosphate, incubated without added carbon and nutrients for 24 hours with agitation, and have the supernatant flocculated with a mixture of calcium and magnesium salts before assay (Lehman *et al.*, 1995). With CLPP, colour development patterns are monitored daily for 7 days. Raw absorbance data is converted to net absorbance after subtraction of the control. An average value for each well is calculated by averaging all 95 responses for all samples and replicates and plotting against time. A classical sigmoid growth curve results. A reading corresponding to the maximum of the linear colour development is chosen and this value for each of the 95 components is normalized by division by the average well colour development for each plate reading.

Principal components and other statistical analyses have been performed on these responses (Garchow *et al.*, 1993; Garland and Mills, 1994; Zak *et al.*, 1994; Lehman *et al.*, 1995). The results provide both a pattern of the presence or absence of positive responses as well as a comparison of the same substrates used by different communities. Garland and Mills (1994) explored temporal trends within four sets of inoculated hydroponically grown plants (each with a different nutrient system), demonstrating consistent changes in the distribution of microbial communities. In particular, they showed a convergence in community-level response and a decrease in the rate of change over time which suggested that the rhizosphere microbial community tended to approach a stable state. Zak *et al.* (1994) showed that by using the Biolog system it was possible to detect considerable variation in the substrate utilization of microbial communities of soils taken from six different plant communities whereby differences in functional diversity were found to be dependent on the class of carbon sources. By application of this methodology on subsurface sediments, Lehman *et al.* (1995) showed that subsurface microbial communities had a much stronger preference for acetate, Tween 40 and amino acids than did surface or groundwater microbial communities and that the subsurface community was clearly different from those communities present in drilling muds and surface sediments.

The community level physiological profile analyses were also found to agree with the phospholipid ester-linked fatty acid (PLFA) analyses in the differentiation of surface and drilling muds from subsurface microbial communities. Utilizing responses of isolates, model communities of known composition, and soils, Haack *et al.* (1995) showed that non-linear substrate oxidation rates were delayed by dilution of the inoculum. Patterns of substrate utilization were reproducible for model communities but the extent of substrate utilization was not. In addition, replicate soil communities from the same pots varied considerably.

MIDI

A highly successful bacterial identification system has been developed by Microbial ID, Inc., Newark, Delaware (known as the MIDI system). The MIDI system measures fatty acids (FAME) from both membrane lipids and lipopolysaccharide. It is most effectively applied to isolated microorganisms cultured with standard media under standardized conditions. Microbial ID, Inc., has commercialized the MIDI system worldwide and sells a comprehensive database. Bacterial isolates are identified by comparing their fatty acid profiles to this MIDI database, which contains over 8000 bacteria (Welch, 1991). The utilization of this system for identification of clinical isolates has been remarkably successful. However, the application of the MIDI system to the analysis of environmental samples has significant drawbacks. The MIDI system was developed to identify clinical microorganisms and requires their isolation and culture on trypticase soy agar at 25°C. Since many isolates are unable to grow at these restrictive growth conditions, the system does not lend itself to identification of many environmental organisms. The isolation and culture of individual organisms for MIDI analysis is somewhat time consuming and expensive so it is rarely applied to the analysis of soil microbiota (Haack *et al.*, 1994). Cavigelli *et al.* (1995) did apply this form of FAME analysis to hydrolysed lipid extracts in defining microbial biomass and community composition in soils. Since many of the FAMEs are common to different microbes, a straight forward interpretation of the results from these soil extracts was complicated. With the application of a principal components analysis it was possible to define similarities and differences between differing soil communities. When a 'signature' FAME for a particular taxa was detected, the analysis became more specific. Comparisons with organisms which grew on R2A agar recovered from the same soils showed distinct differences indicating that many significant organisms were not cultured. However, the patterns of some of the isolates were recognizable in the soil community FAME profiles.

MIDI is most effective with analysis of cellular components on the relatively clean samples derived from isolated bacterial, fungal, and actinomycete cultures. Although the great majority of cellular fatty acids in bacteria are ester-linked in the phospholipids, some are ester-linked in the lipopolysaccharide within the Gram-negative classification of bacteria. Amide-linked fatty acids in the lipids or lipopolysaccharides will not be detected by the MIDI procedure. By applying the whole cell saponification approach used in this type of analysis, specificity is lost due to the lack of knowledge as to where the FAMEs reside in the microbial cell.

Phospholipid Ester-linked Fatty Acid

Increasing the specificity of the lipid analysis by extracting the intact lipids, fractionating them into major lipid classes (usually neutral lipids, glycolipids and polar lipids) and derivatizing by use of a mild alkaline transesterification has provided an analytical system for the direct analysis of soil lipid extracts with improved selectivity. When polar lipid fractions were examined for patterns of fatty acids it proved possible to readily detect differences in soil microbiota as a result of different agricultural practices (Zelles *et al.*, 1992), with exposure to alkaline dust (Bååth *et al.*, 1992), and with differences in exposure to heavy metal dust (Frostegård *et al.*, 1993a,b). Shifts in viable microbial biomass and community composition as a result of exposure to pollutants in the subsurface (Smith *et al.*, 1986) were shown to reflect an increase in the viable biomass and in increased proportions of PLFA characteristic of Gram-negative heterotrophs, and in type II methane-oxidizing bacteria in soil columns gassed with methane and air (Nichols *et al.*, 1987). Addition of different fatty acid substrates to anaerobic sediment cores induced marked and expected changes in the bacterial community structure (Parkes *et al.*, 1992). Subsurface sediments perfused with methane, propane, and air showed shifts in community structure which correlated with trichloroethylene (TCE) biodegradation (Ringelberg *et al.*, 1988; Cox *et al.*, 1994). The polar lipid fatty acid analysis was not, however, sufficiently specific to reflect differences in agricultural soil management practices in a long-term farming system trial (Wander *et al.*, 1995). In this study, variation between sample replicates obscured any significant differences related to differing soil practices. Trends in the analysis did show that organic cover cropped soils contained the largest biomass and most heterogeneous communities whereas organic-manure amended soils showed the least amount of heterogeneity but the highest level of metabolically active organisms. The phospholipid ester-linked fatty acid analyses (PLFA) of an inoculated soil has been compared to the FAME analyses (MIDI system) of the individual isolates that were used to inoculate the soil (Haack *et al.*, 1994). This study showed that not all the isolates in the community could be detected in the composite profile since there was overlap of 'signature' PLFA. Although the PLFA analysis did not allow for the detection of all of the species present in the soil, it did provide an overview of the community as a whole and did allow for detections based on specific physiological traits.

PLFA profiles of isolates have been shown to shift with temperature, substrate sources, and growth conditions (Lechevalier, 1977). However, when bacteria are grown in communities under natural conditions their cells contain a relatively constant proportion of their biomass as phospholipids (White *et al.*, 1979, 1980). Experience has shown that organisms added to communities, or that are induced to proliferate in that community by manipulations, can maintain a sufficiently characteristic PLFA profile to be recognized (White, 1988). The marked changes in PLFA induced in some microbes in monoculture by shifts in the environment apparently are not detected in the environment where competition with better adapted species restricts survival of specific strains to much narrower conditions of growth.

The specificity of the lipid analysis can be increased further by use of a more comprehensive examination, namely the signature lipid biomarker (SLB) analysis. This analysis involves the assay of a larger proportion of the cellular components of the microbiota. The cellular components chosen for assessment should be reasonably labile in the microbes so that dead 'fossil' organisms can be readily differentiated from viable cells. They should be sufficiently discriminatory so various subsets of the microbial community can be determined and, if possible, should provide indications of the nutritional/physiological status of the cells.

Signature Lipid Biomarkers

The rationale for this analysis is based on the hypothesis that soil microbial community composition is a faithful reflection of the soil's recent history and provides predictable effects on sustainable crop yields and other measures of soil quality/health (Tunlid and White, 1992). Waxman (1927) made a prediction that quantitative analysis of soil microorganisms could indicate the actual or potential fertility of a particular soil. Visser and Parkinson (1992) indicated that soil quality correlations be tested at three levels: the population of individual species, the community level, and at the soil ecosystem level through measures of organic matter decomposition rates, soil respiration as CO_2 efflux, and microbial biomass carbon. The community level assessment seemed most problematic as methods based on classical microbiology with microscopic or cultural methods were wholly inadequate. The great majority of soil microbes are not culturable and their direct microscopic analysis in the soil or after attempts at quantitative release from the soil matrix are time consuming and not very effective. Domsch *et al.* (1983) summarized results of many studies using these methods and suggested various culturable microbes could be classified into those with high, medium, and low sensitivities to side-effects from agrochemicals. Nitrifiers, *Rhizobium*, actinomycetes, and rates of organic matter degradation showed the highest sensitivity to the side-effects; *Azotobacter*, total culturable microbes, ammonifiers, and aerobic nitrogen fixing capacity were the least sensitive predictors; algae, bacteria, fungi, soil respiration, CO_2 evolution, oxygen uptake, denitrification and ammonification were of medium sensitivity in predicting the side effects. None of these tests were sufficiently comprehensive in themselves to serve as the sole quantitative analysis of soil microbes needed to test Waxman's predictions. A powerful quantitative assessment method for soil microbiota has been developed over the past 20 years in our laboratory which defines the viable biomass, community composition and nutritional/physiological status of environmental microbial communities. This is based on the application of SLB analysis (White, 1988, 1993; Vestal and White, 1989; Tunlid and White, 1992).

The SLB analysis can provide *in situ* indications of starvation, exposure to toxicity, unbalanced growth, phosphate availability, microniche pH, moisture, aerobic/anaerobic metabolic balance, growth rate, and the proportion of the recently lysed (dead) microorganisms all of which provide significant insight into the phenotypic activity of the microbial community (White, 1995a). Soil and rhizosphere microbial biodiversity is still considered to be largely undefined (Aldhous, 1994), although an application of SLB analysis with its increased specificity has yet to be utilized on a significant scale. Since the SLB analysis includes other classes of lipids in addition to the PLFA, it is also very important that each of the components being analysed be verified through the use of mass spectrometry (White, 1988, 1993; Vestal and White, 1989; Tunlid and White, 1992). We have found that it is critical that each component be structurally verified and that this often entails forming different derivatives, such as that used in the confirmation of the position and configuration of double bonds (Nichols *et al.*, 1986). Often, analysis of fragmentation patterns at the front and tail of a chromatographic peak will confirm the presence of multiple components which, without mass spectrometry, would go undetected.

SLB analysis provides quantitative insight into three important attributes of microbial communities.

Viable biomass

The determination of the total phospholipid ester-linked fatty acids (PLFA) provides a quantitative measure of the viable or potentially viable biomass. Viable microbes have an intact membrane which contains PLFA. The cellular enzymes hydrolyse the phosphate group within minutes to hours of cell death (White *et al.*, 1979). The remaining lipid is diglyceride (DG). The resulting DG contains the same signature fatty acids as the phospholipids, allowing for a comparison of the ratio of phospholipid fatty acids to diglyceride fatty acids (viable to non-viable microbes). The ratio of diglyceride fatty acids/PLFA was shown to generally increase with depth in the soils and subsurface sediments of the semi arid northwest of the USA (White and Ringelberg, 1996). A careful study of subsurface sediments showed that the viable biomass as determined by PLFA was equivalent (but with a much smaller standard deviation) to that estimated by intercellular ATP, cell wall muramic acid, and very carefully done acridine orange direct counts (AODC) (Balkwill *et al.*, 1988). Conversion factors between lipid analysis and cell numbers have been discussed (White *et al.*, 1995).

Community structure

The SLB analysis provides a quantitative definition of the microbial community structure. Specific groups of microbes often contain unusual lipids (White, 1988, 1993; Vestal and White, 1989; Tunlid and White, 1992). For example, specific PLFA are prominent in the hydrogenase-containing *Desulfovibrio* sulphate-reducing bacteria, whereas the *Desulfobacter* type of sulphate-reducing bacteria contain distinctly different PLFA (Edlund *et al.*, 1985; Dowling *et al.*, 1986). Hierarchical cluster analysis of PLFA patterns have shown that relationships between species of isolated methane-oxidizing and sulphate-reducing bacteria almost exactly parallel the phylogenetic relationships based on sequence similarities of the 16S rRNA (Guckert *et al.*, 1991b; Kohring *et al.*, 1994). Hierarchical cluster analyses of PLFA patterns of total microbial communities can be used to quantitatively define relatedness between different microbial communities. This has been done with deep subsurface sediments in which the microbial communities of permeable strata were shown to be distinct from surface soil, clay aqualude, and drilling fluid communities (White *et al.*, 1991, 1996; White and Ringelberg, 1995, 1996).

The analysis of other lipids such as sterols (for the microeukaryotes such as fungi, algae and protozoa) (White *et al.*, 1980), glycolipids (phototrophs, Gram-positive bacteria), or hydroxy fatty acids from the LPS Lipid A (Gram-negative bacteria) (Parker *et al.*, 1982; Walket *et al.*, 1993; Ringelberg *et al.*, 1994) can provide an even more detailed community structure analysis.

Nutritional status

The formation of poly β-hydroxyalkanoic acid (PHA) in bacteria (Nickels *et al.*, 1979; Findlay and White, 1983) or triglyceride in microeukaryotes (Gehron and

White, 1982) relative to the PLFA provides a measure of nutritional status. For example, bacteria grown with adequate carbon and terminal electron acceptors form PHA when they cannot divide because some essential component is missing (phosphate, nitrate, trace metal, etc.). Furthermore, specific patterns of PLFA can indicate physiological stress (Guckert *et al.*, 1986). Exposure to toxic environments can lead to minicell formation and a relative increase in specific *trans* monoenoic PLFA compared to the *cis* homologues. It has also been shown that for increasing concentrations of phenol toxicants, the bacteria *Pseudomonas pudita* forms increasing proportions of *trans* PLFA (Heipieper *et al.,* 1992). Some useful insights come from the analysis of organisms like the *Pseudomonas* species which have been shown to form acyl-ornithine lipids when grown with limited bioavailable phosphate (Minnikin and Abdolrahimzadeh, 1974). Some Gram-positive bacteria form increased levels of acylamino acid phosphatidylglycerols when grown at sub-optimal acidic pH levels (Lennarz, 1970). Aerobic growth conditions with high-potential electron donors induce facultative Gram-negative heterotrophic bacteria to form respiratory benzoquinones (White, 1988, 1993; Vestal and White, 1989; Tunlid and White, 1992). Prolonged exposure to conditions sustaining stationary growth phase induce the formation of cyclopropane PLFA (Guckert *et al.*, 1986; Tunlid and White, 1992). The distribution of organisms showing these specific responses in most soils is sufficiently universal that they can be utilized to define the conditions in the microniches they occupy. This has been established by manipulating environments and then subjecting them to SLB analysis (White, 1988).

There are additional insights that can be gained by further extending the SLB analysis but it is important to differentiate between free fatty acids, ester-linked fatty acids, amide linked fatty acids and vinyl ether linked fatty components. The alkaline hydrolysis used in the analysis of FAME or PLFA does not liberate the amide linked fatty acids that could be present in lipopolysaccharide or other extractable lipids. The use of acid hydrolysis is almost always necessary for the quantitative recovery of amide linked fatty acids.

DNA

Two approaches to DNA extraction from soil have been utilized, namely direct extraction of the whole soil and indirect extraction of a crude bacterial fraction of the soil. Of these two methods the direct approach yields approximately one order of magnitude more DNA than the indirect approach (Ogram *et al.*, 1987; Steffan *et al.*, 1988). Torsvik *et al.* (1994) determined that soils contained thousands of species of microbes by determining the heterogeneity of DNA. Ritz and Griffiths (1994) applied a total DNA community hybridization to the analysis of shifts in the soil community structure whereby similarity was determined by measuring the relative extent to which DNA from one community cross-hybridized with that from another (Lee and Fuhrman, 1990). The method showed great potential for the analysis of shifts in the isolated bacterial fraction of the soil although the more heterogeneous DNA isolated from the whole soil proved too complex to analyse using this technique (Ritz and Griffiths, 1994).

Once protocols were established for the recovery of DNA from soils (Ogram *et al.*, 1987; Steffan *et al.*, 1988), the technology of gene probing was applied toward the detection of specific genes in the environment (Ogram and Sayler, 1988; Steffan *et al.*, 1989; Sayler and Layton, 1990; Sayler, 1991; Sanserverino *et al.*, 1994). Specific genes for specific enzymes can be detected by DNA hybridization, although the presence of a gene does not necessarily indicate its activity. Theoretically, it is possible to detect one gene in a sample (if no inhibitors are present and all controls are negative) by enhancing the sensitivity of nucleic acid analysis via the use of the polymerase chain reaction (PCR). Specific mRNA can often be recovered from environmental samples as again pioneered in the Sayler laboratory (Flemming and Sayler, 1993) and the community structure can be determined by recovery of the gene coding for ribosomal RNA (rDNA) (Ward *et al.*, 1992). Highly conserved portions of the small and large subunit rRNA sequences can be used as universal probes with kingdom specific sequences used to determine the proportions of the eukaryotes, Archaea and Bacteria, and more variable sequences used as species specific probes (Ward *et al.*, 1992).

One of the major problems encountered when utilizing these powerful molecular techniques for analysis of environmental samples is the lack of quantitative recovery of nucleic acids from various complex environmental matrices. DNA extraction depends on quantitative lysis and it has been repeatedly demonstrated that components of the soil microbiota are differentially lysed by chemical and physical methods (Moré *et al.*, 1994). Gram-positive bacteria such as actinomycetes, various cocci, spores and yeasts may be particularly difficult to lyse and subsequently extract (Johnson, 1991). In addition, fractionation of the microbial biomass prior to DNA extraction introduces a bias towards those organisms which are easily dislodged from the soil resulting in non-representative sampling. Polyphenols, tannins, iron chelates and clays make the recovery, amplification, and detection of DNA (by DNA:DNA hybridization) difficult. Where enzyme action is required in the analysis, as in the amplification by Tac polymerase in the PCR or restriction digestion, the concurrent extraction of enzyme inhibitors is a serious problem. For example, although PCR is theoretically capable of detecting one target molecule in a sample, impurities present with DNA extracted from soil can lower the sensitivity of PCR by 10^4–10^8 cells g^{-1} of soil (Picard *et al.*, 1992). Despite the large efforts of several laboratories, quantitative recoveries have been restricted to environments lacking in tannins, clays or enzyme inhibitors such as thermal spring microbial mats or pelagic seawater communities (Ward *et al.*, 1992). Where RNA (either mRNA or rRNA) is the target molecule, the recovery problems are magnified due to the presence of RNA degrading enzymes (Flemming *et al.*, 1993).

The DNA probe analysis offers powerful insights because of its exquisite specificity for the detection of functional genes for enzyme processes and their control, and for the rDNA (coding for the bacterial small and large subunit rRNA and microeukaryote small and large subunit rRNA) for organism identification at the kingdom, family, genus or species levels. However, at present there is a lack of knowledge regarding the effects of all parameters involved in DNA analysis which precludes obtaining truly quantitative results. As such, care should be taken when interpreting results of community structure analysis solely based on nucleic acid analysis. Although several methods have been developed for DNA extraction there is no guarantee that they extract all of the naturally occurring DNA or RNA. For example, sampling bias can result from cell recovery procedures which are selective against specific types of microorganism (Hahn *et al.*, 1990), while failure to lyse all members of a microbial community also negatively impacts the DNA extraction (Ward *et al.*, 1992). Furthermore, during enzymic processes such as PCR other restrictions such as primer choice (Cariello *et al.*, 1991), PCR conditions (Mayerhans *et al.*, 1990; Cariello *et al.*, 1991; Weller *et al.*, 1991; Reysenbach *et al.*, 1992), PCR drift and selection (Wagner *et al.*, 1994) and cloning (Cariello *et al.*, 1991) are introduced, again biasing results. PCR selection occurs when a reaction favours certain members of a gene family (i.e. separate reactions produce results skewed towards the same genes) while PCR drift is a bias that occurs as a result of random events in the early cycles of the reaction. PCR drift can be countered by carrying out several reactions and pooling the products to reduce any skewness of the results caused by the independent random PCR events (Wagner *et al.*, 1994). In addition, PCR selection can be countered by carrying out the reaction using the smallest possible number of amplification cycles (Wagner *et al.*, 1994). Recent evidence has shown that genome size and the number of rRNA genes (rDNA) present per cell will impact on the species represented in clonal libraries of environmental samples (Farrelly *et al.*, 1995). Unless these variables are known for the species present it will be impossible to truly quantify the number of species present in a sample from such libraries.

Another major disadvantage of PCR when applied to community structure analysis and organism identification is the formation of chimeric sequences, that is hybrid molecules made up of two different 16S rDNA types (Leisack *et al.*, 1991). The formation of chimera introduces risk when using PCR to examine microbial community structures. As such, critical analysis of sequences using a computer program such as Check Chimera (Olsen *et al.*, 1992) is vital to any interpretation of the data. These factors should all be taken into account when interpreting nucleic acid data in the absence of corroborating analyses.

Recovery of DNA from Lipid Extraction

Recent evidence indicates that the lipid extraction used in the SLB analysis also liberates cellular DNA (Kehrmeyer *et al.*, 1995). Over 99% of ^{32}P-labelled DNA added to soil was recovered in the residue and aqueous phase of the lipid extraction. Over 50% of the gene *nahH-lux* added in intact *Pseudomonas fluorescens* to the soil recovered by the standard techniques utilized in the Sayler Laboratory (Ogram *et al.*, 1987) was recovered in the aqueous phase of the lipid extraction (Kehrmeyer *et al.,* 1996). The DNA recovered from the lipid extraction was amplified successfully by PCR. The SLB analysis can now be expanded to also provide DNA for gene probing and enzymatic amplification on the same samples. The powerful, quantitative assessment method developed over the past 20 years by our laboratories to define the viable biomass, community structure and nutritional/physiological status of environmental microbial communities based on signature lipid analysis can be expanded to include defining the proportions of specific genes or microbial species based on the extracted DNA.

Concomitant DNA/lipid analysis readily provides quantitative recoveries independent of the ability to isolate or culture the microbes. Whereas the actual biomass can be calculated from the lipid analysis, the gene probes enable comparisons of specific proportions of the community to the more general universal, bacterial, archaeal and 'eukaryote' sequences in the rDNA. The second advantage of the combined SLB/DNA analysis comes from the fact that detection of a functional gene shows only the potential for its activity. The lipid analysis provides an indication of physiological status reflective of the activity indicated in the detection of the enzyme(s) by gene probe analysis. Most genes are not expressed (White, 1995a). Since the SLB analysis provides *in situ* indications of a number of physiological traits, the analysis provides a basis by which a better prediction of the metabolic activity of the genes detected by probe analysis can be made. The concomitant SLB and DNA gene probe analysis of soils and rhizospheres should provide a powerful new analysis system for determining the ecology and community dynamics of soil microbiota. The signature lipid biomarker analysis is both quantitative and provides insights into the nutritional/physiologic status of a microbial community. The addition of a DNA extraction into the technique greatly extends its specificity which will make the utility in toxicity assessment and the determination of 'how clean is clean' much more rational and scientifically defensible. An analysis of currently available signature biomarkers for soil microbiota is illustrated in a diagram of the extractions, fractionations, acid, and alkaline hydrolysates with detection/identification by GC/LC/MS in Fig. 15.1.

Application of SLB Analysis to the Rhizosphere Microbiota

One application of the SLB analysis compared the effects of adding bacteria isolated from the rhizosphere of the rape plant (*Brassica napus*) to surface sterilized seeds of the same plant, and then compared the resulting root systems post germination and growth (Tunlid *et al.*, 1985). From these studies it was shown that the PLFA of sterile roots are very simple which allows for the presence of bacteria to be easily determined. When incubated in sterile sand, elaborations of

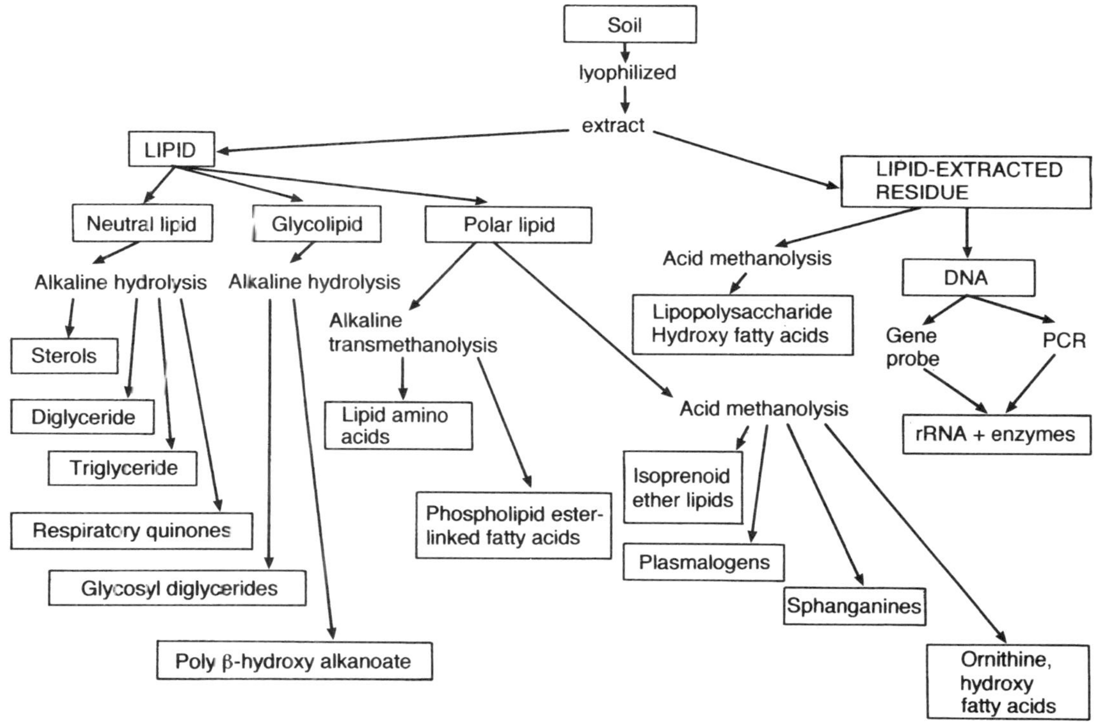

Fig. 15.1. Diagram of the combined SLB and DNA analysis method. The lipid is extracted and fractionated into neutral lipids, glycolipids and phospholipids on silicic acid columns. After acid or alkaline hydrolysis (or methanolysis) the components are derivatized and analysed by GC/MS. The lipid-extracted residue is extracted for DNA recovery and acid hydrolysed and extracted to recover covalently bound lipids.

the roots greatly stimulates the growth of bacteria. A comparison of the nutritional status of the bacteria attached to the roots with those not associated with the roots showed that the rhizosphere bacteria contained no evidence of starvation (increases in the ratio of cyclopropane PLFA to the monoenoic PLFA precursors) or toxicity (increases in the ratio of monoenoic PLFA in the *trans* configuration relative to PLFA in the *cis* configuration) whereas the non-rhizosphere associated bacteria showed high ratios of PHA/PLFA indicating a state of unbalanced growth (growth without cell division). The effects of exudates, like those from roots, on the nutritional status of attached and unattached bacteria has been demonstrated in a model with a filter apparatus which supplies nutrients through the filter. The stimulation of a biofilm on the membrane surface with starving bacteria in the bulk phase was demonstrated (Odham *et al.*, 1986).

The effects of increased temperature induced shifts in the rhizosphere microbiota (Zac *et al.*, 1996). In addition, it has been shown that exposure of plants to increased CO_2 increased the total rhizosphere microbiota associated with the fine root mass without affecting the community composition in white oaks exposed for 3 years in nitrogen limited forest soils in open topped enclosures (Ringelberg *et al.*, 1997). With SLB we showed wood chip compost can be manipulated so that it can either be conducive or suppressive to the growth of the damping off fungal pathogen *Rhizoctonia* (Tunlid *et al.*, 1989a). The attached rhizosphere microbiota of cucumber seedlings (*Cucumis sativus* L.) in compost which suppressed the infection with *Rhizoctonia* showed a lower viable biomass (about 60% of the PLFA associated with the roots of conducive compost grown plants), lower proportions of the short terminal branched PLFA characteristic of Gram-positive bacteria like *Arthrobacter*, decreased proportions of the tuberculostearic acid PLFA characteristic of *Actinomycetes*, and higher proportions of the monoenoic PLFA *cis*-vaccenic acid, (16:1ω7c), including a branched 17 carbon monoenoic PLFA, (i17:1ω8c) all formed by a bacterial biosynthetic pathway characteristic of Gram-negative bacteria. These root-associated Gram-negative bacteria showed greater evidence of starvation (increased cyclopropane to precursor monoenoic PLFA ratio) in the suppressive compost as compared to the conducive compost. The root-associated microbiota had at least a 5-fold greater biomass than the substrate in both suppressive and conducive composts. It proved possible to generate suppressive conditions in conducive compost by growing cucumbers with seeds dipped in cultures of the bacterium *Flavobacterium balustinum* 299. Rhizosphere microbiota recovered from the plants grown in conducive compost exposed as seeds to the *F. balustinum* 299 showed a similar pattern to those in suppressive compost by exhibiting high levels of *cis*-vaccenic acid and the iso-branched 17 carbon monoenoic PLFA characteristic of the bacterium. We have shown differences in sterols and PLFA profiles between 3 and 4 year old oaks grown in the same forest soil. The 4 year old oaks were grown from acorns and inoculated with *Pisolithus tinctorius* in soil-free mix and then planted in the enclosures and 3 year old trees were transplanted as seedlings into the enclosures as nursery stock.

A series of experiments with long needle pines grown in sand and exposed to ambient, ambient + 360 μmol mol^{-1}, and ambient + 720 μmol mol^{-1} CO_2, with either 40 or 400 kg ha^{-1} $year^{-1}$ reduced nitrogen show significant increases in the viable rhizosphere biomass when grown with high reduced nitrogen (G.B. Runion and H.H. Rogers, Auburn University). The greatest increase was observed when both high reduced nitrogen and elevated CO_2 were present. Changes in the rhizosphere microbial community structure were detected. The principal components most influential in the separation between treatments are those associated with the fungal mycorrhiza (White, 1995b).

Detection of Disturbance with SLB Analysis

Since the SLB technique involves the separation and assay of distinctive lipid biomarkers, rates of formation from precursors or turnover during growth may be determined (White and Tucker, 1969). Radioactive or mass labelled precursors are added to the sediment and the incorporation or metabolic activity determined at timed intervals after isolation of the product. Adding the labelled precursor to sediments in slurries, or by injection with various degrees of disturbance resulted in progressively greater apparent metabolic rates (Findlay *et al.*, 1990a,b).

Natural disturbances by benthic sediment processing invertebrates, sting ray feeding, or wave action on tidal flats can be detected both as increased rates of incorporation of labelled precursors and/or shifts in PHA/PLFA ratios if the gentlest methods of labelled precursor application are utilized (Findlay *et al.*, 1990a,b). With these methods, the sedimentary microbiota have been shown to be remarkably responsive to disturbances that allow metabolic activity. Microbes in sediments exist like coiled springs awaiting metabolic opportunities thereby creating opportunities for determinations of activity that are much greater than the actual basal rates. Measurements at the surface with recovered subsurface sediment slurries gave metabolic rates 5 orders of magnitude higher than could possibly maintain the oxygen measured in the ground water at that depth and at the known minimal recharge from the surface (Phelps *et al.*, 1994). Since estimates of carbon dioxide and methane production by soil and benthic microbes are important in calculations of greenhouse effects, these disturbance artefacts have possibly introduced serious errors.

Despite the complexities of the sedimentary microbiota, the SLB technique allows *in situ* determinations that provide insights into sedimentary processes. By sampling just ahead and just behind echinoderm sand dollars *Mellita quinquiesperforata* as they slowly move through the sediments processing the sand, it was possible to show with the SLB that the feeding was selective for protozoans and bacteria to some extent (Findlay and White, 1983). Diatoms passed through the sand dollar alimentary tract intact. Excluding the top predators (fish and crabs) from an estuary by caging induced changes in the sedimentary microbiota (Federle *et al.*, 1983). There was an immediate overgrowth of the opportunistic polychaete *Mediomastus ambiseta* with markedly decreased polyenoic PLFA characteristic of nematodes and algae. There was a concomitant increase in bacteria, especially anaerobic sulphate-reducing bacteria, as detected in specific PLFA patterns. Shortly after the increase in bacteria, the proportions of linoleic acid characteristic of bacteriovorous protozoa increased. The overgrowth of the polychaetes grazed the nematodes and algae which allowed the bacterial overgrowth. The change in bioturbation decreased the aeration of the sediment with an increase in the proportions of anaerobes.

Microcosms isolated from the environment they are designed to mimic are created for assessing the toxicity of xenobiotics on benthic biota. The quantitative definition of the sedimentary microbial community structure in these microcosms can be compared directly with that in the field by using the SLB technique. Experiments showed microbial biomass and community structure were detectably different. However, the degree of difference was not large and did not increase with time when compared with microcosms from a shallow, turbid, highly disturbed bay enriched with riparian runoff that is characterized by low macroscopic species diversity and high biomass (Federle *et al.*, 1986). Microcosms prepared from a more stable, higher salinity system with much greater diversity that is controlled by epibenthic predators showed great differences between replicate microcosms themselves as well as to the field; moreover, the divergences between microcosms increased with time (Federle *et al.*, 1986).

Utility of SLB as a Toxicity Assessment Monitor

The responsiveness of the microbiota to manipulations of the environment can provide a multi-species multi-trophic level toxicity assessment assay. At Oak Ridge National Laboratory, this multi-trophic level, multi-species assay system based on periphyton (the slime covering rocks in streams) to monitor the disturbance involved with pollution abatement in streams was used (Guckert *et al.*, 1991a). Unglazed tiles or rocks were incubated in an unpolluted stream for a month and then transferred to three sites in East Fork Poplar Creek which had different levels of toxicity as estimated by the responses to *Ceriodaphnia* and *Pimephales promelas* larval assays. After a month the tiles were recovered and a portion incubated in ^{14}C-acetate for an hour with the rest of the tiles subjected to SLB analysis. The ratio of rates of PLFA synthesis (membrane) to PHA (storage lipid) synthesis showed an increase as the system was more highly impacted. Toxicity increased the formation of membrane lipids without much effect on the storage lipids. Principal components analysis of the PLFA showed three distinct clusters in which signature PLFA of diatoms were associated with the least impacted cluster and signature PLFA of green algae were associated with the most impacted cluster. The intermediate site was intermediate. The experiment was repeated three times at different seasons with identical results.

The periphyton analysis of wastewater abatement was repeated in a different stream with identical results (Napolitano *et al.*, 1994). Outdoor artificial streams made from plastic rain gutters, seeded with tiles and over which water was pumped from ponds and the toxicant added showed a similar response (Kohring, 1994). We have proposed applying the same system of monitoring the environmental health in the subsurface by analysis of shifts in the subsurface microbes in studies sponsored by the EPA (White and Wilson, 1989). Studies are underway to test natural attenuation and the response of the subsurface microbiota in a deliberate implantation of petroleum hydrocarbons at an Air Force Base in Mississippi (Stapleton *et al.*, 1994). Active biodegradation of petroleum hydrocarbons in subsurface sediments results in increases in viable biomass, shifts to largely aerobic Gram-negative heterotrophic bacterial communities, decrease in biomarkers indicative of stationary phase growth, decrease in PHA/PLFA ratio, and increases in the proportion of benzoquinone respiratory quinones indicative of aerobic electron transport activity (Ringelberg and White, 1992). The degree of these shifts parallel the effectiveness of the bioremediation in many cases.

These results indicate that the SLB analysis of soil microbiota is a reasonable target for defining effects of manipulations on soil quality/health and the subsequent predictions of productivity. There is ample evidence that agricultural practices such as crop rotations, residue management, fertilization, cultural and other management practices can significantly affect soil quality by changing the soil physical, chemical and biological parameters. These changes are reflected in changes in the soil microbiota (Eash *et al.*, 1994; Fauci and Dick, 1994; Karlen *et al.*, 1994; Rice and Garcia, 1994).

Application of SLB/DNA Analysis

If the SLB analysis, particularly with the added specificity of concomitant DNA gene probing, is so comprehensive and useful why is it not universally applied? The principal reasons why the SLB methodology is not widely utilized for microbial characterizations are because lipid extraction, fractionation and derivatization procedures are time consuming and labour intensive. The current SLB extraction procedures also require extensive attention to detail in the purification of solvents, reagents and glassware. The interpretation of the SLB analysis requires extensive understanding of a widely dispersed data-base. These factors have prevented the wide usage of this quantitative analytical definition of the soil and sediment microbiota in long term ecological studies. Signature lipid biomarker analytical techniques are just too difficult for graduate students or investigators interested in ecological interactions involving the soil or sediment microbiology to take up as a task secondary to their primary ecological interest.

Microbial Insights, Inc. (Knoxville, Tennessee, USA) has made the SLB and selected gene probe analysis commercially available to users. Currently, intense research efforts are being aimed at producing an automated SLB analytical system to lower the cost, increase its sensitivity, selectivity and specificity by creating derivatives (Tunlid *et al.*, 1989b), and speed up the analysis with supercritical fluid extraction (Anderson *et al.*, 1993, Hawthorne *et al.*, 1992).

Acknowledgements

This work has been supported by grant DE-FG05-90ER60988 from the Subsurface Science program, administered by F.W. Wobber, grant 94UOT001S from the National Institute for Global Environmental Change, South East Regional Center, from the US Department of Energy and grants N00014-94-1-0961, N00014-94-1-0765 and N00014-93-1-1317 from the Office of Naval Research, US Department of Defense, and grants HRA 699-510-94 and WQI 699-524-94 from the National Water Research Institute, Fountain City, California.

References

Albrechtsen, H.-J. and Winding, A. (1992) Microbial biomass and activity in subsurface sediments from Vejen, Denmark. *Microbial Ecology* 23, 303–317.

Aldhous, P. (1994) Biodiversity: Ecologists draft plan to dig in the dirt. *Science* 264, 1521.

Anderson, T.R., Nivens, D.E. and White, D.C. (1993) Use of supercritical fluid extraction to obtain bacterial polar lipid fatty acids and neutral lipids from environmental samples. *93rd American Society for Microbiology General Meeting.* Atlanta, Georgia., p. 391 (Abstract).

Bååth E., Frostegård, Å. and Fritze, H. (1992) Soil bacterial biomass, activity, phospholipid fatty acid pattern, and pH tolerance in an area polluted with alkaline dust deposition. *Applied and Environmental Microbiology* 58, 4026–4031.

Balkwill, D.L., Leach, F.R., Wilson, J.T., McNabb, J.F. and White, D.C. (1988) Equivalence of microbial biomass measures based on membrane lipid and cell wall components, adenosine triphosphate, and direct counts in subsurface sediments. *Microbial Ecology* 16, 73–84.

Boucher, B.R. and Savageau, M. (1977) Generalized indicator plate for genetic, metabolic, and taxonomic studies with microorganisms. *Applied and Environmental Microbiology* 33, 434–444.

Cariello, N.F., Thilly, W.G., Swenberg, J.A. and Skopek, T.R. (1991) Deletion mutagenesis during polymerase chain reaction: Dependence on DNA polymerase. *Gene* 99, 105–108.

Cavigelli, M.A., Robertson, G.P. and Klug, M.J. (1995) Fatty acid methyl ester (FAME) profiles as measures of soil microbial community structure. In: Collins, H.P., Robertson, G.P. and Klug, M.J. (eds) *The Significance and Regulation of Soil Biodiversity.* Kluwer Academic Publishers, Boston, Massachusetts, pp. 97–113.

Cox, E.E., Major, D.W., Acton, D.W., Phelps, T.J. and White, D.C. (1994) Evaluating trichloroethyene biodegradation by measuring the *in situ* status and activities of microbial populations. In: Hinchee, R.E., Leeson, A., Semprini, L. and Ong, S.K. (eds) *Bioremediation of Chlorinated Polycyclic Aromatic Compounds.* Lewis Publishers, Ann Arbor, Michigan, pp. 37–49.

Domsch, K.H., Jagnow, G. and Anderson, T.-H. (1983) An ecological concept for the assessment of side-effects of agrochemicals on soil microorganisms. *Residue Reviews* 86, 65–105.

Doran, J.W. and Parkin, T.B. (1994) Defining and assessing soil quality. In: Doran, J.W., Coleman, D.C. and Bezdicek, D.F. (eds) *Defining Soil Quality for a Sustainable Environment.* Serial Publications Number 35, Soil Science Society of America, Madison, Wisconsin, pp. 221–228.

Dowling, N.J.E., Widdel, F. and White, D.C. (1986) Phospholipid ester-linked fatty acid biomarkers of acetate-oxidizing sulfate reducers and other sulfide forming bacteria. *Journal of General Microbiology* 132, 1815–1825.

Eash, N.S., Karlen, D.L. and Parkin, T.B. (1994) Fungal contributions to soil aggregation and soil quality. In: Doran, J.W., Coleman, D.C. and Bezdicek, D.F. (eds) *Defining Soil Quality for a Sustainable Environment.* Serial Publications Number 35, Soil Science Society of America, Madison, Wisconsin, pp. 221–228.

Edlund, A., Nichols, P.D., Roffey, R. and White, D.C. (1985) Extractable and lipopolysaccharide fatty acid and hydroxy acid profiles from *Desulfovibrio* species. *Journal of Lipid Research* 26, 982–988.

Farrelly, V., Rainey, F.A. and Stackebrandt, E. (1995) Effect of genome size and *rrn* gene copy number on PCR amplification of 16S rRNA genes from a mixture of bacterial species. *Applied and Environmental Microbiology* 61, 2798–2801.

Fauci, M. and Dick, R.P. (1994) Microbial biomass as an indicator of soil quality: Effects of long-term management and recent soil amendments. In: Doran, J.W., Coleman, D.C. and Bezdicek, D.F. (eds) *Defining Soil Quality for a Sustainable Environment.* Serial Publication Number 35, Soil Science Society of America, Madison, Wisconsin, pp. 229–234.

Federle, T., Livingston, R.J., Meeter, D.A and White, D.C. (1983) Modification of estuarine sedimentary microbiota by exclusion of epibenthic predators. *Journal of Experimental Marine Biology and Ecology* 73, 81–94.

Federle, T.W., Livingston, R.J., Wolfe, L.E. and White, D.C. (1986) A quantitative comparison of microbial community structure of estuarine sediments from microcosms and the field. *Canadian Journal of Microbiology* 32, 319–325.

Findlay, R.H. and White, D.C. (1983) The effects of feeding by the sand dollar *Mellita quinquiesperforata* on the benthic microbial community. *Journal of Experimental Marine Biology and Ecology* 72, 1068–1074.

Findlay, R.H. and White, D.C. (1993) Polymeric beta-hydroxyalkanoates from environmental samples of *Bacillus megaterium. Applied and Environmental Microbiology* 45, 71–78.

Findlay, R.H., Trexler, M.B., Guckert, J.B. and White, D.C. (1990a) Laboratory study of disturbance in marine sediments: Response of a microbial community. *Marine Ecological Progress Series* 61, 121–133.

Findlay, R.H., Trexler, M.B. and White, D.C. (1990b) Response of a benthic microbial community to biotic disturbance. *Marine Ecology Progress Series* 61, 135–148.

Flemming, J. and Sayler, G.S. (1993) Quantitative analysis of naphthalene dioxygenase mRNA in polyaromatic hydrocarbon contaminated soils. *Environmental Science and Technology* 27, 1068–1074.

Flemming, J., Sanseverino, J. and Sayler, G.S. (1993) Quantitative relationship between napthalene catabolic frequency and expression in predicting PAH degradation at town gas manufacturing sites. *Environmental Science and Technology* 27, 1068–1074.

Frostegård Å., Tunlid, A. and Bååth, E. (1993a) Phospholipid fatty acid composition, biomass and activity of microbial communities from two soil types experimentally exposed to different heavy metals. *Applied and Environmental Microbiology* 59, 3605–3617.

Frostegård Å., Bååth E. and Tunlid, A. (1993b) Shifts in the structure of soil microbial communities in limed forests as revealed by phospholipid fatty acid analysis. *Soil Biology and Biochemistry* 25, 723–730.

Garchow, H.L., Collins, H.P. and Klug, M.J. (1993) Characterization of whole soil microbial communities using Biolog GN microplates. *Third Annual Center for Microbiology Ecology Forum,* Michigan State University, East Lansing, Michigan, September 1993.

Garland, J.L and Mills, A.L. (1991) Classification and characterization of heterotrophic microbial communities on the basis of patterns of community-level sole-carbon-source utilization. *Applied and Environmental Microbiology* 57, 2351–2359.

Garland, J.L. and Mills, A.L. (1994) A community level physiological approach for studying microbial communities. In: Ritz, K., Dighton, J. and Giller, K.E. (eds) *Beyond the Biomass, Compositional and Functional Analysis of Soil Microbial Communities.* John Wiley & Sons, New York, pp. 77–83.

Gehron, M.J. and White, D.C. (1982) Quantitative determination of the nutritional status of detrital microbiota and the grazing fauna by triglyceride glycerol analysis. *Journal of Experimental Marine Biology* 64, 145–158.

Guckert, J.B., Hood, M.A. and White, D.C. (1986) Phospholipid, ester-linked fatty acid profile changes during nutrient deprivation of *Vibrio cholerae*: Increases in the *trans/cis* ratio and proportions of cyclopropyl fatty acids. *Applied and Environmental Microbiology* 52, 794–801.

Guckert, J.B., Ringelberg, D.B., White, D.C., Henson, R.S. and Bratina, B.J. (1991a) Membrane fatty acids as phenotypic markers in the polyphasic taxonomy of methylotrophs within the proteobacteria. *Journal of General Microbiology* 137, 2631–2641.

Guckert, J.B., Nold, S.C., Boston, H.L. and White, D.C. (1991b) Periphyton response along an industrial effect gradient: lipid-based physiological stress analysis and pattern recognition of microbial community structure. *Canadian Journal of Fish Aquatic Science* 49, 2579–2597.

Haack, S.K., Garchow, H., Odelson, D.A., Forney, L.J. and Klug, M.J. (1994) Accuracy, reproducibility and interpretation of fatty methyl ester profiles of model bacterial communities. *Applied and Environmental Microbiology* 60, 2483–2493.

Haack, S.K., Garchow, H., Klug, M.J. and Forney, L.J. (1995) Analysis of factors affecting the accuracy, reproducibility, and interpretation of microbial community carbon source utilization patterns. *Applied and Environmental Microbiology* 61, 1458–1468.

Hahn, D., Kester, R., Starrenburg, M.J.C. and Akkermans, A.D.L. (1990) Extraction of ribosomal RNA from soil for detection of *Frankia* with oligonucleotide probes. *Archiv für Mikrobiologie* 154, 329–335.

Hawthorne, S.B., Miller, D.J., Nivens, D.E. and White, D.C. (1992) Supercritical fluid extraction of polar analytes using *in situ* chemical derivatization. *Analytical Chemistry* 64, 405–412.

Heipieper, H.-J., Diffenbach, R. and Keweloh, H. (1992) Conversion of *cis* unsaturated fatty acids to *trans*, a possible mechanism for the protection of phenol degrading *Pseudomonas putida* P8 from substrate toxicity. *Applied and Environmental Microbiology* 58, 1847–1852.

Johnson, J.L. (1991) Isolation and purification of nucleic acids. In: Stackebrandt and Goodfellow, M. (eds) *Nucleic Acid Techniques in Bacterial Systematics.* John Wiley & Sons, New York, pp. 1–19.

Karlen, D.L. Wollenhaupt, N.C., Erbach, D.C., Berry, E.C., Swan, N.S. and Jordahl, J.L. (1994) Crop residue effects on soil quality following 10 years of no-till corn. *Soil and Tillage Research* 31, 149–167.

Kehrmeyer, S.R., Applegate, B.M., Pinkart, H., Hedrick, D.B., White, D.C. and Sayler, G.S. (1996) Combined lipid/DNA extraction method for environmental samples. *Journal of Microbiological Methods* 25, 153–163.

Kohring, L.L. (1994) Quantitative multi-species toxicity assessment by analysis of freshwater stream periphyton pigments and lipids. Master thesis in environmental toxicology, University of Tennessee, Knoxville, Tennessee.

Kohring, L.L., Ringelberg, D.B, Devereux, R., Stahl, D.A., Mittelman, M.W. and White, D.C. (1994) Comparison of phylogenetic relationships based on phospholipid fatty acid profiles and ribosomal RNA sequence similarities among dissimilatory sulfate-reducing bacteria. *FEMS Microbiology Letters* 119, 303–308.

Lechevalier, M.P. (1977) Lipids in bacterial taxonomy – a taxonomist's view. *CRC Reviews of Microbiology* 7, 109–210.

Lee, S. and Fuhrman, J.A. (1990) DNA hybridization to compare species compositions of natural bacterioplankton assemblages. *Applied and Environmental Microbiology* 56, 739–746.

Lehman, R.M., Colwell, F.S., Ringelberg, D.B. and White, D.C. (1995) Combined microbial community-level analyses for quality assurance of terrestial subsurface cores. *Journal of Microbiological Methods* 22, 263–281.

Lennarz, W.J. (1970) Bacterial lipids. In: Wakil, S. (ed.) *Lipid Metabolism.* Academic Press, New York, pp. 155–183.

Liesack, W., Weyland, H. and Stackebrandt, E. (1991) Potential risks of gene amplification by PCR as determined by 16S rDNA analysis of a mixed-culture of strict barophilic bacteria. *Microbial Ecology* 21, 191–198.

Mayerhans, A., Vartanian, J. and Wain-Hobson, S. (1990) DNA recombination during PCR. *Nucleic Acids Research* 18, 1687.

Minnikin, D.E. and Abdolrahimzadeh, H. (1974) The replacement of phosphatidylethanlolamine and acidic phospholipids by ornithine-amid lipid and a minor phosphorus-free lipid in *Pseudomonas fluorescens* NCMB129. *FEMS Letters* 43, 257–260.

Montagna, P.A. (1982) Sampling design and enumeration statistics for bacteria extracted from marine sediments. *Applied and Environmental Microbiology* 43, 1366–1372.

Moré, M.I., Herrick, J.B., Silva, M.C., Ghiorse, W.C. and Madsen, E.L. (1994) Quantitative cell lysis of indigenous microorganisms and rapid extraction of microbial DNA from sediment. *Applied and Environmental Microbiology* 60, 1572–1580.

Napolitano, G.E, Hill, W.R., Guckert, J.B, Stewart, A.J., Nold, S.C. and White, D.C. (1994) Changes in periphyton fatty acid composition in chlorine-polluted streams. *Journal of the North American Benthological Society* 13, 237–249.

Nichols, P.D., Guckert, J.B. and White, D.C. (1986) Determination of monounsaturated fatty acid double-bond position and geometry for microbial monocultures and complex consortia by capillary GC-MS of their dimethyl disulphide adducts. *Journal of Microbiological Methods* 5, 49–55.

Nichols, P.D., Henson, J.M., Antworth, C.P., Parsons, J., Wilson, J.T. and White, D.C. (1987) Detection of a microbial consortium including type II methanotrophs by use of phospholipid fatty acids in an aerobic halogenated hydrocarbon-degrading soil column enriched with natural gas. *Environmental Toxicology and Chemistry* 6, 89–97.

Nickels, J.D., King, J.D. and White, D.C. (1979) Poly-beta-hydroxybutyrate accumulation as a measure of unbalanced growth of the estuarine detrital microbiota. *Applied and Environmental Microbiology* 37, 459–465.

Odham, G., Tunlid, A., Valeur, A., Sundin, P. and White, D.C. (1986) Model system for studies of microbial dynamics at exuding surfaces such as the rhizosphere. *Applied and Environmental Microbiology* 52, 191–196.

Ogram, A. and Sayler, G.S. (1988) The use of gene probes in the rapid analysis of natural microbial communities. *Journal of Industrial Microbiology* 3, 281–292.

Ogram, A., Sayler, G.S. and Barkay, T. (1987) The extraction and purification of microbial DNA from sediments. *Journal of Microbiological Methods* 7, 57–66.

Olsen R.A. and Bakken, L.R. (1987) Viability of soil bacteria: Optimization of plate-counting technique and comparison between total counts and plate counts within different size groups. *Microbial Ecology* 13, 59–74.

Olsen, G.L., Overbeek, R., Larsen, N., Marsh, T.L., McCaughey, M.J., Maciukenas, M.A., Kuan, W.M., Make, T.J., Xing, Y. and Woese, C.R. (1992) The ribosomal database project. *Nucleic Acids Research* 20, 2199–2200.

Parker, J.H., Smith, G.A., Fredrickson, H.L., Vestal, J.R. and White, D.C. (1982) Sensitive assay, based on hydroxy-fatty acids from lipopolysaccharide lipid A for gram negative bacteria in sediments. *Applied and Environmental Microbiology* 44, 1170–1177.

Parkes, R.J., Dowling, N.J.E., White, D.C., Herbert, A. and Gibson, G.R. (1992) Characterization of sulfate-reducing bacterial populations within marine and estuarine sediments with different rates of sulfate reduction. *FEMS Microbial Ecology* 102, 235–250.

Phelps, T.J., Murphy, E.M., Pfiffner, S.M. and White, D.C. (1994) Comparison of geochemical and biological estimates of subsurface microbial activities. *Microbial Ecology* 28, 335–349.

Picard, C.C., Ponsonnet, C., Paget, E., Nesme, X. and Simonet, P. (1992) Detection and enumeration of bacteria in soil by direct DNA extraction and polymerase chain reaction. *Applied and Environmental Microbiology* 58, 2717–2722.

Reysenbach, A., Giver, L.J., Wickham, G.S. and Rice, N.R. (1992) Differential amplification of rRNA genes by polymerase chain reaction. *Applied and Environmental Microbiology* 58, 3417–3418.

Rice, C.W. and Garcia, F.O. (1994) Biologically active pools of carbon in tallgrass prairie soil. In: Doran, J.W., Coleman, D.C. and Bezdicek, D.F. (eds) *Defining Soil Quality for a Sustainable Environment.* Serial Publication Number 35, Soil Science Society of America, Madison, Wisconsin, pp. 201–208.

Ringelberg, D.B. and White, D.C. (1992) Fatty acid profiles. In: Adler, H.I., Jolley, R.L. and Donaldson, T.L. (eds) *Bioremediation of Petroleum-contaminated Soil on Kwajalein Island: Microbial Characterization and Biotreatability Studies.* Oak Ridge National Laboratory, Oak Ridge, Tennessee, ORNL/TM-11925, pp. 31–36.

Ringelberg, D.B, Davis, J.D., Smith, G.A., Pfiffner, S.M., Nichols, P.D., Nickels, J.B., Hensen, J.M., Wilson, J.T., Yates, J.T., Kampbell, D.H., Reed, H.W., Stocksdale, T.T. and White, D.C. (1988) Validation of signature polarlipid fatty acid biomarkers for alkane-utilizing bacteria in soils and subsurface aquifer materials. *FEMS Microbial Ecology* 62, 39–50.

Ringelberg, D.B., Townsend, T., DeWeerd, A., Suflita, J.M. and White, D.C. (1994) Detection of the anaerobic declorinator *Desulfomonile tiedjei* in soil by its signature lipopolysaccharide branched-long-chained hydroxy fatty acids. *FEMS Microbial Ecology* 114, 9–18.

Ringelberg, D.B., Stair, J.O. Norby, R.J., O'Neill, E.G., Zac, D.R. and White, D.C. (1997) Elevated atmospheric CO_2 and the below-ground microbiota of *Quercus alba* L. *Microbial Ecology,* (in press).

Ritz, K. and Griffiths, B.S. (1994) Potential application of a community hybridization technique for assessing changes in the population structure of soil microbial communities. *Soil Biology and Biochemistry* 26, 963–971.

Sanseverino, J., Flemming, J.T., Heitzer, A., Applegate, B.M. and Sayler, G.S. (1994) Application of environmental biotechnology to Bioremediation. In: Wise, D. and Ttantolo, D. (eds) *Remediation of Hazardous Waste Contaminated Soils.* Marcel Dekker, Inc. pp. 97–103.

Sayler, G.S. (1991) Contribution of molecular biology to bioremediation. *Journal of Hazardous Materials* 28, 13–27.

Sayler, G.S. and Layton, A.C. (1990) Environmental application of nucleic acid hybridization. *Annual Review of Microbiology* 44, 625–648.

Smith, G.A., Nickels, J.S., Kerger, B.D., Davis, J.D., Collins, S.P., Wilson, J.T., McNabb, J.F. and White, D.C. (1986) Quantitative characterization of microbial biomass and community structure in subsurface material: a prokaryotic consortium responsive to organic contamination. *Canadian Journal of Microbiology* 32, 104–111.

Stapleton, R.D., Applegate, B.M., Johnson, W.H. and Sayler, G.S. (1994) Microbial characterization of shallow, heterogeneous aquifer system. *Annual Meeting of the American Society for Microbiology* Q-347, 499 (Abstract).

Steffan, R.J., Goksøyr, J., Bej, A.K. and Atlas, R.M. (1988) Recovery of DNA from soils and sediments. *Applied and Environmental Microbiology* 54, 2908–2915.

Steffan, R.J., Breen, A., Atlas, R.M. and Sayler, G.S. (1989) Application of gene probe methods for monitoring specific microbial populations in freshwater ecosystems. *Canadian Journal of Microbiology* 35, 681–685.

Torsvik, V., Goksøyr, J., Daae, F.L., Sørheim, R., Michalsen, J. and Sale, K. (1994) Use of DNA analysis to determine the diversity of microbial communities. In: Ritz, K., Dighton, J. and Giller, K.E. (eds) *Beyond the Biomass. Compositional and Functional Analysis of Soil Microbial Communities.* John Wiley and Sons, New York, pp. 39–48.

Tunlid, A., Baird, B.H., Trexler, M.B., Olsoon, S., Findlay, R.H., Odham, G. and White, D.C. (1985) Determination of phospholipid ester-linked fatty acids and poly beta hydroxybutyrate for the estimation of bacterial biomass and activity in the rhizosphere of the rape plant *Brassica napus* (L.). *Canadian Journal of Microbiology* 31, 1113–1119.

Tunlid, A. and White, D.C. (1992) Biochemical analysis of biomass, community structure, nutritional status, and metabolic activity of the microbial community in soil. *Soil Biochemistry* 7, 229–262.

Tunlid A., Hoitink, H.A.J., Low, C. and White, D.C. (1989a) Characterization of bacteria that suppress *Rhizoctonia* damping-off in bark compost media by analysis of fatty acid biomarkers. *Applied and Environmental Microbiology* 55, 1368–1374.

Tunlid, A., Ringelberg, D.B., Phelps, T.J., Low, C. and White, D.C. (1989b) Measurement of phospholipid fatty acids at picomolar concentrations in biofilms and deep subsurface sediments using gas chromatography and chemical ionization mass spectrometry. *Journal of Microbiological Methods* 10, 139–153.

Vestal, J.R. and White, D.C. (1989) Lipid analysis in microbial ecology: Quantitative approaches to the study of microbial communities. *Bioscience* 39, 535–541.

Visser, S. and Parkinson, D. (1992) Soil biological criteria as indicators of soil quality: Soil microorganisms. *American Journal of Alternative Agriculture* 7, 33–37.

Wagner, A., Blackstone, N., Cartwright, P., Dick, M., Misof, B., Snow, P., Wagner, G.P., Bartels, J., Murtha, M. and Pendleton, J. (1994) Surveys of gene families using polymerase chain reaction: PCR selection and PCR drift. *Systematic Biology* 42, 250–261.

Walker, J.T., Sonesson, A., Keevil, C.W. and White, D.C. (1993) Detection of *Legionella pneumophila* in biofilms containing a complex microbial consortium by gas chromatography-mass spectrometric analysis of genus-specific hydroxy fatty acids. *FEMS Microbiology Letters* 113, 139–144.

Wander, M.M., Hedrick, D.S., Kaufman, D., Traina, S.J., Skinner, B.R., Kehrmeyer, S.R. and White, D.C. (1995) The functional significance of the microbial biomass in organic and conventionally managed soils. In: Collins, H.P, Robertson, G.P. and Klug, M.J. (eds) *The Significance and Regulation of Soil Biodiversity.* Kluwer Academic Publishers, Boston, Massachusetts, pp. 87–97.

Ward, D.M., Bateson, M.M., Weller, R. and Ruff-Roberts, A.L. (1992) Ribosomal RNA analysis of microorganisms as they occur in nature. *Advances in Microbial Ecology* 12, 219–286.

Waxman, S.A. (1927) *Principles of Soil Microbiology,* The Williams and Wilkins Co., Baltimore, Maryland.

Welch, D.F. (1991) Applications of cellular fatty acid analysis. *Clinical Microbiology Reviews* 4, 422–438.

Weller, R., Walsh-Weller, J. and Ward, D.M. (1991) 16S rRNA sequences of uncultivated hot spring cyanobacterial mat inhabitants retrieved as randomly primed cDNA. *Applied and Environmental Microbiology* 57, 1146–1151.

White, D.C. (1988) Validation of quantitative analysis for microbial biomass, community structure, and metabolic activity. *Advances in Limnology* 31, 1–18.

White, D.C. (1993) *In situ* measurement of microbial biomass, community structure, and nutritional status. *Philosophical Transactions of the Royal Society of London, Series* 344A, 59–67.

White, D.C. (1995a) Chemical ecology: possible linkage between macro- and microbial ecology. *Oikos* 74, 177–184.

White, D.C. (1995b) Increasing accuracy of climate change models by application of lipid molecular techniques to assess belowground biomass, prokaryotic and eukaryotic community shifts, and nutritional/physiological status (for estimating activity) in the rhizosphere and soils of important Southeastern US plants exposed to elevated CO_2. *Annual Report, Southeast Regional Center National Institute for Global Environmental Change for Fiscal 1994.* University of Alabama, Tuscaloosa, Alabama, pp. 70–74.

White, D.C. and Tucker, A.N. (1969) Phospholipid metabolism during bacterial growth. *Journal of Lipid Research* 10, 220–223.

White, D.C. and Ringelberg, D.B. (1995) Utility of signature lipid biomarker analysis in determining *in situ* viable biomass, community structure, and nutritional/physiological status of the deep subsurface microbiota. In: Amy, P.S. and Halderman, D.L. (eds) *The Microbiology of the Terrestrial Subsurface.* CRC Press, Boca Raton, Florida.

White, D.C. and Ringelberg, D.B. (1996) Monitoring deep subsurface microbiota for assessment of safe long-term nuclear waste disposal. *Canadian Journal of Microbiology,* 47, 375–381.

White, D.C. and Wilson, J.W. (1989) Subsurface microbiota as monitors of contaminant migration and mitigation. In: Molz, J.F. (ed.) *Symposium on New Field Techniques for Quantitating the Physical and Chemical Properties of Heterogeneous Aquifers.* National Water Well Association, Dublin, Ohio.

White, D.C., Davis, W.M., Nickels, J.S., King, J.D. and Bobbie, R.J. (1979) Determination of the sedimentary microbial biomass by extractable lipid phosphate. *Oecologia* 400, 51–62.

White, D.C., Bobbie, R.J., Nickels, J.S., Fazio, S.D. and Davis, W.M. (1980) Nonselective biochemical methods for the determination of fungal mass and community structure in estuarine detrital microflora. *Botanica Marina* 23, 239–250.

White, D.C., Ringelberg, D.B., Guckert, J.B. and Phelps, T.J. (1991) Biochemical markers for *in situ* microbial community structure. In: Fliermans, C.B. and Hazen, T.C. (eds) *Proceedings of the First International Symposium on Microbiology of the Deep Subsurface.* WSRC Information Service, Aiken, Orlando, pp. 445–456.

White, D.C., Ringelberg, D.C. and Palmer, R.J. (1996) Quantitative comparison of the *in situ* microbial communities in different biomes. *Seventh International Symposium on Microbial Ecology,* Santos, Brazil.

White, D.C., Pinkart, H.C. and Ringelberg, D.B. (1997) Biomass measurements: biochemical approaches. In: Hurst, C.H., Knudsen, G., McInerney, M., Stetzenbach, L.D. and Walter, M. (eds) *Manual of Environmental Microbiology, 1st edn.* American Society for Microbiology Press, Washington, DC, pp. 91–101.

Zac, D.R., Ringelberg, D.B., Pregitzer, K.S., Randlett, D.L., White, D.C and Curtis, P.S. (1996) Soil microbial communities beneath *Populus grandidentata* Michx grown under elevated atmospheric CO_2. *Ecological Applications* 6, 257–262.

Zak, J.C., Willig, M.R., Moorhead, D.L. and Wildman, H.G. (1994) Functional diversity of microbial communities: a quantitative approach. *Soil Biology and Biochemistry* 26, 1101–1108.

Zelles L., Bai, Q.Y., Beck, T. and Breese, F. (1992) Signature fatty acids in phospholipids and lipopolysaccharides as indicators of microbial biomass and community structure in agricultural soils. *Soil Biology and Biochemistry* 24, 317–323.

16

Use of Genetically Modified Microbial Biosensors for Soil Ecotoxicity Testing

G.I. Paton[1,2,3], E.A.S. Rattray[1], C.D. Campbell[3], M.S. Cresser[1], L.A. Glover[2], J.C.L. Meeussen[3] and K. Killham[1]

[1]*Department of Plant and Soil Science, Cruickshank Building, University of Aberdeen AB9 2UE, UK;*
[2]*Department of Molecular and Cell Biology, Marischal College, University of Aberdeen, Aberdeen AB9 1AS, UK;*
[3]*Macaulay Land Use Research Institute, Craigiebuckler, Aberdeen AB9 2QJ, UK*

Introduction

Worldwide contamination of soils by both organic and inorganic pollutants is becoming increasingly apparent. The problems of use and management of these contaminated soils necessitate the development of reliable ecotoxicity tests (dose-response assays) to assess the significance of the contamination and the potential for bioremediation. Biosensors, whether or not involving genetic modification, offer a very promising way forward in the provision of these ecotoxicity tests.

Biosensors are biological materials which, when exposed to an analyte (e.g. air, soil, water), provide an information linked response via a suitable transducer (van der Leslie *et al.*, 1994). The biological material can comprise plants (cells/organs or whole plants), vertebrates, microorganisms/microbial tissue, enzymes, nucleic acid probes, antibodies and various other types of biological receptor. In soil ecotoxicity testing, the analyte to which the biological material is exposed may be the soil itself or some representative soil fraction (e.g. soil solution/extract) and the type of transducer (e.g. acoustic, colorimetric, optical and electrochemical) will vary depending on the nature of the biological material used.

Microbial biosensors appear to have numerous advantages in soil ecotoxicity testing. Microorganisms are generally simple and cheaper to culture than higher organisms, can be freeze-dried for storage and respond rapidly to toxins. Furthermore, the selection of microorganisms for biosensors can be made to ensure environmental relevance. Genetic modification can greatly enhance detection of microbial biosensor response, primarily through use of reporter genes (e.g. *lux* genes). The reporter genes may either be fused to genes involved in the response to a particular toxin or used to indicate overall metabolic status.

Biosensor detection of specific heavy metals has been achieved through transcriptional fusion of *lux* reporter genes to appropriate heavy metal resistance promoters (i.e. light output is switched on by the presence of the particular heavy metal). These systems are available for detection of mercury (Mölders, 1990; Selifonova *et al.*, 1993), copper (Corbisier *et al.*, 1992), nickel and cobalt (Liesegang *et al.*, 1993), chromium (Nies *et al.*, 1989, 1990), zinc (Mergeay *et al.*, 1985; Nies, 1992), cadmium and arsenic (Corbisier *et al.*, 1993).

Detection of specific xenobiotics (i.e. pollutants), through fusion of reporter genes such as *lux*, with catabolic genes encoding degradation of the organic compound in question, provides a potentially powerful role for genetically modified microbial biosensors. *Lux* gene reporting of naphthalene (King *et al.*, 1990) and polychlorinated biphenyls (PCB) (Springael *et al.*, 1991) degradation has provided promising results, although the degree of compound specificity in xenobiotic detection is unlikely to be as strict as for heavy metals, with a range of related organic substrates generally being recognized by any one biosensor.

Use of reporter genes to indicate the general metabolic status of the biosensor microorganism has considerable advantages, particularly where recognition of specific components of an analyte is not a priority. Paton *et al.* (1995a,b) have reported on the overall toxicity of soil and water samples contaminated with heavy metals and xenobiotics as assessed by the bioluminescence response of a *lux*-marked (chromosomal integration) pseudomonad. The recent use of genetically modified microbial (GMM) biosensors in this way is analogous to the approach used in a number of commercially available toxicity tests (e.g. Microtox, Lumintox) based on the response of naturally luminescent marine bacteria (e.g. *Photobacterium phosphoreum*) (Beckman, 1982). Clearly, however, the use of marine microorganisms for ecotoxicity testing in terrestrial and freshwater environments presents a number of problems, mainly related to the necessity of maintaining high saline concentrations in the analyte under test.

Probably the greatest single advantage in using biosensors including GMM-based biosensors in soil ecotoxicity testing is that the biosensor can indicate bioavailability in a way that conventional chemical analyses cannot. Biosensors can directly assess heavy metal bioavailability, effectively integrating the complexity of environmental factors (pH, Eh (redox potential), exchangeable cations, biological activity, etc.) that contribute to bioavailability (Corbisier *et al.*, 1993; Holmes *et al.*, 1993). Of course, consideration of bioavailability is equally relevant to organic pollutants as to heavy metals. However, many recalcitrant organics are not toxic to microorganisms. Fusion of reporter genes to specific catabolic activity (i.e. switching reporter genes on in the presence of particular organic pollutants), therefore, rather than reporting of general metabolic activity, will be the preferred option in biosensor development because the former strategy provides for greater sensitivity of detection.

The use of GMM biosensors to assess toxin bioavailability in soils necessitates appropriate and representative sampling and exposure of the soil to the biosensor in a way that does not compromise assessment of bioavailability. Most toxins will be bioavailable when in the soil solution, and so procedures to obtain representative samples of soil solution will be an essential prerequisite of biosensor use. Paton *et al.* (1995a) obtained representative soil solution by centrifugation to determine bioavailability of contaminant heavy metals with a bioluminescence-marked bacterial biosensor. In some instances, however, bioavailability of compounds can only be assessed *in situ* in the presence of soil particulates. In these cases, the exposure of soil directly to biosensor microorganisms immobilized onto a recoverable surface may offer the way forward (Brouwer *et al.*, 1990).

Although the use of GMM-based biosensors in soil ecotoxicity testing potentially offers great sensitivity and rapidity of detection, this is only advantageous if the genetic modification has not affected its response to potentially toxic elements (PTES) or its ecological fitness (i.e. the GMM still behaves like its wild-type counterpart). The latter is important to ensure that the GMM biosensors remain representative of their ecological niches and accurately reflect the response of the wild type to stress even when inserted genes are expressed. These assumptions can be checked by comparing the overall growth kinetics and toxic responses of the GMM and its wild type using conventional growth inhibition tests (Amin-Hanjani *et al.*, 1993). This approach, however, is only likely to measure large perturbations to the partitioning of metabolically derived energy between normal and introduced growth processes. Checks on other important functions such as effective nodulation with *Rhizobium* spp. and uptake of PTEs dependent on active transport mechanisms may also be necessary. In addition, more rigorous and sensitive testing of any debilitating modification or reduction in ecological fitness might be achieved using techniques such as metabolic fingerprinting (Bochner, 1989), which assesses the ability of the organism to degrade a wide range of substrates.

Whilst the ecological relevance of GMM biosensors is major step forward, it will also be important in their future development to relate them to more traditional assays involving higher organisms (Kaiser, 1993) and to situations in which long term chronic effects have been found, which may not be predicted using acute bioassays (e.g. microbial biosensors). Only by doing this can GMM biosensors provide the early warning system of potential environmental damage that is needed for soils. The research described in this chapter is tackling this by using GMM biosensors in long term experiments in which a range of other microbial measurements are being made.

The aim of this chapter is to discuss the potential of genetically modified microbial biosensors (involving *lux* gene reporting of overall metabolic activity of two soil bacteria *Pseudomonas fluorescens* and *Rhizobium leguminosarum* biovar *trifolii*) for soil ecotoxicity testing. Comparisons were made with a naturally bioluminescent biosensor (*Photobacterium phosphoreum* – Microtox) and with two other bioassays (dehydrogenase and ATP luciferin–luciferase) which indicate the status of microbial growth activity.

Development of *lux*-Marked GMM Biosensors

Construction of GMM biosensors and preparation and storage of inocula

Pseudomonas fluorescens 10586s FAC510 (Amin-Hanjani *et al.*, 1993) and *Rhizobium leguminosarum* biovar *trifolii* TA1 Tn5*luxAB* contain the *Vibrio fischeri lux A* and *B* genes encoding the luciferase enzyme, integrated into the chromosome. *Pseudomonas fluorescens* 10586s pUCD607 (Amin-Hanjani *et al.*, 1993) contains the *V. fischeri luxABCDE* genes present as a multi-copy plasmid. The *lux*-marked biosensors were prepared and stored as freeze-dried cultures, using batch grown cultures (Amin-Hanjani *et al.*, 1993) and lyophilized using standard procedures (Rudge, 1990). The use of freezing drying for producing inocula for ecotoxicity testing has previously been used for the production of the Microtox organism (Beckman, 1982). The method allows for the preservation of bacteria over a considerable period of time (Lapage *et al.*, 1970), which are genetically stable, have a high degree of viability and can be rapidly resuscitated. These aspects of preparation and storage are of paramount importance when developing a rapid screening biosensor. For optimal assay conditions, the freeze dried cultures were resuscitated over a 2 h period in growth media, washed in potassium chloride (KCl) to remove any growth media and utilized within 30 min of washing.

Colony enumeration and bioluminescence response of washed and unwashed cells of *P. fluorescens* 10586s pUCD607 were monitored over a period of 6 months at weekly intervals. It was shown that maximum bioluminescence was expressed 2 h after the commencement of resuscitation. The relative light units (RLUs) per cell were shown to remain constant for cells resuscitated over a 6 month storage period (unpublished data). Assessment of the EC_{50} value of *P. fluorescens* 10586s pUCD607 for Zn and dichlorophenol (DCP) demonstrated that freeze-drying and resuscitation had no effect on the Zn response. The storage of the freeze-dried culture in a moisture free environment at –20°C for a period of 6 months was also shown to have no effect on the EC_{50} values (unpublished results). It was concluded that freeze-drying of a large batch of cells offered an ideal method to provide a rapid and reproducible assay over a period of time and ensured that the cells represent a known genetic pool for which the toxicity response to a range of potential toxins had been well characterized.

Assay procedures

Bioluminescence assay

To perform the *lux*-based GMM assay, a freeze-dried culture was resuscitated and washed in KCl (0.1 M). The assay itself includes a washed cell suspension (100 μl; 0.1 M KCl) and 900 μl of the test solution (Paton *et al.*, 1995b), which was incubated at a constant temperature (25°C) for the required time. The two chromosomal constructs required the exogenous addition of decanal for light production, which was added prior to luminescence measurements. Luminescence was measured using a Bio-Orbit 1251 luminometer (Rattray *et al.*, 1990; Paton *et al.*, 1995b).

Dehydrogenase assay

The 2,3,5 triphenyltetrazolium chloride (TTC) dehydrogenase assay of Bitton and Koopman (1986) was modified and used to determine dehydrogenase activity of *P. fluorescens* 10586s and *R. leguminosarum* bv. *trifolii* TA1 in response to toxicants. It should be noted, however that its use is limited when testing samples that contain Cu (Chander and Brookes, 1991). The assay included the use of 0.5 M TES buffer with 1% tetrazolium chloride (pH 7.80) and resuscitated, unwashed cells of *P. fluorescens* 10586s or *R. leguminosarum* bv. *trifolii* TA1 to which the test pollutant was added and vortexed. The assay was incubated and dehydrogenase activity determined as described by Bitton and Koopman (1986).

ATP luciferin–luciferase assay

The adenosine triphosphate (ATP) luciferin–luciferase assay is a rapid detection technique for the non-specific quantification of microbial biomass, the assay measurement reflecting the total intracellular volume of ATP (Lundin, 1989). The assay used pure cultures of *P. fluorescens* 10586s to determine the effect of pollutants on microbial ATP content (Paton *et al.*, 1995a). Resuscitated cells of *P. fluorescens* 10586s were exposed to the test pollutants for 5 h, after which cellular ATP was extracted and quantified using the luciferin–luciferase reaction (Lundin, 1989). Luminescence was measured at 25°C on a Bio-Orbit 1251 luminometer.

Microtox and Photobacterium phosphoreum *(844)*

The Microtox assay was carried out as described by Beckman Instruments Inc. (1982) and comprises of freeze-dried culture of *Photobacterium phosphoreum*. As a comparison to the Microtox organism, a culture collection strain of *P.*

phosphoreum (844) was also used. Freeze-dried cultures of *P. phosphoreum* (844) were prepared (Rudge, 1990) and the assay procedure was as described for the Microtox assay (Beckman, 1982).

Data interpretation

The plotted graph of bioluminescence as a percentage of the maximum against toxin concentration provided a rapid method of assessing the EC_{50} value. Analysis was carried out as described by Paton *et al.* (1995a) using the statistical package Genstat 5 Rel 2.2 (Rothamsted Experimental Station, 1991). For the other assays (Microtox, *P. phosphoreum* (844), ATP luciferin-luciferase and dehydrogenase) linear regression analysis was carried out on the triplicate data sets to predict the EC_{50} value for each toxin value (using Genstat 5 Rel 2.2.). A two-way analysis of variance was then carried out to calculate mean EC_{50} values and LSDs to allow a cross comparison between the significance of different toxins and bioassays.

Effect of pH

The bioluminescence response of washed cells of both *lux*-marked *Pseudomonas fluorescens* and *R. leguminosarum* bv. *trifolii* across a range of pH values was

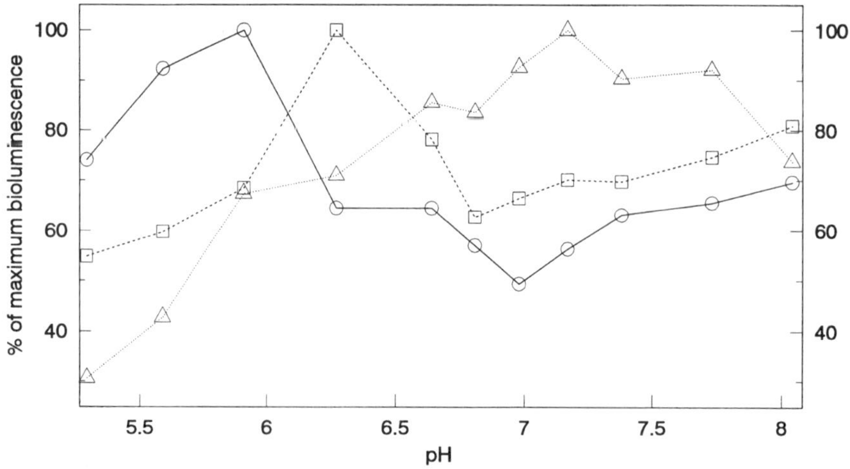

Fig. 16.1. The effect of pH on the bioluminescence response of washed cells of *lux*-marked biosensors (*Pseudomonas fluorescens* 10586s FAC510, ○, *P. fluorescens* 10586s pUCD607, □, and *Rhizobium leguminosarum* bv. *trifolii* TA1 Tn5*luxAB*, △) in solution buffered with Na_2HPO_4 and KH_2PO_4. Least significant difference ($P \leq 0.05$) values are 13%, 8% and 17% for 10586s FAC510, 10586s pUCD607 and TA1 Tn5*luxAB*, respectively.

measured. This was carried out both with a non-buffered solution using sterile deionized water adjusted with NaOH and HCl to within the pH range 4.0–9.0 and with a buffered solution using Na_2HPO_4 and KH_2PO_4 in the range 5.2–7.4.

A pH optimum of 5.5 was found for bioluminescence for both of the *P. fluorescens* constructs in unbuffered solution (unpublished data). In buffered solution, the pH optimum for both strains of *P. fluorescens* was between pH 5.8 and 6.3 (Fig. 16.1). The light output was 50% or greater across the whole pH range from 4.0–9.0. The pH optimum for *R. leguminosarum* bv. *trifolii* TA1 Tn5*luxAB* was found to be 6.4 in unbuffered solution (data not shown) and 7.2 in buffered solution (Fig. 16.1). The light output was greater than 50% of the maximum luminescence across a pH range of 5.7–8.0 and was not less than 30% of the maximum in the pH range of 4.0–5.7. Because of the variable bioluminescence response across the pH range, the strategy used was to carry out all experiments at pH 5.5 as this was considered to represent a relevant soil pH for north-east Scotland.

Comparison of different lux*-marked biosensors in aqueous solutions*

Aqueous solutions of heavy metals (Zn, Cu, Cd, Cr and Ni prepared from sulphate salts at pH 5.5) and 3,5-dichlorophenol (DCP) were used to compare the bioluminescent response of the different *lux*-marked biosensors by calculating EC_{50} values (Paton *et al.*, 1995b). Paton *et al.* (1995b) described a decline in bioluminescence with increasing concentration of toxins for *lux*-marked *P. fluorescens* and suggested some of the reasons for a significant difference in the sensitivity of light output response depended on the *lux* construct used. The order of toxicity (from most to least toxic) was Cu = Zn > Cd > Ni > Cr > DCP for the plasmid construct and Cu = Zn > Cd > Cr > Ni > DCP for the chromosomal construct. For *R. leguminosarum* bv. *trifolii* TA1 Tn5*luxAB*, the order of toxicity was Zn > Cu > Cd > DCP > Cr.

Comparison of lux*-marked biosensors with standard assays*

The EC_{50} values for the range of bioassays for 3 PTEs and DCP are shown in Table 16.1. The bioluminescence based assays (Microtox, *Photobacterium phosphoreum* 844 and the *lux*-marked biosensors) proved to be the most rapid and to offer the greatest degree of sensitivity. The ATP luciferin–luciferase assay and the dehydrogenase assays were carried out in growth media at pH 7.8. This may cause the binding of toxins to substrates and thus render them less available and reduce the toxicity. A reduction in toxicity will accordingly increase the EC_{50} value. The ATP and dehydrogenase assays using *Pseudomonas fluorescens* 10586s were found to provide EC_{50} values of a similar order of magnitude. With the exception of EC_{50} values for Cd, it was not possible to distinguish between

Table 16.1. Comparison of the EC_{50} values (ppm) for a range of bioassays using three PTEs and DCP and a range of microorganisms

Assay	Microorganism	Zn	Cu	Cd	DCP	LSD	CV
ATP	*Pseudomonas fluorescens* 10586s	6.33[a]	7.19[a]	8.21[a]	8.31[a]	3.05	23
Dehydrogenase	*P. fluorescens* 10586s	7.38[a]	NA	12.59[b]	8.46[a]	1.20	20
Dehydrogenase	*Rhizobium trifolii* TA1	19.72[c]	NA	16.42[c]	10.11[c]	9.35	62
Bioluminescence	Microtox[1]	2.35[d]	1.89[d]	9.78[e]	1.68[d]	0.78	12
Bioluminescence	*Photobacterium phosphoreum*[1]	3.45[d]	2.34[d]	NA	2.46[d]	0.92	17
Bioluminescence	*Pseudomonas fluorescens* 10586s FAC510[1]	0.89[f]	0.76[f]	0.98[g]	3.42[d]	0.30	18
Bioluminescence	*P. fluorescens* 10586s pUCD607[1]	0.09[h]	0.09[h]	0.17[i]	1.86[d]	0.13	12
Bioluminescence	*R. trifolii* TA1 Tn5*luxAB*[2]	0.48[f]	0.78[f]	2.14[d]	2.40[d]	0.46	22

LSD, Least significant difference ($P \leq 0.05$) within the assay; CV, coefficient of variance as a percentage; NA, data not available; [1]EC_{50} calculated after 20 min exposure; [2]EC_{50} calculated after 30 min exposure; Values followed by the same letter were not significantly ($P \leq 0.05$) different as determined by ANOVA for assay and metal.

the EC_{50} values and accordingly between the sensitivity of these two assays for *P. fluorescens* (ATP and dehydrogenase). EC_{50} values for Cu, using the dehydrogenase assay were not determined for either *P. fluorescens* 10586s or *R. leguminosarum* bv. *trifolii* TA1, as the presence of Cu in samples is known to interfere with assay (Chander and Brookes, 1991). The *P. fluorescens* 10586s dehydrogenase and ATP assays were less reproducible, having a higher coefficient of variance than associated with the bioluminescence-based bacterial assays. The coefficients of variance determined in this study were at levels that agreed with literature values (Bitton and Koopman, 1986).

The dehydrogenase assay using the *R. leguminosarum* bv. *trifolii* TA1 was found to be the least sensitive and was associated with the highest coefficient of variance. This may have been due to greater production of extracellular polysaccharides by the rhizobium compared to pseudomonads during the relatively long incubation period (18 h) required by the dehydrogenase assay, causing complexation of the toxins, thus further reducing the toxicity of the pollutant. DCP was found to be more toxic to *R. leguminosarum* bv. *trifolii* TA1 than either Zn or Cd when assessed by the dehydrogenase method.

ATP and dehydrogenase assays measured the effect of a toxin on cell activity over a period of time, during which time cell division occurs so that the use of these assays may reflect inhibition of both cell activity and division.

The Microtox assay and the assay using *Photobacterium phosphoreum* 844 were found not to be significantly ($P \leq 0.05$) different for any of the PTE or DCP studied. The three *lux* GMM assays were found to be more sensitive for PTEs than the assays using *P. phosphoreum* and there was no difference in the sensitivity of the assays for the detection of DCP. The *Pseudomonas fluorescens* 10586s pUCD607 assay was the most sensitive assay for detecting inorganic PTEs and identified the same order of toxicity as the *P. fluorescens* 10586s FAC510 assay which was found to be significantly less sensitive. The *P. fluorescens* 10586s FAC510 assay was as sensitive as *R. leguminosarum* bv. *trifolii* TA1 Tn5*luxAB* at detecting Zn, Cu and DCP but was more sensitive to Cd.

Cellular accumulation of Zn by lux*-marked biosensor*

In addition to studying the decline in bioluminescence with increasing PTE concentration, it was also important to assess the cellular concentration of PTEs. Zinc was selected for study because of the excellent detection sensitivity using flame atomic absorption spectroscopy (FAAS). Resuscitated *lux*-marked cells were exposed to varying concentrations of filter sterilized $ZnSO_4$ (pH 5.5), after which the cells were collected onto a filter membrane, digested in concentrated HNO_3 and the Zn concentration determined by FAAS. It was found that there was no significant ($P \leq 0.05$) difference in the cellular concentration between the different strains of *P. fluorescens* and *R. leguminosarum* bv. *trifolii* across the Zn aqueous concentration range tested (unpublished data) suggesting that the differences in sensitivity (as determined by bioluminescence) between *P. fluorescens* 10586s FAC510 and pUCD607 strains were probably caused by the expression of bioluminescence and not by a difference in the availability, and hence the cellular accumulation, of the Zn. These results have also been observed for Cu (unpublished data). Table 16.2 shows the mean data for cellular

Table 16.2. Cellular accumulation of Zn by *Pseudomonas fluorescens* and *Rhizobium leguminosarum* biovar *trifolii* in a range of Zn concentrations.

[1]Strains	Zn solution conc. (mg ml^{-1})	Zn (fg cfu^{-1})	l.s.d. ($P \leq 0.05$)
P. fluorescens 10586s	0	1.0	0.4
P. fluorescens 10586s	0.5	2.2	
P. fluorescens 10586s	2.0	3.4	
R. trifolii TA1	0	0.5	0.9
R. trifolii TA1	1	1.5	
R. trifolii TA1	5	3.5	
R. trifolii TA1	10	4.7	
R. trifolii TA1	20	6.9	

[1]The average values of Zn accumulation for the different strains of *P. fluorescens* and *R. trifolii* are given.

concentration of Zn (on a *fg* Zn per cfu basis) associated with *P. fluorescens* 10586s and *R. leguminosarum* bv. *trifolii*. Cellular concentration of Zn for both *P. fluorescens* and *R. leguminosarum* bv. *trifolii* was found to increase as Zn aqueous concentration increased (Table 16.2). It should be pointed out, however, that the distribution of PTEs associated with cells was not studied.

Effect of complexing agents

In anticipation of applying the assay to soil extracts, it was considered important to investigate the role of chelators in complexing metals. Paton *et al.* (1995a) demonstrated that the EC_{75} values for Cu, Zn, Ni and Cd were less than 10 μM and the decline in bioluminescence with increase metal concentration was linear around concentrations of this value. This range was therefore used to test the complexation of metals with different complexants present in varying molar equivalents.

Metal standards of 20 μM were prepared for each of the divalent metals (M^{2+}) Cu, Zn, Ni and Cd and 40 μM solutions of citric acid, EDTA (disodium salt) and acetic acid were made. All of the solutions were adjusted to pH 5.50 (± 0.02) using HCl/NaOH. The GMM bioassay was then carried out using *P. fluorescens* 10586s pUCD607 over a range of metal and chelator concentrations as follows; 0 M^{2+}, 10 μM M^{2+}, 10 μM M^{2+} + 2.5 μM chelator, 10 μM M^{2+} + 5.0 μM chelator, 10 μM M^{2+} + 7.5 μM chelator, 10 μM M^{2+} + 10 μM chelator, 10 μM M^{2+} + 15 μM chelator, 10 μM M^{2+} + 20 μM chelator.

The results show that acetate or citrate, unlike EDTA, had no effect on metal toxicity (Fig. 16.2). Acetate and citrate are much less effective in binding metals than EDTA, although Cu complexation, and hence some reduction in toxicity, might have been expected under these conditions. EDTA markedly decreased the bioavailability of all metals (Fig. 16.3). For Cd and Cu, the toxic response related to the amount of non-complexed metal present. This suggests that the metal–EDTA complex was less toxic to the organisms and that toxicity was governed by the free metal concentration. For Zn the results were somewhat different, in that it continued to depress bioluminescence above a 1:1 metal: EDTA ratio. This may have been due to different stability of the Zn–EDTA complex compared to stabilities of Cu and Cd complexes, but this is not very likely as the stabilities of Zn–EDTA and Cd–EDTA are comparable. EDTA, at the concentration studied, itself was not toxic (unpublished results), so possibly the Zn–EDTA complex was toxic to the organisms, caused by direct uptake of the Zn-complex or by the ability of the microorganisms to (actively) release Zn from the complex.

In addition, cellular accumulation of Zn was investigated for EDTA complexed solutions at concentrations of 0 Zn, 100 μM Zn, 100 μM Zn + 50 μM EDTA, 100 μM Zn + 100 μM EDTA, 100 μM Zn + 200 μM EDTA, all at pH

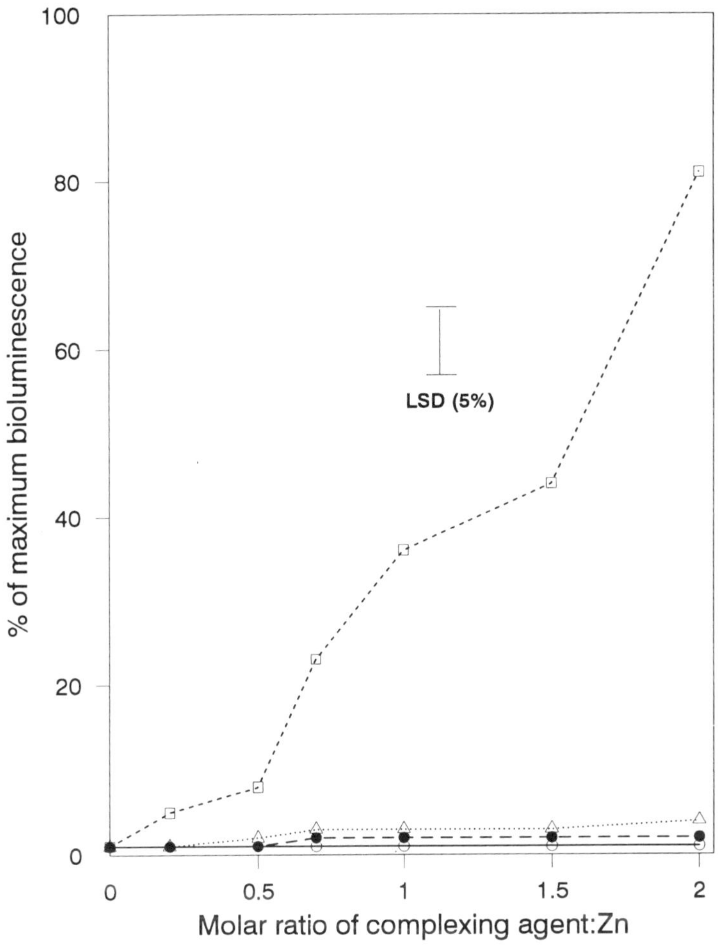

Fig. 16.2. The effect on the bioluminescence response of *Pseudomonas fluorescens* 10586s pUCD607 of the addition of chelating agents (EDTA, □, citrate, △, acetate, ●, and aqueous solution, ○) to a 10 μM Zn solution (as $ZnSO_4$).

5.50 (± 0.02). At 100 μM Zn cellular accumulation of Zn was 12.0 *fg* Zn cfu^{-1} (Fig. 16.4). With the addition of 1 molar equivalent of EDTA, bioluminescence was 38% of maximum corresponding to 6.1 *fg* Zn cfu^{-1}. At 2 molar equivalents cellular Zn in *P. fluorescens* 10586s pUCD607 was found to be 3.2 *fg* Zn cfu^{-1}, suggesting a deleterious effect of Zn on bioluminescence via binding or uptake even in the presence of EDTA.

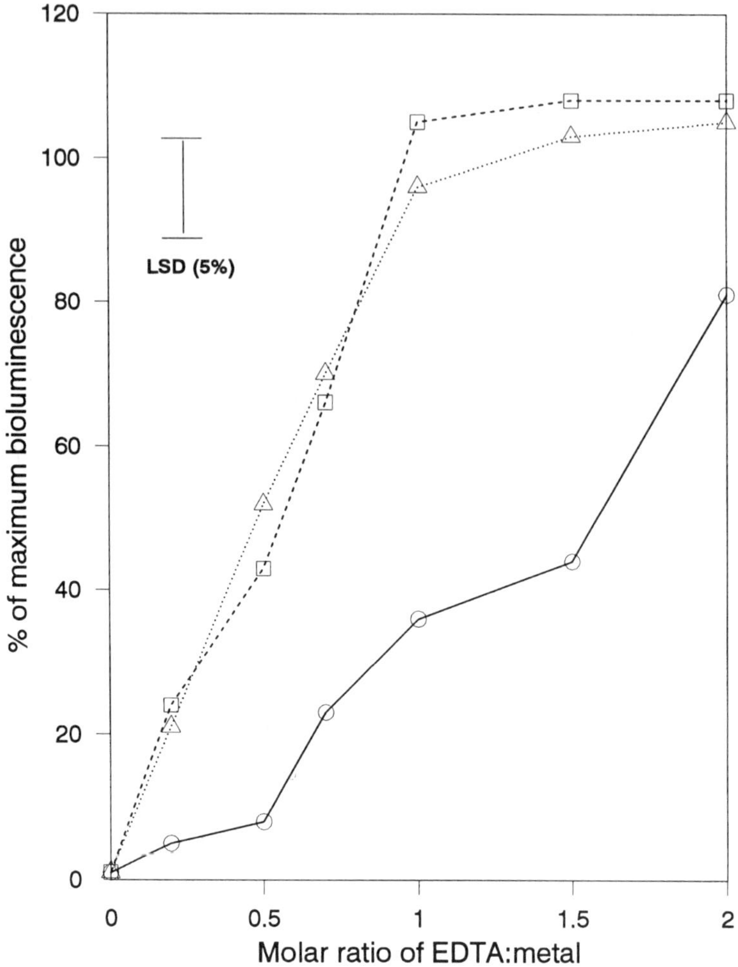

Fig. 16.3. The effect of EDTA chelation on the bioavailability of three PTEs (each at a concentration of 10 μM PTE (as $M^{2+}SO_4$)) (Zn, ○, Cu, □, and Cd, △) as determined by the bioluminescence response of *Pseudomonas fluorescens* 10586s pUCD607.

Effect of pH on Cu toxicity

One of the most important factors determining the availability of metal cations and in particular, Cu (Kratgen, 1978) is pH. Cu rapidly becomes bound in the hydroxyl form from pH 4 upwards. Figure 16.5 illustrates the change in bioluminescence with increased Cu concentration at pH 5.5, 7.0 and 8.5 for *P. fluor-*

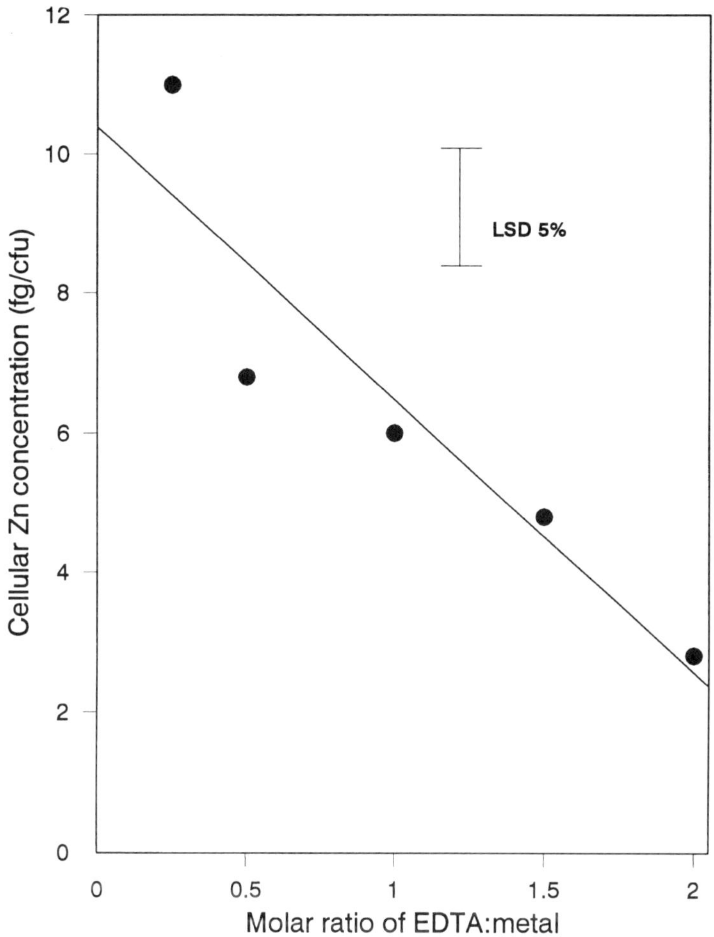

Fig. 16.4. The cellular concentration of Zn in *Pseudomonas fluorescens* 10586s pUCD607 after the addition of a range of concentrations of EDTA to a 10 μM Zn solution (as $ZnSO_4$).

escens 10586s pUCD607 and at pH 7.0 (with 2% NaCl) for *Photobacterium phosphoreum* 844. This figure illustrates two points; firstly, *P. fluorescens* 10586s pUCD607 is more sensitive to Cu than *P. phosphoreum* and, secondly, as pH increases, the EC_{50} of Cu increases. There is a significant ($P \leq 0.05$) difference in the EC_{50} between pH 8.5 and the other two pH values. When the more sensitive EC_{60} value was used to compare the pH effects, there were significant ($P \leq 0.05$) differences between all pH values when assessing Cu toxicity.

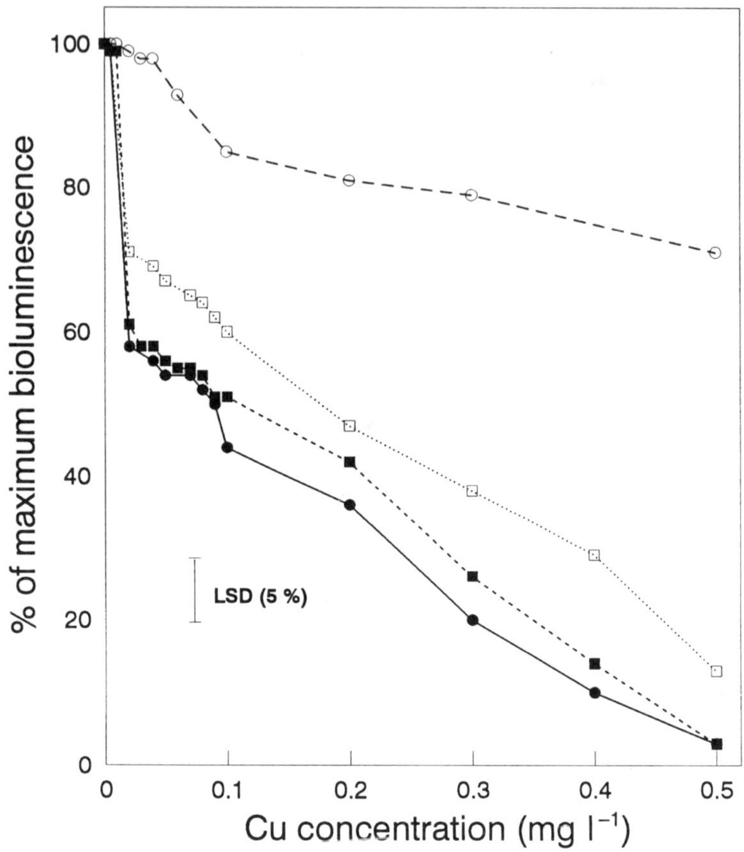

Fig. 16.5. The effect of a range of Cu concentrations on the bioluminescence response of *Pseudomonas fluorescens* 10586s pUCD607 at three different pHs (pH 5.5, ●, pH 7.0, ■, and pH 8.5, □) compared with the bioluminescence response of *Photobacterium phosphoreum* 844 at pH 7.0 and in 2% NaCl (○).

It is essential therefore that comparisons are carried out with samples that are at the same pH. The response of *P. phosphoreum* was not sufficiently sensitive at the EC_{50} within the range of EU drinking water limits for Cu (Paton *et al.*, 1995a).

Application of GMM Biosensors to Ecotoxicity Testing in Soil

Extraction of soil water

Soil samples were taken from 0–10 cm depth from four contrasting soils from north-east Scotland (Table 16.3). The soils were sieved through a 2 mm sieve

Table 16.3. Characteristics of the soils from which extract was removed.

Location	Soil type	Series	Texture	pH (H_2O)
Craigiebuckler	Freely drained podzol	Dess	sandy silt loam	5.9
Craibstone	Imperfectly drained podzol	Countesswells	loamy sand	6.2
Insch	Freely drained brown forest soil	Insch	sandy loam	5.8
Cruden Bay	Poorly drained brown forest soil	Tipperty	clay loam	5.3

and visible root and faunal material were removed. The soils were stored field moist at 4°C. Deionized water was used to increase water contents prior to extraction by wetting the soil to field capacity and allowing equilibration in a sealed plastic bag for 24 h. Soils were extracted by centrifugation as described by Linehan *et al.* (1989). The water extracts obtained for each soil were combined, and used immediately or else filter sterilized (through a Whatman cellulose acetate 0.22 μm filter) and stored at 4°C. Preliminary investigations demonstrated that unless the soil water was filter sterilized, microbial growth took place rapidly, changing the chemistry of the extract.

Determination of EC_{50} values in spiked soil water extracts

Fresh soil water extract of an imperfectly drained humus iron podzol (Dess Series) was sequentially filtered through a 0.8 μm, 0.45 μm, 0.22 μm and aliquots of each of the filtrates retained. 10 ml aliquots of the 0.22 μm filtrate were then spiked with either Zn or Cu (as sulphate salts) at 10 mg l^{-1} and the pH adjusted to 5.50 (± 0.02). Standards were made up for each of the metals in different filtrates (varying filter pore size) by diluting the spiked stock into soil extracts. The GMM bioassay was then carried out using *Pseudomonas fluorescens* 10586s pUCD607 to assess the EC_{50} value of soil extract calculated for different pore sizes and for different PTEs.

For the Dess Series soil, the EC_{50} for *P. fluorescens* 10586s pUCD607 declined with a reduction in membrane filter pore size (Table 16.4). The reproducibility of the assay also increased with decreasing pore size, demonstrated by the reduction of the % coefficient of variance (% CV) (Table 16.4). The % CV declined from 36% and 52% at 2.5 μm to 21% and 18% at 0.22 μm for Cu and Zn, respectively. Total bioluminescence (RLU) was found to decline slightly (12%) between 2.5 μm and 0.22 μm. The decline in bioluminescence was quantified against non-contaminated standards for each of the appropriate pore sizes and, accordingly, the difference in the total bioluminescence was not important. The effect on the EC_{50} values indicated that there were changes in

Table 16.4. EC_{50} values in Dess Series soil water extract after pH adjustment and filtration and a 30 min exposure time.

	Filter membrane pore diameter (μm)				
	2.5	0.8	0.45	0.22	LSD
Zn	1.24 (52)	1.20 (31)	1.14 (20)	0.08 (18)	0.62
Cu	0.56 (36)	0.34 (27)	0.42 (23)	0.28 (21)	0.25

Figure in brackets after the EC_{50} value is the mean coefficient of variance for each data set; LSD at 5% confidence limit.

the biology and the chemistry of the sample due to the sequential filtration procedure. Whether the filtration process removed material (e.g. colloids) that would form complexes with the metals, or the biota, thus reducing competition for the pollutants is, however, unclear and requires further study. The effect was greatest in the filtrate < 0.22 μm for Zn. The decline in bioluminescence may be indicative of osmotic and substrate fluctuations that are associated with different size components of the extract.

The cellular accumulation of Zn was measured in Zn spiked soil extracts using the *P. fluorescens* 10586s pUCD607 at 0, 0.5, 1.0 and 2.0 mg l^{-1} Zn. *Pseudomonas fluorescens* 10586s pUCD607 tended to have lower cellular metal concentrations when in a soil extract than in an aqueous solution. At a Zn concentration of 2 mg l^{-1} in water and Dess soil extract, cellular accumulation of Zn by *P. fluorescens* 10586s pUCD607 was significantly ($P \leq 0.05$) greater in water at 3.3 *fg* cfu^{-1} compared to 2.3 *fg* cfu^{-1} in soil water extract, confirming that the metal was less available. A proportion of the metal spike was therefore chemically bound to other components within the soil extract and not available.

Testing of lux-marked biosensors with exposure to Cu spikes in four different soils

Soil water extracts from four contrasting soils (Dess, Countesswells, Insch and Tipperty Series) were spiked with 0, 20, 40, 50, 60, 75 and 100 μM Cu (as $CuSO_4$) and the pH adjusted to 5.50 (± 0.03). The GMM bioassay was then carried out using *P. fluorescens* 10586s pUCD607 and bioluminescence was expressed as a percentage of that for the unspiked extract for each of the individual soils.

The soils tested are representative of some soil types in north-east Scotland (Table 16.3), varying in the texture from loamy sand to a clay loam. Two of the soils have low base saturation (the two podzols) and the two brown forest soils have a higher base saturation. The soil extracts were filtered through a 0.22 μm filter before the assay was carried out and the pH of all of the samples was

Table 16.5. Bioluminescence in relation to the total organic carbon (TOC) of the 0.22 membrane filtered soil extracts.

				EC_{50}	
Soil	pH of filtrate	RLU*	TOC (mg l^{-1})	μM	mg l^{-1}
Deionised H_2O	5.48	10200	0	0.1	0.09
Dess	5.44	10100	5	7.3	0.45
Countesswells	5.52	15600	194	45.0	2.86
Insch	5.46	19600	163	73.1	6.35
Tipperty	5.40	28000	148	66.5	4.19
l.s.d. (5%)	0.06	6631	32	6.8	0.43

* Relative light units (Rattray *et al.*, 1990) after 15 min.

adjusted to between pH 5.40 and 5.52 to allow cross comparison between different soils. The maximum bioluminescence in the non-contaminated samples varied between the soils (Table 16.5). *Lux*-marked *P. fluorescens* 10586s pUCD607 showed highest bioluminescence in the Tipperty extract. Bioluminescence was not correlated with the total organic carbon in solution and may therefore be determined by the availability of substrates in the extracts. Variability in turbidity and colour, which may affect light quenching was not assessed, and whether this would cause differences in the maximum bioluminescence detected remains unclear.

When the copper spikes were added to the soil water extract, the EC_{50} values were calculated relative to maximum bioluminescence for each individual non-contaminated soil extract (Table 16.5). These were calculated on a micromolar basis and have also been expressed in a mg l^{-1} form to allow cross comparisons. It can be seen that all the spiked soil extracts provided a significantly ($P \leq 0.05$) higher EC_{50} value than the spiked deionized water. The EC_{50} values, however, do not reflect ideally the decline in bioluminescence which is shown in the soil samples (Fig. 16.6). In Fig. 16.5, the decline in bioluminescence due to exposure to Cu at pH 5.50 was rapid and the gradient of the decline lessened as the concentration increased. This immediate decline was also found in the Dess extract, although the gradient was less steep than that found for the aqueous sample suggesting that the soil had a greater PTE buffering capacity. In the case of the other three soil extracts, the decline in bioluminescence occurred at approximately 30 μM for Countesswells, 40 μM for Tipperty and 50 μM for the Insch soil extract. This demonstrates that the use of the EC_{50} value for these samples may not convey the maximum amount of useful information as it does not indicate how the toxicity declines with increasing Cu concentration. It may prove to be more suitable therefore to list the EC_{10}, EC_{25}, EC_{75} and EC_{90} values to allow the user to rapidly interpret important characteristics in the bioluminescence decline.

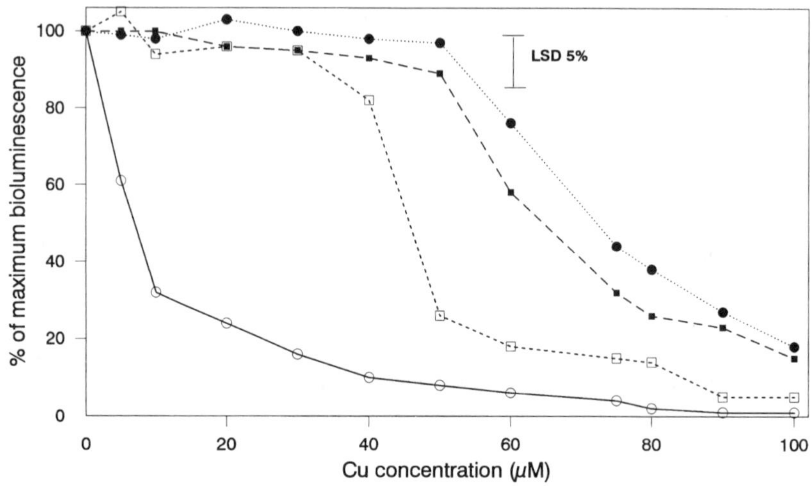

Fig. 16.6. The effect of a range of Cu spikes in four different soil extracts (Dess, ○, Countesswells, □, Insch, ● and Tipperty, ■), on the bioluminescence response of *Pseudomonas fluorescens* 10586s pUCD607.

Figure 16.6 demonstrates that spiked Insch and Tipperty soil water extracts have a greater capacity to render Cu less available to the *lux*-marked *P. fluorescens* than spiked Countesswells soil water extract. The Dess soil water extract was associated with the least capacity to render Cu unavailable, although this capacity was still significantly ($P \leq 0.05$) greater than that for the aqueous solution of Cu.

The inorganic chemical characteristics of the Dess and Countesswells extracts were determined by inductively coupled plasma mass spectroscopy (ICPMS) and the TOC by a TOC Sin II (aqueous carbon analyser). The two samples were shown to have similar chemical characteristics (unpublished data) except that the TOC concentration of the Dess extract was 5 mg l^{-1}, while for the Countesswells extract it was 194 mg l^{-1}. To study this characteristic in more detail and to try to assess if it was an important property for rendering Cu unavailable, composites of the two extracts were made and spiked accordingly with Cu. The results (Fig. 16.7) showed that the EC_{50} was raised as the relative contribution of the Countesswells soil extract was increased. TOC did not solely account for the buffering capacity of the soil against the Cu (data not shown) and it may be concluded that there are many chemical and biological factors that will alter the availability of the PTE to the *lux*-based biosensor. TOC is one of these factors, but the form of the organic components must also be considered. The binding affinity of Cu to EDTA, for example, has been demonstrated to be much greater than that found for acetate or citrate.

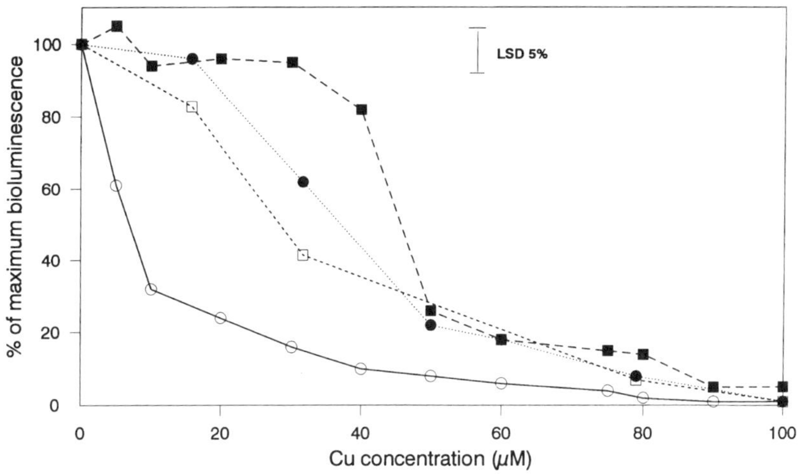

Fig. 16.7. The effect of varying the TOC concentrations by using composites of Dess and Countesswells soil extracts on the bioluminescence response of *Pseudomonas fluorescens* 10586s pUCD607 (5 mg l^{-1} TOC, ○, 25 mg l^{-1} TOC, □, 98 mg l^{-1} TOC, ●, 194 mg l^{-1} TOC, ■) at a range of Cu concentrations.

Conclusions

High levels of bioluminescence over the wide pH range associated with soils demonstrated that use of *lux*-marked soil bacteria as biosensors offers a very powerful approach to ecotoxicity testing in the terrestrial environment. Bioavailability of heavy metals and other potentially toxic pollutants in soil is particularly dependent upon pH and therefore assay conditions must reflect field pH if reliable assessment of ecotoxicity is to be made.

Results from ecotoxicity testing of aqueous standards showed the use of *lux*-marked biosensor bacteria provides appropriate sensitivity for ecotoxicity testing of a broad spectrum (organic and inorganic) of pollutants with the additional flexibility in sensitivity being available through the type of *lux* gene modifications of the bacteria. Differences in sensitivity were found to be due to variations in light output rather than the characteristics of metal accumulation in the bacterial cell.

Reductions in EC_{50} of heavy metals due to the presence of EDTA indicated that complexing ligands will often reduce the bioavailability and hence ecotoxicity of many potentially toxic elements in the soil. Nevertheless, the *lux*-based biosensor approach was found to be very effective in testing contaminated soil extracts. The differences in ecotoxicity of inorganic pollutants in soil extracts due to membrane filter pore diameter highlight the need for careful consideration of soil sampling/extraction protocols. Ultimately, fibre optic linked membrane bound biosensor probes may facilitate *in situ* ecotoxicity monitoring of soil and overcome the problems associated with soil extraction.

The primary attraction of the development and use of *lux*-marked biosensors in soil ecotoxicity testing is the flexibility to select bacteria which are representative of the environment under test. Furthermore, battery testing using *lux*-marked soil bacteria from a variety of contrasting ecological niches may well provide information predictive of a broader range of ecological effects of soil contamination than could be obtained from a single test organism.

Microbial biosensors in ecotoxicity testing represent an essential part of the suite of methods needed in diagnosing soil health. Their main contribution will be assessment of the bioavailability of potentially toxic contaminants in soil, a concept vital to the effective management and use of polluted soils.

Acknowledgements

This research was funded by the Aberdeen Research Consortium (G.I. Paton), BBSRC, Yorkshire Water Services and Siemens Environmental Systems Limited (E.A.S. Rattray) and Scottish Office Agriculture and Fisheries Department (C.D. Campbell). The authors gratefully acknowledge S. Sousa and C. Duffy for technical assistance.

References

Amin-Hanjani, S., Meikle, A., Glover, L.A., Prosser, J.I. and Killham, K. (1993) Plasmid and chromosomally encoded luminescence marker systems for detection of *Pseudomonas fluorescens* in soil. *Molecular Ecology* 2, 47–54.

Beckman Instruments Inc. (1982) *Beckman Microtox Operating Manual.* Microbics Corporation, Carlsbad.

Bitton, G. and Koopman, B. (1986) Effect of toxicants on dehydrogenases. In: Bitton, G. and Dutka, B.J. (ed.) *Toxicity Testing Using Microorganisms.* Volume 1. CRC Press, Inc., Boca Raton, Florida, pp. 32–41.

Bochner, B.R. (1989) Sleuthing out bacterial identities. *Nature* 339, 157–158.

Brouwer, H., Murphy, Y. and McArdle, L. (1990) A sediment-contact bioassay with *Photobacterium phosphoreum. Environmental Toxicology and Chemistry* 9, 1353–1358.

Chander, K. and Brookes, P.C. (1991) Is the dehydrogenase assay invalid as a method to estimate microbial activity in copper contaminated soils? *Soil Biology and Biochemistry* 23, 909–915.

Corbisier, P., Duels, L., Nuyts, G., Baeyers, W. and Mergeay, M. (1992) Copper resistance in large plasmids of *Alcaligenes eutrophus. Proceedings, American Society for Microbiology, New Orleans.* American Society for Microbiology, Washington.

Corbisier, P., Ji, G., Nuyts, G., Mergeay, M. and Silver, S. (1993) *LuxAB* fusions with the arsenic and cadmium resistance operons of *Staphylococcus aureus* plasmid p1258. *FEMS Microbiology Letters* 110, 231–238.

Holmes, D.S., Dubey, S.K. and Gangolli, S. (1993) Development of biosensors to measure metal ion bioavailability in mining and metal waste. In: Torma, A.E., Apel, M.L. and Brierley, C.L. (eds) *Biohydrometallurgical Technologies.* The Minerals, Metals and Materials Society, pp. 659–666.

Kaiser, K.L.E. (1993) Qualitative and quantitative relationships of microtox data with toxicity for other aquatic species. In: Richardson, M. (ed.) *Ecotoxicity Monitoring.* VCH, Weinheim, pp. 197–211.

King, J.M.H., Digrazia, P.M., Applegate, B., Burlage, R., Sansererino, J., Dunbar, P., Larimer, F. and Sayler, G.S. (1990) Rapid, sensitive bioluminescent reporter technology of naphthalene exposure and biodegradation. *Science* 249, 778–781.

Kratgen, J. (1978) Atlas of Metal–Ligand Equilibria in Aqueous Solution. Ellis Horwood Ltd, Chichester, p. 782.

Lapage, S.P., Shelton, J.E., Mitchell, T.G. and Mackenzie, A.R. (1970) Culture collections and the preservation of bacteria. In: Norris, J.R. and Ribbons, D.W. (eds) *Methods in Microbiology.* Vol 3A, Academic Press, London, pp. 135–228.

Liesegang, H., Lemke, K., Siddique, R.A. and Schlegel, H.G. (1993) Characterisation of the inducible nickel and cobalt resistance determinant *cnr* from PMOD28 of *Alcaligenes eutrophus* CH34. *Journal of Bacteriology* 175, 767–778.

Linehan, D.J., Sinclair, A.H. and Mitchell, M.C. (1989) Seasonal changes in Cu, Mn, Zn and Co concentrations in soil in the root zone of barley (*Hordeum vulgare* L.). *Journal of Soil Science* 40, 103–115.

Lundin, A. (1989) ATP assays in routine microbiology: from visions to realities in the 1980s. In: Stanley, P.E., McCarthy, P.L. and Smither, R. (eds) *ATP luminescence: Rapid Methods in Microbiology.* Blackwell, Oxford, pp. 11–30.

Mergeay, M., Nies, D.H., Schlegel, H.G., Gents, J., Charles, P. and van Gysegen, F. (1985) *Alcaligenes eutrophus* OH34 is a facultative chemoautotroph with plasmid-bound resistance to heavy metals. *Journal of Bacteriology* 162, 328–334.

Mölders, H. (1990) Methods for detecting the presence of mercury using microorganisms with mercury enhanced bioluminescence (PCT/DE90/00063), Patent *n.z.*

Nies, A., Nies, D.H. and Silver, S. (1989) Cloning and expression of plasmid genes encoding resistances to chromate and cobalt in *Alcaligenes eutrophus. Journal of Bacteriology* 171, 5065–5070.

Nies, A., Nies, D.H. and Silver S. (1990) Nucleotide sequence and expression of a plasmid-encoded chromate resistance determinant from *Alcaligenes eutrophus. Journal of Biological Chemistry* 265, 5648–5653.

Nies, D.H. (1992) *CzcR* and *CzcD,* gene products affecting regulation of resistance to cobalt, zinc and cadmium (*czc* system) in *Alcaligenes eutrophus. Journal of Bacteriology* 174, 8102–8110.

Paton, G.I., Campbell, C.D., Cresser, M.S., Glover, L.A., Rattray, E.A.S. and Killham, K. (1995a) Bioluminescence-based ecotoxicity testing of soil and water. *Pro-*

ceedings of the OECD International Workshop on Bioremediation, Tokyo, Japan. OECD Publication, Paris, Cedex, France.

Paton, G.I., Campbell, C.D., Glover, L.A. and Killham, K. (1995b) A novel method to assess ecotoxicity using a *lux*-modified strain of *Pseudomonas fluorescens*. *FEMS Letters in Microbiology* 20, 52–57.

Rattray, E.A.S., Prosser, J.I., Killham, K. and Glover, L.A. (1990) Luminescence-based non-extractive techniques for *in situ* detection of *Escherichia coli* in soil. *Applied and Environmental Microbiology* 56, 3368–3374.

Rudge, R.H. (1990) Maintenance of bacteria by freeze-drying In: Kirsop, B.E. and Doyle, A. (eds.) *A Manual of Laboratory Methods*. Academic Press, London. pp. 31–38.

Selifonova, O., Burlage, R. and Barkay, T. (1993) Bioluminescent sensors for the detection of bioavailable Hg (II) in the environment. *Applied and Environmental Microbiology* 59, 3083–3090.

Springael, D., Baeyers, W., Hannes, L., De Wilde, K. and Mergaey, M. (1991) Use of Tn 4431 containing the promoter-less *lux* genes of *Vibrio fischeri* to detect genes involved in PCB degradation of *Alcaligenes* sp. AS. *Archives Internationales de Physiologie et de Biochimie* 99, 77.

van der Leslie, D., Corbisier, P., Baeyers, W., Wuertz, S., Diels, L. and Mergeay, M. (1994) The use of biosensors in environmental monitoring. In: *Bioremediation: Scientific and Technological Issues*, 10th Forum in Microbiology.

Biological Indicators of Soil Health: Synthesis

17

C.E. Pankhurst[1], B.M. Doube[1]
and V.V.S.R. Gupta[2]
[1]*CSIRO Land and Water and* [2]*Cooperative Research Centre for Soil and Land Management, Private Bag No. 2, Glen Osmond, South Australia 5064, Australia*

Introduction

The increasing human demand for resources (food, water, soil, structural material, space) is outgrowing the capacity of the earth's natural systems to provide them. Evidence of the damage to the earth's ecological infrastructure can be seen in eroding soils, drying rivers, falling water tables, shrinking forests, climatic instability, and the disappearance of species (Brown, 1995). Each year, over 90 million extra mouths appear on the face of the globe. The most urgent global problem over the next decade will be food shortages. Nearly all the potentially arable land is already in production and the total area under production is shrinking each year. Increased production can only come via increased production from current crop lands, and this must persist in perpetuity.

What has the biological health of soils to do with such a catalogue of gloom?

Maintaining and improving the productive capacity of our soils is essential to human survival and healthy soils are an essential element of this process. Because the physical and chemical components of soils are largely fixed by geographic constraints, the flexibility in soil ecosystems is primarily due to their biological composition and the way this is modified by human inputs (fertilizers, cultivation, plant species, etc.). Furthermore, biological systems are highly sensitive to incipient degradation; hence a change in the biological status of the system may provide an 'early warning' of environmental collapse and so allow us to react before irreversible damage occurs.

The exploitative nature of human agriculture has resulted in a common belief that natural systems are more 'healthy' than managed (agricultural) systems, but the reality of this needs to be questioned. The converse may be true.

(eds C.E. Pankhurst, B.M. Doube and V.V.S.R. Gupta)

Certainly the adoption of conservation management practices (e.g. minimum tillage with plant residue retention) and organic agriculture are attempts to return to a more natural, less disturbed ecosystem but again, are such soils healthier than those of conventional agriculture? Can the world afford the production penalty which can accompany conservation management (e.g. Kirkegaard, 1995). Will international economics and social issues allow us the luxury of using 'natural' systems of modest productivity in order to conserve soil biodiversity? Many believe that high input systems are the only way to meet the growing global demand for food, but are these sustainable?

Obviously such matters are ethical and social, as well as scientific, but a reliable basis for predicting the capacity of the world's soil to remain healthy and productive will be a useful contribution to resolving the ecological crisis facing mankind. This book is directed towards that aim.

In the following sections, we attempt to synthesize the issues concerning the concept and assessment of soil health canvassed by the authors of this volume. We also summarize the arguments which consider the usefulness of a range of biological indicators and discuss whether these indicators might be used as part of a package of indicators for the assessment and monitoring of soil health and sustainable productivity in managed systems (especially agroecosystems).

The Importance of Soil

Soil is a dynamic, living resource that plays many key roles in terrestrial ecosystems (Doran and Safely, Chapter 1, this volume). The health of a soil is largely defined in terms of its function in terrestrial ecosystems: the character of a healthy soil is given by a composite of its physical, chemical and biological properties. The major soil functions are: (i) to provide a medium for plant growth and a habitat for many animals and microorganisms; (ii) to regulate and partition water flow in the environment; and (iii) to serve as an environmental buffer in the formation, attenuation, and degradation of environmentally hazardous compounds (Larson and Pierce, 1994).

Soil serves as a medium for plant growth by providing physical support, water and essential nutrients. Its suitability for sustaining plant growth is a function of physical properties (e.g. structure, water holding capacity), chemical properties (e.g. pH, nutrient supplying ability, salt content) and biological properties (e.g. nutrient mineralization capacity, mutualistic associations with microorganisms). Microbial decomposition of organic residues in soil plays a key role in the cycling of essential plant nutrients (N, P, S and trace elements). The soil's capacity to store water is a major regulator of water supply to plants and also influences the transport of pollutants to the surface and into ground water. Soil has several additional functions which are linked to human activity and require a different set of qualities; these functions include the use of soil as a physical medium to support housing, for recreation, refuse disposal, a source of raw materials (minerals), and the use of soil as a source of cultural heritage (Blum and Santelises, 1994).

Traditional agricultural practices (including monoculture plant production, mechanical cultivation, excessive and indiscriminate use of chemical fertilizers and pesticides) used to maintain and increase crop and fibre production, are reducing the soil's capacity to maintain its function. This has numerous facets including decreased soil organic matter content, loss of soil structure, loss of soil through wind and water erosion, development of acidic, saline and sodic soils, and soil contamination with pesticide residues and heavy metals (Doran and Parkin, 1994). Soil degradation is not a new process, as the deserts of North Africa and the Middle East testify, but the rate of the process is greater than ever seen before.

Soil Biota in Perspective

Biological activity in soils is largely concentrated in the topsoil, the depth of which may vary from a few to 30 cm. In that soil, the biological components occupy a tiny fraction (<0.5%) of the total soil volume and make up less than 10% of the total organic matter in soil (Fig. 17.1). This living component of the soil organic matter consists of 5–15% plant roots and 85–95% soil organisms. The soil organism component generally contains 15–30% macro- and mesofauna and 60–80% microorganisms (Fig. 17.1). Soil microorganisms (predominantly bacteria and fungi) thus make up the bulk of the biomass of organisms in the soil and are responsible for 80–90% of the soil's biological activity (Reichle, 1977). Much of this biological activity is associated with processes which regulate nutrient cycling (e.g. mineralization, denitrification, nitrogen fixation) and the decomposition of organic residues. Collectively, the soil biota are a vital force which serves to maintain the health of soils.

Very few species, if any, occur in all environments, but commonly species with similar niche requirements will progressively replace each other across environmental gradients. One consequence of this is that sampling species which occupy similar niches (i.e. sampling within functional groups) across environmental gradients may provide a measure of the status of that functional group independent of the environmental gradient, and independent of the species composition *per se*.

The ability to define soil health is essential to the development of indicators which can be used to assess and monitor the sustainability of soil and land management systems.

Defining Soil Health

A largely unresolved problem encountered by all the authors in this volume and in numerous other forums (e.g. the Soil Science Society of America Annual

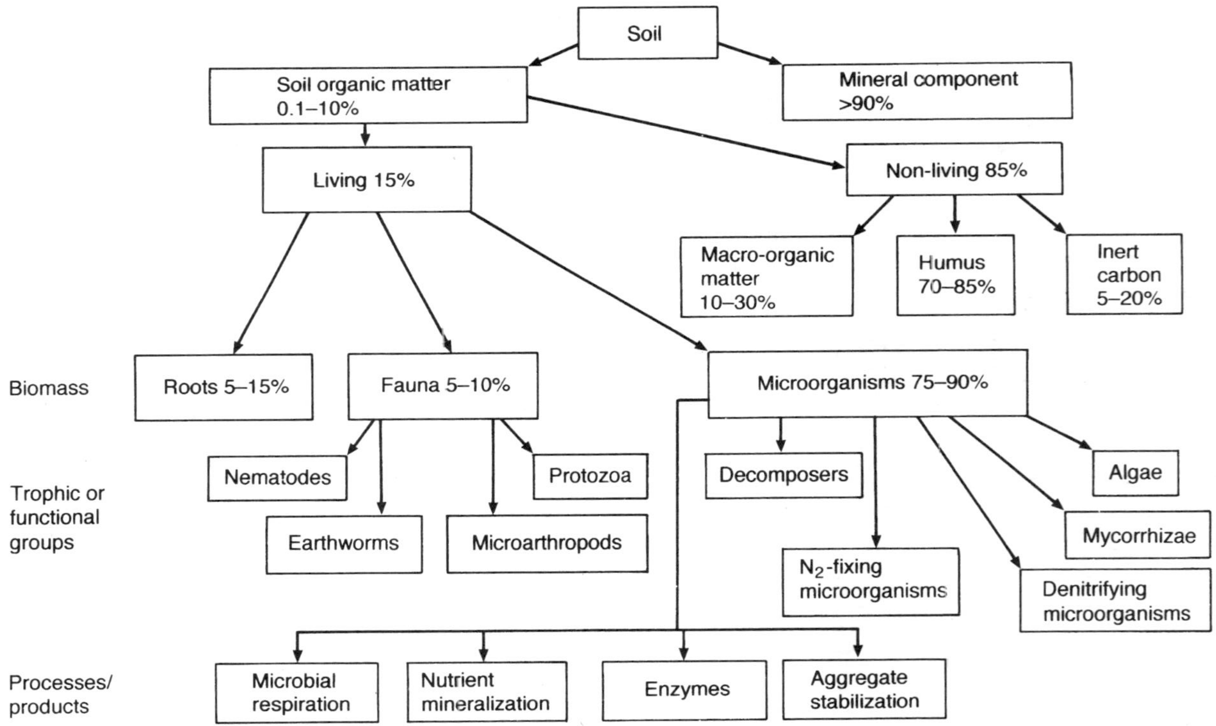

Fig. 17.1. Composition of a typical fertile soil in terms of its biota, functional groups of microorganisms and processes/products carried out by microorganisms. Numbers are percentage dry weight (after Theng *et al.*, 1989).

Meeting, Minneapolis, USA, 1992; the OECD Soil Biota Meeting, Adelaide, Australia, 1994) has been the definition of soil health and its distinction, if any, from soil quality. Doran and Safley (Chapter 1, this volume) have defined soil health as 'the continued capacity of soil to function as a vital living system, within ecosystem and land-use boundaries, to sustain biological productivity, promote the quality of air and water environments, and maintain plant, animal, and human health'. Two elements of this definition of soil health distinguish it from the definition of soil quality proposed by Doran and Parkin (1994). They are: (i) the inclusion of a time component (i.e. 'the continued capacity of' – reflecting the importance of the soil in being able to continue to function over time); and (ii) recognition of soil 'as a vital living system' (i.e. emphasizing the importance of the soil biota to soil functioning).

There appears to be agreement in the current literature that the use of the term soil quality is appropriate when the intended use of the soil (purpose) is specified (e.g. Doran and Parkin, 1994; Larson and Pierce, 1994). This viewpoint is supported strongly by Rapport *et al.* (Chapter 2, this volume), Sparling (Chapter 5, this volume) and Doube and Schmidt (Chapter 11, this volume). One important implication of this is that the quality of a soil required for a particular purpose, e.g. to produce a wheat crop, may be different to that required for another purpose, e.g. to produce a vegetable crop. Further, the attributes of healthy soils for annual crops may well be quite different from those in long-lived productive systems (e.g. forestry) and within these there may be distinctions between monocultures (e.g. pine plantations) and diverse natural systems (e.g. regrowth eucalyptus forest). Soil quality is thus closely linked to the production of individual plant species or plant communities. The intrinsic properties of the soil (e.g. its mineral composition, texture and its depth) are therefore important attributes of soil quality.

In contrast, the concept of soil health captures the ecological attributes of the soil which have implications beyond its quality or capacity to produce a particular crop. These attributes are chiefly those associated with the soil biota; its biodiversity, its food web structure, its activity and the range of functions it performs. For example, biodiversity is not a soil property that is critical for the production of a wheat crop *per se*, but it is a property that may be important for the continued capacity of the soil to produce wheat. Without maintenance of biodiversity, the soil's capacity to recover from natural or anthropogenic perturbations may well be reduced. Similarly, maintenance of the soil's capacity to perform functional processes such as those associated with nutrient cycling and the breakdown of xenobiotics are important if the soil is to be able to sustain plant growth in the long-term.

There is a dichotomy amongst the authors of this volume, some of whom search for an objective measure of soil health (e.g. Sparling, Chapter 5; Doube and Schmidt, Chapter 11), while most emphasize its holistic nature and accept subjective assessments of what is healthy. Inconsistencies between definitions of soil health create conceptual problems but it is necessary to have a functional definition of soil health so that we may attempt to develop indicators which report its status. All authors consider that the term soil health encompasses the living and dynamic nature of soil, and that this differentiates it from soil quality.

We therefore adopt the view that although the concepts of soil quality and soil health overlap to a major degree and that in many instances the two terms are used synonymously (e.g. Doran *et al.*, 1994; Harris and Bezdicek, 1994), soil quality focuses more on the soil's capacity to meet defined human needs such as the growth of a particular crop, whilst soil health focuses more on the soil's continued capacity to sustain plant growth and maintain its functions.

Indicators of Soil Health

Indicators are measurable surrogates for environmental processes or endpoints such as plant productivity, soil pollution and soil degradation, and sustained biodiversity. As discussed by Elliott (Chapter 3, this volume) indicators of soil health should: (i) be linked and/or correlated with ecosystem processes; (ii) integrate soil physical, chemical, and biological properties or processes; (iii) be easy and cost effective to measure and calculate; (iv) be responsive to variations in management and climate at an appropriate time scale; and (v) be components of existing soil data bases (Doran and Safley, Chapter 1, this volume). Because no single indicator will possess all of these desirable properties, a set or package of complementary indicators is required.

Soil properties which meet many of the desirable criteria for indicators of soil health are discussed by Doran and Safley (Chapter 1) and listed in Table 17.1. The relative importance of these properties as indicators will depend, in part, on understanding their role in the ecosystem. It is important to note that the indicators will have more relevance to the measurement of soil health: (i) if they are considered as part of an indicator package of soil health and (ii) if the soil health indicator package is interpreted with respect to soil function, e.g. plant biomass production. In Table 17.1 we also draw attention to distinctions between soil health and soil quality on the basis that some indicators have more relevance to one concept than the other (as discussed above).

It should be emphasized that indicators of soil health are simply 'indicators'; i.e. they are not measures of soil health, but rather measures that collectively tell us whether the soil is functioning normally. As discussed by Elliott (Chapter 3, this volume) individual indicators will tell us if a particular component or process in the soil is present or is functioning and, given sufficient testing and analysis of different indicators, we may then be able to construct a set of probabilities that the soil is healthy based on these indicators.

Table 17.1. Soil physical, chemical and biological properties of relevance as indicators of soil quality and soil health[1].

	Relevance to:	
Indicator	Soil quality	Soil health
Physical indicators		
Mineral composition	+	–
Texture	+	–
Depth	+	–
Bulk density	+	+
Water holding capacity	+	+
Porosity	+	+
Chemical indicators		
pH	+	+
Electrical conductivity	+	+
Cation exchange capacity	+	+
Organic matter	+	+
Major elements	+	+
Heavy metals	+	+
Biological indicators		
Microbial biomass	+	+
Soil respiration	+	+
Mineralizable N	+	+
Enzyme activity	+	+
Abundance of microflora	+	+
Abundance of soil fauna	+	+
Root disease	+	+
Soil biodiversity	–	+
Food web structure	–	+
Plant growth	+	+
Plant biodiversity	–	+

[1] –, of little or no relevance; +, relevant.

Biological Indicators of Soil Health

Soil organisms contribute to the maintenance of soil health by controlling the decomposition of plant and animal residues, biogeochemical cycling (including N_2 fixation), the formation and maintenance of soil structure, and the fate of agrochemicals and pollutants applied to soil. In addition to their critical role in soil functioning, soil organisms are potentially useful indicators of soil health because they respond to soil management in time scales (months/years) that are relevant to land managers (Pankhurst, 1994). For example, changes in microbial biomass, or abundance of selected functional groups of microorganisms (e.g. mycorrhizal fungi), may be detected well in advance of changes in soil organic matter contents or other soil physical or chemical properties (Powlson *et al.*, 1987; Gupta *et al.*, 1994; Miller and Jastrow, 1994; Pankhurst, 1994; Sparling, Chapter 5, this volume).

While the biological health of soils is largely determined by above-ground vegetative processes (inputs from organic residues, plant species, etc.) as well as from a whole host of anthropogenic influences (tillage, agrochemicals, etc.), soil biological factors may also be important determinants of the composition of the plant communities. For example, the presence of specific symbionts (e.g. mycorrhizal fungi, root nodule bacteria, *Frankia*) or plant pathogens can determine whether particular plant species establish and grow (Pankhurst, 1994).

One of the major difficulties in the use of soil organisms *per se*, or soil processes mediated by soil organisms, as indicators of soil health has been methodological (what to measure and how and when to measure it) and how to interpret changes in terms of soil function (Pankhurst, 1994; Turco *et al.*, 1994). Despite these difficulties there have been major advances in our understanding of the soil biota and its functioning at the community level in recent years (Brussaard *et al.*, 1990; Beare *et al.*, 1993; Pankhurst *et al.*, 1995; Roper and Gupta, 1995). A summary of the current status of potential biological indicators of soil health examined in the previous chapters is presented below.

Microbial biomass

Sparling (Chapter 5, this volume) considers that soil microbial biomass can provide a more sensitive measure of change in soil health than total organic matter content, and can usefully reveal trends within 1–5 years rather than decades after changes in land management. By comparing levels of microbial biomass and soil respiration under a particular land management practice with a reference soil (of similar type and under a reference management practice), an estimate of the rate of change under that land management can be obtained. Derived indices such as the microbial quotient and qCO_2 can provide additional information about the microbial processes in soil. Major difficulties with using microbial biomass as a bioindicator of soil health include: (i) the large natural range in microbial biomass contents in different soil types and ecosystems, seasonal fluctuations and inconsistent trends in relation to soil fertility and plant production; (ii) lack of accepted baseline or reference values; and (iii) no recognized threshold above or below which microbial biomass indices could be justifiably regarded as healthy or unhealthy.

Soil microflora

Roper and Ophel-Keller (Chapter 7, this volume) emphasize that many studies show that the abundance of soil microorganisms (bacteria, fungi, actinomycetes, algae) respond to changes in soil health. However, problems with the poor culturability of most microorganisms and spatial and temporal heterogeneity make it nearly impossible to use absolute values of microfloral populations or processes as direct bioindicators of soil health. However, changes in microbial populations, or in specific functional groups (e.g. nitrifying bacteria, nitrogen fixing bacteria, root nodule bacteria, mycorrhizal fungi) (Mhatre and Pankhurst, Chapter 14, this volume), have been used as measures of the impact of chemical pollution in soils. New techniques for measuring the structure and functional diversity of microbial communities (e.g. GC-FAME, BIOLOG, DNA techniques) (White and Macnaughton, Chapter 15, this volume) offer new and largely unexplored dimensions for using microbes as bioindicators.

At a more specific level, Paton *et al.* (Chapter 16, this volume) have illustrated how soil bacteria equipped with special reporter genes can be used as biosensors to detect the presence of toxic elements in the soil. The reporter genes may be fused directly to genes involved in the response to a particular toxin or used to indicate overall metabolic status. Of particular importance is the capacity of the biosensor to indicate the bioavailability of contaminant molecules in the soil. This impressive technology requires validation and currently can only be used in sophisticated laboratories.

Root pathogens

Hornby and Bateman (Chapter 8, this volume) correctly point out that root pathogens currently have no reputation *per se* as bioindicators of soil health, although no farmer would regard his soil as healthy if a disease problem was present. However, the presence of disease may indicate the existence of other soil health problems (e.g. inadequate residue inputs, nutrient imbalance), and therefore provide a strategy for using disease as a bioindicator. The development of new techniques for rapid detection of root pathogens directly in soil (Saylor, *et al.*, 1992) will help to promote this strategy.

Biodiversity

Reduction in biological diversity is one of the most profound ecological consequences of modern agriculture, with the species composition of many agroecosystems being largely determined by immigration and climate. Diverse local plant assemblages have been replaced by monocultures of about a dozen grain producing species. Mammals have been reduced to about a dozen species of domestic stock with man as the primary predator. Native earthworm species have largely disappeared from agricultural soils and have been replaced by about a dozen peregrine species (Doube and Schmidt, Chapter 11, this volume).

Whether the same is true for microbial communities in agricultural soils is largely unknown but it is a key question for functional soil ecology.

There is some good evidence that links biodiversity with ecosystem function and stability (Tilman and Downing, 1994; Naeem *et al.*, 1995). Other evidence links the biodiversity of soil organisms with the maintenance of soil functional processes (Pankhurst, Chapter 12, this volume). However, it is still uncertain how much biodiversity within a given functional group of organisms (i.e. those capable of carrying out a specific soil function) is needed to ensure continuance of that function in the face of perturbations. Use of biodiversity as a bioindicator of soil health has many attractions but our knowledge of how to put this into practice is in its infancy.

Soil microfauna

Soil protozoa and nematodes play critical roles in nutrient cycling in soils. Protozoa, with their rapid growth rates and delicate external membranes, respond rapidly to environmental changes. However, as with nematodes, measurement of total soil populations of protozoa is only likely to be useful as an indicator of soil health in extreme situations, such as in contaminated soils (Gupta and Yeates, Chapter 9, this volume). Functional or trophic diversity would appear to be the most useful attribute of these organisms that could be used as a bioindicator of soil health. This diversity is largely governed by the available food source in the soil, e.g. in soils subjected to minimum tillage and residue retention, the soil microflora is fungal dominated, which is reflected in a predominance of mycophagous protozoa and fungivorous nematodes. A major constraint to using either protozoa or nematodes as bioindicators is the technical expertise required to identify trophic groups and species.

Soil arthropods

The composition (species distribution) of collembola and mites has been used successfully in many studies as an indicator of soil health, particularly in relation to soil pH, soil C/N ratio and heavy metal contamination of soils (van Straalen, Chapter 10, this volume). Community composition analysis was shown to be more useful than single species analysis as an indicator of changed soil health. Other ways of classifying soil arthropods based on the diversity within their life-histories, feeding type and physiotype (ranking in response to a particular ecological factor, e.g. pH) were considered. It was concluded that a combination of physiotype approaches and multivariate statistical analysis showed the greatest promise.

Soil macrofauna

Doube and Schmidt (Chapter 11, this volume) considered that amongst the soil macrofauna (termites, ants, soil-dwelling insect larvae, etc.), only earthworms are of sufficiently wide distribution in natural and agricultural soils to act as indicators of soil health. Their discussion centred on agricultural soils around the world. These contain few indigenous earthworm species and have been colonized by a suite of about a dozen peregrine species, with individual soils commonly containing two to five species. The biology of the peregrine species suggests that they have many of the attributes of useful indicators of soil health, but their abundance varies widely with soil type and climate. Nevertheless, their abundance is promoted by conservation management practices, which also promote soil health. Their capacity to act as indicators of soil health was examined in a series of ten case studies of nine long-term trials around the globe with disappointing results. Even within the constraints of soil type and climate (i.e. within long-term trials) there was rarely any consistent association between earthworm abundance and crop yield, although the presence of earthworms was commonly associated with improved soil structure. They conclude that earthworm abundance is substantially more variable than plant production and that the factors which determine crop yield are not necessarily those which determine earthworm abundance. Earthworms, therefore, offer only limited potential to act as bioindicators of soil health. However, they do show considerable potential as bioaccumulators of enivronmental contaminants.

Soil enzymes

Soil enzymes are the mediators and catalysts of most soil processes, and hence have potential to provide an integrative assessment of soil health (Dick, Chapter 6, this volume). Methods for measuring the activity of over 50 different enzymes are available. Certain enzymes can exist only in viable cells and thus provide assessments of the activity of the biological component of the soil. However, the majority of enzymes are present in both viable cells and as extracellular enzymes (abiontic) in soil solution or complexed to the soil matrix. This latter characteristic provides the opportunity to incorporate an 'historical' component into enzyme assays that reflects cumulative changes in soil health. Soil enzyme activities have been used successfully to discriminate between a wide range of soil management practices, but only when a reference (baseline) value was available for comparison. They have been particularly useful in determining the impact of pollution or severe perturbations on soil health and to evaluate the success of remediation activities. More research is needed across a range of soil types, ecosystems, and soil management practices to calibrate soil enzyme activities or to develop relative soil enzyme indices that are interpretable and independent of soil type and environment.

Plants

Some would argue that plant growth is the ultimate indicator of soil health. Whilst plant growth is extremely sensitive to many of the chemical, physical and biological attributes of the soil, these attributes can easily be masked by anthropogenic inputs such as fertilizers, pesticides and water. It may thus be difficult to use plant growth *per se* as an indicator of soil health in cropping systems. However, in natural ecosystems the composition of plant communities or the presence of particular plant species may be useful as indicators of the chemical and physical condition of the soil (Pandolfini *et al.*, Chapter 13, this volume). Of particular interest is the use of both lower plants (lichens and bryophytes) and higher plants as accumulative bioindicators of trace elements, heavy metals and radionucleotides in soils. Genetically uniform material and standardized methods of exposure and sampling are now available for several species of higher plants employed as accumulative bioindicators of heavy metals. Biochemical markers such as stress marker enzymes, e.g. peroxidase, also have potential as indicators of soil contamination.

What Can We Conclude?

Most authors have considered the constraints which determine whether particular soil biological properties have potential to act as indicators of soil health. The attributes of these biological properties, whether they be populations of organisms or functional processes, have generally been evaluated against a spectrum of management practices imposed on the soil and across a range of soil types. Problems associated with sampling, measurement and patterns of temporal and spatial variability have been described and reviewed. It is clear that some potential indicators are too variable in many situations to be of consistent use as indicators, e.g. populations of microorganisms.

There remain substantial conceptual and practical difficulties associated with the development of indicators of soil health. These difficulties are discussed by some authors in this volume. However, it is clear that these difficulties are not unique to biological indicators and apply equally well to the chemical and physical indicators of soil health/quality discussed by other authors (e.g. Doran and Parkin, 1994).

The conceptual difficulties with the selection and use of biological indicators of soil health include:

1. On-going difficulties with the definition of soil health.
2. The absence of any clear base line data which might act as a reference point for soils of known (or defined) health levels.
3. Identification of what are the most suitable indicators, because individually measured components represent the summed response of the whole system.

4. How to assess health in view of the multitude of components that contribute to it.
5. How to deal with systems (e.g. different soil types) that show no consistency in their responses to perturbations.

The practical difficulties with the selection and use of biological indicators of soil health include:

1. Complex methodologies and technical expertise are commonly required.
2. High levels of temporal and spatial heterogeneity affect all measurements in most systems.
3. Lack of validation of biological indicators in diverse situations (e.g. soil types and climatic zones).
4. The requirements for estimates of soil health may vary between end-users (e.g. farmers and researchers, see Lynch and Elliott, Chapter 4, this volume).

If we accept that there remain substantial impediments to developing useful and widely applicable bioindicators of soil health, is there anything that can be proposed to facilitate our quest for methods which allow evaluation of whether we have sustainable agricultural ecosystems?

At a very minimum, for example within one soil type within one region, are there biological indicators which might indicate soil health? In an attempt to evaluate the wide range of indicators which have been examined by the authors in this volume, we have developed a ranking system which allows us to compare the intrinsic and realised potential of the different biological indicators (Table 17.2). Whilst there is necessarily a degree of subjectivity inherent in this analysis, some interesting observations can be made. Firstly, most of the bioindicators rank high for spatial and temporal variability and also for their sensitivity or responsiveness to perturbations (e.g. management practices). In nearly all cases they are seen as integrative, that is they are known to reflect or respond to a wide range of chemical, physical and biological influences in the soil. In terms of their realized potential, most require a high level of technical expertise and in many instances currently available methodology (especially for the microflora and micro- and mesofauna) requires further development. Furthermore, it is suggested that the available knowledge and our capacity to interpret responses of many bioindicators to soil perturbations is inadequate, this applies particularly to the soil mesofauna. In terms of overall ranking we conclude that at present, microbial biomass, soil respiration (and derived parameters), selected functional groups of soil microorganisms (e.g. mycorrhizal fungi), microbial community composition and functional diversity, soil enzymes, functional diversity among microfaunal and mesofaunal populations and plant growth offer the most potential as bioindicators. Clearly, much more research is needed and this has been emphasized by all authors in this volume. Not only do we require more information about the specific bioindicators but we also need information on how they can be used and interpreted by the potential clients, farmers, land managers, environmental groups and researchers.

Table 17.2. Biological indicators of soil health – intrinsic and realized potential.

	Intrinsic potential				Realized potential				
	Variability								
Biological indicator	Spatial	Temporal	Sensitivity to perturbation	Integrative	Methods available	Skills required	Knowledge available	Ease of interpretation	Overall ranking
Microbial biomass									
Absolute	++	+	+++	+++	+++	+++	+++	++	+++
Microbial quotient	++	+	++	+++	+++	+++	+	++	+++
Microbial activity									
Soil respiration	+++	+++	+++	+++	+++	++	+++	+++	++
Respiratory quotient	+++	+++	+++	+++	+++	++	+	+	+++
N mineralization	++	+++	++	++	++	+++	++	++	++
Populations of microflora									
Bacteria	+++	+++	++	+	++	+	++	+	+
Fungi	+++	+++	++	+	++	++	++	+	+
Actinomycetes	+++	+++	++	+	+++	++	+	+	+
Algae	+++	+++	++	+	+++	++	+	+	+
Functional groups of microflora									
Cellulose decomposers	+++	+	++	++	+++	++	++	++	++
Nitrifying bacteria	+++	++	++	+	+++	+++	++	++	++
Denitrifying bacteria	+++	+++	++	+	+++	+++	++	+	+
Rhizobium	++	+	+++	+	+++	++	+++	+++	++
Free living N-fixers	+++	+++	+	+	++	+++	++	+	+
Mycorrhizae	+++	+	+++	++	++	+++	++	++	+++
Root pathogens	+++	+++	+++	+++	+++	+++	++	+++	+++
Microbial communities									
Composition	++	+	+++	+++	++	+++	+	++	+++
Functional diversity	+++	++	+++	+++	++	+++	+	++	+++
Soil enzymes									
Dehydrogenases	++	++	++	+++	+++	++	+++	++	+++
Phosphatases	++	++	++	++	+++	++	+++	++	+++
Sulphatases	++	++	++	++	+++	++	+++	++	+++

Peptidases	++	++	++	++	+++	++	+++	++	+++
Enzyme index	++	+–	++	++	–	+++	+	++	+++
Microfauna									
Protozoa									
Abundance	+++	+++	++	++	++	+++	++	+++	++
Biodiversity	++	++	+++	+++	++	+++	+	+++	+++
Bioassays	–	–	+++	+++	+	+++	+	++	++
Nematodes									
Abundance	+++	–++	++	++	+++	++	+++	+++	++
Trophic diversity	++	++	+++	+++	++	+++	++	+++	+++
Maturity index	++	++	+++	+++	++	+++	+	++	+++
Plant parasitic index	++	–+	+++	+++	++	+++	+	++	+++
Mesofauna									
Collembola									
Abundance	+++	+++	+++	+++	++	++	++	++	++
Diversity	++	++	+++	+++	+	+++	+	++	++
Physiotype	+–	++	+++	+++	+	+++	+	++	+++
Mites									
Abundance	+++	+++	+++	+++	+++	++	++	++	++
Diversity	+++	+++	+++	+++	+	+++	+	++	++
Physiotype	+++	+++	+++	++	+	+++	+	++	+++
Macrofauna									
Earthworms									
Abundance	++	+++	+++	++	+++	+	+++	+++	++
Diversity	++	++	+++	+++	+++	+++	++	++	++
Plants									
Lower plants									
Lichens	–+	+	+++	+++	+	+++	++	++	+++
Bryophytes	++	+	+++	+++	+	+++	+	++	++
Higher plants									
Diversity	+++	+	+++	+++	++	+++	+	+	+++
Growth	+	+++	+++	+++	+++	+	+++	+++	+++
Biomarkers	+	++	+++	+++	+	+++	+	++	++

The scoring system used is as follows: –, absent; +, low; ++, medium; +++, high.

Whilst it is clear to most soil scientists that the soil biota plays a crucial role in the functioning of soils and hence in the maintenance of the health of all terrestrial ecosystems, it is probably true to say that public perception of this role is very poor indeed. More needs to be done to increase the awareness of the importance of the soil biota and the processes it carries out. Similarly, more needs to be done to emphasize the potential value bioindicators have in terms of their capacity to detect dysfunctioning or disease in our soils and how attentive husbandry of the soil biota can provide benefits which are compatible with the need to have sustainable food production systems, conserve non-renewable resources, recycle wastes, remove harmful pollutants from the environment, preserve water quality and provide a pleasant living environment.

References

Beare, M.H., Pohlad, B.R., Wright, D.H. and Coleman, D.C. (1993) Residue placement and fungicide effects on fungal communities in conventional and no-tillage soils. *Soil Science Society of America Journal* 57, 392–399.

Blum, W.E.H. and Santelises, A.A. (1994) A concept of sustainability and resilience based on soil functions: the role of ISSS in promoting sustainable land use. In: Greenland, D.J. and Szabolcs, I. (eds) *Soil Resilience and Sustainable Land Use.* CAB International, Wallingford, UK, pp. 535–542.

Brown, L.R. (ed.) (1995) *State of the World 1996. A Worldwatch Institute Report on Progress Towards a Sustainable Society.* W.W. Norton, New York.

Brussaard, L., Bouwman, L.A., Geurs, M., Hassink, J. and Zwart, K.B. (1990) Biomass, composition and temporal dynamics of soil organisms of a silt loam soil under conventional and integrated management. *Netherlands Journal of Agricultural Science* 38, 283–302.

Doran, J.W. and Parkin, T.B. (1994) Defining and assessing soil quality. In: Doran, J.W., Coleman, D.C., Bezdicek, D.F. and Stewart, B.A. (eds) *Defining Soil Quality for a Sustainable Environment.* Soil Science Society of America, Inc. Special Publication No. 35, Madison, Wisconsin, USA, pp. 3–21.

Doran, J.W., Sarrantonio, M. and Janke, R. (1994) Strategies to promote soil quality and soil health. In: Pankhurst, C.E., Doube, B.M., Gupta, V.V.S.R. and Grace, P.R. (eds) *Soil Biota: Management in Sustainable Farming Systems.* CSIRO Press, Melbourne, pp. 230–237.

Gupta, V.V.S.R., Roper, M.M., Kirkegaard, J.A. and Angus, J.F. (1994) Changes in microbial biomass and organic matter levels during the first year of modified tillage and stubble management practices on a red earth. *Australian Journal of Soil Research* 32, 1339–1354.

Harris, R.F. and Bezdicek, D.F. (1994) Descriptive aspects of soil quality/health. In: Doran, J.W., Coleman, D.C., Bezdicek, D.F. and Stewart, B.A. (eds) *Defining Soil Quality for a Sustainable Environment.* Soil Science Society of America, Inc. Special Publication No. 35, Madison, Wisconsin, USA, pp. 23–35.

Kirkegaard, J.A. (1995) A review of trends in wheat yield responses to conservation cropping in Australia. *Australian Journal of Experimental Agriculture* 35, 835–848.

Larson, W.E. and Pierce, F.J. (1994) The dynamics of soil quality as a measure of sustainable management. In: Doran, J.W., Coleman, D.C., Bezdicek, D.F. and Stewart, B.A. (eds) *Defining Soil Quality for a Sustainable Environment.* Soil Science Society of America, Inc. Special Publication No. 35, Madison, Wisconsin, USA, pp. 37–52.

Miller, M.A. and Jastrow, J.D. (1994) Vesicular-arbuscular mycorrhizae and biogeochemical cycling. In: Pfleger, F.L. and Lindeman, R.G. (eds) *Mycorrhizae in Plant Health.* American Phytopathological Society, St Paul, Minnesota, pp. 189–212.

Naeem, S., Thompson, L.J., Lawler, S.P., Lawton, J.H. and Woodfin, R.M. (1995) Declining biodiversity can alter the performance of ecosystems. *Nature* 368, 734–737.

Pankhurst, C.E. (1994) Biological indicators of soil health and sustainable productivity. In: Greenland, D.J. and Szabolcs, I. (eds) *Soil Resilience and Sustainable Land Use.* CAB International, Wallingford, UK, pp. 331–351.

Pankhurst, C.E., Hawke, B.G., McDonald, H.J., Kirkby, C.A., Buckerfield, J.C., Michelson, P., O'Brien, K.A., Gupta, V.V.S.R. and Doube, B.M. (1995) Evaluation of soil biological properties as potential bioindicators of soil health. *Australian Journal of Agricultural Resesearch* 35, 1015–1028.

Powlson, D.S., Brookes, P.C. and Christensen, B.T. (1987). Measurement of soil microbial biomass provides an early indication of changes in total soil organic matter due to straw incorporation. *Soil Biology and Biochemistry* 19, 159–164.

Reichle, D.E. (1977) The role of soil invertebrates in nutrient cycling. In: Lohm, U. and Persson, T. (eds) *Soil Organisms as Components of Ecosystems.* Ecological Bulletin, Stockholm 25, 145–156.

Roper, M.M. and Gupta, V.V.S.R. (1995) Management practices and soil biota. *Australian Journal of Soil Research* 33, 321–339.

Saylor, G.S., Nikbakht, K. and Fleming, J.T. (1992) Application of molecular techniques to soil biochemistry. In: Stotzky, G. and Bollag, J.-M. (eds) *Soil Biochemistry* 7, Marcel Dekker, Inc., New York, pp. 131–172.

Theng, B.K.G., Tate, K.R., Sollins, P., Moris, N., Nadkarni, N. and Tart, R.L. III (1989) Constituents of organic matter in temperate and tropical soils. In: Coleman, D.C., Oades, J.M. and Uehara, G. (eds) *Dynamics of Soil Organic Matter in Tropical Ecosystems.* University of Hawaii Press, Honolulu, pp. 5–32.

Tilman D. and Downing, J.A. (1994) Biodiversity and stability in grasslands. *Nature* 367, 363–365.

Turco, R.F., Kennedy, A.C. and Jawson, M.D. (1994) Microbial indicators of soil quality. In: Doran, J.W., Coleman, D.C., Bezdicek, D.F. and Stewart, B.A. (eds) *Defining Soil Quality for a Sustainable Environment.* Soil Science Society of America, Special Publication Number 35, Madison, Wisconsin, pp. 73–90.

Index